Catalysis for Green Chemistry

Catalysis for Green Chemistry

Editor

Lu Liu

Basel • Beijing • Wuhan • Barcelona • Belgrade • Novi Sad • Cluj • Manchester

Editor
Lu Liu
School of Chemistry and
Molecular Engineering
East China Normal
University
Shanghai
China

Editorial Office
MDPI
St. Alban-Anlage 66
4052 Basel, Switzerland

This is a reprint of articles from the Special Issue published online in the open access journal *Molecules* (ISSN 1420-3049) (available at: www.mdpi.com/journal/molecules/special_issues/28KQL4N98R).

For citation purposes, cite each article independently as indicated on the article page online and as indicated below:

Lastname, A.A.; Lastname, B.B. Article Title. *Journal Name* **Year**, *Volume Number*, Page Range.

ISBN 978-3-7258-0950-9 (Hbk)
ISBN 978-3-7258-0949-3 (PDF)
doi.org/10.3390/books978-3-7258-0949-3

© 2024 by the authors. Articles in this book are Open Access and distributed under the Creative Commons Attribution (CC BY) license. The book as a whole is distributed by MDPI under the terms and conditions of the Creative Commons Attribution-NonCommercial-NoDerivs (CC BY-NC-ND) license.

Contents

Xiaoli Zhu, Xunshen Liu, Fei Xia and Lu Liu
Theoretical Study on the Copper-Catalyzed *ortho*-Selective C-H Functionalization of Naphthols with α-Phenyl-α-Diazoesters
Reprinted from: *Molecules* 2023, 28, 1767, doi:10.3390/molecules28041767 1

Tong-Tong Xu, Jin-Lan Zhou, Guang-Yuan Cong, Jiang-Yi-Hui Sheng, Shi-Qi Wang, Yating Ma, et al.
Boron Lewis Acid Catalysis Enables the Direct Cyanation of Benzyl Alcohols by Means of Isonitrile as Cyanide Source
Reprinted from: *Molecules* 2023, 28, 2174, doi:10.3390/molecules28052174 14

Xin-Zhang Yu, Wen-Long Wei, Yu-Lan Niu, Xing Li, Ming Wang and Wen-Chao Gao
Homocouplings of Sodium Arenesulfinates: Selective Access to Symmetric Diaryl Sulfides and Diaryl Disulfides
Reprinted from: *Molecules* 2022, 27, 6232, doi:10.3390/molecules27196232 29

Xiao-Yu Miao, Yong-Ji Hu, Fu-Rao Liu, Yuan-Yuan Sun, Die Sun, An-Xin Wu, et al.
Synthesis of Diversified Pyrazolo[3,4-*b*]pyridine Frameworks from 5-Aminopyrazoles and Alkynyl Aldehydes via Switchable C≡C Bond Activation Approaches
Reprinted from: *Molecules* 2022, 27, 6381, doi:10.3390/molecules27196381 43

Naaser A. Y. Abduh, Abdullah A. Al-Kahtani, Mabrook S. Amer, Tahani Saad Algarni and Abdel-Basit Al-Odayni
Fabricated Gamma-Alumina-Supported Zinc Ferrite Catalyst for Solvent-Free Aerobic Oxidation of Cyclic Ethers to Lactones
Reprinted from: *Molecules* 2023, 28, 7192, doi:10.3390/molecules28207192 55

Marcin Muszyński, Janusz Nowicki, Mateusz Zygadło and Gabiela Dudek
Comparsion of Catalyst Effectiveness in Different Chemical Depolymerization Methods of Poly(ethylene terephthalate)
Reprinted from: *Molecules* 2023, 28, 6385, doi:10.3390/molecules28176385 72

György Orsy, Sayeh Shahmohammadi and Enikő Forró
A Sustainable Green Enzymatic Method for Amide Bond Formation
Reprinted from: *Molecules* 2023, 28, 5706, doi:10.3390/molecules28155706 98

Bojie Li, Wu Wen, Wei Wen, Haifeng Guo, Chengpeng Fu, Yaoyao Zhang, et al.
Application of Chitosan/Poly(vinyl alcohol) Stabilized Copper Film Materials for the Borylation of α, β-Unsaturated Ketones, Morita-Baylis-Hillman Alcohols and Esters in Aqueous Phase
Reprinted from: *Molecules* 2023, 28, 5609, doi:10.3390/molecules28145609 109

Cheng Zuo and Qian Su
Research Progress on Propylene Preparation by Propane Dehydrogenation
Reprinted from: *Molecules* 2023, 28, 3594, doi:10.3390/molecules28083594 120

Mengting Chen, Yun Wang, Limin Jiang, Yuran Cheng, Yingxin Liu and Zuojun Wei
Highly Efficient Selective Hydrogenation of Cinnamaldehyde to Cinnamyl Alcohol over CoRe/TiO$_2$ Catalyst
Reprinted from: *Molecules* 2023, 28, 3336, doi:10.3390/molecules28083336 133

Yu Liang, Xinyan Chen, Jianli Zeng, Junqing Ye, Bin He, Wenjin Li, et al.
Mesoporous Polymeric Ionic Liquid via Confined Polymerization for Laccase Immobilization towards Efficient Degradation of Phenolic Pollutants
Reprinted from: *Molecules* **2023**, *28*, 2569, doi:10.3390/molecules28062569 . . . . **146**

Nuray Uzunlu, Péter Pongrácz, László Kollár and Attila Takács
Alkyl Levulinates and 2-Methyltetrahydrofuran: Possible Biomass-Based Solvents in Palladium-Catalyzed Aminocarbonylation
Reprinted from: *Molecules* **2023**, *28*, 442, doi:10.3390/molecules28010442 . . . . **161**

Iuliana Porukova, Vadim Samoilov, Dzhamalutdin Ramazanov, Mariia Kniazeva and Anton Maximov
In Situ-Generated, Dispersed Cu Catalysts for the Catalytic Hydrogenolysis of Glycerol
Reprinted from: *Molecules* **2022**, *27*, 8778, doi:10.3390/molecules27248778 . . . . **176**

Insaf Abdouli, Frederic Dappozze, Marion Eternot, Chantal Guillard and Nadine Essayem
TiO_2 Catalyzed Dihydroxyacetone (DHA) Conversion in Water: Evidence That This Model Reaction Probes Basicity in Addition to Acidity
Reprinted from: *Molecules* **2022**, *27*, 8172, doi:10.3390/molecules27238172 . . . . **196**

Anna Wolny and Anna Chrobok
Silica-Based Supported Ionic Liquid-like Phases as Heterogeneous Catalysts
Reprinted from: *Molecules* **2022**, *27*, 5900, doi:10.3390/molecules27185900 . . . . **213**

Article

Theoretical Study on the Copper-Catalyzed *ortho*-Selective C-H Functionalization of Naphthols with α-Phenyl-α-Diazoesters

Xiaoli Zhu [1], Xunshen Liu [1,2], Fei Xia [1,3,*] and Lu Liu [1,2,*]

[1] School of Chemistry and Molecular Engineering, East China Normal University, 500 Dongchuan Road, Shanghai 200241, China
[2] Shanghai Engineering Research Center of Molecular Therapeutics and New Drug Development, East China Normal University, Shanghai 200062, China
[3] NYU-ECNU Center for Computational Chemistry at New York University, East China Normal University, 3663 Zhongshan Road, Shanghai 200062, China
* Correspondence: fxia@chem.ecnu.edu.cn (F.X.); lliu@chem.ecnu.edu.cn (L.L.)

Abstract: The aromatic C(sp^2)-H functionalization of unprotected naphthols with α-phenyl-α-diazoesters under mild conditions catalyzed by CuCl and CuCl$_2$ exhibits high efficiency and unique *ortho*-selectivity. In this study, the combination of density functional theory (DFT) calculations and experiments is employed to investigate the mechanism of C-H functionalization, which reveals the fundamental origin of the site-selectivity. It explains that CuCl-catalyzed *ortho*-selective C-H functionlization is due to the bimetallic carbene, which differs from the reaction catalyzed by CuCl$_2$ via monometallic carbene. The results demonstrate the function of favourable H-bond interactions on the site- and chemo-selectivity of reaction through stabilizing the rate-determining transition states in proton (1,3)-migration.

Keywords: copper catalysis; *ortho*-C(sp^2)-H bond functionalization; naphthols; diazo compounds; metal carbene; density functional theory (DFT) calculations

1. Introduction

The metal-carbenes, normally generated from diazo compounds via the catalysis of transition-metals, have been widely used in organic synthesis as one of the most significant reactive intermediates due to the versatile transformations [1–16], such as cycloaddition, cyclopropanation, ylide formation and rearrangement, O-H bond insertion, N-H bond insertion, and C-H bond functionalization [17–31]. Thus, it is highly desirable to develop a new reactional methodology by using metal-carbenes as the key species in synthetic chemistry. On the other hand, because phenyl rings, such as benzene, phenol, aniline and their derivatives, occur widely in natural products, bioactive molecules and drugs are important platforms in organic synthesis. The highly site-selective aromatic C(sp^2)-H bond functionalization of phenyl ring without the directing groups is still challenging, especially for the phenol derivatives [32–38]. In this field, due to its high activity in organic reactions, metal-carbene species have the advantage in the activation of inert C-H bond, while still meeting the chemo- and region-selectivities. For example, the phenolic O-H insertion occurred during the reactions of phenol derivatives with diazoesters in the presence of various metal catalysts such as Rh, Fe, Ru, Cu, and Pd (Scheme 1a) [39–44]. In 2014, we disclosed the first gold-catalyzed aromatic C(sp^2)-H bond functionalization of free phenols with diazo compounds in high efficiency with excellent *para*-selectivity [45]. Later, we reported the challenging *ortho*-selective aromatic C(sp^2)-H bond alkylation of phenols with diazoesters by using B(C$_6$F$_5$)$_3$ and naphthols with diazoesters by using gold catalyst (Scheme 1b) [46,47]. Nemoto disclosed the similar gold catalyzed C(sp^2)-H bond functionalization/cyclization of β-naphthols with diazoesters [48]. However, the catalysts used in these reactions, like gold complexes and B(C$_6$F$_5$)$_3$, are expensive. Thus, the development of inexpensive and abundant catalysts to replace the noble catalysts is a long-term need. Furthermore, understanding the origins of new catalytic systems in carbene chemistry is also very important, which could guide the design of new reactions and new catalysts.

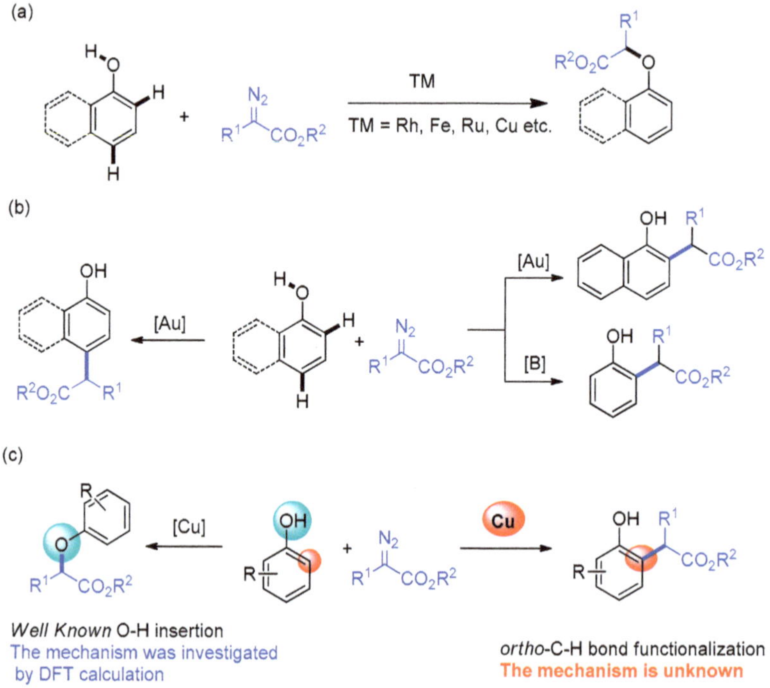

Scheme 1. The reactions of phenol derivatives with diazoesters. (a) O-H insertion of phenols; (b) C-H bond functionalization of phenols by gold and boron; (c) Copper-catalyzed reaction of phenols with diazoesters.

Compared to other commonly used transition-metals in organic synthesis, copper represents a type of ideal catalyst for chemical reactions. It attracts much attention due to its low-toxicity, low cost, ready availability, and its benign environmental impact [49–53]. It has been used in carbene transfer reactions for half a century. However, the O-H insertion was the major reaction when phenol derivatives reacted with diazo compounds in the presence of copper [54–57]. In 1952, Yates reported the reaction of phenols and diazo compounds under copper catalyst, delivering O-H insertion product as the primary one along with the side product via the *ortho*-selective C-H bond functionalization/cycliztion [54]. Recently, Zhou disclosed an elegant asymmetric copper-catalyzed O-H insertion of phenols [56]. Apart from experimental investigations, mechanistic investigation of the O-H insertion by copper-catalyzed carbenes transferred from diazo compounds were also performed [58–62]. Pérez et al. examined the mechanism of copper-catalyzed O-H insertion reaction, in which they investigated the ligand effects of Tp^X (hydrotris(3,5-dimethylpyrazolyl)borate and its derivatives) on chemoselectivity by experiments [58]. Yu et al. explored the detailed mechanism of O-H insertion of diazoacetates under copper (I) by DFT calculation [59]. They discovered that the (1,2)-H migration favored the copper-associated ylide pathway, in which the water molecule acted as an effective proton shuttle for the (1,2)-H shift in Cu-catalyzed O-H insertion. In contrast to the above cases, we recently reported a copper-catalyzed *ortho*-selective C-H bond alkylation of naphthols and phenols that provided an important route for the synthesis of *ortho*-substituted phenol derivatives [63]. However, the detailed mechanism of copper-catalyzed C(sp^2)-H bond functionalization is still unknown due to the more complex structures and versatile valence states of copper (Scheme 1c). In our previous work, we studied the competitive pathways of gold catalyzed C-H bond functionalization and O-H bond insertions of phenols with diazo compounds via the combination of DFT calculations and experiments [61,62]. In this report, both CuCl and CuCl$_2$ were efficient catalysts for this reaction, providing a good yield of the *ortho*-selective products with excellent site-selectivity (Table 1). In this study, we investigated the mechanism of site-selective C(sp^2)-H bond functionalization of 1-naphthol and α-diazoacetate catalyzed by CuCl and CuCl$_2$ by combining DFT calculations and experiments.

Table 1. Copper-catalyzed C(sp^2)-H bond functionalization of naphthols with α-diazoesters [63].

Catalyst	Yield of 3 (%) [1]	Yield of 4 (%) [1]
CuCl	76	0
CuCl$_2$	87	0

[1] The yield was determined by ^{1}H NMR of crude product by using CH$_2$Br$_2$ as internal standard.

Previously, we reported a mechanistic study on how to yield the active Cu-carbenes from diazo compounds with CuCl and CuCl$_2$ [64]. The DFT calculations revealed that the most stable structure of CuCl in the solution was dimer, while that of CuCl$_2$ was monomer. Then, the decomposition of α-phenyl-α-diazoester under CuCl generated the bimetallic carbene, while the monometallic Cu(II)-carbene was obtained in the presence of CuCl$_2$ (Figure 1). Therefore, the two types of Cu-carbenes will be used as the precursors for the site-selective C(sp$_2$)-H bond functionalization of naphthols.

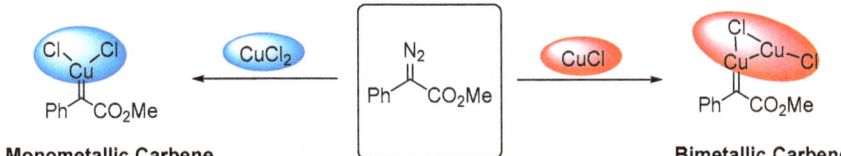

Figure 1. The reactions of the formation of the monometallic and bimetallic Cu carbenes [64].

2. Results and Discussion

Under the catalysis of transition metal Cu, we studied the mechanisms of C-H bond functionalization between diazo compounds and naphthols to generate the *ortho*-substituted products. Previous experimental studies reported that the transition metal complexes initially reacted with the methyl phenyl diazoacetates to form metal carbenes by releasing the N$_2$ molecules [65–67]. The metal carbenes and naphthols were regarded as the precursors for the C-H insertion. Thus, the Cu-carbenes and naphthols are used as the reaction precursors in our DFT calculations. The relative energies of all intermediates and transition states to the sum of precursors are calculated and presented in all figures. The red and pink lines denote the lowest energy pathways of *ortho*-C-H and *para*-C-H insertions, respectively.

2.1. The C-H Insertion of Naphthols Catalyzed by the (CuCl)₂ Dimer

As discussed, the DFT calculations revealed that CuCl preferred the dimer to the monomer in the solution [64]. Assuming the (CuCl)$_2$ dimer as an effective catalyst, it reacts with the diazo compound to yield the bimetallic Cu carbene. Figure 2 shows the calculated free energy profiles for the detailed reaction pathways of the C(sp^2)-H bonds of naphthols inserted by bimetallic Cu carbenes. The bimetallic Cu carbenes and naphthols undergo the electrophilic addition to form the intermediate **1-Int-o2** through the ortho-substituted transition state **1-TS-o1** with a barrier of 11.9 kcal mol^{-1}. Another reaction pathway of the addition at the para-C(sp^2) of naphthols via **1-TS-p1** has a higher barrier of 12.7 kcal mol^{-1}. In the two addition pathways, the protons of the aromatic C-H bonds transfer to the carboxyl oxygen atoms via the optimized five-membered ring transition states **1-TS-o3/p3**. In this case, the protons of **1-Int-o2/p2** at the ortho- and para-carbons of naphthols move to the carbonyl oxygens to form the enols **1-Int-o4/p4**, which overcome the activation barriers of 8.5 and 8.6 kcal mol^{-1}, respectively.

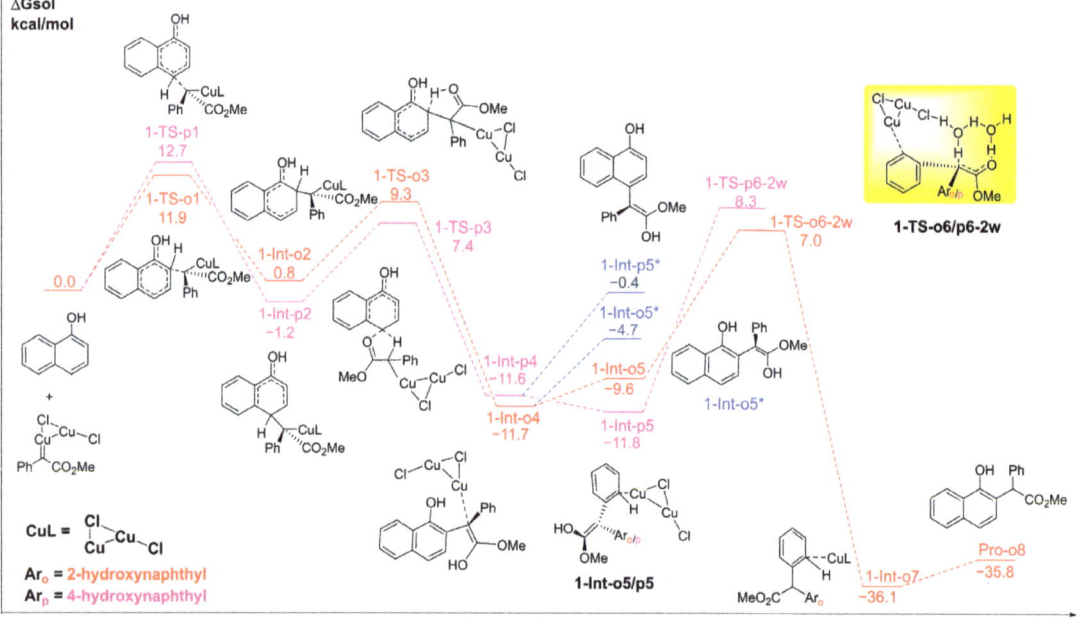

Figure 2. The calculated free energy profiles ΔG_{sol} and the corresponding structures of intermediates and transition states along different pathways of the C–H bonds of naphthols inserted by the bimetallic carbenes of the (CuCl)$_2$ dimer. The reasonable ortho-selective pathway is shown in red, the para-selective pathway is pink, and other possible pathways are in blue. The free energies are given in kcal mol^{-1}.

There are two possible reaction pathways from **1-Int-o4** to yield the final product **Pro-o8**. One is the conversion of **1-Int-o4** into **1-Int-o5*** through the (CuCl)$_2$ dimer dissociating into the solution by absorbing an energy of 7.0 kcal mol^{-1}. The second pathway is that the (CuCl)$_2$ dimer in **1-Int-o4** undergoes the (1,3)-migration to the phenyl group to yield the less stable enol **1-Int-o5**. However, the formation of **1-Int-o5** is more favorable than **1-Int-o5*** by 4.9 kcal mol^{-1} in energy. For the para-C-H insertion of naphthols, the reaction process is like the ortho-C-H insertion. The enol complex **1-Int-p4** could be further stabilized through the intramolecular (1,3)-migration of the (CuCl)$_2$ dimer, leading to the enol intermediate **1-Int-p5** rather than to **1-Int-p5***. Subsequently, the metal catalysts participate in the two-water assisted (1,3)-H migration via the eight-membered ring transition states **1-TS-o6-2w** and **1-TS-p6-2w**. Comparison of the calculated energies of the optimized TSs for proton transfer suggests that the energy of **1-TS-o6-2w** is lower than that of **1-TS-p6-2w** by 1.4 kcal mol^{-1}. In this case, the proton transfer from the hydroxyl group to the ortho-position carbon of naphthol through **1-TS-o6-2w** leads to the formation of **1-Int-o7** with a barrier of 18.7 kcal mol^{-1}. Finally, the (CuCl)$_2$ dimer of **1-Int-o7** dissociates into the solution to yield the final product **Pro-o8** with the calculated energy of −35.8 kcal mol^{-1}.

2.2. The Key H-Bond Interactions Formed with (CuCl)$_2$

To further explain the site-selectivity of the bimetallic Cu carbene catalyzed C(sp^2)-H functionalization of naphthols with diazo compounds, we focus on the two important elementary steps during the activation of the ortho-C-H bond of naphthols by the bimetallic Cu carbenes: the electrophilic addition of bimetallic Cu carbenes and naphthols, and the two water-assisted (1,3)-proton transfer with the participation of Cu catalyst, which is the rate-determining step of the reaction. It is found that both the electrophilic addition and the (1,3)-H migration at the ortho-sites of naphthols are superior to that of the para-sites. The optimized structures and energies of **1-TS-1o**, **1-TS-1p**, **1-TS-o6-2w**, and **1-TS-6p-2w** are shown in Figure 3. It is noted that the electrophilic addition at the ortho-C(sp^2) of naphthol has a lower barrier of 11.9 kcal mol^{-1} relative to that at the para-C(sp^2), which is consistent with the known experiment results [63]. The **1-TS-o1** is stabilized by the O-H···Cl H-bond interaction between the hydroxyl group of naphthol and the Cl atom of bimetallic Cu carbene, with the H···Cl distance of 2.14 Å, as shown in Figure 3.

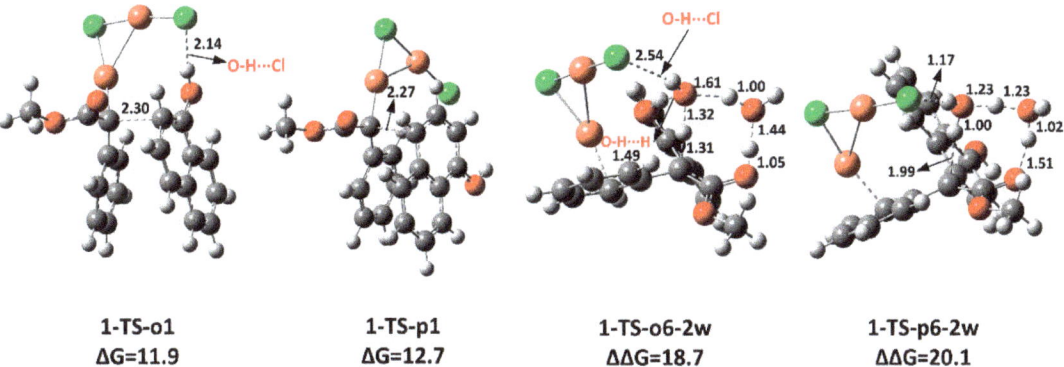

Figure 3. Optimized geometries of four transition states **1-TS-1o**, **1-TS-1p**, **1-TS-6-2w**, and **1-TS-p6-2w** corresponding to Figure 2. The distances are in the units of angstroms and the values of free energy barriers are in kcal mol^{-1}.

Additionally, the H-bond interactions also play an important role in the two water-assisted proton transfer. The remote (1,3)-H migration needs two water molecules as a shuttle rather than one, which has been demonstrated in our previous study on the C-H insertion of phenols by Au-carbenes [61,62]. There are two kinds of H-bonds formed in the transition state **1-TS-o6-2w**, including the O-H···Cl H-bond interaction formed from the shuttle water and the Cl atom of the (CuCl)$_2$ dimer, and the O-H···O H-bond interaction between the hydroxyl group of naphthol and the shuttle water. Due to the stabilization of the formed H-bonds in **1-TS-o6-2w**, the barrier of the (1,3)-proton shift is only 18.7 kcal mol^{-1}, lower than that of 20.1 kcal mol^{-1} of **1-TS-p6-2w**, leading to the ortho-substituted products. Thus, the presence of H-bond interactions in the ortho-C-H functionalization reduces the energy barriers of the electrophilic addition of the metal carbenes with naphthols and the (1,3)-proton transfer process.

2.3. The C-H Insertion of Naphthols Catalyzed by the CuCl Monomer

We also studied the $C(sp^2)$-H bond insertion mechanism of naphthols catalyzed by the CuCl monomer instead of the $(CuCl)_2$ dimer. Figure 4 displays the calculated free energy profiles of reaction pathways of the C-H bonds of naphthols inserted by the monometallic Cu carbenes. The first step remains the addition of the monometallic carbenes and naphthols and the calculated results indicate that the electrophilic addition at the para-$C(sp^2)$ of naphthols occurs via the transition state **1′-TS-p1** with a lower energy barrier of 15.6 kcal mol^{-1}. This is lower than that of 17.6 kcal mol^{-1} at the ortho-$C(sp^2)$ via **1′-TS-o1**. Thus, the addition through **1′-TS-p1** is more kinetically favorable and not consistent with the experimental observations. If **1′-Int-o4** releases the moiety of the CuCl monomer into the solution, it yields a more stable intermediate **1′-Int-o5*** with the exothermicity of 2.8 kcal mol^{-1}. However, the (1,3)-migration of CuCl to the phenyl group in **1′-Int-o4** leads to the enol **1′-Int-o5** with an exothermic energy of 5.9 kcal mol^{-1}. Because **1′-Int-o5** is more stable than **1′-Int-o5*** by 3.1 kcal mol^{-1}, the catalyst CuCl participates in the proton transfer. Also, **1′-Int-p4** transforms to the **1′-Int-p5** through the (1,3)-migration of the monomer CuCl instead of the dissociation of CuCl to form **1′-Int-p5***. Furthermore, the barrier from **1′-Int-o5** to **1′-TS-o6-2w** is high (up to 28.2 kcal mol^{-1}) and considered as the rate-determining step, which is higher than that of **1′-Int-p5** to **1′-TS-p6-2w** by 5.8 kcal mol^{-1}. Such a high barrier is counter to the experimental results. As such, the reaction pathways catalyzed by the monometallic carbenes of CuCl are excluded [63]. In summary, the calculated reaction pathways of C-H bond functionalization of naphthols catalyzed by the bimetallic and monometallic carbenes account for the site-selectivity of the C-H bond functionalization by the mild catalyst CuCl. The calculated results of these two steps show that the Cu-catalyzed C-H bond functionalization of naphthols with diazo esters is more inclined to obtain the ortho-substituted products.

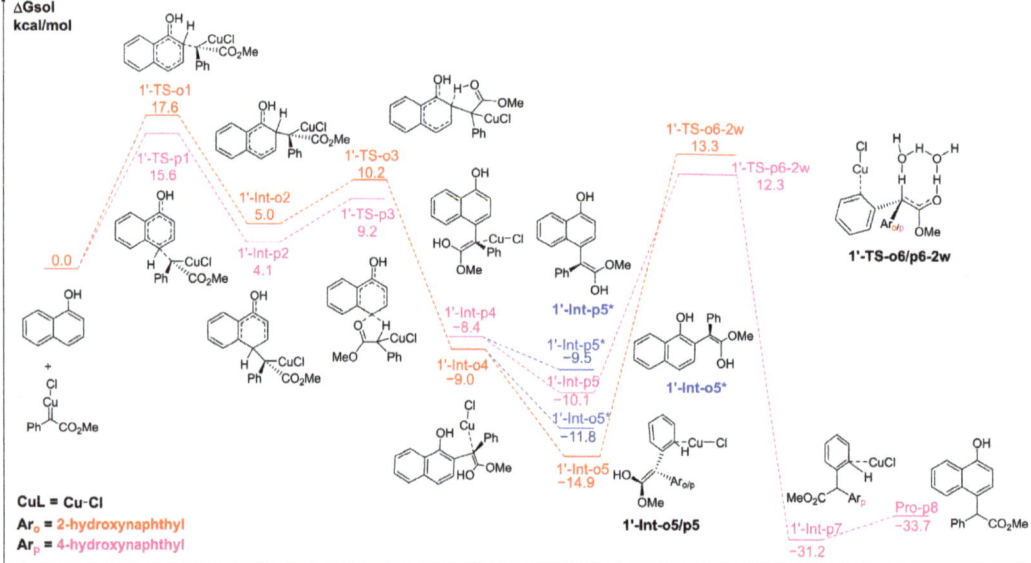

Figure 4. The calculated free energy profiles ΔG_{sol} and the corresponding structures of intermediates and transition states along different pathways of the C–H bonds of naphthols inserted by the monometallic carbenes of the CuCl monomer. The reasonable ortho-selective pathway is shown in red, the para-selective pathway is pink, and other possible pathways are in blue. The free energies are given in kcal mol^{-1}.

2.4. The C-H Insertion of Naphthols Catalyzed by the CuCl$_2$ Monomer

Figure 5 shows the calculated free energy profiles for the possible C-H bonds insertion pathways of naphthols catalyzed by the monometallic carbenes of CuCl$_2$. The difference, when compared to Cu(I), is the energy barriers of the electrophilic addition at the ortho-C(sp^2) of naphthols, which is higher than that at the para-C(sp^2) by 2.8 kcal mol^{-1}. For ortho-C-H insertion, it is an endothermic process when the CuCl$_2$ moiety of **2-Int-o4** migrates from the C=C double bond to the phenyl group to generate **2-Int-o5*** via (1,3)-migration of the monomer CuCl$_2$. The **2-Int-o4** can transform to the key intermediate **2-Int-o5** through the CuCl$_2$ monomer as it dissociates into the solution by releasing an energy of 4.6 kcal mol^{-1}, which is relatively more feasible and stable compared to the formation of **2-Int-o5***. Like the process of ortho-C-H insertion, the enol complex **2-Int-p4** can be further stabilized by releasing CuCl$_2$ into the solution, leading to a free enol **2-Int-p5** rather than **2-Int-p5***. Subsequently, the two water-assisted (1,3)-H migration without the participation of CuCl$_2$ via the eight-membered ring transition states **2-TS-o6-2w** and **2-TS-p6-2w** are the pivotal steps in the ortho-C-H and para-C-H insertions, with the calculated barriers of 19.8 and 22.2 kcal mol^{-1}, respectively. The activation energy barrier of **2-TS-p6-2w** is higher than that of **2-TS-o6-2w** by 2.4 kcal mol^{-1}, implying that the C(sp^2)-H insertion catalyzed by Cu(II) prefers the ortho-selective product **Pro-o8** to the para-substituted product **Pro-p8**, which is in accordance with previous experimental results [58–62].

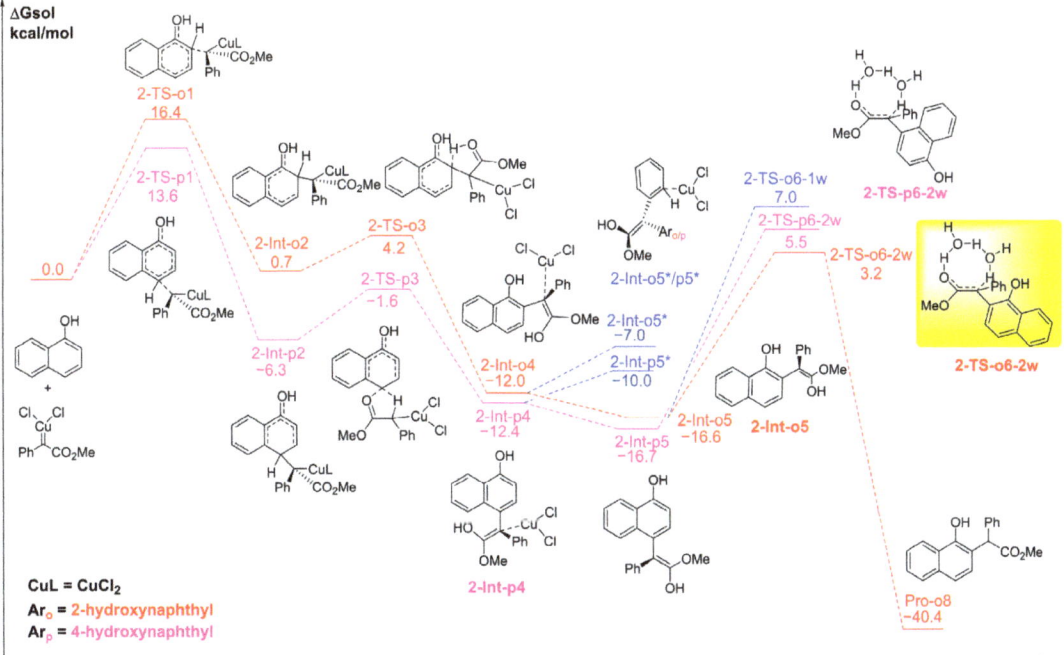

Figure 5. The calculated free energy profiles ΔG$_{sol}$ and the corresponding structures of intermediates and transition states along different pathways of the C–H bonds of naphthols inserted by the monometallic carbenes of CuCl$_2$. The reasonable ortho-selective pathway is shown in red, the para-selective pathway is pink, and other possible pathways are in blue. The free energies are given in kcal mol^{-1}.

The structures of the key transition states **2-TS-1o** and **2-TS-1p** of the electrophilic additions in the two pathways are shown in Figure 6. The calculated distance between the C1 atom of the copper-carbene and the C2 atom of naphthol in **2-TS-1p** is longer than that of **2-TS-1o**. Although neither **2-TS-1o** nor **2-TS-1p** has H-bonds formed, the H-bond interaction of O-H···O between the hydroxyl group of naphthol and the water molecule in **2-TS-o6-2w** plays a crucial role in stabilizing the proton transfer through the Cu-free pathways, with a distance of 1.60 Å. The **2-TS-o6-2w** has a lower barrier than that of **2-TS-p6-2w**, which could be the key to control the chemo- and site-selectivity of the C-H bond insertion of naphthols catalyzed by Cu(II).

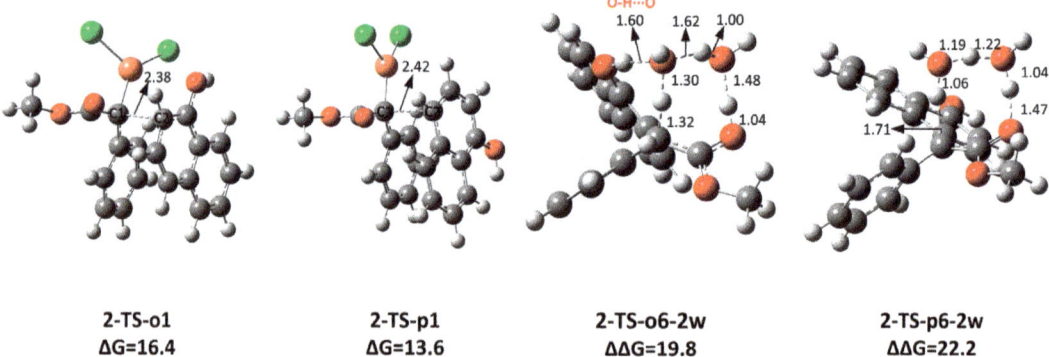

Figure 6. Optimized geometries of four transition states **2-TS-1o**, **2-TS-1p**, **2-TS-o6-2w**, and **2-TS-6p-2w** corresponding to Figure 5. The distances are in the units of angstroms and the values of free energy barriers are in kcal mol^{-1}.

2.5. Experimental Reactivity of 1-Methoxynaphthalene Catalyzed by Cu Catalysts

To further prove the Cu-catalyzed ortho-C-H insertion mechanism of naphthols we proposed, we performed the experiments of the 1-methoxynaphthalene **5** and methyl α-diazoacetate **2** catalyzed by Cu catalysts and obtained a trace amount of the ortho-selective C-H products **6** (Scheme 2). Due to the high structural similarity of the reactants, we presuppose that it also follows the same reaction mechanism of naphthols inserted by the Cu carbenes. Since we have discussed the significance of the electrophilic addition and the hydrogen transfer assisted by two water molecules, the DFT calculations are primarily performed to investigate the two crucial elementary steps.

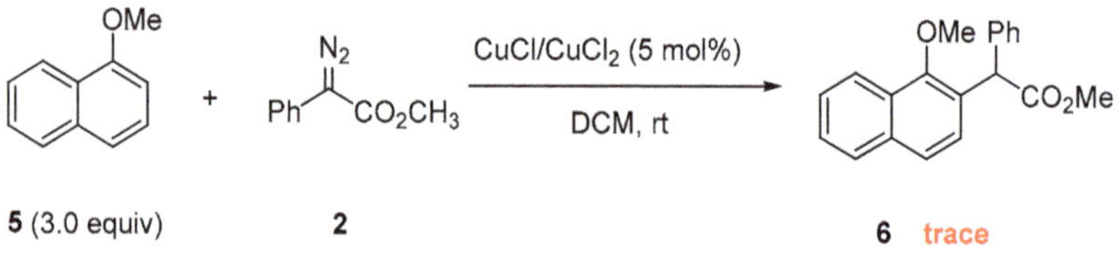

Scheme 2. C(sp^2)-H functionalization of 1-methoxynaphthalene with α-diazoesters catalyzed by CuCl or CuCl$_2$.

Figure 7 shows the calculated energy profiles of the ortho- and para-C-H bonds of 1-methoxynaphthalenes inserted by the bimetallic and monometallic Cu carbenes. The calculated reaction pathways of the C-H insertion of 1-methoxynaphthalenes are similar to that of naphthols. With the catalysts CuCl or CuCl$_2$, the calculated free energy barriers of the electrophilic addition at the ortho-positions of 1-methoxynaphthalenes are higher than the counterparts at the para-positions. It has been emphasized that the H-bonds formed by the hydroxyl groups of naphthols and the bimetallic carbenes are the key factors to determine the site-selectivity in the addition step. Nevertheless, such a specific O-H···Cl H-bond interaction does not exist in the transition states **3-TS-o1** and **4-TS-o1** due to the steric effect of methoxy groups of 1-methoxynaphthalenes.

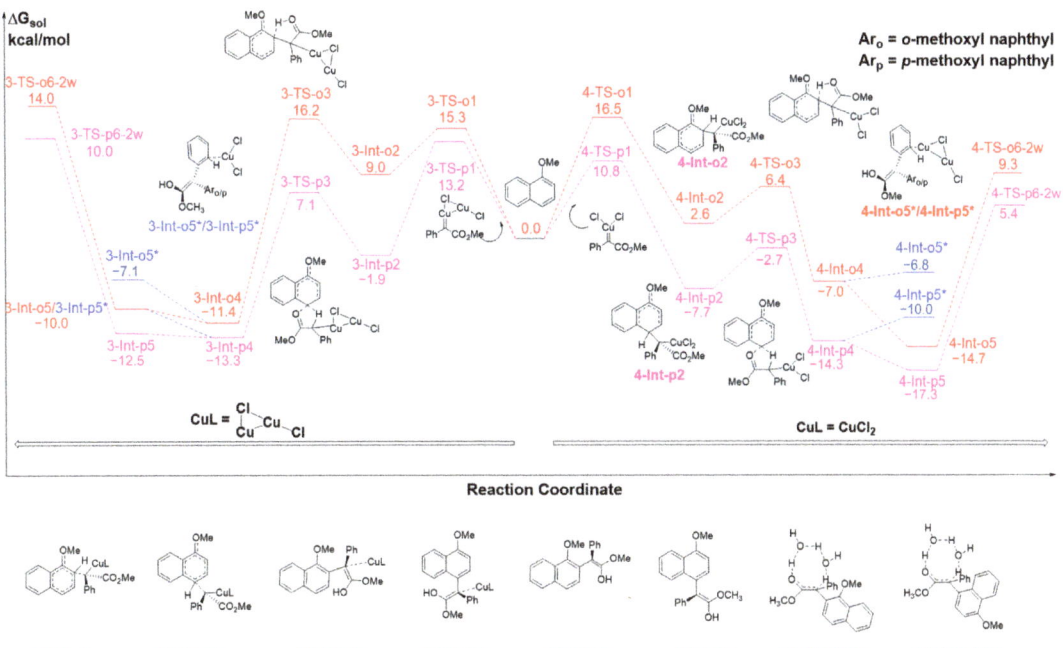

Figure 7. The calculated free energy profiles ΔG$_{sol}$ and the corresponding structures of intermediates and transition states along different pathways of the C–H bonds of 1-methoxynaphthalenes inserted by the bimetallic and monometallic Cu carbenes. The reasonable ortho-selective pathways are shown in red, the para-selective pathways are pink, and other possible pathways are in blue. The free energies are given in kcal mol^{-1}. Ar$_o$ = o-methoxyl naphthyl, Ar$_p$ = p-methoxyl naphthyl.

The energy barriers of the subsequent rate-determining steps, namely the proton transfer steps at the ortho-positions, are relatively higher than that of the para-positions. Specifically, the crucial barriers of the (1,3)-H transfer in the transition states **3-TS-o6-2w** and **3-TS-p6-2w** at the ortho-C-H and para-C-H insertion catalyzed by the (CuCl)$_2$ dimer reach high energies of 25.4 and 23.3 kcal mol^{-1}, respectively. In another case, the (1,3)-H transfer of the ortho-C-H and para-C-H insertion catalyzed by the CuCl$_2$ monomer also have high activation barriers of 24.0 and 22.7 kcal mol^{-1}, respectively. The high barriers of both **3-TS-o6-2w** and **4-TS-o6-2w** mean that it is difficult to pass through them and obtain the product at the room temperature when using the trace amount of the product **6** obtained in our experiments. The optimized structures of the transition states **3/4-TS-o6-2w** and **3/4-TS-p6-2w** are remarkably different from that of naphthols due to the lack of the crucial H-bonds formed between the chemical groups of reactants and the metal catalysts.

3. Materials and Methods

3.1. General Information

^{1}H NMR spectra were recorded on a BRUKER 500 spectrometer (Billerica, MA, USA) in CDCl$_3$. Chemical reagents were purchased from Leyan (Shanghai, China). Anhydrous dichloromethane (DCM) was distilled from calcium hydride to use. Catalysts CuCl and CuCl$_2$ were purchased from Alfa-Aesar Company (Haverhill, MA, USA) and used directly.

3.2. Synthetic Procedure for the Reaction of 1-Methoxynaphthalene and Diazoester

In a dried glass tube, copper catalyst CuCl (0.02 mmol, 5 mol%), 1-methoxynaphthalene **5** (189.6 mg, 1.2 mmol, 3 equiv), and DCM (1 mL) was added at room temperature. Then a solution of methyl phenyl diazaester **2** (76.2 mg, 0.4 mmol) dissolved in 1 mL DCM was introduced into the reaction mixture by a syringe. The resulting mixture was continually stirred at room temperature until product **6** was consumed completely, determined by TLC analysis. After being filtrated through celite and concentrated, the residue was purified by column chromatography on silica gel to obtain the desired product. The yield was determined by ^{1}H NMR of crude product by using CH_2Br_2 as internal standard.

3.3. Computational Methods

All DFT calculations are performed using the Gaussian09 program package [68]. The geometric structures of intermediates and transition states are directly optimized in the solution phase by using the ωB97XD functional [69,70]. The SDD basis set [71] combined with the effective core potential is used to describe the metal element Cu, and the large 6-31 + G** basis set [72] is utilized to describe the nonmetallic elements C, H, O, N and Cl. Frequency analyses are also performed at the same computational level to confirm that the intermediates are local minima and the transition states have only one imaginary frequency. The intrinsic reaction coordinate (IRC) [73,74] calculations are performed to make sure that all transition state structures connect the correct reactants and products in the forward and backward reaction directions. The solvent effect of dichloromethane is evaluated using the SMD [75] model with a dielectric constant $\varepsilon = 8.93$ in Gaussian09. All the calculated energies refer to the Gibbs free energies in the units of kcal mol^{-1} at the temperature of 298.15 K. Further structure details about the intermediates and transition states are provided in Supporting Information.

4. Conclusions

The detailed mechanisms of the C(sp^2)-H bond functionalization of naphthols and α-aryl-α-diazoacetates by the catalysts CuCl and CuCl$_2$ are studied through the combined experimental and computational methods. The DFT calculations reveal that the *ortho*-selective products catalyzed by CuCl are obtained from the C-H insertion of naphthols by the bimetallic carbenes. Also, the optimized TS structures of the steps of addition and (1,3)-H transfer reveal that the H-bonds formed by the OH groups of naphthols and the Cl atoms of metal catalysts play an important role in stabilizing the TSs and lowering their energies. In the reaction catalyzed by CuCl$_2$, the DFT results indicate that the monometallic carbenes insert into the C(sp^2)-H bonds of naphthols, rather than the bimetallic species. It is proposed that the H-bond interactions between the Cu carbenes and substrates play an essential role in stabilizing the site-selectivity-determining TSs in all cases, resulting in a lower energy barrier and generating the experimentally observed *ortho*-selective products. The proposed H-bonds assisted insertion by Cu catalysts are supported by our further experiments of the C-H insertion of 1-methoxynaphthalenes catalyzed by the CuCl/CuCl$_2$ catalysts as well as the corresponding DFT calculations. Our studies systematically provide the mechanistic insights into the unprecedented C-H functionalization by the CuCl/CuCl$_2$ catalysts, which is instructive in designing Cu-catalyzed chemo- and site-selective transformations.

Supplementary Materials: The following supporting information can be downloaded at: https://www.mdpi.com/article/10.3390/molecules28041767/s1, Figure S1: Calculated reaction pathways of diazoacetates and copper catalyst CuCl monomer, the (CuCl)$_2$ dimer and CuCl$_2$ monomer; Scheme S1 C(sp^2)-H bond functionalization of 1-methoxynaphthalene with α-diazoesters catalyzed by CuCl or CuCl$_2$

Author Contributions: F.X. and L.L. conceived the idea; X.Z. performed the most DFT calculations; X.L. assisted in some experiments; X.L. and X.Z. collected and analyzed the data; F.X. guided the DFT calculations; F.X. and L.L. guided this project and wrote the manuscript. All authors have read and agreed to the published version of the manuscript.

Funding: This research was funded by the National Natural Science Foundation of China (Nos. 21971066, 22171088) and the Science and Technology Commission of Shanghai Municipality (18JC1412300) for financial support.

Institutional Review Board Statement: Not applicable.

Informed Consent Statement: Not applicable.

Data Availability Statement: The data are available on request from the corresponding authors.

Conflicts of Interest: The authors declare no conflict of interest.

Sample Availability: Samples of the compounds **1–6** are available from the authors.

References

1. Doyle, M.P. Catalytic Methods for Metal Carbene Transformations. *Chem. Rev.* **1986**, *86*, 919–939. [CrossRef]
2. Davies, H.M.L.; Denton, J.R. Application of donor/acceptor-carbenoids to the synthesis of natural products. *Chem. Soc. Rev.* **2009**, *38*, 3061–3071. [CrossRef] [PubMed]
3. Maas, G. New syntheses of diazo compounds. *Angew. Chem. Int. Ed.* **2009**, *48*, 8186–8195. [CrossRef] [PubMed]
4. Ford, A.; Miel, H.; Ring, A.; Slattery, C.N.; Maguire, A.R.; McKervey, M.A. Modern Organic Synthesis with α-Diazocarbonyl Compounds. *Chem. Rev.* **2015**, *115*, 9981–10080. [CrossRef] [PubMed]
5. Liu, L.; Zhang, J. Gold-catalyzed transformations of α-diazocarbonyl compounds: Selectivity and diversity. *Chem. Soc. Rev.* **2016**, *45*, 506–516. [CrossRef]
6. Zhu, D.; Chen, L.; Fan, H.; Yao, Q.; Zhu, S. Recent progress on donor and donor-donor carbenes. *Chem. Soc. Rev.* **2020**, *49*, 908–950. [CrossRef]
7. Yadagiri, D.; Anbarasan, P. Catalytic Functionalization of Metallocarbenes Derived from α-Diazocarbonyl Compounds and Their Precursors. *Chem. Rec.* **2021**, *21*, 3872–3883. [CrossRef] [PubMed]
8. Yu, Z.; Qiu, H.; Liu, L.; Zhang, J. Gold-Catalyzed Construction of Two Adjacent Quaternary Stereocenters via Sequential C-H Functionalization and Aldol Annulation. *Chem. Commun.* **2016**, *52*, 2257–2260. [CrossRef]
9. Ma, B.; Wu, Z.; Huang, B.; Liu, L.; Zhang, J. Gold-catalysed facile access to indene scaffold via sequential C-H functionalization and 5-endo-dig carbocyclization. *Chem. Commun.* **2016**, *52*, 9351–9354. [CrossRef]
10. Ma, B.; Wu, J.; Liu, L.; Zhang, J. Gold-catalyzed para-selective C–H bond alkylation of benzene derivatives with donor/acceptor-substituted diazo compounds. *Chem. Commun.* **2017**, *53*, 10164–10167. [CrossRef]
11. Li, Y.; Tang, Z.; Zhang, J.; Liu, L. Gold-Catalyzed Intermolecular [4+1] Spiroannulation via Site Selective Aromatic C(sp^2)-H Functionalization and Dearomatization of Phenol Derivatives. *Chem. Commun.* **2020**, *56*, 8202–8205. [CrossRef] [PubMed]
12. Yu, Z.; Li, G.; Zhang, J.; Liu, L. Iron-catalysed chemo- and ortho-selective C-H bond functionalization of phenols with α-aryl-α-diazoacetates. *Org. Chem. Front.* **2021**, *8*, 3770–3775. [CrossRef]
13. Liu, X.; Tang, Z.; Li, Z.; Li, M.; Xu, L.; Liu, L. Modular and stereoselective synthesis of tetrasubstituted vinyl sulfides leading to a library of AIEgens. *Nat. Commun.* **2021**, *12*, 7298. [CrossRef] [PubMed]
14. Liu, X.; Li, M.; Dong, K.; Peng, S.; Liu, L. Highly Stereoselective Synthesis of Tetrasubstituted Vinyl Selenides via Rhodium-Catalyzed [1,4]-Acyl Migration of Selenoesters and Diazo Compounds. *Org. Lett.* **2022**, *24*, 2175–2180. [CrossRef]
15. Liu, X.; Tang, Z.; Si, Z.-Y.; Zhang, Z.; Zhao, L.; Liu, L. Enantioselective para-C(sp^2)-H Functionalization of Alkyl Benzene Derivatives via Cooperative Catalysis of Gold/Chiral Brønsted Acid. *Angew. Chem. Int. Ed.* **2022**, *61*, e202208874. [CrossRef]
16. Dong, K.; Liu, X.-S.; Wei, X.; Zhao, Y.; Liu, L. Borane-catalysed S–H insertion reaction of thiophenols and thiols with α-aryl-α-diazoesters. *Green Synth. Catal.* **2021**, *2*, 385–388. [CrossRef]
17. Maas, G. Ruthenium-catalysed carbenoid cyclopropanation reactions with diazo compounds. *Chem. Soc. Rev.* **2004**, *33*, 183–190. [CrossRef]
18. Morandi, B.; Carreira, E.M. Iron-catalyzed cyclopropanation in 6M KOH with in situ generation of diazomethane. *Science* **2012**, *335*, 1471–1474. [CrossRef]
19. Davies, H.M.L.; Beckwith, R.E. Catalytic enantioselective C–H activation by means of metal-carbenoid-induced C–H insertion. *Chem. Rev.* **2003**, *103*, 2861–2904. [CrossRef]
20. Davies, H.M.L.; Manning, J.R. Catalytic C–H functionalization by metal carbenoid and nitrenoid insertion. *Nature* **2008**, *451*, 417–424. [CrossRef]
21. Diaz-Requejo, M.M.; Perez, P.J. Coinage metal catalyzed C-H bond functionalization of hydrocarbons. *Chem. Rev.* **2008**, *108*, 3379–3394. [CrossRef] [PubMed]
22. Doyle, M.P.; Duffy, R.; Ratnikov, M.; Zhou, L. Catalytic carbene insertion into C-H bonds. *Chem. Rev.* **2010**, *110*, 704–724. [CrossRef] [PubMed]
23. Wencel-Delord, J.; Droge, T.; Liu, F.; Glorius, F. Towards mild metal-catalyzed C-H bond activation. *Chem. Soc. Rev.* **2011**, *40*, 4740–4761. [CrossRef]
24. Zheng, C.; You, S. Recent development of direct asymmetric functionalization of inert C-H bonds. *RSC Adv.* **2014**, *4*, 6173–6214. [CrossRef]

25. Arockiam, P.B.; Bruneau, C.; Dixneuf, P.H. Ruthenium(II)-catalyzed C-H bond activation and functionalization. *Chem. Rev.* **2012**, *112*, 5879–5918. [CrossRef] [PubMed]
26. Ma, B.; Liu, L.; Zhang, J. Gold-Catalyzed Site-Selective C−H Bond Functionalization with Diazo Compounds. *Asian J. Org. Chem.* **2018**, *7*, 2015–2025. [CrossRef]
27. Ma, B.; Chu, Z.; Huang, B.; Liu, Z.; Liu, L.; Zhang, J. Highly *para*-selective C-H Alkylation of Benzene Derivatives with 2,2,2-Trifluoroethyl α-Aryl-α-diazoesters. *Angew. Chem. Int. Ed.* **2017**, *56*, 2749–2753. [CrossRef]
28. Ji, X.; Zhang, Z.; Wang, Y.; Han, Y.; Peng, H.; Li, F.; Liu, L. Catalyst-free synthesis of α,α-disubstituted carboxylic acid derivatives under ambient conditions via a Wolff rearrangement reaction. *Org. Chem. Front.* **2021**, *8*, 6916–6922. [CrossRef]
29. Li, Y.; Wang, H.; Su, Y.; Li, R.; Li, C.; Liu, L.; Zhang, J. Phosphine-Catalyzed [3+2] Cycloaddition reaction of α-Diazoacetates and β-Trifluoromethyl Enones: A Facile Access to Multisubstituted 4-CF_3 Pyrazolines. *Org. Lett.* **2018**, *20*, 6444–6448. [CrossRef]
30. Gillingham, D.; Fei, N. Catalytic X-H insertion reactions based on carbenoids. *Chem. Soc. Rev.* **2013**, *42*, 4918–4931. [CrossRef]
31. Shi, Y.; Gulevich, A.V.; Gevorgyan, V. Rhodium-catalyzed NH insertion of pyridyl carbenes derived from pyridotriazoles: A general and efficient approach to 2-picolylamines and imidazo[1,5-a]pyridines. *Angew. Chem. Int. Ed.* **2014**, *53*, 14191–14195. [CrossRef]
32. Rousseau, G.; Breit, B. Removable directing groups in organic synthesis and catalysis. *Angew. Chem. Int. Ed.* **2011**, *50*, 2450–2494. [CrossRef] [PubMed]
33. Lee, D.; Kwon, K.; Yi, C.S. Dehydrative C-H Alkylation and Alkenylation of Phenols with Alcohols: Expedient Synthesis for Substituted Phenols and Benzofurans. *J. Am. Chem. Soc.* **2012**, *134*, 7325–7328. [CrossRef] [PubMed]
34. Yu, D.G.; de Azambuja, F.; Glorius, F. Direct functionalization with complete and switchable positional control: Free phenol as a role model. *Angew. Chem. Int. Ed.* **2014**, *53*, 7710–7712. [CrossRef] [PubMed]
35. Dai, J.; Shao, N.; Zhang, J.; Jia, R.; Wang, D. Cu(II)-Catalyzed ortho-Selective Aminomethylation of Phenols. *J. Am. Chem. Soc.* **2017**, *139*, 12390–12393. [CrossRef]
36. Huang, Z.; Lumb, J. Phenol-Directed C–H Functionalization. *ACS Catal.* **2018**, *9*, 521–555. [CrossRef]
37. Yu, C.; Patureau, F.W. Cu-Catalyzed Cross-Dehydrogenative ortho-Aminomethylation of Phenols. *Angew. Chem. Int. Ed.* **2018**, *57*, 11807–11811. [CrossRef]
38. Murai, M.; Yamamoto, M.; Takai, K. Rhenium-Catalyzed Regioselective ortho-Alkenylation and [3+2+1] Cycloaddition of Phenols with Internal Alkynes. *Org. Lett.* **2019**, *21*, 3441–3445. [CrossRef]
39. Miller, D.J.; Moody, C.J. Synthetic applications of the O-H insertion reactions of carbenes and carbenoids derived from diazocarbonyl and related diazo compounds. *Tetrahedron* **1995**, *51*, 10811–10843. [CrossRef]
40. Zhu, S.; Zhou, Q. Transition-metal-catalyzed enantioselective heteroatom-hydrogen bond insertion reactions. *Acc. Chem. Res.* **2012**, *45*, 1365–1377. [CrossRef]
41. Guo, X.; Hu, W. Novel multicomponent reactions via trapping of protic onium ylides with electrophiles. *Acc. Chem. Res.* **2013**, *46*, 2427–2440. [CrossRef]
42. Xie, X.; Zhu, S.; Guo, J.; Cai, Y.; Zhou, Q. Enantioselective palladium-catalyzed insertion of α-aryl-α-diazoacetates into the O-H bonds of phenols. *Angew. Chem. Int. Ed.* **2014**, *53*, 2978–2981. [CrossRef]
43. Gao, X.; Wu, B.; Huang, W.; Chen, M.; Zhou, Y. Enantioselective palladium-catalyzed C-H functionalization of indoles using an axially chiral 2,2′-bipyridine metal. *Angew. Chem. Int. Ed.* **2015**, *54*, 11956–11960. [CrossRef]
44. Zhang, Y.; Yao, Y.; He, L.; Liu, Y.; Shi, Y. Rhodium(II)/Chiral Phosphoric Acid-Cocatalyzed Enantioselective O-H Bond Insertion of α-Diazo Esters. *Adv. Synth. Catal.* **2017**, *359*, 2754–2761. [CrossRef]
45. Yu, Z.; Ma, B.; Chen, M.; Wu, H.; Liu, L.; Zhang, J. Highly site-selective direct C-H bond functionalization of phenols with α-aryl-α-diazoacetates and diazooxindoles via gold catalysis. *J. Am. Chem. Soc.* **2014**, *136*, 6904–6907. [CrossRef] [PubMed]
46. Yu, Z.; Li, Y.; Shi, J.; Ma, B.; Liu, L.; Zhang, J. $B(C_6F_5)_3$ Catalyzed Chemoselective and *ortho*-Selective Substitution of Phenols with α-Aryl α-Diazoesters. *Angew. Chem. Int. Ed.* **2016**, *55*, 14807–14811. [CrossRef] [PubMed]
47. Yu, Z.; Li, Y.; Zhang, P.; Liu, L.; Zhang, J. Ligand and counteranion enabled regiodivergent C-H bond functionalization of naphthols with α-aryl-α-diazoesters. *Chem. Sci.* **2019**, *10*, 6553–6559. [CrossRef]
48. Harada, S.; Sakai, C.; Tanikawa, K.; Nemoto, T. Gold-catalyzed chemoselective formal (3þ2)-Annulation reaction between β-naphthols and methyl aryldiazoacetate. *Tetrahedron* **2019**, *75*, 3650. [CrossRef]
49. Zhang, C.; Tang, C.; Jiao, N. Recent advances in copper-catalyzed dehydrogenative functionalization via a single electron transfer (SET) process. *Chem. Soc. Rev.* **2012**, *41*, 3464–3484. [CrossRef] [PubMed]
50. Zhao, X.; Zhang, Y.; Wang, J. Recent developments in copper-catalyzed reactions of diazo compounds. *Chem. Commun.* **2012**, *48*, 10162–10173. [CrossRef]
51. Guo, X.; Gu, D.; Wu, Z.; Zhang, W. Copper-catalyzed C-H functionalization reactions: Efficient synthesis of heterocycles. *Chem. Rev.* **2015**, *115*, 1622–1651. [CrossRef]
52. Zhu, X.; Chiba, S. Copper-catalyzed oxidative carbon-heteroatom bond formation: A recent update. *Chem. Soc. Rev.* **2016**, *45*, 4504–4523. [CrossRef]
53. Yang, Y.; Gao, W.; Wang, Y.; Wang, X.; Cao, F.; Shi, T.; Wang, Z. Recent Advances in Copper Promoted Inert C(sp^3)–H Functionalization. *ACS Catal.* **2021**, *11*, 967–984. [CrossRef]
54. Yates, P. The Copper-Catalyzed Decomposition of Diazoketones. *J. Am. Chem. Soc.* **1952**, *74*, 5376–5381. [CrossRef]
55. Maier, T.C.; Fu, G.C. Catalytic enantioselective O-H insertion reactions. *J. Am. Chem. Soc.* **2006**, *128*, 4594–4595. [CrossRef]

56. Chen, C.; Zhu, S.; Liu, B.; Wang, L.; Zhou, Q. Highly enantioselective insertion of carbenoids into O-H bonds of phenols: An efficient approach to chiral α-aryloxycarboxylic esters. *J. Am. Chem. Soc.* **2007**, *129*, 12616–12617. [CrossRef] [PubMed]
57. Song, X.; Zhu, S.; Xie, X.; Zhou, Q. Enantioselective copper-catalyzed intramolecular phenolic O-H bond insertion: Synthesis of chiral 2-carboxy dihydrobenzofurans, dihydrobenzopyrans, and tetrahydrobenzooxepines. *Angew. Chem. Int. Ed.* **2013**, *52*, 2555–2558. [CrossRef]
58. Morilla, M.E.; Molina, M.J.; Díaz-Requejo, M.M.; Belderraín, T.R.; Nicasio, M.C.; Trofimenko, S.; Pérez, P.J. Copper-catalyzed carbene insertion into O-H bonds: High selective conversion of alcohols into ethers. *Organometallics* **2003**, *22*, 2914–2918. [CrossRef]
59. Liang, Y.; Zhou, H.; Yu, Z. Why is copper(I) complex more competent than dirhodium(II) complex in catalytic asymmetric O-H insertion reactions? A computational study of the metal carbenoid O-H insertion into water. *J. Am. Chem. Soc.* **2009**, *131*, 17783–17785. [CrossRef]
60. Cai, Y.; Lu, Y.; Yu, C.; Lyu, H.; Miao, Z. Combined C-H functionalization/O-H insertion reaction to form tertiary β-alkoxy substituted β-aminophosphonates catalyzed by [Cu(MeCN)$_4$]PF$_6$. *Org. Biomol. Chem.* **2013**, *11*, 5491–5499. [CrossRef]
61. Liu, Y.; Luo, Z.; Zhang, J.Z.; Xia, F. DFT Calculations on the Mechanism of Transition-Metal-Catalyzed Reaction of Diazo Compounds with Phenols: O-H Insertion versus C-H Insertion. *J. Phys. Chem. A* **2016**, *120*, 6485–6492. [CrossRef] [PubMed]
62. Liu, Y.; Yu, Z.; Zhang, J.Z.; Liu, L.; Xia, F.; Zhang, J. Origins of unique gold-catalysed chemo- and site-selective C-H functionalization of phenols with diazo compounds. *Chem. Sci.* **2016**, *7*, 1988–1995. [CrossRef]
63. Ma, B.; Tang, Z.; Zhang, J.; Liu, L. Copper-catalysed ortho-selective C-H bond functionalization of phenols and naphthols with α-aryl-α-diazoesters. *Chem. Commun.* **2020**, *56*, 9485–9488. [CrossRef] [PubMed]
64. Li, F.; Zhang, J.Z.; Xia, F. How CuCl and CuCl$_2$ Insert into C-N Bonds of Diazo Compounds: An Electronic Structure and Mechanistic Study. *J. Phys. Chem. A* **2020**, *124*, 2029–2035. [CrossRef]
65. Fructos, M.R.; Belderrain, T.R.; Fremont, P.; Scott, N.M.; Nolan, S.P.; Diaz-Requejo, M.M.; Perez, P.J. A gold catalyst for carbene-transfer reactions from ethyl diazoacetate. *Angew. Chem. Int. Ed.* **2005**, *44*, 5284–5288. [CrossRef] [PubMed]
66. Barluenga, J.; Lonzi, G.; Tomas, M.; Lopez, L.A. Reactivity of stabilized vinyl diazo derivatives toward unsaturated hydrocarbons: Regioselective gold-catalyzed carbon-carbon bond formation. *Chem. Eur. J.* **2013**, *19*, 1573–1576. [CrossRef]
67. Liu, Y.; Yu, Z.; Luo, Z.; Zhang, J.; Liu, L.; Xia, F. Mechanistic Investigation of Aromatic C(sp^2)-H and Alkyl C(sp^3)-H Bond Insertion by Gold Carbenes. *J. Phys. Chem. A* **2016**, *120*, 1925–1932. [CrossRef]
68. Frisch, M.J.; Trucks, G.W.; Schlegel, H.B.; Scuseria, G.E.; Robb, M.A.; Cheeseman, J.R.; Scalmani, G.; Barone, V.; Mennucci, B.; Petersson, G.A.; et al. *Gaussian 09, Revision B.01*; Gaussian, Inc.: Wallingford, CT, USA, 2010.
69. Chai, J.; Head-Gordon, M. Long-range corrected hybrid density functionals with damped atom-atom dispersion corrections. *Phys. Chem. Chem. Phys.* **2008**, *10*, 6615–6620. [CrossRef]
70. Burns, L.A.; Vazquez-Mayagoitia, A.; Sumpter, B.G.; Sherrill, C.D. Density-functional approaches to noncovalent interactions: A comparison of dispersion corrections (DFT-D), exchange-hole dipole moment (XDM) theory, and specialized functionals. *J. Chem. Phys.* **2011**, *134*, 084107. [CrossRef]
71. Dolg, M.; Wedig, U.; Stoll, H.; Preuss, H. Energy-adjusted ab initio pseudopotentials for the first row transition elements. *J. Chem. Phys.* **1987**, *86*, 866–872. [CrossRef]
72. Hehre, W.J.; Ditchfield, R.; Pople, J.A. Self-Consistent Molecular Orbital Methods. XII. Further Extensions of Gaussian-Type Basis Sets for Use in Molecular Orbital Studies of Organic Molecules. *J. Chem. Phys.* **1972**, *56*, 2257–2261. [CrossRef]
73. Fukui, K. A Formulation of the reaction coordinate. *J. Phys. Chem.* **1970**, *74*, 4161–4163. [CrossRef]
74. Fukui, K. The path of chemical reactions-the IRC approach. *Acc. Chem. Res.* **1981**, *14*, 363–368. [CrossRef]
75. Marenich, A.V.; Cramer, C.J.; Truhlar, D.G. Universal solvation model based on solute electron density and on a continuum model of the solvent defined by the bulk dielectric constant and atomic surface tensions. *J. Phys. Chem. B* **2009**, *113*, 6378–6396. [CrossRef] [PubMed]

Disclaimer/Publisher's Note: The statements, opinions and data contained in all publications are solely those of the individual author(s) and contributor(s) and not of MDPI and/or the editor(s). MDPI and/or the editor(s) disclaim responsibility for any injury to people or property resulting from any ideas, methods, instructions or products referred to in the content.

Article

Boron Lewis Acid Catalysis Enables the Direct Cyanation of Benzyl Alcohols by Means of Isonitrile as Cyanide Source

Tong-Tong Xu, Jin-Lan Zhou, Guang-Yuan Cong, Jiang-Yi-Hui Sheng, Shi-Qi Wang, Yating Ma and Jian-Jun Feng *

State Key Laboratory of Chemo/Biosensing and Chemometrics, Advanced Catalytic Engineering Research Center of the Ministry of Education, College of Chemistry and Chemical Engineering, Hunan University, Changsha 410082, China
* Correspondence: jianjunfeng@hnu.edu.cn

Abstract: The development of an efficient and straightforward method for cyanation of alcohols is of great value. However, the cyanation of alcohols always requires toxic cyanide sources. Herein, an unprecedented synthetic application of an isonitrile as a safer cyanide source in $B(C_6F_5)_3$-catalyzed direct cyanation of alcohols is reported. With this approach, a wide range of valuable α-aryl nitriles was synthesized in good to excellent yields (up to 98%). The reaction can be scaled up and the practicability of this approach is further manifested in the synthesis of an anti-inflammatory drug, naproxen. Moreover, experimental studies were performed to illustrate the reaction mechanism.

Keywords: boron Lewis acid; α-aryl nitrile; cyanation; isonitrile; green chemistry

1. Introduction

The need for the development of new reactions that are based on applying the atom-economy concept [1] and avoiding the use of toxic reagents has become a consensus. Alcohols are highly attractive starting materials for synthesis because they are stable, have low toxicity, and are available. Direct nucleophilic substitution of an alcohol is attractive since water is, in principle, the only by-product [2–5]. However, this reaction is difficult because hydroxide is such a poor leaving group and therefore alcohols are classically derivatized to halides or pseudohalides prior to substitution, which results in the formation of vast amounts of waste. Thus, the development of new catalytic methodologies for dehydrative substitutions of alcohols was considered a central issue, as demonstrated by the inclusion of the "direct substitution of alcohols" in the ACS Green Chemistry Institute® Pharmaceutical Roundtable's 2018 update on key green chemistry research areas [6]. In this context, the deoxygenative cyanation of readily available benzyl alcohols represent one of the most powerful methods for preparing α-aryl nitriles [7–20], an important class of core structures found in bioactive moleculars [21] and functional materials [22], and precursors that have applications in the synthesis of well-known drugs such as verapamil [23], naproxen [24], and cytenamide [25] as shown in Figure 1.

As early as 1967, the one-pot method for the conversion of alcohols into cyanides based on the concept of the Mitsunobu reaction using NaCN as the cyanide source has been described [26]. Subsequently, there are a few reports on one-pot transformations of alcohols to α-aryl nitrile using $Me_3SiCl/NaI/NaCN$ [27], $PPh_3/nBu_4NCN/DDQ$ [28], N-(p-toluenesulfonyl)imidazole (TsIm)/NaCN [29], and $PPh_3/DEAD$/acetone cyanohydrin [30]. However, these methods suffer from major disadvantages such as the presence of hazardous and toxic cyanide sources and the use of stoichiometric activating reagents.

Recently, catalytic synthesis of α-aryl nitrile from benzyl alcohols was successfully developed. Ding's group performed pioneering work on the direct cyanation of α-aryl alcohols with trimethylsilyl cyanide (TMSCN) by indium halide catalysis [31]. Later, other Lewis acids, such as $FeCl_3·6H_2O$ [32], $Zn(OTf)_2$ [33], Brønsted acid montmorillonite catalysts [34], and others [35] were also used to catalyze this transformation. Besides these, ruthenium-catalyzed cyanation of benzyl alcohol with cuprous cyanide (CuCN) was also reported [36]. Again, these methods are typically plagued by notorious toxic cyanide source issues. To solve the severe safety issues associated with the handling of traditional cyanide sources, such as metal cyanides, ketone cyanohydrins [37–39], or TMSCN, several safer alternatives have been introduced, including: DMF [40], DMSO/ammonium ion [41], azobisisobutyronitrile (AIBN) [42], TsN(Ph)CN [43], isocyanides [44–51], and so on [52,53]. However, no direct cyanation of alcohols has been described so far for the synthesis of α-aryl nitriles from these safer organic CN surrogates (Scheme 1A).

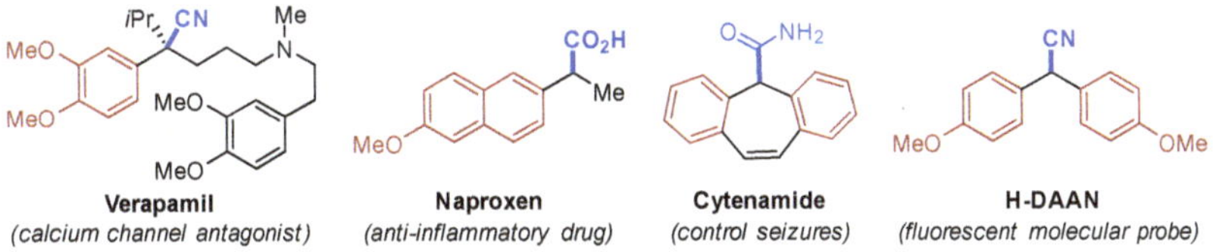

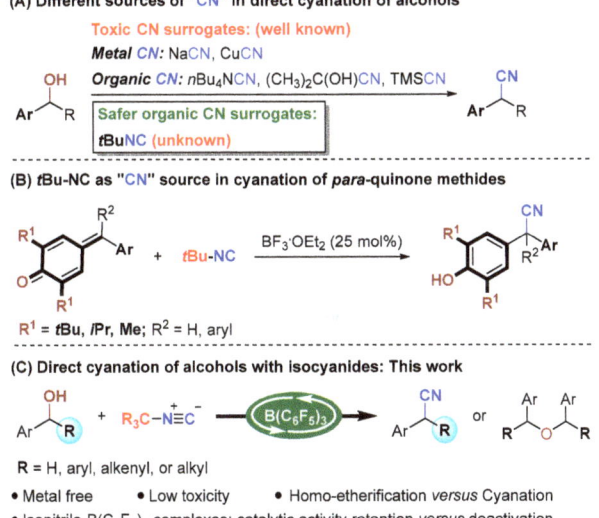

Figure 1. Verapamil, H-DAAN, and some derivatives of α-aryl nitriles.

Scheme 1. Our catalytic strategy to access α-aryl nitriles and its scientific context.

Isocyanides, which are isoelectronic with carbon monoxide, have emerged as powerful C1 building blocks in organic synthesis [54–60]. The distinctive reactivity of isocyanides makes them well-known in Passerini and Ugi multicomponent reactions and others [61]. In 1982, Saegusa and co-workers performed seminal work on the conjugate hydrocyanation and 1,2-addition reactions with *tert*-butyl isocyanide in the presence of stoichiometric Lewis acids [44]. Subsequently, commercially available *tert*-butyl isocyanide as a safer cyanide alternative in C-H cyanation [45–48], noble-metal-catalyzed cyanation of aryl iodides [49], and cyanothiolation of alkynes [50] has been reported, which further broadened the application of isocyanides in organic synthesis. To date, the catalytic method to obtain valuable α-aryl nitriles relied upon the usage of isocyanides as a cyanide source; however, these are rarely known. As a rare example, Muthukrishnan and co-workers reported a BF₃·OEt₂-catalyzed 1,6-conjugate addition reaction of *p*-quinone methides (*p*-QMs) with *tert*-butyl isocyanide for synthesis of α-diaryl and α-triaryl nitriles (Scheme 1B) [51]. Nonetheless, the substrates of the reaction are limited to *p*-QMs that feature bulky *tert*-butyl substituents at the 2- and 6-positions. Therefore, the development of new reactions with isocyanides for synthesis of α-aryl nitriles is highly desirable.

In recent years, B(C$_6$F$_5$)$_3$ as a non-metallic Lewis acid has received widespread attention because of its strong Lewis acidity, commercial availability, and environmental friendliness [62–68]. Although still limited in its success, it mainly involves the B(C$_6$F$_5$)$_3$-catalyzed activation of hydroxyl groups, as reported by Meng, Zhao and Chan [69,70], Marek [71], Tang [72], Maji [73], Gevorgyan [74], and Moran [75]. Inspired by these reports and building on our ongoing interest in the developing atom-economic reactions [76–79], we questioned if the direct cyanation of alcohols with isocyanides in the presence of B(C$_6$F$_5$)$_3$ could be realized to meet the requirements of atom economy and green chemistry (Scheme 1C). However, this hypothesis may face considerable challenges, such as the following: (a) the catalyst should be stable in wet and Lewis basic conditions, and (b) tert-butylisocyanide/B(C$_6$F$_5$)$_3$ and nitrile/B(C$_6$F$_5$)$_3$ adducts can be easily formed, as reported by Berke and Erker and co-workers [80]. It is unknown whether B(C$_6$F$_5$)$_3$ can maintain its catalytic activity during the current cyanation reaction, and (c) the catalyst should be able to dissociate from nitrile products. Finally, (d) another challenge is to suppress the B(C$_6$F$_5$)$_3$-catalyzed homo-etherification of alcohols reported by Chan and co-workers [70].

2. Results

We initiated our investigation with the optimization of the reaction of benzhydryl alcohol (**1a**) and *tert*-butyl isocyanide (**2a**). Initially, **1a** and 1.5 equivalents of **2a** were subjected to a solution of FeCl$_3$ (10 mol%) in toluene at 100 °C (Table 1, entry 1). However, FeCl$_3$ showed almost no catalytic activity. Other commonly used Lewis acids, including AlCl$_3$, Cu(OTf)$_2$, AgClO$_4$, and BF$_3$·(OEt)$_2$ were also tested, but led to low yield (see entries 2–5 in Table 1). Of note, using AlCl$_3$, BF$_3$·(OEt)$_2$, or TsOH·H$_2$O as a catalyst, a mixture of **3a** and some homo-etherification side product **4a** was obtained (entries 2, 5 and 7). With the use of Brønsted acid Tf$_2$NH, the cyanation provided the desired **3a** in 72% NMR yield. Interestingly, the reaction favors the formation of etherification product **4a** rather than **3a** when using diphenyl phosphate as the catalyst (entry 8 versus 9). Gratifyingly, we found that B(C$_6$F$_5$)$_3$ affords the desired α-aryl nitrile **3a** in >99% NMR yield at 100 °C (entry 6). No reaction occurred when B(C$_6$F$_5$)$_3$ was used as the catalyst at 50 °C (entry 11). Increasing the temperature to 80 °C improved both the yield of **3a** (20%) and ether **4a** (30%). Different solvents such as 1,2-dichloroethane (DCE), THF, and hexafluoroisopropanol (HFIP) (entries 12–15) were also tested, and the results revealed that toluene was superior to other solvents.

With the optimized conditions identified, we then proceeded to explore the scope of isocyanides (Scheme 2). Besides *t*Bu-NC (**2a**), other tertiary amine-derived isocyanides, such as **2b** and **2c**, can also be used as a novel cyano source in the current direct cyanation of α-aryl alcohols. In contrast, when secondary amine- and aniline-derived isocyanides were used as substrates, only the corresponding ether products were obtained (not shown). Of note, treatment of **1a** and **2c** with the standard conditions afforded **3a** in 81% NMR yield together with internal alkene **5** in 53% NMR yield and terminal alkene **6** in 38% NMR yield (Scheme 2b), indicating a tertiary carbon cation might be an intermediate.

Next, we turned to explore the generality of this cyanation reaction with a variety of alcohols with *t*Bu-NC (**2a**). As shown in Scheme 3, a wide range of benzylic alcohols can smoothly react with **2a** under the optimized conditions, giving the corresponding α-aryl nitriles in good to excellent yields. Diarylsubstituted alcohols (R^2 = aryl) underwent reaction with *t*Bu-NC to furnish the corresponding products (**3a**–**3k**) in 19–98% yields.

Table 1. Optimization of reaction conditions [a].

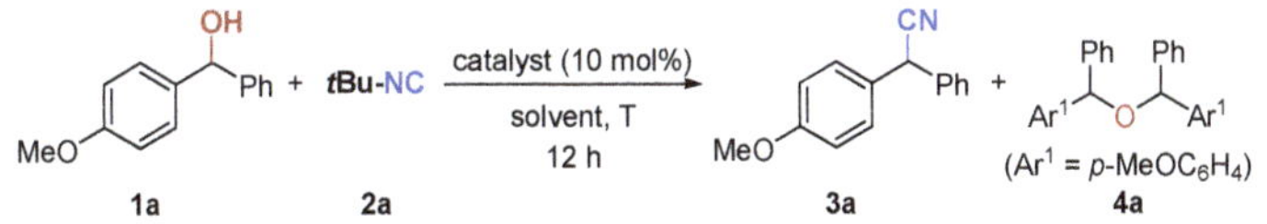

Entry	Catalyst	Solvent	T [°C]	Conv. (%)	3a [%] [b]	4a [%] [b]
1	FeCl$_3$	toluene	100	10	0	0
2	AlCl$_3$	toluene	100	86	16	28
3	Cu(OTf)$_2$	toluene	100	100	57	0
4	AgClO$_4$	toluene	100	21	7	0
5	BF$_3$·Et$_2$O	toluene	100	48	24	11
6	B(C$_6$F$_5$)$_3$	toluene	100	100	>99(98)	0
7	TsOH·H$_2$O	toluene	100	84	32	29
8	Tf$_2$NH	toluene	100	100	72	0
9	(PhO)$_2$P(=O)OH	toluene	100	100	0	45
10	B(C$_6$F$_5$)$_3$	toluene	80	70	20	30
11	B(C$_6$F$_5$)$_3$	toluene	50	0	0	0
12	B(C$_6$F$_5$)$_3$	DCE	reflux	100	93	0
13	B(C$_6$F$_5$)$_3$	THF	reflux	100	82	0
15	B(C$_6$F$_5$)$_3$	HFIP	reflux	50	20	0

[a] Reaction conditions: **1a** (0.20 mmol), **2a** (0.30 mmol, 1.5 equiv), catalyst (10 mol%), solvent (2 mL), T °C, under N$_2$ for 12 h. [b] The yields were determined by ^{1}H NMR spectroscopy using CH$_2$Br$_2$ as the internal standard. The number in the parentheses is the isolated yield of **3a**.

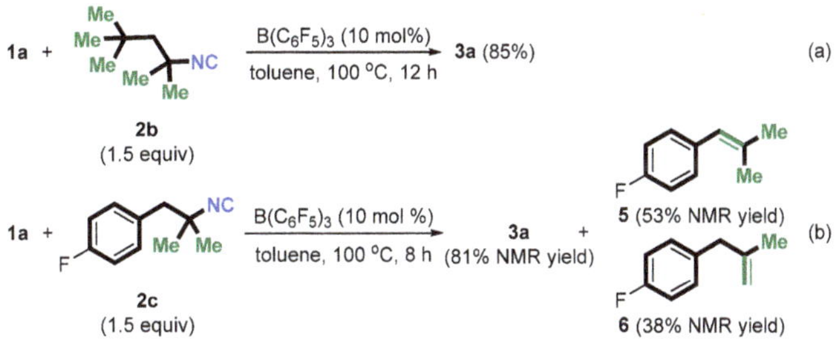

Scheme 2. Survey of the scope of isocyanides (a,b).

Diarylsubstituted alcohols bearing electron-donating (methoxy, alkyl) groups at the *para-*, *ortho-* or *meta-*positions of the benzene rings reacted smoothly (**1a–1f**). A variety of electron-withdrawing functional groups at the *para-*positions of the benzene rings such as -Br, -CF$_3$, and -CO$_2$Me were tolerated, affording the desired products in moderate to good yields (**1g–1i**). However, a low yield of **3j** was isolated from **1j** having an *ortho*-nitro group on the aromatic substituent. The naphthyl-containing alcohol **1k** and heteroaryl-containing alcohol **1l** also afforded the corresponding products in good to excellent yields. In addition, alkyl-substituted alcohols (R^2 = methyl, ethyl, and *tert*-butyl) were also well-tolerated to afford the desired products **3m–3o**. Of note, a competitive side reaction encountered in the reaction with **1m** or **1n** is the formation of styrene derivatives via the dehydration reactions of alcohols. Besides R^1 = methoxy (**1p–1q**), the substituent R^1 can be benzamide (**1r**). Benzhydrol **1s** was tested but afforded the corresponding product **3s** in low yield together with (oxybis(methanetriyl))tetrabenzene in 64% NMR yield. However, 1,2,3,4-tetrahydronaphthalen-1-ol (**1t**) and allylic alcohol (**1u**) underwent smooth cyanation. It is worth noting that the reported indium halide-catalyzed protocol with TMSCN as the cyanide source is not amenable to primary alcohol **1v** for cyanation reaction [31]. Our system, however, gives reasonable yield for the same substrate. To our delight, the precursors for naproxen and cytenamide, respectively, can also be efficiently obtained in high yields by this protocol (**3w** and **3x**).

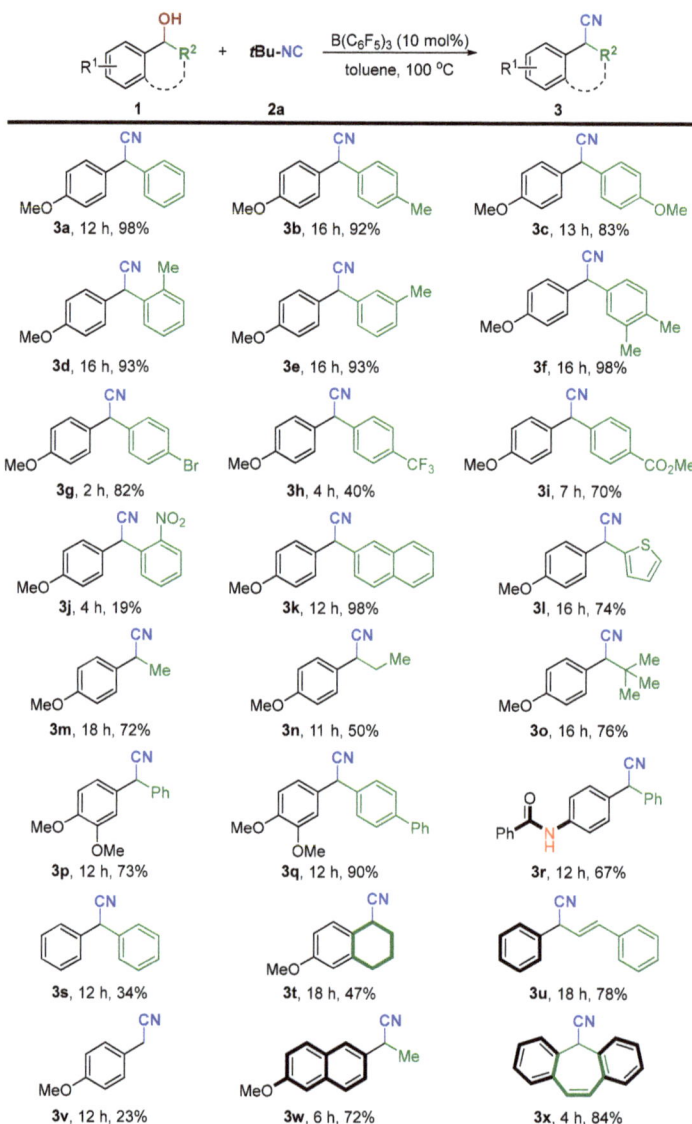

Scheme 3. Survey of the scope of α-aryl alcohols [a,b]. [a] Standard conditions: **1** (0.20 mmol), **2a** (0.30 mmol, 1.5 equiv), B(C₆F₅)₃ (10 mol%), toluene (2 mL), at 100 °C for 2–18 h. [b] Isolated yield of **3**.

To disclose the synthetic practicability of the developed method, we studied a gram-scale reaction, and 66% yield of **3w** was obtained, which might provide potential value in synthetic chemistry. Having established a protocol for the efficient synthesis of α-aryl nitrile **3w**, (±)-naproxen, a nonsteroidal anti-inflammatory drug [24], was prepared in three steps from commercially available materials (Scheme 4).

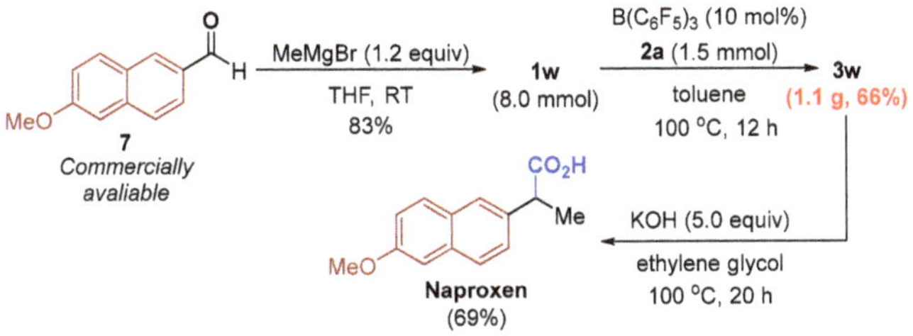

Scheme 4. Scale-up synthesis and synthetic transformations.

To gain insight into the reaction mechanism, several control experiments were conducted. When an enantiomerically pure sample of (R)-**1y** was subjected to standard conditions, the resulting nitrile **3y** was obtained in racemic form (Scheme 5a). Moreover, (R)-**1y** was also employed to perform the etherification reaction under the standard condition. The ^{1}H NMR spectrum showed the ether **4y** was obtained in 50% NMR yield with a 48/52 ratio of the two diastereomers (Scheme 5b). These control experiments support an S_N1 pathway and rule out a concerted S_N2 mechanism. The *tert*-Butylisocyanide-B(C$_6$F$_5$)$_3$ adduct (**8**) was easily prepared [80] and used as a catalyst in the current reaction, affording the corresponding product **3a** in 99% NMR yield (Scheme 5c). As shown in Table 1, entry 10, the reaction did form **3a** in 20% NMR yield along with ether **4a** in 30% NMR yield at 80 °C. Treatment of **4a** with the standard setup then gave the desired **3a** in 70% NMR yield (Scheme 5d).

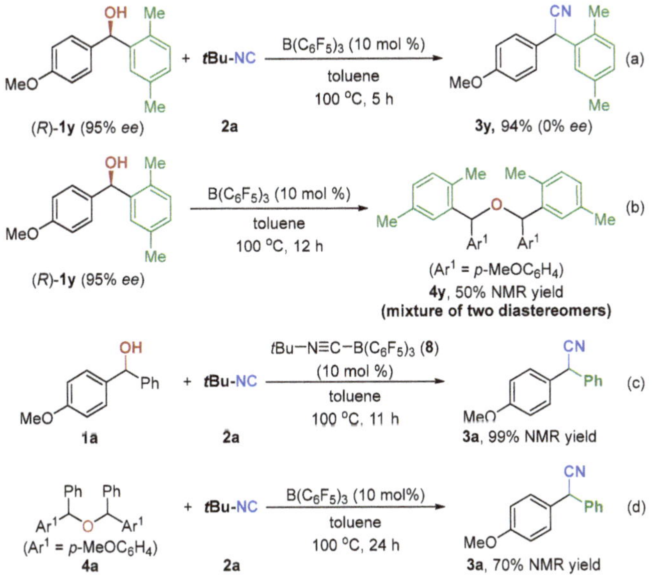

Scheme 5. Control experiments (a–d).

Based on the results above, a plausible mechanism is proposed in Scheme 6. First, *tert*-butylisocyanide **2a** forms a reversibly Lewis adduct **8** with B(C$_6$F$_5$)$_3$ [80]. The homo-etherification of alcohol in the presence of the B(C$_6$F$_5$)$_3$ quickly delivers the ether **4** [70], which could furnish an adduct **9** with B(C$_6$F$_5$)$_3$ through the oxygen center. Subsequently, the adduct **9** could break into an intermediate **10** and carbocation **11**. However, an alternative reaction path for the formation of the carbocation **11** directly from alcohol in the presence of the in situ-generated strong Brønsted acid B(C$_6$F$_5$)$_3$·nH$_2$O or boron Lewis acid B(C$_6$F$_5$)$_3$

(not shown, see Supporting Information for details) cannot be ruled out [69,70]. The carbocation **11** could then be intercepted instantaneously by the *t*Bu-NC (**2a**) to afford an intermediate **12** with a borate anion **10** as the counteranion. The stability of the tertiary carbon cation is the driving force to break the C-N bond in **12**, leading to the α-aryl nitrile **3** and 2-methylpropene by proton elimination via a *tert*-butyl carbon cation intermediate (supported by Scheme 2b) [81]. The borate anion **10**, on the other hand, transforms into alcohol **1** with the regeneration of the $B(C_6F_5)_3$ catalyst.

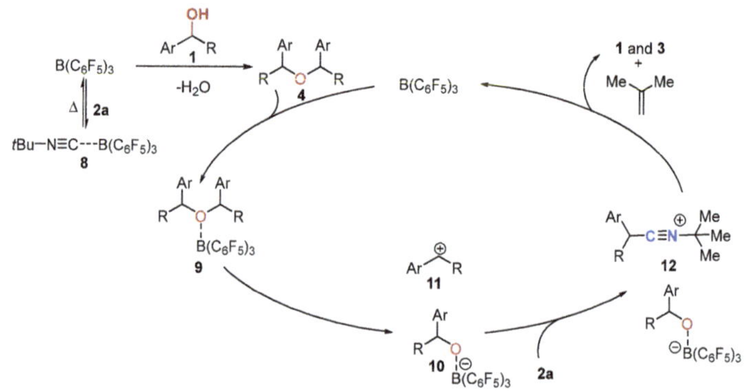

Scheme 6. Proposed mechanism.

3. Materials and Methods

3.1. General Information

All reactions were performed in flame-dried glassware using conventional Schlenk techniques under a static pressure of nitrogen unless otherwise stated. Liquids and solutions were transferred with syringes. The known alcohols **1** [31,35] and *t*Bu-NC-B(C_5F_5)$_3$ adduct **8** [80] were prepared according to reported procedures. (*R*)-**1y** was prepared in 80% yield according to the known procedure [82] (95% *ee* of (*R*)-**1y** was determined by HPLC: OJ-H Column, 5/95 iPrOH/hexane, 0.5 mL/min, 254 nm, 35 °C; retention time = 75.36 min (minor), 81.66 min (major)). Tris(pentafluorophenyl)borane (B(C_5F_5)$_3$, 98%, *Energy Chemical*) and *tert*-butyl isocyanate (97%, *Energy Chemical*) were purchased from commercial suppliers and used as received. Other commercially available reagents were purchased from *Sigma-Adrich*, *Leyan* and *Bide* Chemical Company. All solvents (tetrahydrofuran, toluene, and 1,2-dichloroethane etc.) were dried and purified following standard procedures. Technical grade solvents for extraction or chromatography (petroleum ether, CH_2Cl_2, and ethyl acetate) were distilled prior to use. Analytical thin layer chromatography (TLC) was performed on silica gel 60 F254 glass plates by *Merck*. Flash column chromatography was performed on silica gel 60 (40–63 μm, 230–400 mesh, ASTM) by *Grace* using the indicated solvents. ^1H, ^{13}C NMR spectra (Supplementary Materials) were recorded in $CDCl_3$ on Bruker AV400 instruments. Chemical shifts are reported in parts per million (ppm) and are referenced to the residual solvent resonance as the internal standard ($CHCl_3$: δ = 7.26 ppm for ^1H NMR and $CDCl_3$: δ = 77.0 ppm for ^{13}C NMR). Data are reported as follows: chemical shift, multiplicity (s = singlet, d = doublet, t = triplett, q = quartet, m = multiplet), coupling constants (Hz), and integration. Mass spectra were recorded on a THERMO FINNIGAN LTQ-XL. The MS inlet capillary temp was always maintained at 275 °C and capillary voltage at 5 kV. No other source gases were used when digestion was performed in microdroplets. The samples were dissolved in 1:1 methanol:water.

3.2. Typical Procedure for Direct Cyanation of Alcohols

In a glove box, alcohol **1** (0.2 mmol), isocyanide **2** (0.3 mmol, 1.5 equiv), $B(C_6F_5)_3$ (10.2 mg, 20 μmol, 10 mol%), and toluene (2.0 mL) were added to an oven-dried 10 mL pressure vial. The vial was sealed and removed from the glove box. The reaction mixture was stirred at 100 °C (oil bath) for 2–18 h. After the reaction was completed, the reaction mixture was purified by silica gel column chromatography by using petroleum ether/ethyl acetate mixture to obtain the desired nitrile 3.

3.3. Procedure for the Preparation of Naproxen

To a solution of 6-methoxy-2-naphthaldehyde (1.86 g, 10.0 mmol, 1.0 equiv) in THF (20 mL, 0.5 M), methylmagnesium bromide (4.0 mL, 12 mmol, 3.0 M, 1.2 equiv) was added. When the reaction was judged to have reached completion (as determined by TLC), sat. NH$_4$Cl was added slowly at 0 °C, and the mixture was extracted with EtOAc. The combined organic layers were washed with brine, dried over MgSO$_4$, and purified by column chromatography on silica gel to obtain 1w (1.60 g, 80% yield).

To an oven-dried 200 mL Schlenk flask containing a magnetic stir bar under an atmosphere of nitrogen, alcohol 1w (1.62 g, 8.0 mmol), isocyanide 2a (1.0 g, 12 mmol, 1.5 equiv), B(C$_6$F$_5$)$_3$ (0.41 g, 0.8 mmol, 10 mol%), and toluene (80 mL) were added. The reaction mixture was stirred at 100 °C (oil bath) for 12 h. After the reaction was completed, the reaction mixture was purified by silica gel column chromatography by using petroleum ether/ethyl acetate mixture to obtain the desired nitrile 3w (1.12 g, 66% yield).

To a suspension of 3w (42.3 mg, 0.2 mmol, 1.0 equiv) in ethylene glycol (0.6 mL), KOH (0.1 mL, 10 M solution in H$_2$O, 5.0 equiv) was added. The reaction vessel was sealed with a rubber septum and submerged in an oil bath at 100 °C for 20 h. We then added 1M HCl (2.5 mL) dropwise, and the mixture was extracted with EtOAc. The combined organic layers were washed with brine, dried over MgSO$_4$, and concentrated under reduced pressure. The residue was purified by column chromatography on silica gel to obtain Naproxen (31.8 mg, 69% yield).

3.4. Characterization Data of the Products

2-(4-methoxyphenyl)-2-phenylacetonitrile (3a) [83]. White solid (43.8 mg, 98% yield); mp 130–132 °C; R$_f$ = 0.50 (petroleum ether/EtOAc = 20/1). ^{1}H NMR (400 MHz, CDCl$_3$): δ 7.39–7.29 (m, 5H), 7.25 (d, J = 8.4 Hz, 2H), 6.88 (d, J = 8.4 Hz, 2H), 5.09 (s, 1H), 3.79 (s, 3H) ppm. ^{13}C NMR (100 MHz, CDCl$_3$): δ 159.4, 136.2, 129.1, 128.9, 128.1, 127.9, 127.6, 119.9, 114.5, 55.3, 41.8 ppm.

2-(4-methoxyphenyl)-2-(p-tolyl)acetonitrile (3b) [83]. White solid (43.9 mg, 92% yield); mp 85–87 °C; R$_f$ = 0.50 (petroleum ether/EtOAc = 20/1). ^{1}H NMR (400 MHz, CDCl$_3$): δ 7.25–7.15 (m, 6H), 6.87 (d, J = 8.8 Hz, 2H), 5.05 (s, 1H), 3.79 (s, 3H), 2.33 (s, 3H) ppm. ^{13}C NMR (100 MHz, CDCl$_3$): δ 159.3, 137.9, 133.2, 129.8, 128.8, 128.2, 127.5, 120.0, 114.4, 55.3, 41.4, 21.0 ppm.

2,2-bis(4-methoxyphenyl)acetonitrile (3c) [83]. White solid (41.9 mg, 83% yield); mp 156–158 °C; R$_f$ = 0.50 (petroleum ether/EtOAc = 20/1). ^{1}H NMR (400 MHz, CDCl$_3$): δ 7.23 (d, J = 8.4 Hz, 4H), 6.88 (d, J = 8.8 Hz, 4H), 5.04 (s, 1H), 3.79 (s, 6H) ppm. ^{13}C NMR (100 MHz, CDCl$_3$): δ 159.3, 128.7, 128.2, 120.1, 114.4, 55.3, 41.0 ppm.

2-(4-methoxyphenyl)-2-(o-tolyl)acetonitrile (3d) [83]. Colorless oil (44.4 mg, 93% yield); R$_f$ = 0.50 (petroleum ether/EtOAc = 20/1). ^{1}H NMR (400 MHz, CDCl$_3$): δ 7.40–7.38 (m, 1H), 7.25 (t, J = 4.8 Hz, 2H), 7.20–7.17 (m, 3H), 6.87 (d, J = 8.8 Hz, 2H), 5.23 (s, 1H), 3.78 (s, 3H), 2.26 (s, 3H) ppm. ^{13}C NMR (100 MHz, CDCl$_3$): δ 159.3, 135.8, 133.8, 131.2, 128.9, 128.5, 128.4, 126.8, 126.7, 119.8, 114.4, 55.3, 39.1, 19.4 ppm. MS (ESI) m/z: [M+H]$^+$ calcd. for C$_{16}$H$_{16}$NO: 238.12; found: 238.18

2-(4-methoxyphenyl)-2-(m-tolyl)acetonitrile (3e) [83]. Colorless oil (44.4 mg, 93% yield); R$_f$ = 0.50 (petroleum ether/EtOAc = 20/1). ^{1}H NMR (400 MHz, CDCl$_3$): δ 7.25–7.22 (m, 3H), 7.14–7.10 (m, 3H), 6.88 (d, J = 8.4 Hz, 2H), 5.04 (s, 1H), 3.78 (s, 3H), 2.33 (s, 3H) ppm. ^{13}C NMR (100 MHz, CDCl$_3$): δ 159.3, 139.0, 136.1, 128.9, 128.84, 128.81, 128.2, 128.0, 124.6, 120.0, 114.4, 55.3, 41.7, 21.3 ppm. MS (ESI) m/z: [M+H]$^+$ calcd. for C$_{16}$H$_{16}$NO: 238.12; found: 238.82.

2-(3,4-dimethylphenyl)-2-(4-methoxyphenyl)acetonitrile (3f) [83]. Colorless oil (55.1 mg, 98% yield); R$_f$ = 0.50 (petroleum ether/EtOAc = 20/1). ^{1}H NMR (400 MHz,

CDCl$_3$): δ 7.24 (d, J = 6.8 Hz, 2H), 7.12–7.03 (m, 3H), 6.87 (d, J = 6.4 Hz, 2H), 5.02 (s, 1H), 3.78 (s, 3H), 2.24 (s, 6H) ppm. ^{13}C NMR (100 MHz, CDCl$_3$): δ 159.3, 137.5, 136.6, 133.6, 130.2, 128.8, 128.7, 128.3, 124.9, 120.1, 114.4, 55.3, 41.4, 19.8, 19.4 ppm. MS (ESI) m/z: [M+H]$^+$ calcd. for C$_{17}$H$_{18}$NO: 252.14; found: 252.18.

2-(4-bromophenyl)-2-(4-methoxyphenyl)acetonitrile (3g) [83]. Colorless oil (49.4 mg, 82% yield); R$_f$ = 0.50 (petroleum ether/EtOAc = 20/1). ^{1}H NMR (400 MHz, CDCl$_3$): δ 7.48 (d, J = 8.4 Hz, 2H), 7.20 (t, J = 7.5 Hz, 4H), 6.88 (d, J = 8.4 Hz, 2H), 5.04 (s, 1H), 3.79 (s, 3H) ppm. ^{13}C NMR (100 MHz, CDCl$_3$): δ 159.6, 135.3, 132.2, 129.2, 128.8, 127.2, 122.2, 119.3, 114.6, 55.3, 41.2 ppm. MS (ESI) m/z: [M+H]$^+$ calcd. for C$_{15}$H$_{13}$BrNO: 302.02; found: 302.36.

2-(4-methoxyphenyl)-2-(4-(trifluoromethyl)phenyl)acetonitrile (3h). Colorless oil (23.2 mg, 40% yield); R$_f$ = 0.40 (petroleum ether/EtOAc = 10/1). ^{1}H NMR (400 MHz, CDCl$_3$): δ 7.63 (d, J = 8.4 Hz, 2H), 7.47 (d, J = 8.0 Hz, 2H), 7.24 (d, J = 8.4 Hz, 2H), 6.91 (d, J = 8.4 Hz, 2H), 5.14 (s, 1H), 3.80 (s, 3H) ppm. ^{13}C NMR (100 MHz, CDCl$_3$): δ 159.8, 140.1, 130.6 (q, J = 32.8 Hz), 128.9, 128.0, 126.9, 126.1 (q, J = 3.7 Hz), 123.8 (q, J = 270.5 Hz, 1H), 119.1, 114.8, 55.3, 41.6 ppm. ^{19}F NMR (376 MHz, CDCl$_3$) δ −62.75 (s) ppm. HRMS (MALDI-TOF/TOF) for C$_{16}$H$_{13}$F$_3$NO [M+H]$^+$: calculated 292.0944, found 292.0947.

Methyl 4-(cyano(4-methoxyphenyl)methyl)benzoate (3i). Yellow oil (39.4 mg, 70% yield); R$_f$ = 0.50 (petroleum ether/EtOAc = 5/1). ^{1}H NMR (400 MHz, CDCl$_3$): δ 8.03 (d, J = 8.2 Hz, 2H), 7.42 (d, J = 8.2 Hz, 2H), 7.23 (d, J = 8.6 Hz, 2H), 6.89 (d, J = 8.6 Hz, 2H), 5.14 (s, 1H), 3.91 (s, 3H), 3.79 (s, 3H) ppm. ^{13}C NMR (100 MHz, CDCl$_3$): δ 166.3, 159.6, 141.0, 130.4, 130.1, 128.9, 127.6, 127.1, 119.2, 114.7, 55.3, 52.2, 41.7 ppm. HRMS (ESI) for C$_{17}$H$_{16}$NO$_3$ [M+H]$^+$: calculated 282.1125, found 282.1126.

2-(4-methoxyphenyl)-2-(2-nitrophenyl)acetonitrile (3j) [84]. Yellow oil (10.1 mg, 19% yield); R$_f$ = 0.40 (petroleum ether/EtOAc = 10/1). ^{1}H NMR (400 MHz, CDCl$_3$): δ 8.04 (d, J = 8.0 Hz, 1H), 7.75 (d, J = 7.6 Hz, 1H), 7.69 (t, J = 7.2 Hz, 1H), 7.53 (t, J = 7.6 Hz, 1H), 7.22 (d, J = 8.4 Hz, 2H), 6.87 (d, J = 8.4 Hz, 2H), 6.11 (s, 1H), 3.79 (s, 3H) ppm. ^{13}C NMR (100 MHz, CDCl$_3$): δ 159.8, 147.7, 134.0, 130.9, 130.7, 129.5, 129.1, 126.0, 125.7, 118.8, 114.7, 55.3, 37.6 ppm.

2-(4-methoxyphenyl)-2-(naphthalen-2-yl)acetonitrile (3k) [85]. Colorless oil (53.5 mg, 98% yield). R$_f$ = 0.50 (petroleum ether/EtOAc = 20/1). ^{1}H NMR (400 MHz, CDCl$_3$): δ 7.87–7.81 (m, 4H), 7.53–7.48 (m, 2H), 7.34 (d, J = 8.4 Hz, 1H), 7.29 (d, J = 8.8 Hz, 2H), 6.89 (d, J = 8.4 Hz, 2H), 5.25 (s, 1H), 3.79 (s, 3H) ppm. ^{13}C NMR (100 MHz, CDCl$_3$): δ 159.5, 133.4, 133.2, 132.7, 129.2, 129.0, 128.0, 127.8, 127.7, 126.7, 126.6, 126.5, 125.2, 119.8, 114.5, 55.3, 42.0 ppm.

2-(4-methoxyphenyl)-2-(thiophen-2-yl)acetonitrile (3l). Colorless oil (33.8 mg, 74% yield); R$_f$ = 0.50 (petroleum ether/EtOAc = 20/1). ^{1}H NMR (400 MHz, CDCl$_3$): δ 7.31 (d, J = 8.4 Hz, 2H), 7.25 (d, J = 5.2 Hz, 1H), 7.06 (d, J = 3.2 Hz, 1H), 6.97–6.95 (m, 1H), 6.91 (d, J = 8.4 Hz, 2H), 5.30 (s, 1H), 3.80 (s, 3H) ppm. ^{13}C NMR (100 MHz, CDCl$_3$): δ 159.7, 139.1, 128.7, 127.5, 127.0, 126.5, 126.3, 119.0, 114.5, 55.3, 37.2 ppm. HRMS (MALDI-TOF/TOF) for C$_{13}$H$_{12}$NOS [M+H]$^+$: calculated 230.0634, found 230.0631.

2-(4-methoxyphenyl)propanenitrile (3m) [48]. Colorless oil (23.3 mg, 72% yield); R$_f$ = 0.70 (petroleum ether/EtOAc = 20/1). ^{1}H NMR (400 MHz, CDCl$_3$): δ 7.27 (d, J = 8.4 Hz, 2H), 6.90 (d, J = 8.4 Hz, 2H), 3.85 (q, J = 7.6 Hz, 1H), 3.81 (s, 3H), 1.61 (d, J = 7.3 Hz, 3H) ppm. ^{13}C NMR (100 MHz, CDCl$_3$): δ 159.2, 129.0, 127.8, 121.8, 114.4, 55.3, 30.4, 21.5 ppm.

2-(4-methoxyphenyl)butanenitrile (3n) [86]. Colorless oil (17.5 mg, 50% yield); R$_f$ = 0.50 (petroleum ether/EtOAc = 20/1). ^{1}H NMR (400 MHz, CDCl$_3$): δ 7.23 (d, J = 8.4 Hz, 2H), 6.90 (d, J = 8.4 Hz, 2H), 3.81 (s, 3H), 3.68 (t, J = 7.2 Hz, 1H), 1.95–1.86 (m, 2H), 1.06 (t, J = 7.2 Hz, 3H) ppm. ^{13}C NMR (100 MHz, CDCl$_3$): δ 159.3, 128.4, 127.7, 121.0, 114.3, 55.3, 38.1, 29.2, 11.4 ppm.

2-(4-methoxyphenyl)-3,3-dimethylbutanenitrile (3o) [87]. White solid (31.1 mg, 76% yield); mp 50–52 °C; R$_f$ = 0.70 (petroleum ether/EtOAc = 20/1). ^{1}H NMR (400 MHz, CDCl$_3$): δ 7.19 (d, J = 6.8 Hz, 2H), 6.88 (d, J = 8.4 Hz, 2H), 3.82 (s, 3H), 3.51 (s, 1H), 1.04 (s, 9H) ppm. ^{13}C NMR (100 MHz, CDCl$_3$): δ 159.3, 130.4, 125.4, 120.6, 113.6, 55.3, 48.9, 35.1, 27.3 ppm.

2-(3,4-dimethoxyphenyl)-2-phenylacetonitrile (**3p**) [88]. Colorless oil (37.0 mg, 73% yield); R_f = 0.50 (petroleum ether/EtOAc = 20/1). ^{1}H NMR (400 MHz, CDCl$_3$): δ 7.38–7.31 (m, 5H), 6.90–6.80 (m, 3H), 5.10 (s, 1H), 3.86 (s, 3H), 3.84 (s, 3H) ppm. ^{13}C NMR (100 MHz, CDCl$_3$): δ 149.4, 148.9, 136.0, 129.1, 128.13, 128.12, 127.5, 120.1, 119.8, 111.3, 110.7, 55.89, 55.88, 42.1 ppm.

2-([1,1′-biphenyl]-4-yl)-2-(3,4-dimethoxyphenyl)acetonitrile (**3q**) [48]. Colorless oil (59.3 mg, 90% yield); R_f = 0.50 (petroleum ether/EtOAc = 20/1). ^{1}H NMR (400 MHz, CDCl$_3$): δ 7.60–7.55 (m, 4H), 7.45–7.39 (m, 4H), 7.37–7.33 (m, 1H), 6.92 (d, J = 8.4 Hz, 1H), 6.85 (d, J = 7.2 Hz, 2H), 5.13 (s, 1H), 3.86 (s, 3H), 3.85 (s, 3H) ppm. ^{13}C NMR (100 MHz, CDCl$_3$): δ 149.5, 149.0, 141.1, 140.1, 135.0, 128.8, 128.1, 128.0, 127.8, 127.6, 127.0, 120.2, 119.8, 111.4, 110.7, 56.0, 55.9, 41.8 ppm. HRMS (MALDI-TOF/TOF) for C$_{22}$H$_{20}$NO$_2$ [M+H]$^+$: calculated 330.1489, found 330.1386.

N-(4-(cyano(phenyl)methyl)phenyl)benzamide (**3r**). White solid (41.6 mg, 67% yield); mp 196–198 °C; R_f = 0.50 (petroleum ether/EtOAc = 20/1). ^{1}H NMR (400 MHz, DMSO-d6): δ 10.35 (s, 1H), 7.98 (d, J = 7.6 Hz, 2H), 7.85 (d, J = 8.4 Hz, 2H), 7.65–7.61 (m, 1H), 7.58–7.55 (m, 2H), 7.48–7.39 (m, 7H), 5.81 (s, 1H) ppm. ^{13}C NMR (100 MHz, DMSO-d6): δ 166.5, 139.8, 137.7, 135.6, 132.5, 130.1, 129.3, 128.8, 128.7, 128.5, 128.3, 121.8, 121.3, 41.2 ppm. HRMS (ESI) m/z: [M+H]$^+$ calcd. for C$_{21}$H$_{17}$N$_2$O: 313.1336; found: 313.1334.

2,2-diphenylacetonitrile (**3s**) [83]. White solid (13.1 mg, 34% yield); mp 70–72 °C; R_f = 0.50 (petroleum ether/EtOAc = 15/1). ^{1}H NMR (400 MHz, CDCl$_3$): δ 7.39–7.30 (m, 10H), 5.14 (s, 1H) ppm. ^{13}C NMR (100 MHz, CDCl$_3$): δ 135.9, 129.2, 128.2, 127.7, 119.6, 42.6 ppm.

6-methoxy-1,2,3,4-tetrahydronaphthalene-1-carbonitrile (**3t**) [48]. Colorless oil (17.5 mg, 47% yield); R_f = 0.50 (petroleum ether/EtOAc = 20/1). ^{1}H NMR (400 MHz, CDCl$_3$): δ 7.26 (d, J = 8.8 Hz, 1H), 6.78–6.75 (m, 1H), 6.64 (s, 1H), 3.92 (t, J = 6.4 Hz, 1H), 3.78 (s, 3H), 2.87–2.71 (m, 2H), 2.13–2.09 (m, 2H), 2.07–1.96 (m, 1H), 1.87–1.77 (m, 1H) ppm. ^{13}C NMR (100 MHz, CDCl$_3$): δ 159.1, 137.7, 129.9, 121.9, 114.2, 112.8, 55.2, 30.1, 28.7, 27.5, 20.7 ppm.

(E)-2,4-diphenylbut-3-enenitrile (**3u**) [89]. Colorless oil (34.0 mg, 78% yield); R_f = 0.50 (petroleum ether/EtOAc = 20/1). ^{1}H NMR (400 MHz, CDCl$_3$): δ 7.42–7.26 (m, 10H), 6.81 (d, J = 15.6 Hz, 1H), 6.19 (dd, J = 16.0, 6.4 Hz, 1H), 4.69 (d, J = 6.4 Hz, 1H) ppm. ^{13}C NMR (100 MHz, CDCl$_3$): δ 135.4, 134.5, 133.2, 129.2, 128.7, 128.4, 127.5, 126.7, 123.2, 118.8, 40.0 ppm.

2-(4-methoxyphenyl)acetonitrile (**3v**) [90]. Colorless oil (17.0 mg, 23% yield, 0.5 mmol scale); R_f = 0.50 (petroleum ether/EtOAc = 15/1). ^{1}H NMR (400 MHz, CDCl$_3$): δ 7.23 (d, J = 8.4 Hz, 2H), 6.90 (d, J = 8.4 Hz, 2H), 3.80 (s, 3H), 3.68 (s, 2H) ppm. ^{13}C NMR (100 MHz, CDCl$_3$): δ 159.3, 129.0, 121.8, 118.1, 114.5, 55.3, 22.8 ppm.

2-(6-methoxynaphthalen-2-yl)propanenitrile (**3w**) [91]. White solid (42.3 mg, 72% yield); mp 72–74 °C; R_f = 0.70 (petroleum ether/EtOAc = 20/1). ^{1}H NMR (400 MHz, CDCl$_3$): δ 7.76–7.72 (m, 3H), 7.38 (dd, J = 8.4, 2.0 Hz, 1H), 7.18 (dd, J = 8.8, 2.4 Hz, 1H), 7.13 (d, J = 2.4 Hz, 1H), 4.02 (q, J = 7.2 Hz, 1H), 3.92 (s, 3H), 1.71 (d, J = 7.2 Hz, 3H) ppm. ^{13}C NMR (100 MHz, CDCl$_3$): δ 158.1, 134.0, 132.0, 129.3, 128.8, 127.9, 125.4, 124.9, 121.7, 119.6, 105.7, 55.3, 31.2, 21.4 ppm.

5H-dibenzo[a,d][7]annulene-5-carbonitrile (**3x**) [25]. Colorless oil (36.6 mg, 84% yield); R_f = 0.50 (petroleum ether/EtOAc = 20/1). ^{1}H NMR (400 MHz, CDCl$_3$): δ 7.68 (s, 2H), 7.40 (t, J = 7.2 Hz, 2H), 7.35–7.28 (m, 4H), 7.11 (s, 2H), 4.72 (s, 1H) ppm. ^{13}C NMR (100 MHz, CDCl$_3$): δ 133.9, 132.3, 131.3, 129.2, 128.6, 127.8, 125.3, 118.3, 41.1 ppm.

2-(2,5-dimethylphenyl)-2-(4-methoxyphenyl)acetonitrile (**3y**) [92]. Colorless oil (47.2 mg, 94% yield); R_f = 0.50 (petroleum ether/EtOAc = 20/1). ^{1}H NMR (400 MHz, CDCl$_3$): δ 7.22–7.17 (m, 3H), 7.07 (s, 2H), 6.87 (d, J = 8.8 Hz, 2H), 5.20 (s, 1H), 3.79 (s, 3H), 2.33 (s, 3H), 2.21 (s, 3H) ppm. ^{13}C NMR (100 MHz, CDCl$_3$): δ 159.3, 136.4, 133.6, 132.6, 131.1, 129.2, 129.1, 128.9, 127.1, 120.0, 114.4, 55.3, 39.1, 21.0, 18.9 ppm.

2,2′-(oxybis((4-methoxyphenyl)methylene))bis(1,4-dimethylbenzene) (**4y**). Colorless oil (17.9 mg, 38% yield); R_f = 0.50 (petroleum ether/EtOAc = 15/1). ^{1}H NMR (400

MHz, CDCl$_3$): δ 7.46 (s, 1H), 7.39 (s, 1H), 7.20–7.18 (m, 4H), 6.98 (d, J = 10.8 Hz, 4H), 6.82 (t, J = 6.4 Hz, 4H), 5.46 (s, 1H), 5.45 (s, 1H), 3.77 (s, 6H), 2.33 (s, 3H), 2.31 (s, 3H), 1.96 (s, 6H) ppm. ^{13}C NMR (100 MHz, CDCl$_3$): δ 158.8, 158.6, 139.9, 139.6, 135.3, 134.0, 133.8, 133.0, 132.3, 130.4, 130.2, 129.1, 128.7, 128.3, 128.0, 127.8, 127.5, 113.6, 113.5, 76.9, 76.5, 55.2, 21.2, 21.2, 18.9, 18.8 ppm. HRMS (ESI) for C$_{32}$H$_{34}$O$_3$Na [M+Na]$^+$: calculated 489.2400, found 489.2414.

2-(6-methoxynaphthalen-2-yl)propanoic acid (Naproxen) [24]. White solid (31.8 mg, 69% yield); mp 156–158 °C; R$_f$ = 0.50 (petroleum ether/EtOAc/MeOH = 5/2/1). ^{1}H NMR (400 MHz, CDCl$_3$): δ 7.70–7.67 (m, 3H), 7.40 (d, J = 8.4 Hz, 1H), 7.14–7.10 (m, 2H), 3.90–3.85 (m, 4H), 1.58 (d, J = 6.8 Hz, 3H) ppm. ^{13}C NMR (100 MHz, CDCl$_3$): δ 179.9, 157.7, 135.1, 133.8, 129.3, 128.9, 127.2, 126.2, 126.1, 118.9, 105.7, 55.3, 45.3, 18.2 ppm.

4. Conclusions

In conclusion, by taking advantage of isonitriles as low-toxic CN surrogates in the metal-free cyanation of alcohols, an efficient and green method for the direct catalytic synthesis of α-aryl nitriles was developed (up to 98% yield). To the best of our knowledge, this is the first B(C$_6$F$_5$)$_3$-catalyzed transformation of isonitriles. Control experiments support an S$_N$1 pathway and rule out a concerted S$_N$2 mechanism. The in situ-generated ether **4** can be converted to the desired α-aryl nitriles under the current catalytic system via cleavage of the C-O bond. The use of readily available starting materials, low catalyst loading, a broad substrate scope, ease of scale-up, and application in the synthesis of the precursors for naproxen and cytenamide make this approach very practical and attractive. With these advantages, we expect that this method will find wide applications in organic synthesis.

Supplementary Materials: The following supporting information can be downloaded at: https://www.mdpi.com/article/10.3390/molecules28052174/s1, ^{1}H, ^{13}C, ^{19}F and HPLC spectra.

Author Contributions: Conceptualization, J.-J.F.; methodology, T.-T.X. and J.-J.F.; investigation, T.-T.X., J.-L.Z., G.-Y.C. and J.-Y.H.S.; data curation, S.-Q.W. and Y.M.; writing—original draft preparation, T.-T.X. and J.-J.F.; writing—review and editing, T.-T.X. and J.-J.F.; project administration, J.-J.F. All authors have read and agreed to the published version of the manuscript.

Funding: This project was supported by the Fundamental Research Funds for the Central Universities.

Institutional Review Board Statement: Not applicable.

Informed Consent Statement: Not applicable.

Data Availability Statement: Not applicable.

Acknowledgments: The authors also thank the Fundamental Research Funds for the Central Universities. We thank Zhongyan Zhou and Lei Yue (Hunan University, China) for experimental assistance and spectroscopic characterization.

Conflicts of Interest: The authors declare no conflict of interest.

Sample Availability: Samples of the compounds **3** are available from the authors.

References

1. Trost, B.M. Atom Economy—A Challenge for Organic Synthesis: Homogeneous Catalysis Leads the Way. *Angew. Chem. Int. Ed. Engl.* **1995**, *34*, 259–281. [CrossRef]
2. Estopina-Duran, S.; Taylor, J.E. Bronsted Acid-Catalysed Dehydrative Substitution Reactions of Alcohols. *Chem. Eur. J.* **2021**, *27*, 106–120. [CrossRef]
3. Huy, P.H. Lewis Base Catalysis Promoted Nucleophilic Substitutions—Recent Advances and Future Directions. *Eur. J. Org. Chem.* **2020**, 10–27. [CrossRef]
4. Moran, J.; Dryzhakov, M.; Richmond, E. Recent Advances in Direct Catalytic Dehydrative Substitution of Alcohols. *Synthesis* **2016**, *48*, 935–959. [CrossRef]
5. Emer, E.; Sinisi, R.; Capdevila, M.G.; Petruzziello, D.; De Vincentiis, F.; Cozzi, P.G. Direct Nucleophilic S$_N$1-Type Reactions of Alcohols. *Eur. J. Org. Chem.* **2011**, 647–666. [CrossRef]

6. Bryan, M.C.; Dunn, P.J.; Entwistle, D.; Gallou, F.; Koenig, S.G.; Hayler, J.D.; Hickey, M.R.; Hughes, S.; Kopach, M.E.; Moine, G.; et al. Key Green Chemistry Research Areas from a Pharmaceutical Manufacturers' Perspective Revisited. *Green Chem.* **2018**, *20*, 5082–5103. [CrossRef]
7. Zhang, W.; Wang, F.; McCann, S.D.; Wang, D.; Chen, P.; Stahl, S.S.; Liu, G. Enantioselective Cyanation of Benzylic C–H Bonds Via Copper-Catalyzed Radical Relay. *Science* **2016**, *353*, 1014–1018. [CrossRef] [PubMed]
8. Wang, D.; Zhu, N.; Chen, P.; Lin, Z.; Liu, G. Enantioselective Decarboxylative Cyanation Employing Cooperative Photoredox Catalysis and Copper Catalysis. *J. Am. Chem. Soc.* **2017**, *139*, 15632–15635. [CrossRef] [PubMed]
9. Yuan, Y.; Yang, J.; Zhang, J. Cu-Catalyzed Enantioselective Decarboxylative Cyanation Via the Synergistic Merger of Photocatalysis and Electrochemistry. *Chem. Sci.* **2023**, *14*, 705–710. [CrossRef] [PubMed]
10. Culkin, D.A.; Hartwig, J.F. Palladium-Catalyzed α-Arylation of Carbonyl Compounds and Nitriles. *Acc. Chem. Res.* **2003**, *36*, 234–245. [CrossRef]
11. Wu, G.; Deng, Y.; Wu, C.; Zhang, Y.; Wang, J. Synthesis of α-Aryl Esters and Nitriles: Deaminative Coupling of α-Aminoesters and α-Aminoacetonitriles with Arylboronic Acids. *Angew. Chem. Int. Ed.* **2014**, *53*, 10510–10514. [CrossRef]
12. Falk, A.; Goderz, A.L.; Schmalz, H.G. Enantioselective Nickel-Catalyzed Hydrocyanation of Vinylarenes Using Chiral Phosphine-Phosphite Ligands and TMS-CN as a Source of HCN. *Angew. Chem. Int. Ed.* **2013**, *52*, 1576–1580. [CrossRef]
13. Singh, D.K.; Prasad, S.S.; Kim, J.; Kim, I. One-Pot, Three-Component Approach to Diarylacetonitriles. *Org. Chem. Front.* **2019**, *6*, 669–673. [CrossRef]
14. Tan, J.-P.; Chen, Y.; Ren, X.; Guo, Y.; Yi, B.; Zhang, H.; Gao, G.; Wang, T. In Situ Phosphonium-Containing Lewis Base-Catalyzed 1,6-Cyanation Reaction: A Facile Way to Obtain α-Diaryl and α-Triaryl Acetonitriles. *Org. Chem. Front.* **2022**, *9*, 156–162. [CrossRef]
15. Kadunce, N.T.; Reisman, S.E. Nickel-Catalyzed Asymmetric Reductive Cross-Coupling between Heteroaryl Iodides and α-Chloronitriles. *J. Am. Chem. Soc.* **2015**, *137*, 10480–10483. [CrossRef] [PubMed]
16. Jiao, Z.; Chee, K.W.; Zhou, J.S. Palladium-Catalyzed Asymmetric α-Arylation of Alkylnitriles. *J. Am. Chem. Soc.* **2016**, *138*, 16240–16243. [CrossRef] [PubMed]
17. Liu, R.Y.; Bae, M.; Buchwald, S.L. Mechanistic Insight Facilitates Discovery of a Mild and Efficient Copper-Catalyzed Dehydration of Primary Amides to Nitriles Using Hydrosilanes. *J. Am. Chem. Soc.* **2018**, *140*, 1627–1631. [CrossRef] [PubMed]
18. Fang, X.; Yu, P.; Morandi, B. Catalytic Reversible Alkene-Nitrile Interconversion through Controllable Transfer Hydrocyanation. *Science* **2016**, *351*, 832–836. [CrossRef]
19. Goswami, P.; Singh, G.; Vijaya Anand, R. N-Heterocyclic Carbene Catalyzed 1,6-Conjugate Addition of Me₃Si-CN to *para*-Quinone Methides and Fuchsones: Access to α-Arylated Nitriles. *Org. Lett.* **2017**, *19*, 1982–1985. [CrossRef] [PubMed]
20. Michel, N.W.M.; Jeanneret, A.D.M.; Kim, H.; Rousseaux, S.A.L. Nickel-Catalyzed Cyanation of Benzylic and Allylic Pivalate Esters. *J. Org. Chem.* **2018**, *83*, 11860–11872. [CrossRef]
21. Fleming, F.F.; Yao, L.; Ravikumar, P.C.; Funk, L.; Shook, B.C. Nitrile-Containing Pharmaceuticals: Efficacious Roles of the Nitrile Pharmacophore. *J. Med. Chem.* **2010**, *53*, 7902–7917. [CrossRef] [PubMed]
22. Yamamoto, T.; Kato, S.; Aoki, D.; Otsuka, H. A Diarylacetonitrile as a Molecular Probe for the Detection of Polymeric Mechanoradicals in the Bulk State through a Radical Chain-Transfer Mechanism. *Angew. Chem. Int. Ed.* **2021**, *60*, 2680–2683. [CrossRef] [PubMed]
23. Sica, D.A.; Prisant, L.M. Pharmacologic and Therapeutic Considerationsin Hypertension Therapy with Calcium Channel Blockers: Focus on Verapamil. *J. Clin. Hypertens.* **2007**, *9*, 1–22. [CrossRef]
24. Harrington, P.J.; Lodewijk, E. Twenty Years of Naproxen Technology. *Org. Process Res. Dev.* **1997**, *1*, 72–76. [CrossRef]
25. Bedford, C.T.M. An Efficient, Large-Scale Synthesis of Cytenamide. *J. Chem. Res.* **2018**, *42*, 153–155. [CrossRef]
26. Brett, D.; Downie, I.M.; Lee, J.B. Sugars with Potential Antiviral Activity. Conversion of Hydroxy Compounds to Nitriles. *J. Org. Chem.* **1967**, *32*, 855–856. [CrossRef] [PubMed]
27. Davis, R.; Untch, K.G. Direct One-Step Conversion of Alcohols into Nitriles. *J. Org. Chem.* **1981**, *46*, 2985–2987. [CrossRef]
28. Iranpoor, N.; Firouzabadi, H.; Akhlaghinia, B.; Nowrouzi, N. Conversion of Alcohols, Thiols, and Trimethysilyl Ethers to Alkyl Cyanides Using Triphenylphosphine/2,3-Dichloro-5,6-Dicyanobenzoquinone/*n*-Bu₄NCN. *J. Org. Chem.* **2004**, *69*, 2562–2564. [CrossRef]
29. Soltani Rad, M.N.; Khalafi-Nezhad, A.; Behrouz, S.; Faghihi, M.A. A Simple One-Pot Procedure for the Direct Conversion of Alcohols into Alkyl Nitriles Using TsIm. *Tetrahedron Lett.* **2007**, *48*, 6779–6784. [CrossRef]
30. Aesa, M.C.; Baán, G.; Novák, L.; Szántay, C. Preparation of Unsaturated Nitriles by the Modification of Mitsunobu Wilk Procedure. *Synth. Commun.* **1995**, *25*, 1545–1550. [CrossRef]
31. Chen, G.; Wang, Z.; Wu, J.; Ding, K. Facile Preparation of α-Aryl Nitriles by Direct Cyanation of Alcohols with TMSCN under the Catalysis of InX₃. *Org. Lett.* **2008**, *10*, 4573–4576. [CrossRef]
32. Trillo, P.; Baeza, A.; Nájera, C. Direct Nucleophilic Substitution of Free Allylic Alcohols in Water Catalyzed by FeCl₃·6H₂O: Which Is the Real Catalyst? *ChemCatChem* **2013**, *5*, 1538–1542. [CrossRef]
33. Theerthagiri, P.; Lalitha, A. Zn(OTf)₂—Catalyzed Direct Cyanation of Benzylic Alcohols—a Novel Synthesis of α-Aryl Nitriles. *Tetrahedron Lett.* **2012**, *53*, 5535–5538. [CrossRef]
34. Wang, J.; Masui, Y.; Onaka, M. Direct Synthesis of Nitriles from Alcohols with Trialkylsilyl Cyanide Using Brønsted Acid Montmorillonite Catalysts. *ACS Catal.* **2011**, *1*, 446–454. [CrossRef]

35. Rajagopal, G.; Kim, S.S. Synthesis of α-Aryl Nitriles through B(C$_6$F$_5$)$_3$-Catalyzed Direct Cyanation of α-Aryl Alcohols and Thiols. *Tetrahedron* **2009**, *65*, 4351–4355. [CrossRef]
36. Bhor, M.D.; Panda, A.G.; Nandurkar, N.S.; Bhanage, B.M. Synthesis of Alkyl Iodides/Nitriles from Carbonyl Compounds Using Novel Ruthenium Tris(2,2,6,6-Tetramethyl-3,5-Heptanedionate) as Catalyst. *Tetrahedron Lett.* **2008**, *49*, 6475–6479. [CrossRef]
37. Yu, R.; Rajasekar, S.; Fang, X. Enantioselective Nickel-Catalyzed Migratory Hydrocyanation of Nonconjugated Dienes. *Angew. Chem. Int. Ed.* **2020**, *59*, 21436–21441. [CrossRef]
38. Gao, J.; Jiao, M.; Ni, J.; Yu, R.; Cheng, G.J.; Fang, X. Nickel-Catalyzed Migratory Hydrocyanation of Internal Alkenes: Unexpected Diastereomeric-Ligand-Controlled Regiodivergence. *Angew. Chem. Int. Ed.* **2021**, *60*, 1883–1890. [CrossRef]
39. Sun, F.; Wang, T.; Cheng, G.-J.; Fang, X. Enantioselective Nickel-Catalyzed Hydrocyanative Desymmetrization of Norbornene Derivatives. *ACS Catal.* **2021**, *11*, 7578–7583. [CrossRef]
40. Ding, S.; Jiao, N. Direct Transformation of N,N-Dimethylformamide to −CN: Pd-Catalyzed Cyanation of Heteroarenes Via C–H Functionalization. *J. Am. Chem. Soc.* **2011**, *133*, 12374–12377. [CrossRef] [PubMed]
41. Ren, X.; Chen, J.; Chen, F.; Cheng, J. The Palladium-Catalyzed Cyanation of Indole C-H Bonds with the Combination of NH$_4$HCO$_3$ and DMSO as a Safe Cyanide Source. *Chem. Commun.* **2011**, *47*, 6725–6727. [CrossRef]
42. Fan, C.; Zhou, Q.-L. Nickel-Catalyzed Group Transfer of Radicals Enables Hydrocyanation of Alkenes and Alkynes. *Chem. Catal.* **2021**, *1*, 117–128. [CrossRef]
43. Cui, J.; Song, J.; Liu, Q.; Liu, H.; Dong, Y. Transition-Metal-Catalyzed Cyanation by Using an Electrophilic Cyanating Agent, N-Cyano-N-Phenyl-p-Toluenesulfonamide (NCTS). *Chem. Asian J.* **2018**, *13*, 482–495. [CrossRef]
44. Ito, Y.; Kato, H.; Imai, H.; Saegusa, T. A Novel Conjugate Hydrocyanation with TiCl$_4$-tert-Butyl Isocyanide. *J. Am. Chem. Soc.* **1982**, *104*, 6449–6450. [CrossRef]
45. Xu, S.; Huang, X.; Hong, X.; Xu, B. Palladium-Assisted Regioselective C–H Cyanation of Heteroarenes Using Isonitrile as Cyanide Source. *Org. Lett.* **2012**, *14*, 4614–4617. [CrossRef] [PubMed]
46. Peng, J.; Zhao, J.; Hu, Z.; Liang, D.; Huang, J.; Zhu, Q. Palladium-Catalyzed C(Sp^2)–H Cyanation Using Tertiary Amine Derived Isocyanide as a Cyano Source. *Org. Lett.* **2012**, *14*, 4966–4969. [CrossRef]
47. Dewanji, A.; van Dalsen, L.; Rossi-Ashton, J.A.; Gasson, E.; Crisenza, G.E.M.; Procter, D.J. A General Arene C-H Functionalization Strategy Via Electron Donor-Acceptor Complex Photoactivation. *Nat. Chem.* **2022**, *15*, 43–52. [CrossRef]
48. Tang, S.; Guillot, R.; Grimaud, L.; Vitale, M.R.; Vincent, G. Electrochemical Benzylic C–H Functionalization with Isocyanides. *Org. Lett.* **2022**, *24*, 2125–2130. [CrossRef] [PubMed]
49. Jiang, X.; Wang, J.-M.; Zhang, Y.; Chen, Z.; Zhu, Y.-M.; Ji, S.-J. Synthesis of Aryl Nitriles by Palladium-Assisted Cyanation of Aryl Iodides Using tert-Butyl Isocyanide as Cyano Source. *Tetrahedron* **2015**, *71*, 4883–4887. [CrossRef]
50. Higashimae, S.; Kurata, D.; Kawaguchi, S.I.; Kodama, S.; Sonoda, M.; Nomoto, A.; Ogawa, A. Palladium-Catalyzed Cyanothiolation of Internal Alkynes Using Organic Disulfides and tert-Butyl Isocyanide. *J. Org. Chem.* **2018**, *83*, 5267–5273. [CrossRef] [PubMed]
51. Shirsath, S.R.; Shinde, G.H.; Shaikh, A.C.; Muthukrishnan, M. Accessing α-Arylated Nitriles Via BF$_3$.OEt$_2$ Catalyzed Cyanation of para-Quinone Methides Using tert-Butyl Isocyanide as a Cyanide Source. *J. Org. Chem.* **2018**, *83*, 12305–12314. [CrossRef]
52. Nauth, A.M.; Opatz, T. Non-Toxic Cyanide Sources and Cyanating Agents. *Org. Biomol. Chem.* **2019**, *17*, 11–23. [CrossRef]
53. Kim, J.; Kim, H.J.; Chang, S. Synthesis of Aromatic Nitriles Using Nonmetallic Cyano-Group Sources. *Angew. Chem. Int. Ed.* **2012**, *51*, 11948–11959. [CrossRef] [PubMed]
54. Wang, W.; Liu, T.; Ding, C.-H.; Xu, B. C(Sp^3)–H Functionalization with Isocyanides. *Org. Chem. Front.* **2021**, *8*, 3525–3542. [CrossRef]
55. Wang, J.; Li, D.; Li, J.; Zhu, Q. Advances in Palladium-Catalysed Imidoylative Cyclization of Functionalized Isocyanides for the Construction of N-Heterocycles. *Org. Biomol. Chem.* **2021**, *19*, 6730–6745. [CrossRef]
56. Giustiniano, M.; Basso, A.; Mercalli, V.; Massarotti, A.; Novellino, E.; Tron, G.C.; Zhu, J. To Each His Own: Isonitriles for All Flavors. Functionalized Isocyanides as Valuable Tools in Organic Synthesis. *Chem. Soc. Rev.* **2017**, *46*, 1295–1357. [CrossRef] [PubMed]
57. Song, B.; Xu, B. Metal-Catalyzed C-H Functionalization Involving Isocyanides. *Chem. Soc. Rev.* **2017**, *46*, 1103–1123. [CrossRef]
58. Zhang, B.; Studer, A. Recent Advances in the Synthesis of Nitrogen Heterocycles via Radical Cascade Reactions Using Isonitriles as Radical Acceptors. *Chem. Soc. Rev.* **2015**, *44*, 3505–3521. [CrossRef]
59. Boyarskiy, V.P.; Bokach, N.A.; Luzyanin, K.V.; Kukushkin, V.Y. Metal-Mediated and Metal-Catalyzed Reactions of Isocyanides. *Chem. Rev.* **2015**, *115*, 2698–2779. [CrossRef] [PubMed]
60. Qiu, G.; Ding, Q.; Wu, J. Recent Advances in Isocyanide Insertion Chemistry. *Chem. Soc. Rev.* **2013**, *42*, 5257–5269. [CrossRef]
61. Dömling, A. Recent Developments in Isocyanide Based Multicomponent Reactions in Applied Chemistry. *Chem. Rev.* **2006**, *106*, 17–89. [CrossRef]
62. Fang, H.; Oestreich, M. Defunctionalisation Catalysed by Boron Lewis Acids. *Chem. Sci.* **2020**, *11*, 12604–12615. [CrossRef] [PubMed]
63. Ma, Y.; Lou, S.J.; Hou, Z. Electron-Deficient Boron-Based Catalysts for C–H Bond Functionalisation. *Chem. Soc. Rev.* **2021**, *50*, 1945–1967. [CrossRef] [PubMed]
64. Basak, S.; Winfrey, L.; Kustiana, B.A.; Melen, R.L.; Morrill, L.C.; Pulis, A.P. Electron Deficient Borane-Mediated Hydride Abstraction in Amines: Stoichiometric and Catalytic Processes. *Chem. Soc. Rev.* **2021**, *50*, 3720–3737. [CrossRef] [PubMed]

65. Kumar, G.; Roy, S.; Chatterjee, I. Tris(Pentafluorophenyl)Borane Catalyzed C-C and C-Heteroatom Bond Formation. *Org. Biomol. Chem.* **2021**, *19*, 1230–1267. [CrossRef]
66. Oestreich, M.; Hermeke, J.; Mohr, J. A Unified Survey of Si-H and H-H Bond Activation Catalysed by Electron-Deficient Boranes. *Chem. Soc. Rev.* **2015**, *44*, 2202–2220. [CrossRef]
67. Melen, R.L. Applications of Pentafluorophenyl Boron Reagents in the Synthesis of Heterocyclic and Aromatic Compounds. *Chem. Commun.* **2014**, *50*, 1161–1174. [CrossRef] [PubMed]
68. Erker, G. Tris(Pentafluorophenyl)Borane: A Special Boron Lewis Acid for Special Reactions. *Dalton Trans.* **2005**, 1883–1890. [CrossRef]
69. Meng, S.-S.; Tang, X.; Luo, X.; Wu, R.; Zhao, J.-L.; Chan, A.S.C. Borane-Catalyzed Chemoselectivity-Controllable N-Alkylation and *ortho* C-Alkylation of Unprotected Arylamines Using Benzylic Alcohols. *ACS Catal.* **2019**, *9*, 8397–8403. [CrossRef]
70. Meng, S.S.; Wang, Q.; Huang, G.B.; Lin, L.R.; Zhao, J.L.; Chan, A.S.C. B(C_6F_5)$_3$ Catalyzed Direct Nucleophilic Substitution of Benzylic Alcohols: An Effective Method of Constructing C-O, C-S and C-C Bonds from Benzylic Alcohols. *RSC Adv.* **2018**, *8*, 30946–30949. [CrossRef]
71. Chen, X.; Patel, K.; Marek, I. Stereoselective Construction of Tertiary Homoallyl Alcohols and Ethers by Nucleophilic Substitution at Quaternary Carbon Stereocenters. *Angew. Chem. Int. Ed.* **2023**. [CrossRef]
72. San, H.H.; Huang, J.; Lei Aye, S.; Tang, X.Y. Boron-Catalyzed Dehydrative Friedel-Crafts Alkylation of Arenes Using β-Hydroxyl Ketone as MVK Precursor. *Adv. Synth. Catal.* **2021**, *363*, 2386–2391. [CrossRef]
73. Guru, M.M.; Thorve, P.R.; Maji, B. Boron-Catalyzed N-Alkylation of Arylamines and Arylamides with Benzylic Alcohols. *J. Org. Chem.* **2020**, *85*, 806–819. [CrossRef]
74. Rubin, M.; Gevorgyan, V. B(C_6F_5)$_3$-Catalyzed Allylation of Secondary Benzyl Acetates with Allylsilanes. *Org. Lett.* **2001**, *3*, 2705–2707. [CrossRef]
75. Dryzhakov, M.; Hellal, M.; Wolf, E.; Falk, F.C.; Moran, J. Nitro-Assisted Bronsted Acid Catalysis: Application to a Challenging Catalytic Azidation. *J. Am. Chem. Soc.* **2015**, *137*, 9555–9558. [CrossRef]
76. Xiao, Y.; Tang, L.; Xu, T.T.; Feng, J.J. Boron Lewis Acid Catalyzed Intermolecular Trans-Hydroarylation of Ynamides with Hydroxyarenes. *Org. Lett.* **2022**, *24*, 2619–2624. [CrossRef]
77. Lin, T.-Y.; Wu, H.-H.; Feng, J.-J.; Zhang, J. Design and Enantioselective Synthesis of β-Vinyl Tryptamine Building Blocks for Construction of Privileged Chiral Indole Scaffolds. *ACS Catal.* **2017**, *7*, 4047–4052. [CrossRef]
78. Feng, J.-J.; Zhang, J. Rhodium-Catalyzed Stereoselective Intramolecular Tandem Reaction of Vinyloxiranes with Alkynes: Atom- and Step-Economical Synthesis of Multifunctional Mono-, Bi-, and Tricyclic Compounds. *ACS Catal.* **2017**, *7*, 1533–1542. [CrossRef]
79. Zhu, C.Z.; Feng, J.J.; Zhang, J. Rhodium(I)-Catalyzed Intermolecular Aza-[4+3] Cycloaddition of Vinyl Aziridines and Dienes: Atom-Economical Synthesis of Enantiomerically Enriched Functionalized Azepines. *Angew. Chem. Int. Ed.* **2017**, *56*, 1351–1355. [CrossRef] [PubMed]
80. Jacobsen, H.; Berke, H.; Döring, S.; Kehr, G.; Erker, G.; Fröhlich, R.; Meyer, O. Lewis Acid Properties of Tris(Pentafluorophenyl)Borane. Structure and Bonding in L–B(C_6F_5)$_3$ Complexes. *Organometallics* **1999**, *18*, 1724–1735. [CrossRef]
81. Tumanov, V.V.; Tishkov, A.A.; Mayr, H. Nucleophilicity Parameters for Alkyl and Aryl Isocyanides. *Angew. Chem. Int. Ed.* **2007**, *46*, 3563–3566. [CrossRef] [PubMed]
82. Zhou, B.; Zhang, Y.Q.; Zhang, K.; Yang, M.Y.; Chen, Y.B.; Li, Y.; Peng, Q.; Zhu, S.F.; Zhou, Q.L.; Ye, L.W. Stereoselective Synthesis of Medium Lactams Enabled by Metal-Free Hydroalkoxylation/Stereospecific [1,3]-Rearrangement. *Nat. Commun.* **2019**, *10*, 3234. [CrossRef] [PubMed]
83. Nambo, M.; Yar, M.; Smith, J.D.; Crudden, C.M. The Concise Synthesis of Unsymmetric Triarylacetonitriles via Pd-Catalyzed Sequential Arylation: A New Synthetic Approach to Tri- and Tetraarylmethanes. *Org. Lett.* **2015**, *17*, 50–53. [CrossRef]
84. Choi, I.; Chung, H.; Park, J.W.; Chung, Y.K. Active and Recyclable Catalytic Synthesis of Indoles by Reductive Cyclization of 2-(2-Nitroaryl)Acetonitriles in the Presence of Co-Rh Heterobimetallic Nanoparticles with Atmospheric Hydrogen under Mild Conditions. *Org. Lett.* **2016**, *18*, 5508–5511. [CrossRef]
85. Asai, K.; Hirano, K.; Miura, M. Divergent Synthesis of Isonitriles and Nitriles by Palladium-Catalyzed Benzylic Substitution with TMSCN. *J. Org. Chem.* **2020**, *85*, 12703–12714. [CrossRef] [PubMed]
86. Shu, X.; Jiang, Y.Y.; Kang, L.; Yang, L. Ni-Catalyzed hydrocyanation of alkenes with formamide as the cyano source. *Green Chem.* **2020**, *22*, 2734–2738. [CrossRef]
87. Weinreb, S.; Sengupta, R. A One-Pot Umpolung Method for Preparation of α-Aryl Nitriles from α-Chloro Aldoximes via Organocuprate Additions to Transient Nitrosoalkenes. *Synthesis* **2012**, *44*, 2933–2937. [CrossRef]
88. Aksenov, A.V.; Aksenov, N.A.; Dzhandigova, Z.V.; Aksenov, D.A.; Rubin, M. Nitroalkenes as surrogates for cyanomethylium species in a one-pot synthesis of non-symmetric diarylacetonitriles. *RSC Adv.* **2015**, *5*, 106492–106497. [CrossRef]
89. Liu, S.W.; Meng, L.L.; Zeng, X.J.; Hammond, G.B.; Xu, B. Synthesis of Acrylonitriles via Mild Base Promoted Tandem Nucleophilic Substitution-Isomerization of α-Cyanohydrin Methanesulfonates. *Chin. J. Chem.* **2021**, *39*, 913–917. [CrossRef]
90. Lindsay-Scott, P.J.; Clarke, A.; Richardson, J. Two-Step Cyanomethylation Protocol: Convenient Access to Functionalized Aryl- and Heteroarylacetonitriles. *Org. Lett.* **2015**, *17*, 476–479. [CrossRef] [PubMed]

91. Chen, Y.; Xu, L.T.; Jiang, Y.W.; Ma, D.W. Assembly of α-(Hetero)aryl Nitriles via Copper-Catalyzed Coupling Reactions with (Hetero)aryl Chlorides and Bromides. *Angew. Chem. Int. Ed.* **2021**, *60*, 7082–7086. [CrossRef] [PubMed]
92. Yuen, O.Y.; Chen, X.; Wu, J.; So, C.M. Palladium-Catalyzed Direct α-Arylation of Arylacetonitriles with Aryl Tosylates and Mesylates. *Eur. J. Org. Chem.* **2020**, 1912–1916. [CrossRef]

Disclaimer/Publisher's Note: The statements, opinions and data contained in all publications are solely those of the individual author(s) and contributor(s) and not of MDPI and/or the editor(s). MDPI and/or the editor(s) disclaim responsibility for any injury to people or property resulting from any ideas, methods, instructions or products referred to in the content.

Article

Homocouplings of Sodium Arenesulfinates: Selective Access to Symmetric Diaryl Sulfides and Diaryl Disulfides

Xin-Zhang Yu [1,2], Wen-Long Wei [2], Yu-Lan Niu [1,*], Xing Li [2], Ming Wang [3] and Wen-Chao Gao [2,3,*]

[1] Department of Chemistry and Chemical Engineering, Taiyuan Institute of Technology, Taiyuan 030008, China
[2] Department of Biomedical Engineering, Taiyuan University of Technology, Taiyuan 030024, China
[3] School of Chemistry and Molecular Engineering, East China Normal University, 3663, Shanghai 200062, China
* Correspondence: niuyulan@163.com (Y.-L.N.); gaowenchao@tyut.edu.cn (W.-C.G.)

Abstract: Symmetrical diaryl sulfides and diaryl disulfides have been efficiently and selectively constructed via the homocoupling of sodium arenesulfinates. The selectivity of products relied on the different reaction systems: symmetrical diaryl sulfides were predominately obtained under the Pd(OAc)$_2$ catalysis, whereas symmetrical diaryl sulfides were exclusively yielded in the presence of the reductive Fe/HCl system.

Keywords: sodium arenesulfinates; homocoupling; symmetric; diaryl sulfides; diaryl disulfides

1. Introduction

Symmetrical sulfides [1–3] and disulfides [4–7] are ubiquitous structural motifs, and their corresponding derivatives have found prevalent existence in many biologically active molecules, pharmaceuticals, ligands, functional materials, and natural products. Owing to their importance, various synthetic methodologies have been developed for the preparation of these two classes of sulfur compounds [8–23]. Although the cross-coupling reactions of various sulfur surrogates with other aromatic reagents, such as aromatic halides, arenediazonium salts, and others, are generally efficient for the construction of these two symmetric structures [24–36], the homocoupling of arylsulfonyl derivatives [37,38] and thiols [39–48] has proven to be the most straightforward and convenient strategy in terms of simplicity. Despite these major advances, the utility of environmentally friendly sulfur sources for symmetric sulfides and disulfides is still highly desirable.

Due to their stable, greener, and inexpensive features, sodium sulfinates have been utilized as ideal sulfur donors and widely applied as the coupling partners in C-S cross-coupling reactions, such as sulfonylation [49,50], thiosulfonylation [51], sulfinylation [52,53], and sulfenylation [54–59]. Especially by reductive coupling, the sodium sulfinates could serve as sulfenylation reagents for the synthesis of unsymmetric sulfides under Pd, Cu, or I$_2$ catalysts (Scheme 1a) [54–59]. However, the application of sodium sulfinates for the construction of symmetric disulfides by homocoupling reactions was still less explored. On the other hand, although several reaction systems, such as EtOP(O)H$_2$, TiCl$_4$/Sm, WCl$_6$/NaI, WCl$_6$/Zn, MoCl$_6$/Zn, Cp$_2$TiCl$_2$/Sm, and Silphos, have been reported to mediate the conversion of sodium arenesulfinates into symmetric diaryl disulfides by reductive coupling (Scheme 1b) [60–63], they generally suffered from limited substrate scope, expensive reagents, or complicated procedures. Herein, we would like to report an efficient strategy for the selective synthesis of symmetrical diaryl sulfides and diaryl disulfides using sodium sulfinates as sulfenylation reagents through homocoupling.

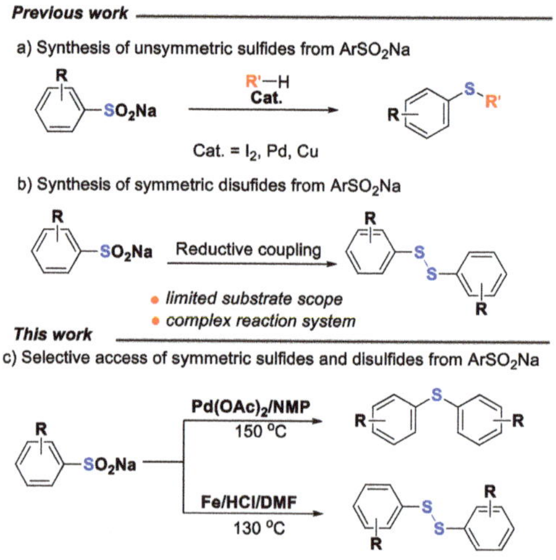

Scheme 1. Sulfides and disulfides construction from sodium sulfinates.

2. Results and Discussion

Our initial study started with sodium phenylsulfinate **1a** as the model substrate to explore the formation of diphenylsulfide **2a**. First, by screening different solvents, NMP was proven to be the most effective out of the others, such as DMF and DMSO (Table 1, entry 3 vs. 1–2). The catalyst played a decisive role in this reaction: among the common metal catalysts, Pd(OAc)$_2$ was the best one to afford diphenylsulfide **2a** in 47% yield (entry 3 vs. 4–6). With CuI and Ni(OAc)$_2$ as catalysts, the product **2a** was not detected at all (entries 4 and 5). Only a trace amount of **2a** was observed when FeCl$_3$ was employed (entry 6). The yield could be improved to 60% by increasing reaction temperature to 150 °C (entry 7 vs. 3 and 8). Most importantly, a decrease of catalyst loading to 2 mol% could further increase the yield to 89%, and no better result was observed by continuously reducing the catalyst loading (entry 10 vs. 7, 9–11). The desired **2a** was not detected when sodium benzenesulfinate was replaced with benzenesulfinic acid in the presence of NaOH under the same conditions (entry 12).

Table 1. Optimization of reaction conditions for sodium benzenesulfinate to diphenylsulfide [a].

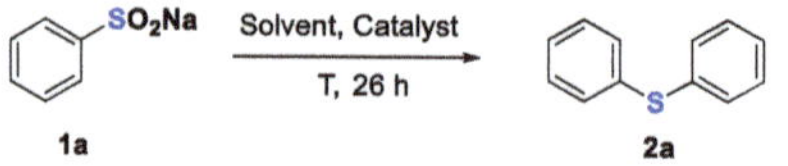

Entry	Solvent (mL)	Catalyst (mol%)	T (°C)	Yield [b] (%)
1	DMF (1.0)	Pd(OAc)$_2$ (2.5)	130	28
2	DMSO (1.0)	Pd(OAc)$_2$ (2.5)	130	trace
3	NMP (1.0)	Pd(OAc)$_2$ (2.5)	130	47
4	NMP (1.0)	CuI (2.5)	130	N.D. [c]
5	NMP (1.0)	Ni(OAc)$_2$ (2.5)	130	N.D. [c]
6	NMP (1.0)	FeCl$_3$ (2.5)	130	trace
7	NMP (1.0)	Pd(OAc)$_2$ (2.5)	150	60
8	NMP (1.0)	Pd(OAc)$_2$ (2.5)	170	52
9	NMP (1.0)	Pd(OAc)$_2$ (2.0)	150	89
10	NMP (1.0)	Pd(OAc)$_2$ (1.5)	150	80
11	NMP (1.0)	Pd(OAc)$_2$ (0.5)	150	71
12 [d]	NMP (1.0)	Pd(OAc)$_2$ (2.0)	150	0

[a] Reactions were performed on a 0.4 mmol scale of sodium benzenesulfinate **1a** (67.0 mg) under identified conditions for 26 h; [b] isolated yield; [c] N.D. = not detected; [d] PhSO$_2$Na was replaced with PhSO$_2$H/NaOH.

Having built the optimal conditions for the construction of diphenylsulfide **2a**, we turned our attention to explore the generality of sodium sulfinates. As shown in Table 2, a variety of substrates could undergo the homocoupling to afford symmetrical diaryl sulfides with high chemoselectivity. It was found that sodium benzenesulfinates with electron-donating groups such as 4-methyl, 3-methyl, 2-methyl, 4-methoxyl, 3-methoxyl, 2-methoxyl, 4-isopropyl, and 4-*tert*-butyl on the phenyl ring gave the corresponding products **2b–2i** in good yields. Electron-withdrawing groups, such as F, Cl, Br, and NO$_2$, were also well-tolerated to provide the desired products **2j–2s** in moderate to good yields that were somewhat lower than the electron-donating groups offered. To our delight, the intramolecular formation of sulfides was also tried, and the desired dibenzothiophene **2t** was produced in a 51% yield.

Table 2. Scope of sodium arylsulfinates to diaryl sulfides [a,b].

[a] Reaction conditions: sodium arylsulfinate **1** (0.4 mmol), Pd(OAc)$_2$ (2 mg, 2 mol%), NMP (1.0 mL), and 150 °C; [b] isolated yield; [c] 4-methylbenzenesulfinic acid (0.4 mmol) was used as a starting material in the presence of Pd(OAc)$_2$ (2 mg, 2 mol%), NaOH (16.0 mg) in NMP (1.0 mL) at 150 °C; [d] N$_2$ was adopted.

During our studies on the synthesis of diphenylsulfide **2a**, 1,2-diphenyldisulfide **3a** was accidentally detected when using CuI as a reductant. This discovery encouraged us to search for optimal conditions for the reductive coupling of sodium benzenesulfinate for 1,2-diphenyldisulfide **3a**. Fortunately, using Fe/HCl as the reductive system, diphenyldisulfide **3a** was isolated as the major product. Subsequently, the investigation of the concentration of hydrochloric acid revealed that increasing the concentration led to the higher yield, and 12 mol/L of hydrochloric acid gave up to 96% yield (Table 3, entry 4 vs. 1−3). The highest yield was obtained when 4.0 equiv. of HCl was used (Table 2, entry 4). Increasing or decreasing the amount resulted in lower yields (entries 5 and 6 vs. 4). It was found that 2.0 equiv. of Fe was suitable for this transformation, and other amounts did not improve the yield further (entry 4 vs. 7 and 8). More notably, the similar high yield was provided when the time was shortened to 9 h (entry 9). However, sodium benzenesulfinate generated in situ by the reaction of NaOH and the equivalent of 4-methylbenzenesulfinic acid only afforded the target product (**3b**) in a 59% yield (entry 10).

Table 3. Optimization of reaction conditions for sodium benzenesulfinate to diphenyldisulfide [a].

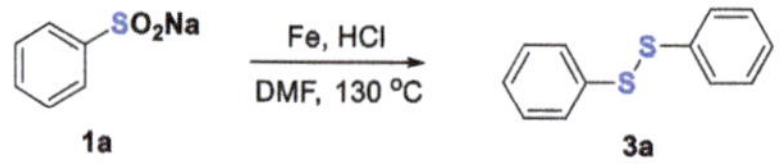

Entry	Concentration of HCl (mol/L)	HCl (equiv.)	Fe (equiv.)	Yield [b] (%)
1	1	4.0	2.0	34
2	4	4.0	2.0	60
3	8	4.0	2.0	78
4	12	4.0	2.0	96
5	12	2.0	2.0	71
6	12	6.0	2.0	90
7	12	4.0	1.5	81
8	12	4.0	2.5	93
9	12	4.0	2.0	96 [c]
10 [d]	12	4.0	2.0	59

[a] Reactions were performed on a 0.2 mmol scale of sodium benzenesulfinate **1a** (33.5 mg) in DMF (1.0 mL) at 130 °C for 20 h. [b] Isolated yield; [c] reaction time: 9 h; [d] PhSO$_2$Na was generated in situ.

With the optimized reaction conditions in hand, we next focused on the evaluation of the scope of the coupling partner to symmetric disulfides, and the results are summarized in Table 4. To our delight, it was found that the reaction could be compatible with a broad range of functional groups, furnishing the corresponding products in good to excellent yields. Although various functional groups, including electronically diverse (**3a–3u**) and sterically hindered (**3d, 3g, 3h, 3i, 3l,** and **3r**) ones are readily tolerated, some substantial influence of electronic properties and steric hindrance of the substituents was observed. The substrates possessing an electron-rich group (Me, MeO, *i*-Pr, and *t*-Bu) showed higher yields than those bearing an electron-poor group (F, Cl, Br, and CF$_3$) (**3b–3i** vs. **3j–3l**, **3p–3s**, and **3u**). Among substrates, sodium *ortho*-substituted arylsulfinates, which are sterically hindered, gave relatively lower yields (**3d** vs. **3b, 3c**, and **3g** vs. **3e, 3f**, and **3l** vs. **3j, 3k**, and **3r** vs. **3p** and **3q**). In addition, an 82% yield was obtained when sodium 2-naphthylsulfinate was employed as a substrate (**3t**). Notably, sodium 3-carboxybenzenesulfinate and sodium thiophene-2-sulfinate could be not transformed into the corresponding disulfides.

To further evaluate the utility of these two protocols, two gram-scale reactions were subsequently carried out (Scheme 2). The corresponding products **2a** and **3a** could be afforded in 88% and 94% yields in a 10 mmol scale, respectively, demonstrating the practicability of the present methodology.

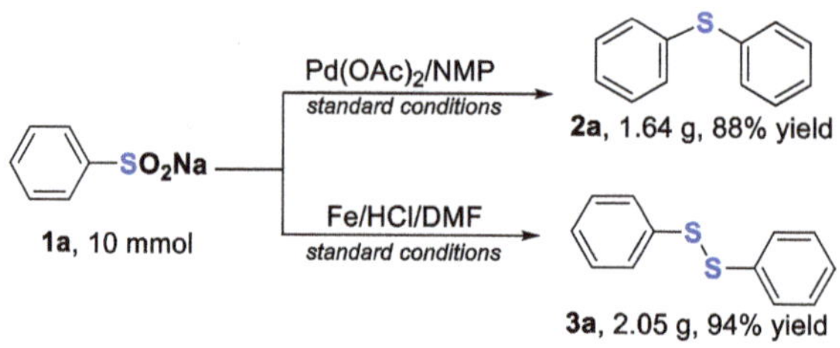

Scheme 2. Gram-scale reactions of sodium benzenesulfinate.

Table 4. Scope of sodium arylsulfinates to diaryl disulfides [a,b].

Reaction scheme: R-C6H4-SO2Na (**1**) → Fe (2 equiv.), Conc. HCl (4 equiv), DMF, 130 °C → (R-C6H4-S-)2 (**3**)

- **3a**, 9 h, 96% — diphenyl disulfide
- **3b**, 9 h, 93% — bis(4-methylphenyl) disulfide
- **3c**, 9 h, 90% — bis(3-methylphenyl) disulfide
- **3d**, 10 h, 85% — bis(2-methylphenyl) disulfide
- **3e**, 9 h, 92% — bis(4-methoxyphenyl) disulfide
- **3f**, 10 h, 88% — bis(3-methoxyphenyl) disulfide
- **3g**, 13 h, 84% — bis(2-methoxyphenyl) disulfide
- **3h**, 13 h, 91% — bis(4-isopropylphenyl) disulfide
- **3i**, 13 h, 87% — bis(4-tert-butylphenyl) disulfide
- **3j**, 13 h, 86% — bis(4-fluorophenyl) disulfide
- **3k**, 13 h, 81% — bis(3-fluorophenyl) disulfide
- **3l**, 15 h, 75% — bis(2-fluorophenyl) disulfide
- **3p**, 9 h, 91% — bis(4-chlorophenyl) disulfide
- **3q**, 10 h, 88% — bis(3-chlorophenyl) disulfide
- **3r**, 13 h, 84% — bis(2-chlorophenyl) disulfide
- **3s**, 9 h, 89% — bis(4-bromophenyl) disulfide
- **3t**, 11 h, 82% — dinaphthyl disulfide
- **3u**, 13 h, 80% — bis(4-trifluoromethylphenyl) disulfide

[a] Reaction conditions: sodium arylsulfinate **1** (0.2 mmol), Fe powder (23.0 mg, 2 equiv.), HCl (12 M, 67.0 uL), DMF (1.0 mL), and 130 °C; [b] isolated yield.

To elucidate the reaction mechanism for the homocoupling of sodium arylsulfinates, several control experiments were conducted (Scheme 3). The formation of both symmetric diphenyl disulfide **3a** and diphenyl sulfide **2a** was not detected after the addition of the radical scavenger 2,2,6,6 tetramethylpiperidine 1 oxyl (TEMPO, 2 equiv.) to the standard reaction systems (Scheme 3a,b), indicating that both of these two transformations underwent a free-radical process. For the synthesis of **2a**, the disulfide **3a** was detected by mass spectrometry. The preparation of disulfides from sodium arylsulfinates under Pd(OAc)$_2$ catalysis was also demonstrated by Xiang and co-workers [55]. In addition, the transformation from **3a** to **2a** could be successfully realized in the presence of a catalytic amount of Pd(OAc)$_2$ and sodium sulfinate **1a** (Scheme 3c,d).

Based on the results of the control experiments and literature reports [64,65], a plausible mechanism for the homocoupling of sodium arylsulfinate **1** to the selective access to symmetric sulfide **2** and disulfide **3** is shown in Scheme 4. First, in the reductive Fe/HCl system, disulfide **3a** could be generated via the homocoupling of the thiyl radical **A**, which comes from the radical reduction of sodium phenylsulfinate **2a** (Scheme 4a). Alternatively, disulfide **3a** could be also formed in the presence of catalytic Pd(OAc)$_2$. After disulfide **3a** was formed, the Pd(II)-insertion to the S-S bond produced the metal-intermediate **B**, which underwent ligand exchange to form intermediate **C**. The thermal extrusion of SO$_2$ of intermediate C resulted in the formation of intermediate **D** [66], which underwent the reductive elimination to give the target sulfide **2a** and regenerate Pd(0) into the next catalytic cycle (Scheme 4b).

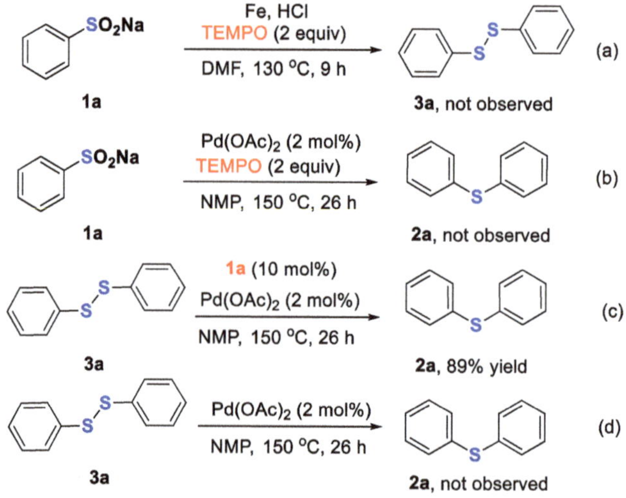

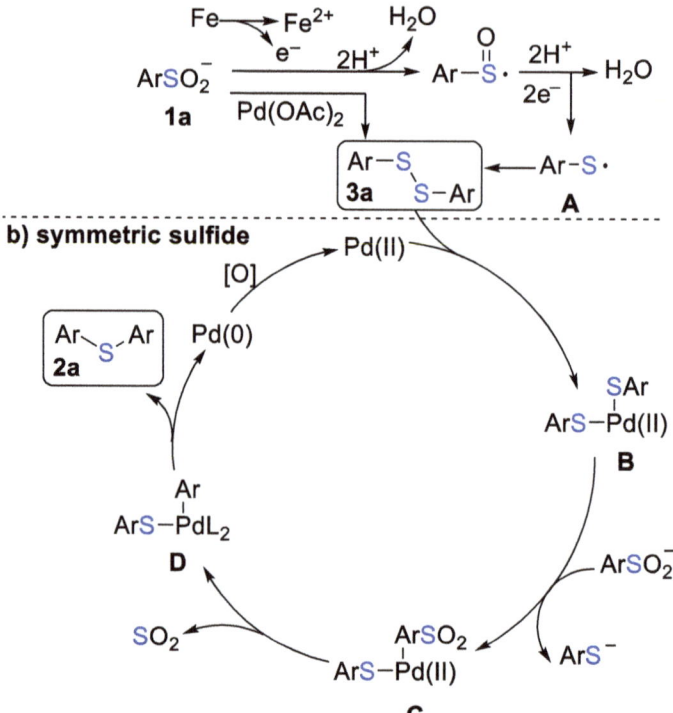

Scheme 3. Control experiments for the homocoupling of sodium phenylsulfinate.

Scheme 4. Proposed mechanism for the homocoupling of sodium arylsulfinates.

3. Materials and Methods

Unless otherwise indicated, all reagents and solvents were purchased from commercial sources and used without further purification. Deuterated solvents were purchased from Sigma–Aldrich(Shanghai, China.). Refinement of the mixed system was achieved through column chromatography, which was performed on silica gel (200–300 mesh) with petroleum ether (solvent A)/ethyl acetate (solvent B) gradients as elution. In addition, all yields were referred to the isolated yields (average of two runs) of the compounds, unless otherwise specified. The known compounds were partly characterized by melting points (for solid samples), 1H NMR, and compared to authentic samples or the literature data. Melting points were measured with an RD-II digital melting point apparatus (Henan, China) and were uncorrected. ^{1}H NMR data were acquired on a Bruker Advance 600 MHz spectrometer (Bruker, Germany). using CDCl$_3$ as solvent. Chemical shifts are reported in ppm from tetramethylsilane, with the solvent CDCl$_3$ resonance as the internal standard (CDCl$_3$ = 7.26). Spectra are reported as follows: chemical shift (δ = ppm), multiplicity (s = singlet, d = doublet, t = triplet, q = quartet, m = multiplet), coupling constants (Hz), integration, and assignment. ^{13}C NMR data were collected at 100 MHz, with complete proton decoupling. The chemical shifts are reported in ppm downfield to the central CDCl$_3$ resonance (δ = 77.0). High-resolution mass spectra were performed on a micrOTOF-Q II instrument (Bruker, Germany), with an ESI source.

3.1. Typical Procedure for Symmetric Diaryl Sulfides 2

The mixture of sodium arylsulfinate **1** (0.4 mmol) and Pd(OAc)$_2$ (2 mg, 2 mol%) in NMP (1.0 mL) was stirred at 150 °C (oil bath) until the substrate was completely consumed, which was determined by TLC. Finally, the reaction mixture was purified by silica gel column chromatography (PE: EA = 40: 1) to afford the desired coupling product diarylsulfides **2**.

3.2. Typical Procedure for Symmetric Diaryl Sisulfides 3

The mixture of sodium arylsulfinate **1** (0.2 mmol), Fe powder (23 mg, 2.0 equiv), and HCl (12 M, 67.0 µL) in DMF (1.0 mL) was stirred at 130 °C (oil bath), until the substrate was completely consumed, which was determined by TLC. Finally, the reaction mixture was purified by silica gel column chromatography (PE: EA = 40: 1) to afford the desired coupling product diaryldisulfides **3**.

3.3. Gram-Scale Reaction of Sodium Benzenesulfinate to Diphenylsulfide

The mixture of sodium benzenesulfinate **1a** (1.64 g, 10 mmol) and the catalyst, Pd(OAc)$_2$ (45 mg, 2 mol%) in NMP (10 mL), was stirred at 150 °C (oil bath) until the substrate was completely consumed, which was determined by TLC. Finally, the reaction mixture was purified by silica gel column chromatography to afford the coupling product diphenylsulfide **2a** (1.640 g, 88% yield).

3.4. Gram-Scale Reaction of Sodium Benzenesulfinate to 1,2-Diphenyldisulfane

The mixture of sodium benzenesulfinate **1a** (1.640 g, 10 mmol), Fe powder (1.150 g, 2.0 equiv.), and 12 mol/L HCl (3.35 mL) in DMF (15 mL) was stirred at 130 °C (oil bath) until the substrate was completely consumed, which was determined by TLC. Finally, the reaction mixture was purified by silica gel column chromatography to afford the coupling product 1,2-diphenyldisulfane **3a** (2.050 g, 94% yield).

3.5. Characterization Data for Homo-Coupling Products of Sodium Arylsulfinates

3.5.1. Characterization Data for the Products of Diaryl sulfides

Diphenyl sulfide (2a) [8]. Colorless liquid (33.1 mg, 89% yield); R_f = 0.6 (petroleum ether); ^{1}H NMR (600 MHz, CDCl$_3$) δ 7.35–7.33 (m, 4H), 7.32–7.27 (m, 4H), 7.26–7.22 (m, 2H) ppm; ^{13}C{^{1}H} NMR (100 MHz, CDCl$_3$) δ 135.9, 131.2, 129.3, and 127.1; HRMS (ESI) m/z [M + H]$^+$ calculated for C$_{12}$H$_{11}$S 187.0576, found 187.0579.

4,4′-Dimethyldiphenyl sulfide (2b) [67]. White solid (36.8 mg, 86% yield); mp 57−58 °C; R_f = 0.6 (petroleum ether); ^{1}H NMR (600 MHz, CDCl$_3$) δ 7.23 (dt, J = 4.8, 2.4 Hz, 4H), 7.10 (d, J = 7.8 Hz, 4H), 2.33 (s, 6H) ppm; ^{13}C{^{1}H} NMR (100 MHz, CDCl$_3$) δ 136.9, 132.7, 131.1, 129.9, and 21.0; HRMS (ESI) m/z [M + H]$^+$ calculated for C$_{14}$H$_{15}$S 215.0889, found 215.0885.

3,3′-Dimethyldiphenyl sulfide (2c) [68]. Colorless liquid (35.5 mg, 83% yield); R_f = 0.6 (petroleum ether); ^{1}H NMR (600 MHz, CDCl$_3$) δ 7.20−7.16 (m, 4H), 7.13 (d, J = 7.8 Hz, 2H), 7.05 (d, J = 7.8 Hz, 2H), and 2.31 (s, 6H) ppm; HRMS (ESI) m/z [M + H]$^+$ calculated for C$_{14}$H$_{15}$S 215.0889, found 215.0885.

Di-o-tolylsulfide (2d) [8]. White solid (33.4 mg, 78% yield); mp 64−65 °C; R_f = 0.6 (petroleum ether); ^{1}H NMR (600 MHz, CDCl$_3$) δ 7.35−7.22 (m, 2H), 7.16 (td, J = 6.0, 1.2 Hz, 2H), 7.11−7.08 (m, 2H), 7.05 (dd, J = 6.6, 1.2 Hz, 2H), and 2.38 (s, 6H) ppm; HRMS (ESI) m/z [M + H]$^+$ calculated for C$_{14}$H$_{15}$S 215.0889, found 215.0885.

Bis(4-methoxyphenyl)sulfide (2e) [8]. White solid (41.8 mg, 85% yield); mp 44−46 °C; R_f = 0.6 (petroleum ether); ^{1}H NMR (600 MHz, CDCl$_3$) δ 7.29−7.26 (m, 4H), 6.83 (dt, J = 9.0, 2.4 Hz, 4H), and 3.79 (s, 6H) ppm; HRMS (ESI) m/z [M + H]$^+$ calculated for C$_{14}$H$_{15}$O$_2$S 247.0787, found 247.0785.

Bis(3-methoxyphenyl)sulfide (2f) [8]. White solid (39.9 mg, 81% yield); mp 45−47 °C; R_f = 0.6 (petroleum ether); ^{1}H NMR (600 MHz, CDCl$_3$) δ 7.24−7.18 (m, 2H), 6.96−6.92 (m, 2H), 6.90−6.88 (m, 2H), 6.81−6.77 (m, 2H), and 3.76 (s, 6H) ppm; HRMS (ESI) m/z [M + H]$^+$ calculated for C$_{14}$H$_{15}$O$_2$S 247.0787, found 247.0785.

Bis(2-methoxyphenyl)sulfide (2g) [8]. White solid (35.9 mg, 73% yield); mp 73−74 °C; R_f = 0.6 (petroleum ether); ^{1}H NMR (400 MHz, CDCl$_3$) δ 7.28−7.21 (m, 2H), 7.06 (dd, J = 5.6, 2.0 Hz, 2H), 6.93−6.90 (dd, J = 7.6, 0.8 Hz, 2H), 6.87−6.85 (dd, J = 6.4, 1.2 Hz, 2H), and 3.87 (s, 6H) ppm; HRMS (ESI) m/z [M + H]$^+$ calculated for C$_{14}$H$_{15}$O$_2$S 247.0787, found 247.0785.

4,4′-Diisopropyldiphenyl sulfide (2h). White solid (45.4 mg, 84% yield); mp 73−75 °C; R_f = 0.6 (petroleum ether); ^{1}H NMR (600 MHZ, CDCl$_3$) δ 7.28−7.25 (m, 4H), 7.17−7.14 (m, 4H), 2.92−2.82 (m, 2H), and 1.24 (d, J = 6.6 Hz, 12H) ppm; ^{13}C{^{1}H} NMR (100 MHz, CDCl$_3$) δ 147.8, 132.8, 131.0, 127.3, 33.7, and 23.9; HRMS (ESI) m/z [M + K]$^+$ calculated for C$_{18}$H$_{22}$SK 309.1074, found 309.1073.

4,4′-Di-tert-butyldiphenyl sulfide (2i) [69]. White solid (48.3 mg, 81% yield); mp 83−84 °C; R_f = 0.6 (petroleum ether); ^{1}H NMR (600 MHz, CDCl$_3$) δ 7.34−7.30 (m, 4H), 7.28 (t, J = 1.8 Hz, 2H), 7.26 (t, J = 2.4 Hz, 2H), and 1.30 (s, 18H) ppm; HRMS (ESI) m/z [M + H]$^+$ calculated for C$_{20}$H$_{27}$S 299.1828, found 299.1821.

4,4′-Difluorodiphenyl sulfide (2j) [8]. Colorless liquid (33.7 mg, 76% yield); R_f = 0.6 (petroleum ether); ^{1}H NMR (600 MHz, CDCl$_3$) δ 7.33−7.28 (m, 4H) and 7.03−6.98 (m, 4H) ppm; HRMS (ESI) m/z [M + H]$^+$ calculated for C$_{12}$H$_9$F$_2$S 223.0388, found 223.0393; ^{19}F NMR (376 MHz, CDCl$_3$) δ -114.3 ppm.

3,3′-Difluorodiphenyl sulfide(2k) [68]. Colorless liquid (32.0 mg, 72% yield); R_f = 0.6 (petroleum ether); ^{1}H NMR (600 MHz, CDCl$_3$) δ 7.33−7.26 (m, 2H), 7.16−7.11 (m, 2H), 7.06−7.01 (m, 2H), and 6.99−6.94 (m, 2H) ppm; HRMS (ESI) m/z [M + H]$^+$ calculated for C$_{12}$H$_9$F$_2$S 223.0388, found 223.0393; ^{19}F NMR (376 MHz, CDCl$_3$) δ -111.5 ppm.

Bis(2-fluorophenyl)sulfide (2l) [70]. Colorless liquid (29.8 mg, 67% yield); R_f = 0.6 (petroleum ether); ^{1}H NMR (600 MHz, CDCl$_3$) δ 7.31−7.22 (m, 4H) and 7.14−7.06 (m, 4H) ppm; HRMS (ESI) m/z [M + H]$^+$ calculated for C$_{12}$H$_9$F$_2$S 223.0388, found 223.0393; ^{19}F NMR (376 MHz, CDCl$_3$) δ -108.7 ppm.

Bis(4-nitrophenyl)sulfide (2m) [8]. White solid (40.3 mg, 73% yield); mp 156−158 °C; R_f = 0.5 (petroleum ether); ^{1}H NMR (600 MHz, CDCl$_3$) δ 8.06 (dt, J = 9.0, 2.4 Hz, 2H), 7.58−7.52 (m, 2H), 7.48−7.44 (m, 2H), and 7.17 (dt, J = 9.0, 2.4 Hz, 2H) ppm; HRMS (ESI) m/z [M + H]$^+$ calculated for C$_{12}$H$_9$N$_2$O$_4$S 277.0278, found 277.0281.

Bis(3-nitrophenyl)sulfide (2n) [71]. White solid (37.5 mg, 68% yield); mp 42−44 °C; R_f = 0.5 (petroleum ether); ^{1}H NMR (400 MHz, CDCl$_3$) δ 8.36 (t, J = 4.0 Hz, 2H), 8.15−8.06 (m, 2H), 7.86−7.77 (m, 2H), and 7.53 (t, J = 12.0 Hz, 2H) ppm; HRMS (ESI) m/z [M + H]$^+$ calculated for C$_{12}$H$_9$N$_2$O$_4$S 277.0278, found 277.0282.

Bis(2-nitrophenyl)sulfide (2o) [8]. Yellow solid (34.8 mg, 63% yield); mp 123−124 °C; R_f = 0.5 (petroleum ether); ^{1}H NMR (400 MHz, CDCl$_3$) δ 8.41 (dd, J = 8.4, 1.2 Hz, 2H), 8.07 (dd, J = 8.4, 1.2 Hz, 2H), 7.84−7.77 (m, 2H), and 7.61−7.54 (m, 2H) ppm; HRMS (ESI) m/z [M + H]$^+$ calculated for C$_{12}$H$_9$N$_2$O$_4$S 277.0278, found 277.0281.

4,4′-Dichlorodiphenyl sulfide (2p) [8]. White solid (42.2 mg, 83% yield); mp 88−89 °C; R_f = 0.6 (petroleum ether); ^{1}H NMR (600 MHz, CDCl$_3$) δ 7.30−7.27 (m, 4H) and 7.26−7.23 (m, 4H) ppm; HRMS (ESI) m/z [M + H]$^+$ calculated for C$_{12}$H$_9$Cl$_2$S 254.9797, found 254.9792.

Bis(3-chlorophenyl)sulfide (2q) [71]. Colorless liquid (39.1 mg, 77% yield); R_f = 0.6 (petroleum ether); ^{1}H NMR (600 MHz, CDCl$_3$) δ 7.32−7.31 (m, 2H), 7.26−7.23 (m, 4H), and 7.22−7.19 (m, 2H) ppm; HRMS (ESI) m/z [M + H]$^+$ calculated for C$_{12}$H$_9$Cl$_2$S 254.9797, found 254.9792.

Bis(2-chlorophenyl)sulfide (2r) [68]. White solid (37.1 mg, 73% yield); mp 68−70 °C; R_f = 0.6 (petroleum ether); ^{1}H NMR (600 MHz, CDCl$_3$) δ 7.47 (dd, J = 7.8, 1.2 Hz, 2H), 7.24 (td, J = 7.8, 1.8 Hz, 2H), 7.19 (td, J = 7.8, 1.2 Hz, 2H), and 7.14 (dd, J = 7.8, 1.8 Hz, 2H) ppm; HRMS (ESI) m/z [M + H]$^+$ calculated for C$_{12}$H$_9$Cl$_2$S 254.9797, found 254.9792.

Bis(4-bromophenyl)sulfide (2s) [8]. White solid (54.7 mg, 80% yield); mp 110−111 °C; R_f = 0.6 (petroleum ether); ^{1}H NMR (600 MHz, CDCl$_3$) δ 7.45−7.40 (m, 4H) and 7.21−7.16 (m, 4H) ppm; HRMS (ESI) m/z [M + H]$^+$ calculated for C$_{12}$H$_9$Br$_2$S 342.8786, found 342.8784.

Dibenzo[b,d]thiophene (2t) [72]. white solid (38 mg, 51% yield); m.p. 95−96 °C. Rf = 0.8 (PE/EA = 20:1); ^{1}H NMR (400 MHz, CDCl$_3$) δ 8.17 (m, 2H), 7.87 (dd, 2 H, J = 8.0, 4.0 Hz), and 7.47 (m, 4 H). The NMR data were consistent with the previous report (see spectra at Supplementary Materials).

3.5.2. Characterization Data for the Products of Diaryl Disulfides

Diphenyl disulfide (3a) [73]. White solid (21.0 mg, 96% yield); mp 61−62 °C; R_f = 0.6 (petroleum ether); ^{1}H NMR (600 MHz, CDCl$_3$) δ 7.54−7.45 (m, 4H), 7.33−7.26 (m, 4H), and 7.25−7.19 (m, 2H) ppm; ^{13}C{^{1}H} NMR (100 MHz, CDCl$_3$) δ 137.0, 129.0, 127.5, and 127.1; HRMS (ESI) m/z [M]$^+$ calculated for C$_{12}$H$_{10}$S$_2$ 218.0224, found 218.0217.

4-Methylphenyl disulfide (3b) [73]. White solid (22.9 mg, 93% yield); mp 47−48 °C; R_f = 0.6 (petroleum ether); ^{1}H NMR (600 MHz, CDCl$_3$) δ 7.39 (d, J = 8.4 Hz, 4H), 7.11 (d, J = 7.8 Hz, 4H), and 2.32 (s, 6H) ppm; ^{13}C{^{1}H} NMR (100 MHz, CDCl$_3$) δ 137.4, 133.9, 129.8, 128.6, and 21.0; HRMS (ESI) m/z [M]$^+$ calculated for C$_{14}$H$_{14}$S$_2$ 246.0537, found 246.0517.

3-Methylphenyl disulfide (3c) [73]. White solid (22.1 mg, 90% yield); mp 112−114 °C; R_f = 0.6 (petroleum ether); ^{1}H NMR (600 MHz, CDCl$_3$) δ 7.31 (d, J = 7.2 Hz, 4H), 7.19 (t, J = 7.4 Hz, 2H), 7.04 (d, J = 7.2 Hz, 2H), and 2.32 (s, 6H) ppm; HRMS (ESI) m/z [M]$^+$ calculated for C$_{14}$H$_{14}$S$_2$ 246.0537, found 246.0519.

Di(o-methylphenyl)disulfide (3d) [73]. White solid (20.9 mg, 85% yield); mp 40−42 °C; R_f = 0.6 (petroleum ether); ^{1}H NMR (600 MHz, CDCl$_3$) δ 7.24−7.21 (m, 2H), 7.19−7.15 (m, 2H), 7.11−7.07 (m, 2H), 7.06 (dd, J = 7.8, 1.8 Hz, 2H), and 2.37 (s, 6H) ppm; HRMS (ESI) m/z [M]$^+$ calculated for C$_{14}$H$_{14}$S$_2$ 246.0537, found 246.0517.

Di(4-methoxyphenyl)disulfide (3e) [34]. White solid (25.6 mg, 92% yield); mp 45−47 °C; R_f = 0.6 (petroleum ether); ^{1}H NMR (600 MHz, CDCl$_3$) δ 7.31−7.27 (m, 4 H), 6.87−6.81 (m, 4 H), and 3.79 (s, 6H) ppm; HRMS (ESI) m/z [M]$^+$ calculated for C$_{14}$H$_{14}$O$_2$S$_2$ 278.0435, found 278.0423.

Di(3-methoxyphenyl)disulfide (3f) [34]. White solid (24.5 mg, 88% yield); mp 106−108 °C; R_f = 0.6 (petroleum ether); ^{1}H NMR (400 MHz, CDCl$_3$) δ 7.18−7.11 (m, 2H), 7.03−6.99 (m, 2H), 6.88−6.81 (m, 2H), 6.73−6.67 (m, 2H), and 3.69 (s, 6H) ppm; HRMS (ESI) m/z [M]$^+$ calculated for C$_{14}$H$_{14}$O$_2$S$_2$ 278.0435, found 278.0423.

Di(2-methoxyphenyl)disulfide (3g) [34]. White solid (23.4 mg, 84% yield); mp 120−121 °C; R_f = 0.6 (petroleum ether); ^{1}H NMR (600 MHz, CDCl$_3$) δ 7.53 (dd, J = 6.6, 1.2 Hz, 2 H), 7.21−7.16 (m, 2 H), 6.93−6.88 (m, 2 H), 6.86 (d, J = 7.8 Hz, 2 H), and 3.90 (s, 6 H) ppm; HRMS (ESI) m/z [M]$^+$ calculated for C$_{14}$H$_{14}$O$_2$S$_2$ 278.0435, found 278.0427.

1,2-bis(4-isopropylphenyl)disulfane (3h) [73]. White solid (27.5 mg, 91% yield); mp 79−81 °C; R_f = 0.6 (Petroleum ether); ^{1}H NMR (600 MHz, CDCl$_3$) δ 7.44 (dt, J = 8.4, 4.8 Hz,

4H), 7.17 (dt, J = 7.8, 4.2 Hz, 4H), 2.93−2.86 (m, 2H), and 1.24 (d, J = 7.2 Hz, 12H) ppm; HRMS (ESI) m/z [M + H]$^+$ calculated for $C_{18}H_{23}S_2$ 303.1236, found 303.1226.

Bis(4-tert-butylphenyl) disulfide (3i) [74]. White solid (28.7 mg, 87% yield); mp 88−89 °C; R_f = 0.6 (petroleum ether); ^{1}H NMR (600 MHz, CDCl$_3$) δ 7.46 (d, J = 9.0 Hz, 4H), 7.34 (d, J = 8.4 Hz, 4H), and 1.31 (s, 18H) ppm; HRMS (ESI) m/z [M]$^+$ calculated for $C_{20}H_{26}S_2$ 330.1476, found 330.1462.

Bis(4-fluorophenyl) disulfide (3j) [34]. White solid (21.8 mg, 86% yield); mp 112−114 °C; R_f = 0.6 (petroleum ether); ^{1}H NMR (600 MHz, CDCl$_3$) δ 7.48−7.41 (m, 4H) and 7.05−6.98 (m, 4H) ppm; HRMS (ESI) m/z [M]$^+$ calculated for $C_{12}H_8F_2S_2$ 254.0035, found 254.0029; ^{19}F NMR (376 MHz, CDCl$_3$) δ -113.4 ppm.

Bis(3-fluorophenyl) disulfide (3k) [75]. White solid (20.6 mg, 81% yield); mp 93−94 °C; R_f = 0.6 (Petroleum ether); ^{1}H NMR (600 MHz, CDCl$_3$) δ 7.31−7.27 (m, 2H), 7.26 (s, 1H), 7.25−7.23 (m, 2 H), 7.22 (t, J = 1.8 Hz, 1H), and 6.96−6.91 (m, 2H) ppm; HRMS (ESI) m/z [M]$^+$ calculated for $C_{12}H_8F_2S_2$ 254.0035, found 254.0030; ^{19}F NMR (376 MHz, CDCl$_3$) δ -111.1 ppm.

Bis(2-fluorophenyl)disulfide (3l) [75]. Slight yellow oil (19.9 mg, 75% yield); R_f = 0.6 (petroleum ether); ^{1}H NMR (600 MHz, CDCl$_3$) δ 7.59 (td, J = 7.8, 1.8 Hz, 2 H), 7.28−7.25 (m, 2H), 7.12 (td, J = 7.8, 1.2 Hz, 2 H), and 7.08−7.04 (m, 2H) ppm; HRMS (ESI) m/z [M]$^+$ calculated for $C_{12}H_8F_2S_2$ 254.0035, found 254.0029; ^{19}F NMR (376 MHz, CDCl$_3$) δ -109.9 ppm.

4,4′-Dichlorodiphenyl disulfide (3p) [73]. White solid (26.0 mg, 91% yield); mp 68−70 °C; R_f = 0.6 (petroleum ether); ^{1}H NMR (600 MHz, CDCl$_3$) δ 7.40 (dt, J = 8.4, 4.8 Hz, 4H) and 7.28 (dt, J = 8.4, 4.8 Hz, 4H) ppm; ^{13}C{^{1}H} NMR (100 MHz, CDCl$_3$) δ 135.1, 133.7, 129.4, and 129.3; HRMS (ESI) m/z [M]$^+$ calculated for $C_{12}H_8Cl_2S_2$ 285.9444, found 285.9421.

Bis(3-clorophenyl)disulfide (3q) [73]. White solid (25.2 mg, 88% yield); mp 80−82 °C; R_f = 0.6 (petroleum ether); ^{1}H NMR (600 MHz, CDCl$_3$) δ 7.48 (d, J = 2.4 Hz, 1 H), 7.37−7.32 (m, 2 H), 7.26 (d, J = 2.4 Hz, 1H), and 7.24−7.20 (m, 4H) ppm; HRMS (ESI) m/z [M]$^+$ calculated for $C_{12}H_8Cl_2S_2$ 285.9444, found 285.9425.

Bis(2-clorophenyl)disulfide (3r) [76]. White solid (24.0 mg, 84% yield); mp 90−91 °C; R_f = 0.6 (petroleum ether); ^{1}H NMR (400 MHz, CDCl$_3$) δ 7.48 (dd, J = 8.0, 1.6 Hz, 2H), 7.30 (dd, J = 7.2, 1.6 Hz, 2H), 7.14 (td, J = 7.6, 1.2 Hz, 2H), and 7.09 (td, J = 7.6, 1.6 Hz, 2H) ppm; HRMS (ESI) m/z [M]$^+$ calculated for $C_{12}H_8Cl_2S_2$ 285.9444, found 285.9421.

Bis(4-bromophenyl)disulfide (3s) [75]. White solid (33.3 mg, 89% yield); mp 110−112 °C; R_f = 0.6 (petroleum ether); ^{1}H NMR (600 MHz, CDCl$_3$) δ 7.44−7.41 (m, 4 H) and 7.35−7.32 (m, 4 H) ppm; HRMS (ESI) m/z [M]$^+$ calculated for $C_{12}H_8Br_2S_2$ 373.8434, found 373.8414.

2,2′-Dinaphthyl disulfide (3t) [77]. White solid (26.1 mg, 82% yield); mp 139−141 °C; R_f = 0.6 (petroleum ether); ^{1}H NMR (600 MHz, CDCl$_3$) δ 7.99 (d, J = 1.8 Hz, 2H), 7.82−7.76 (m, 4H), 7.75−7.71 (m, 2H), 7.62 (dd, J = 9.0, 6.6 Hz, 2H), and 7.49−7.44 (m, 4H) ppm; HRMS (ESI) m/z [M + H]$^+$ calculated for $C_{20}H_{15}S_2$ 319.0610, found 319.0614.

Bis(4-trifluoromethylphenyl)disulfide (3u) [42]. White solid (28.3 mg, 80% yield); mp 119−120 °C; R_f = 0.6 (petroleum ether); ^{1}H NMR (600 MHz, CDCl$_3$) δ 7.60−7.55 (m, 8H) ppm; HRMS (ESI) m/z [M + H]$^+$ calculated for $C_{14}H_9F_6S_2$ 355.0044, found 355.0039; ^{19}F NMR (376 MHz, CDCl$_3$) δ -62.5 ppm.

4. Conclusions

In summary, we have developed an efficient protocol for the selective access to symmetrical diaryl sulfides and disulfides using sodium sulfinates as sulfenylation reagents via homocoupling reaction. The utilization of readily available sodium sulfinates as coupling partners and good functional group tolerance with modest to excellent yields for most substrates enable these two types of novel transformations to become attractive alternatives for the preparation of the corresponding sulfur compounds. More importantly, sodium sulfinates were used for the first time to access symmetrical diaryl sulfides. The convinced mechanism, selectivity, and synthetic application of this transformation are still under investigation.

Supplementary Materials: The following supporting information can be downloaded at: https://www.mdpi.com/article/10.3390/molecules27196232/s1, ^{1}H, ^{13}C, ^{19}F NMR, IR and HPLC spectra.

Author Contributions: Conceptualization, X.L. and W.-C.G.; methodology, X.-Z.Y.; formal analysis, Y.-L.N.; data curation, X.-Z.Y.; writing—original draft preparation, X.L.; writing—review and editing, M.W. and W.-C.G.; superversion, W.-L.W.; funding acquisition, W.-C.G. All authors have read and agreed to the published version of the manuscript.

Funding: This research was funded by the National Natural Science Foundation of China, grant number 21901179, and the Natural Science Foundation of Shanxi Province, grant numbers 20210302123141 and 202103021224067.

Institutional Review Board Statement: Not applicable.

Informed Consent Statement: Not applicable.

Data Availability Statement: Not applicable.

Conflicts of Interest: The authors declare no conflict of interest.

Sample Availability: Samples of the compounds are available from the authors.

References

1. Pasquini, S.; Mugnaini, C.; Tintori, C.; Botta, M.; Trejos, A.; Arvela, R.K.; Larhed, M.; Witvrouw, M.; Michiels, M.; Christ, F.; et al. Investigations on the 4-Quinolone-3-carboxylic Acid Motif. 1. Synthesis and Structure–Activity Relationship of a Class of Human Immunodeficiency Virus type 1 Integrase Inhibitors. *J. Med. Chem.* **2008**, *51*, 5125. [CrossRef] [PubMed]
2. Beletskaya, I.P.; Ananikov, V.P. Transition-Metal-Catalyzed C–S, C–Se, and C–Te Bond Formation via Cross-Coupling and Atom-Economic Addition Reactions. *Chem. Rev.* **2011**, *111*, 1596. [CrossRef] [PubMed]
3. Brigg, S.; Pribut, N.; Basson, A.E.; Avgenikos, M.; Venter, R.; Blackie, M.A.; van Otterlo, W.A.L.; Pelly, S.C. Novel indole sulfides as potent HIV-1 NNRTIs. *Bioorg. Med. Chem. Lett.* **2016**, *26*, 1580. [CrossRef] [PubMed]
4. Srogl, J.; Hývl, J.; Révészb, Á.; Schröder, D. Mechanistic insights into a copper–disulfide interaction in oxidation of imines by disulfides. *Chem. Commun.* **2009**, 3463. [CrossRef] [PubMed]
5. Ge, W.; Wei, Y. Iodine-catalyzed oxidative system for 3-sulfenylation of indoles with disulfides using DMSO as oxidant under ambient conditions in dimethyl carbonate. *Green Chem.* **2012**, *14*, 2066. [CrossRef]
6. Lee, M.H.; Yang, Z.; Lim, C.W.; Lee, Y.H.; Dongbang, S.; Kang, C.; Kim, J.S. Disulfide-Cleavage-Triggered Chemosensors and Their Biological Applications. *Chem. Rev.* **2013**, *113*, 5071. [CrossRef] [PubMed]
7. Feng, M.; Tang, B.; Liang, S.; Jiang, X. Sulfur Containing Scaffolds in Drugs: Synthesis and Application in Medicinal Chemistry. *Curr. Top. Med. Chem.* **2016**, *16*, 1200. [CrossRef]
8. Ke, F.; Qu, Y.; Jiang, Z.; Li, Z.; Wu, D.; Zhou, X. An Efficient Copper-Catalyzed Carbon–Sulfur Bond Formation Protocol in Water. *Org. Lett.* **2011**, *13*, 454. [CrossRef]
9. García, N.; García-García, P.; Fernández-Rodríguez, M.A.; Rubio, R.; Pedrosa, M.R.; Arnáiz, F.J.; Sanz, R. Pinacol as a New Green Reducing Agent: Molybdenum- Catalyzed Chemoselective Reduction of Sulfoxides and Nitroaromatics. *Adv. Synth. Catal.* **2012**, *354*, 321. [CrossRef]
10. Zhao, P.; Yin, H.; Gao, H.; Xi, C. Cu-Catalyzed Synthesis of Diaryl Thioethers and S-Cycles by Reaction of Aryl Iodides with Carbon Disulfide in the Presence of DBU. *J. Org. Chem.* **2013**, *78*, 5001. [CrossRef] [PubMed]
11. Jang, Y.; Kim, K.T.; Jeon, H.B. Deoxygenation of Sulfoxides to Sulfides with Thionyl Chloride and Triphenylphosphine: Competition with the Pummerer Reaction. *J. Org. Chem.* **2013**, *78*, 6328. [CrossRef]
12. García, N.; García-García, P.; Fernández-Rodríguez, M.A.; García, D.; Pedrosa, M.R.; Arnáiz, F.J.; Sanz, R. An unprecedented use for glycerol: Chemoselective reducing agent for sulfoxides. *Green Chem.* **2013**, *15*, 999. [CrossRef]
13. Mitsudome, T.; Takahashi, Y.; Mizugaki, T.; Jitsukawa, K.; Kaneda, K. Hydrogenation of Sulfoxides to Sulfides under Mild Conditions Using Ruthenium Nanoparticle Catalysts. *Angew. Chem. Int. Ed.* **2014**, *53*, 8348. [CrossRef]
14. Touchy, A.S.; Siddiki, S.M.A.H.; Onodera, W.; Kon, K.; Shimizu, K. Hydrodeoxygenation of sulfoxides to sulfides by a Pt and MoOx co-loaded TiO$_2$ catalyst. *Green Chem.* **2016**, *18*, 2554. [CrossRef]
15. Wu, R.; Huang, K.; Qiu, J.; Liu, J.-B. Synthesis of Thioethers from Sulfonyl Chlorides, Sodium Sulfinates, and Sulfonyl Hydrazides. *Synthesis* **2019**, *51*, 3567. [CrossRef]
16. Barba, F.; Ranz, F.; Batanero, B. Electrochemical transformation of diazonium salts into diaryl disulfides. *Tetrahedron Lett.* **2009**, *50*, 6798. [CrossRef]
17. Taniguchi, N. Copper-catalyzed chalcogenation of aryl iodides via reduction of chalcogen elements by aluminum or magnesium. *Tetrahedron* **2012**, *68*, 10510. [CrossRef]
18. Xiao, X.; Feng, M.; Jiang, X. New Design of a Disulfurating Reagent: Facile and Straightforward Pathway to Unsymmetrical Disulfanes by Copper-Catalyzed Oxidative Cross-Coupling. *Angew. Chem. Int. Ed.* **2016**, *55*, 14121. [CrossRef]

19. Abbasi, M.; Nowrouzi, N.; Borazjani, S.G. Conversion of organic halides to disulfanes using KCN and CS_2. *Tetrahedron Lett.* **2017**, *58*, 4251. [CrossRef]
20. Xiao, X.; Xue, J.; Jiang, X. Polysulfurating reagent design for unsymmetrical polysulfide construction. *Nat. Commun.* **2018**, *9*, 2191. [CrossRef]
21. Xue, J.; Jiang, X. Unsymmetrical polysulfidation via designed bilateral disulfurating reagents. *Nat. Commun.* **2020**, *11*, 4170. [CrossRef] [PubMed]
22. Wu, Z.; Pratt, D.A. A Divergent Strategy for Site-Selective Radical Disulfuration of Carboxylic Acids with Trisulfide-1,1-Dioxides. *Angew. Chem. Int. Ed.* **2021**, *60*, 15598. [CrossRef] [PubMed]
23. Wang, F.; Chen, Y.; Rao, W.; Ackermann, L.; Wang, S.-Y. Efficient preparation of unsymmetrical disulfides by nickel-catalyzed reductive coupling strategy. *Nat. Commun.* **2022**, *13*, 2588. [CrossRef] [PubMed]
24. Chen, H.-Y.; Peng, W.-T.; Lee, Y.-H.; Chang, Y.-L.; Chen, Y.-J.; Lai, Y.-C.; Jheng, N.-Y.; Chen, H.-Y. Use of Base Control to Provide High Selectivity between Diaryl Thioether and Diaryl Disulfide for C–S Coupling Reactions of Aryl Halides and Sulfur and a Mechanistic Study. *Organometallics* **2013**, *32*, 5514. [CrossRef]
25. Kamal, A.; Srinivasulu, V.; Murty, J.N.S.R.C.; Shankaraiah, N.; Nagesh, N.; Reddy, T.S.; Rao, A.V.S. Copper Oxide Nanoparticles Supported on Graphene Oxide- Catalyzed S-Arylation: An Efficient and Ligand-Free Synthesis of Aryl Sulfides. *Adv. Synth. Catal.* **2013**, *355*, 2297. [CrossRef]
26. Cai, M.; Yao, R.; Chen, L.; Zhao, H. A simple, efficient and recyclable catalytic system for carbon–sulfur coupling of aryl halides with thioacetamide. *J. Mol. Catal. A Chem.* **2014**, *395*, 349. [CrossRef]
27. Firouzabadi, H.; Iranpoor, N.; Gorginpour, F.; Samadi, A. Dithiooxamide as an Effective Sulfur Surrogate for Odorless High-Yielding Carbon–Sulfur Bond Formation in Wet PEG200 as an Eco-Friendly, Safe, and Recoverable Solvent. *Eur. J. Org. Chem.* **2015**, 2914. [CrossRef]
28. Ghorbani-Choghamarani, A.; Taherinia, Z. The first report on the preparation of peptide nanofibers decorated with zirconium oxide nanoparticles applied as versatile catalyst for the amination of aryl halides and synthesis of biaryl and symmetrical sulfides. *New. J. Chem.* **2017**, *41*, 9414. [CrossRef]
29. Liu, X.; Cao, Q.; Xu, W.; Zeng, M.-T.; Dong, Z.-B. Nickel-Catalyzed C–S Coupling: Synthesis of Diaryl Sulfides Starting from Phenyldithiocarbamates and Iodobenzenes. *Eur. J. Org. Chem.* **2017**, 5795. [CrossRef]
30. Cheng, Y.; Liu, X.; Dong, Z.-B. Phenyldithiocarbamates: Efficient Sulfuration Reagents in the Chan–Lam Coupling Reaction. *Eur. J. Org. Chem.* **2018**, 815. [CrossRef]
31. Dong, Z.-B.; Balkenhohl, M.; Tan, E.; Knochel, P. Synthesis of Functionalized Diaryl Sulfides by Cobalt-Catalyzed Coupling between Arylzinc Pivalates and Diaryl Disulfides. *Org. Lett.* **2018**, *20*, 7581. [CrossRef] [PubMed]
32. Arguello, J.E.; Schmidt, L.C.; Penenory, A.B. "One-Pot" Two-Step Synthesis of Aryl Sulfur Compounds by Photoinduced Reactions of Thiourea Anion with Aryl Halides. *Org. Lett.* **2003**, *5*, 4133. [CrossRef] [PubMed]
33. Soleiman-Beigi, M.; Mohammadi, F. A novel copper-catalyzed, one-pot synthesis of symmetric organic disulfides from alkyl and aryl halides: Potassium 5-methyl-1,3,4-oxadiazole-2-thiolate as a novel sulfur transfer reagent. *Tetrahedron Lett.* **2012**, *53*, 7028. [CrossRef]
34. Li, Z.K.; Ke, F.; Deng, H.; Xu, H.L.; Xiang, H.F.; Zhou, X.G. Synthesis of disulfides and diselenides by copper-catalyzed coupling reactions in water. *Org. Biomol. Chem.* **2013**, *11*, 2943. [CrossRef]
35. Soleiman-Beigi, M.; Hemmati, M. An efficient, one-pot and CuCl-catalyzed route to the synthesis of symmetric organic disulfides via domino reactions of thioacetamide and aryl (alkyl) halides. *Appl. Organometal. Chem.* **2013**, *27*, 734. [CrossRef]
36. Li, X.; Du, J.; Zhang, Y.; Chang, H.; Gao, W.; Wei, W. Synthesis and nano-Pd catalyzed chemoselective oxidation of symmetrical and unsymmetrical sulfides. *Org. Biomol. Chem.* **2019**, *17*, 3048. [CrossRef]
37. Olah, G.A.; Narang, S.C.; Field, L.D.; Karpeles, R. Synthetic methods and reactions. 101. Reduction of sulfonic acids and sulfonyl derivatives to disulfides with iodide in the presence of boron halides. *J. Org. Chem.* **1981**, *46*, 2408. [CrossRef]
38. Zheng, Y.; Qing, F.-L.; Huang, Y.; Xu, X.-H. Tunable and Practical Synthesis of Thiosulfonates and Disulfides from Sulfonyl Chlorides in the Presence of Tetrabutylammonium Iodide. *Adv. Synth. Catal.* **2016**, *358*, 3477. [CrossRef]
39. Hajipour, A.R.; Mallakpour, S.E.; Adibi, H. Selective and Efficient Oxidation of Sulfides and Thiols with Benzyltriphenylphosphonium Peroxymonosulfate in Aprotic Solvent. *J. Org. Chem.* **2002**, *67*, 8666. [CrossRef]
40. Banfield, S.C.; Omori, A.T.; Leisch, H.; Hudlicky, T. Unexpected Reactivity of the Burgess Reagent with Thiols: Synthesis of Symmetrical Disulfides. *J. Org. Chem.* **2007**, *72*, 4989. [CrossRef]
41. Dhakshinamoorthy, A.; Alvaro, M.; Garcia, H. Aerobic oxidation of thiols to disulfides using iron metal–organic frameworks as solid redoxcatalysts. *Chem. Commun.* **2010**, *46*, 6476. [CrossRef] [PubMed]
42. Oba, M.; Tanaka, K.; Nishiyama, K.; Ando, W. Aerobic Oxidation of Thiols to Disulfides Catalyzed by Diaryl Tellurides under Photosensitized Conditions. *J. Org. Chem.* **2011**, *76*, 4173. [CrossRef] [PubMed]
43. Corma, A.; Rodenas, T.; Sabater, M.J. Aerobic oxidation of thiols to disulfides by heterogeneous gold catalysts. *Chem. Sci.* **2012**, *3*, 398. [CrossRef]
44. Li, X.-B.; Li, Z.-J.; Gao, Y.-J.; Meng, Q.-Y.; Yu, S.; Weiss, R.G.; Tung, C.-H.; Wu, L.-Z. Mechanistic Insights into the Interface-Directed Transformation of Thiols into Disulfides and Molecular Hydrogen by Visible-Light Irradiation of Quantum Dots. *Angew. Chem. Int. Ed.* **2014**, *53*, 2085. [CrossRef]

45. Tabrizian, E.; Amoozadeh, A.; Rahmani, S. Sulfamic acid-functionalized nano-titanium dioxide as an efficient, mild and highly recyclable solid acid nanocatalyst for chemoselective oxidation of sulfides and thiols. *RSC Adv.* **2016**, *6*, 21854. [CrossRef]
46. Samanta, S.; Ray, S.; Ghosh, A.B.; Biswas, P. 3,6-Di(pyridin-2-yl)-1,2,4,5-tetrazine (pytz) mediated metal-free mild oxidation of thiols to disulfides in aqueous medium. *RSC Adv.* **2016**, *6*, 39356. [CrossRef]
47. Paul, S.; Islam, S.M. Oxidative dehydrogenation of thiols to disulfides at room temperature using silica supported iron oxide as an efficient solid catalyst. *RSC Adv.* **2016**, *6*, 95753. [CrossRef]
48. Laudadio, G.; Straathof, N.J.W.; Lanting, M.D.; Knoops, B.; Hessel, V.; Noël, T. An environmentally benign and selective electrochemical oxidation of sulfides and thiols in a continuous-flow microreactor. *Green Chem.* **2017**, *19*, 4061. [CrossRef]
49. Ning, Y.; Ji, Q.; Liao, P.; Anderson, E.A.; Bi, X. Silver-Catalyzed Stereoselective Aminosulfonylation of Alkynes. *Angew. Chem. Int. Ed.* **2017**, *56*, 13805. [CrossRef]
50. Yue, H.; Zhu, C.; Rueping, M. Cross-Coupling of Sodium Sulfinates with Aryl, Heteroaryl, and Vinyl Halides by Nickel/Photoredox Dual Catalysis. *Angew. Chem. Int. Ed.* **2018**, *57*, 1371. [CrossRef]
51. Cao, L.; Luo, S.-H.; Jiang, K.; Hao, Z.-F.; Wang, B.-W.; Pang, C.-M.; Wang, Z.-Y. Disproportionate Coupling Reaction of Sodium Sulfinates Mediated by BF3·OEt2: An Approach to Symmetrical/Unsymmetrical Thiosulfonate. *Org. Lett.* **2018**, *20*, 4754. [CrossRef] [PubMed]
52. Miao, T.; Li, P.; Zhang, Y.; Wang, L. A Sulfenylation Reaction: Direct Synthesis of 3-Arylsulfinylindoles from Arylsulfinic Acids and Indoles in Water. *Org. Lett.* **2015**, *17*, 832. [CrossRef] [PubMed]
53. Ji, Y.-Z.; Li, H.-J.; Yang, H.-R.; Zhang, Z.-Y.; Xie, L.-J.; Wu, Y.-C. TMSOTf-Promoted Sulfinylation of Electron-Rich Aromatics with Sodium Arylsulfinates. *Synlett* **2020**, *31*, 349.
54. Xiao, F.; Xie, H.; Liu, S.; Deng, G.-J. Iodine-Catalyzed Regioselective Sulfenylation of Indoles with Sodium Sulfinates. *Adv. Synth. Catal.* **2014**, *356*, 364. [CrossRef]
55. Guo, S.; He, W.; Xiang, J.; Yuan, Y. Palladium-catalyzed direct thiolation of ethers with sodium sulfinates. *Tetrahedron Lett.* **2014**, *55*, 6407. [CrossRef]
56. Lin, Y.-M.; Lu, G.-P.; Cai, C.; Yi, W.-B. Odorless, One-Pot Regio- and Stereoselective Iodothiolation of Alkynes with Sodium Arenesulfinates under Metal-Free Conditions in Water. *Org. Lett.* **2015**, *17*, 3310. [CrossRef]
57. Wang, B.W.; Jiang, K.; Li, J.-X.; Luo, S.-H.; Wang, Z.-Y.; Jiang, H.-F. 1,1-Diphenylvinylsulfide as a Functional AIEgen Derived from the Aggregation-Caused-Quenching Molecule 1,1-Diphenylethene through Simple Thioetherification. *Angew. Chem. Int. Ed.* **2020**, *59*, 2338. [CrossRef]
58. Liu, Y.; Lam, L.Y.; Ye, J.; Blanchard, N.; Ma, C. DABCO-promoted Diaryl Thioether Formation by Metal-catalyzed Coupling of Sodium Sulfinates and Aryl Iodides. *Adv. Synth. Catal.* **2020**, *362*, 2326. [CrossRef]
59. Lam, L.Y.; Ma, C. Chan–Lam-Type C–S Coupling Reaction by Sodium Aryl Sulfinates and Organoboron Compounds. *Org. Lett.* **2021**, *23*, 6164. [CrossRef]
60. Pinnick, H.W.; Reynolds, M.A.; McDonald, R.T., Jr.; Brewster, W.D. Reductive coupling of aromatic sulfinate salts to disulfides. *J. Org. Chem.* **1980**, *45*, 930. [CrossRef]
61. Wang, J.Q.; Zhang, Y.M. The Reduction of Arylsulfonyl Chlorides and Sodium Arylsulfinates with TiCl4/Sm System. A Novel Method for the Preparation of Diaryldisulfides. *Syn. Commun.* **1996**, *26*, 135. [CrossRef]
62. Firouzabadi, H.; Karimi, B. Efficient Deoxygenation of Sulfoxides to Thioethers and Reductive Coupling of Sulfonyl Chlorides to Disulfides with Tungsten Hexachloride. *Synthesis* **1999**, *1999*, 500. [CrossRef]
63. Iranpoor, N.; Firouzabadi, H.; Jamalian, A. Deoxygenation of Sulfoxides and Reductive Coupling of Sulfonyl Chlorides, Sulfinates and Thiosulfonates Using Silphos [PCl3-n(SiO2)n] as a Heterogeneous Phosphine Reagent. *Synlett* **2005**, *9*, 1447. [CrossRef]
64. Still, I.W.J.; Watson, I.D.G. An efficient synthetic route to aryl thiocyanates from arenesulfinates. *Synth. Commun.* **2001**, *31*, 1355. [CrossRef]
65. Emmett, E.J.; Hayter, B.R.; Willis, M.C. Palladium-Catalyzed Synthesis of Ammonium Sulfinates from Aryl Halides and a Sulfur Dioxide Surrogate: A Gas- and Reductant-Free Process. *Angew. Chem. Int. Ed.* **2014**, *53*, 10204. [CrossRef] [PubMed]
66. Wang, Y.; Deng, J.; Chen, J.; Cao, F.; Hou, Y.; Yang, Y.; Deng, X.; Yang, J.; Wu, L.; Shao, X.; et al. Dechalcogenization of Aryl Dichalcogenides to Synthesize Aryl Chalcogenides via Copper Catalysis. *ACS Catal.* **2020**, *10*, 2707. [CrossRef]
67. Li, Y.-M.; Nie, C.-P.; Wang, H.-P.; Verpoort, F.; Duan, C.-Y. A Highly Efficient Method for the Copper-Catalyzed Selective Synthesis of Diaryl Chalcogenides from Easily Available Chalcogen Sources. *Eur. J. Org. Chem.* **2011**, *2011*, 7331–7338. [CrossRef]
68. Li, X.-K.; Yuan, T.-J.; Chen, J.-M. Efficient Copper(I)-Catalyzed S-Arylation of KSCN with Aryl Halides in PEG-400. *Chin. J. Chem.* **2012**, *30*, 651–655. [CrossRef]
69. Carpino, L.A.; Gao, H.S.; Ti, G.S.; Segev, D. Thioxanthene Dioxide Based Amino-Protecting Groups Sensitive to Pyridine Bases and Dipolar Aprotic Solvents. *J. Org. Chem.* **1989**, *54*, 5887–5897. [CrossRef]
70. Zhu, Y.-C.; Li, Y.; Zhang, B.-C.; Yang, Y.-N.; Wang, X.-S. Palladium-Catalyzed Enantioselective C-H Olefination of Diaryl Sulfoxides via Parallel Kinetic Resolution and Desymmetrization. *Angew. Chem. Int. Ed.* **2018**, *57*, 5129–5133. [CrossRef] [PubMed]
71. Hajipour, A.-R.; Karimzadeh, M.; Azizi, G. Highly Efficient and Magnetically Separable Nano-CuFe$_2$O$_4$ Catalyzed S-Arylation of Thiourea by Aryl/Heteroaryl Halides. *Chin. Chem. Lett.* **2014**, *25*, 1382–1386. [CrossRef]
72. García-López, J.-A.; Çetin, M.; Greaney, M.F. Synthesis of Hindered Biaryls via Aryne Addition and in Situ Dimerization. *Org. Lett.* **2015**, *17*, 2649–2651. [CrossRef]

73. Liu, Y.-Y.; Wang, H.; Wang, C.-P.; Wan, J.-P.; Wen, C.-P. Bio-Based Green Solvent Mediated Disufide Synthesis via Thiol Couplings Free of Catalyst and Additive. *RSC Adv.* **2013**, *3*, 21369–21372. [CrossRef]
74. Murahashi, S.I.; Zhang, D.Z.; Iida, H.; Miyawaki, T.; Uenaka, M.; Uenaka, K.; Meguro, K. Flavin-Catalyzed Aerobic Oxidation of Sulfides and Thiols with Formic Acid/Triethylamine. *Chem. Commun.* **2014**, *50*, 10295–10298. [CrossRef] [PubMed]
75. Ruano, J.L.G.; Parra, A.; Alemán, J. Efficient Synthesis of Disulfides by Air Oxidation of Thiols under Sonication. *Green Chem.* **2008**, *10*, 706–711. [CrossRef]
76. Chai, P.J.; Li, Y.S.; Tan, C.X. An Efficient and Convenient Method for Preparation of Disulfides from Thiols Using Air as Oxidant Catalyzed by Co-Salophen. *Chin. Chem. Lett.* **2011**, *22*, 1403–1406. [CrossRef]
77. Wang, L.; Clive, D.L. [[(tert-Butyl)dimethylsilyl]oxy]methyl Group for Sulfur Protection. *Org. Lett.* **2011**, *42*, 1734–1737. [CrossRef]

Article

Synthesis of Diversified Pyrazolo[3,4-b]pyridine Frameworks from 5-Aminopyrazoles and Alkynyl Aldehydes via Switchable C≡C Bond Activation Approaches

Xiao-Yu Miao [1], Yong-Ji Hu [1], Fu-Rao Liu [1], Yuan-Yuan Sun [1], Die Sun [1], An-Xin Wu [2,*] and Yan-Ping Zhu [1,*]

[1] Key Laboratory of Molecular Pharmacology and Drug Evaluation, Ministry of Education, Collaborative Innovation Center of Advanced Drug Delivery System and Biotech Drugs in Universities of Shandong, School of Pharmacy, Yantai University, Yantai 264005, China
[2] Key Laboratory of Pesticide & Chemical Biology, Ministry of Education, College of Chemistry, Central China Normal University, Wuhan 430079, China
* Correspondence: chwuax@mail.ccnu.edu.cn (A.-X.W.); chemzyp@foxmail.com or chemzyp@ytu.edu.cn (Y.-P.Z.)

Abstract: A cascade *6-endo-dig* cyclization reaction was developed for the switchable synthesis of halogen and non-halogen-functionalized pyrazolo[3,4-*b*]pyridines from 5-aminopyrazoles and alkynyl aldehydes via C≡C bond activation with silver, iodine, or NBS. In addition to its wide substrate scope, the reaction showed good functional group tolerance as well as excellent regional selectivity. This new protocol manipulated three natural products, and the arylation, alkynylation, alkenylation, and selenization of iodine-functionalized products. These reactions demonstrated the potential applications of this new method.

Keywords: pyrazolo[3,4-*b*]pyridine; alkyne activation; regional selectivity; *6-endo-dig* cyclization

1. Introduction

A series of natural products and biologically active molecules contain pyrazolo[3,4-*b*]pyridine as a key structural motif [1,2]. Several of these compounds are effective antienterovirals, antimalarials, anticancer agents, and kinase inhibitors (Figure 1) [3–5]. This has inspired the development of efficient methods to construct these compounds and has become a hot topic in modern organic synthesis.

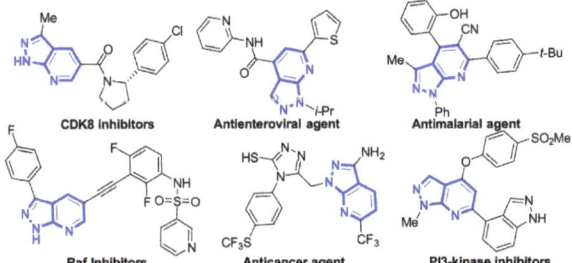

Figure 1. Some biologically active pyrazolo[3,4-*b*]pyridine derivatives.

Recently, catalytic carbon-carbon bond activation has emerged as a useful tool to build complex molecules rapidly and efficiently [6–9]. There are versatile intermediates involved in these reactions, which could be trapped in situ by a second molecule that triggers subsequent tandem reactions [10–15]. The nucleophilic/electrophilic addition reactions of alkynes are well-known and provide a convenient way to synthesize functionalized molecules [16–20]. The high reactivity, good selectivity, excellent functional-group tolerance, and mild reaction conditions of these reactions have inspired significant research over the past few decades. Generally, this process forms highly active intermediates using transition metals, such as Ag, Au, Rh, Cu, and Co [21–27], or electrophiles like I$_2$, NXS (X = I, Br, Cl), Se, S, and P (Scheme 1a) [28–34]. Each reagent type sees significant use in the development of synthetic methodologies and applications to prepare bioactive compounds or complex naturally occurring skeletons. However, to our knowledge, among those strategies, a direct and efficient protocol for the selective synthesis of polysubstituted and functionalized fused heterocycles, such as halogen-functionalized pyrazolo[3,4-b]pyridine frameworks, by C≡C bond activation has seldom been described. Thus, developing convenient and sustainable synthetic methods to build these high-value compounds merits attention.

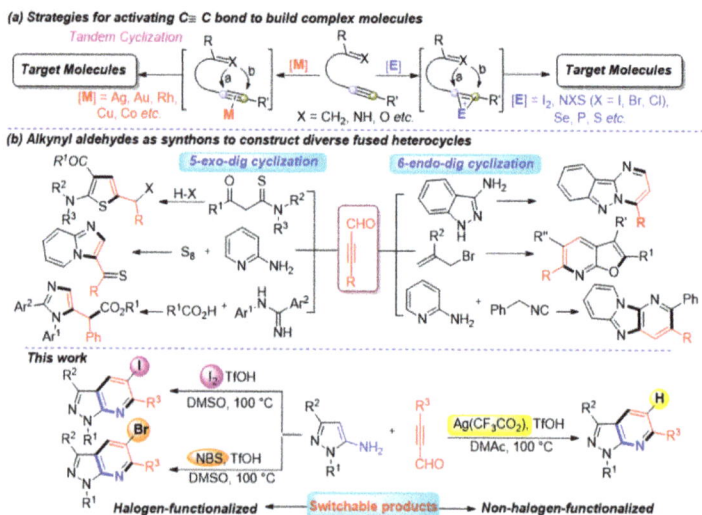

Scheme 1. Strategies for the synthesis of diverse molecules via activating C≡C bond.

As a kind of synthetic block with bifunctional groups (C≡C and carbonyl), alkynyl aldehydes are essential synthons with rich and unexpected chemical properties [35–38]. Tandem cyclization reactions using alkynyl aldehydes as synthons yields a variety of heterocycles. Generally, tandem cyclization occurs in one of two ways, 5-exo-dig or 6-endo-dig cyclizations. For example, some efficient synthesis strategies have been reported for the synthesis of multi-substituted thiazoles, imidazo[1,2-a]pyridines, and imidazoles by using alkynyl aldehydes as synthons via 5-exo-dig cyclization [39–41]. The 6-endo-dig cyclization of alkynyl aldehydes is an alternative method to construct complex fused ring systems (Scheme 1b) [42–44]. These protocols typically use simple starting materials, with good functional tolerance and high yields. Inspired by these achievements, we sought to selectively activate the C≡C bond by changing the reaction conditions to obtain a series of compounds with a pyrazolo[3,4-b]pyridine structure core.

2. Results and Discussion

To evaluate our idea, we chose 3-methyl-1-phenyl-1H-pyrazol-5-amine (**1a**) and 3-phenylpropiolaldehyde (**2a**) as the model substrates for the optimization of the conditions. Through optimization of the catalyst, additive, solvent and temperature, the optimal reaction conditions can be summarized as follows: **1a** (0.2 mmol) and **2a** (0.2 mmol) in DMAc (1.5 mL) with Ag(CF$_3$CO$_2$) (10 mol%), TfOH (30 mol%), at 100 °C for 2 h (details appear in Supplementary Materials Tables S3–S6).

Having optimized the reaction conditions, we determined the versatility of this reaction. We examined a series of 5-aminopyrazole derivatives to test the generality of this method and evaluate the electronic influence of aromatic ring substitutions. As shown in

Scheme 2, pyrazole rings bearing electron-donating groups (e.g., 3-Me, 3-(t-Bu), 1-Ph, 3-Ph) led to good yields (74–84%) of the corresponding products (**3a**–**3d** and **3f**–**3k**). Notably, the structure of compound **3a** was confirmed by X-ray single-crystal diffraction (Scheme 2). The substrates of aromatic rings attached to halogen atoms (e.g., 4-F, 4-Br) also led to their corresponding products (**3e** and **3l**–**3p**) in yields between 68–81%. A strongly electron-deficient substrate was applied and afforded its product in 63% yield for the corresponding product **3q**. Pyrazole rings only bearing alkyl groups were used as starting materials and yielded the expected products (**3r**–**3t**) in moderate to good yields (66–75%). Additionally, 5-aminoisoxazoles also readily reacted with 3-phenylpropiolaldehyde, yielding the desired products (**3u**–**3x**) in good yields (70–75%). However, 3-methylisothiazol-5-amine did not yield the desired product **3y**.

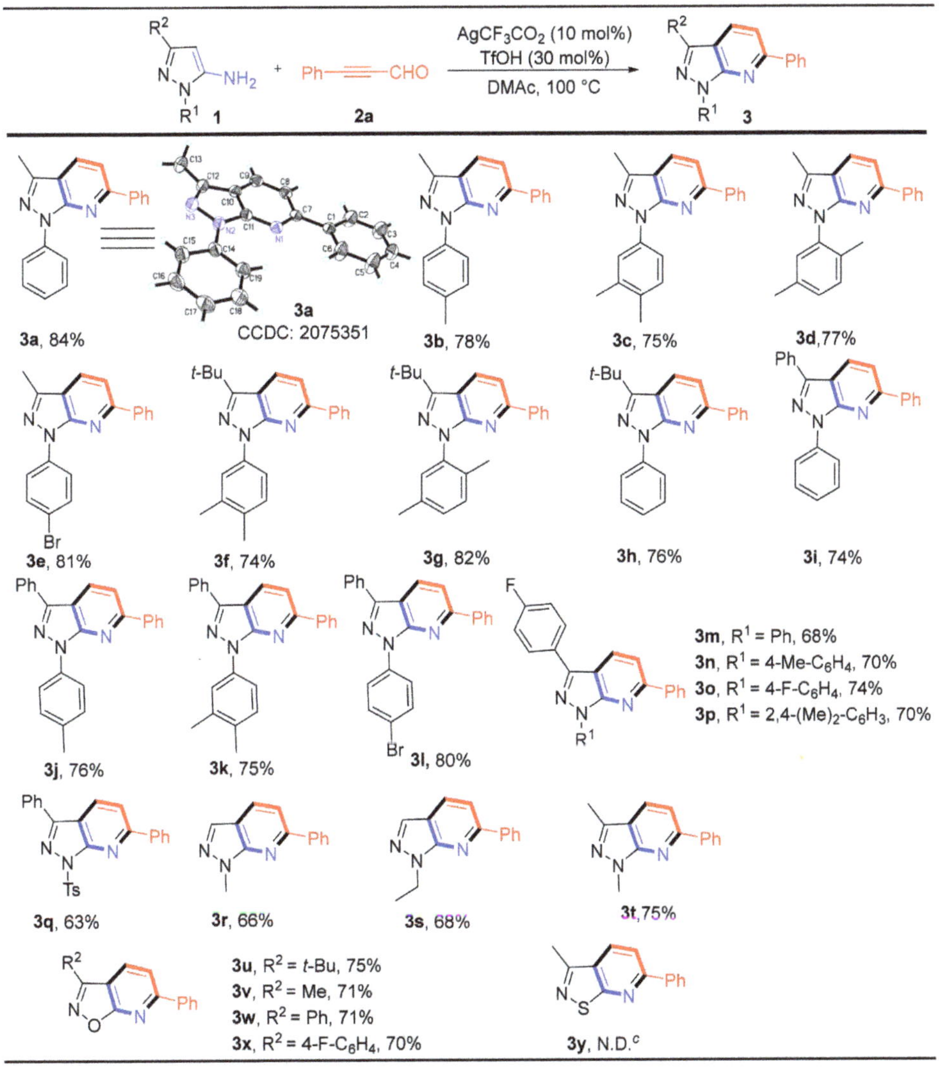

Scheme 2. Substrate scope and Isolated yield of substituted 5-aminopyrazoles and derivatives. Reaction conditions: **1** (0.2 mmol), **2a** (0.2 mmol), Ag(CF$_3$CO$_2$) (10 mol%), TfOH (30 mol%) in DMAc (1.5 mL) at 100 °C for 2 h. [c] N.D. = not detected.

We next investigated the scope of alkynyl aldehydes derivatives for this reaction (Scheme 3). First, we examined 3-phenylpropiolaldehydes with phenyl rings containing electron-rich substituents (e.g., 4-Me, 4-Et, 4-OMe, 4-OEt, 3,4-(OMe)$_2$). Annulation reactions occurred smoothly to deliver products (**4b–4f**) in 64–78% yields. For 3-phenylpropiolaldehyde containing electron-withdrawing groups (e.g., 4-Ac, 4-CO$_2$Me, 4-CF$_3$, 4-F, 3-F, 3-Cl, 4-Br), the reaction proceeded smoothly and afforded products (**4g–4i** and **4l–4o**) in moderate to good yields (66–81%). In addition, when 4-phenylbut-2-ynal and 3-(trimethylsilyl)propiolaldehyde were used as starting materials, products **4j** and **4k** were obtained in good yields (67% and 73% respectively). It is worth noting that when using 3-(trimethylsilyl)propiolaldehyde, compound **4k** was the product of the trimethylsilyl group removal. Furthermore, different heterocyclic aldehydes were also investigated including furan, thiophene, and pyridine to generate products **4p–4s** in 65–72% yields. We were delighted to find that alkyl alkynyl aldehydes gave the corresponding products **4t-4u** in moderate yields as well.

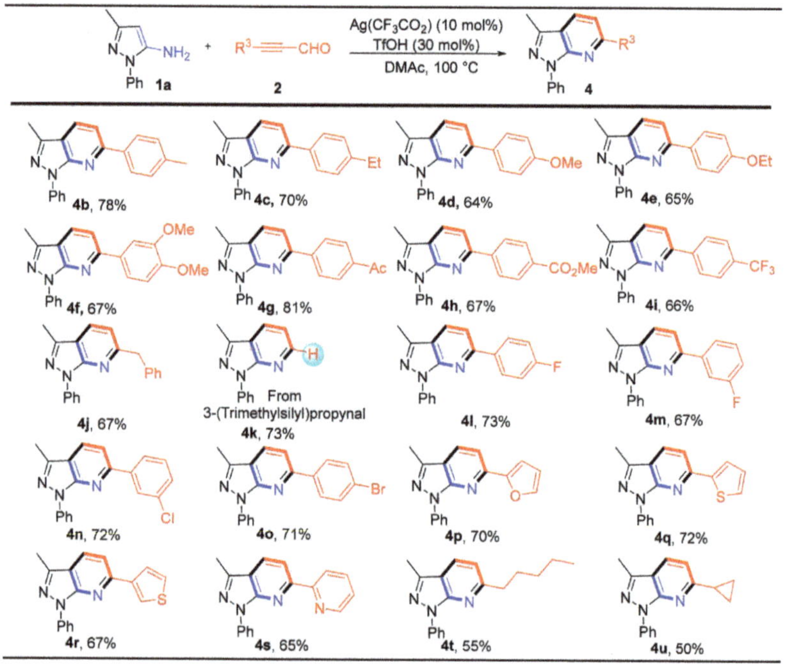

Scheme 3. Substrate scope and Isolated yield of substituted alkynyl aldehydes and derivatives. Reaction conditions: **1a** (0.2 mmol), **2** (0.2 mmol), Ag(CF$_3$CO$_2$) (10 mol%), TfOH (30 mol%) in DMAc (1.5 mL) at 100 °C for 2 h.

The iodinated product was detected when 1.0 equivalent of iodine was added to the reaction system (control experiment, Scheme 7d). We chose **1a** and **2a** as model substrates to investigate the optimal conditions to synthesize iodinated products (more details appear in Supplementary Materials Tables S7 and S8).

Next, various 5-aminopyrazoles and alkynyl aldehydes were tested to determine the scope of iodine-functionalized products (Scheme 4). These reactions produced the corresponding products **5a–5j** 58–68% yields. Meanwhile, 3-methylisoxazol-5-amine tolerated the reaction conditions and reacted with 3-phenylpropiolaldehyde (**2a**) to generate **5k** in moderate yield. When iodine was replaced by NBS, the expected compounds **5l–5r** were obtained in moderated yields (53–66%). However, after many trials, the Cl-functionalized product **5s** was not obtained.

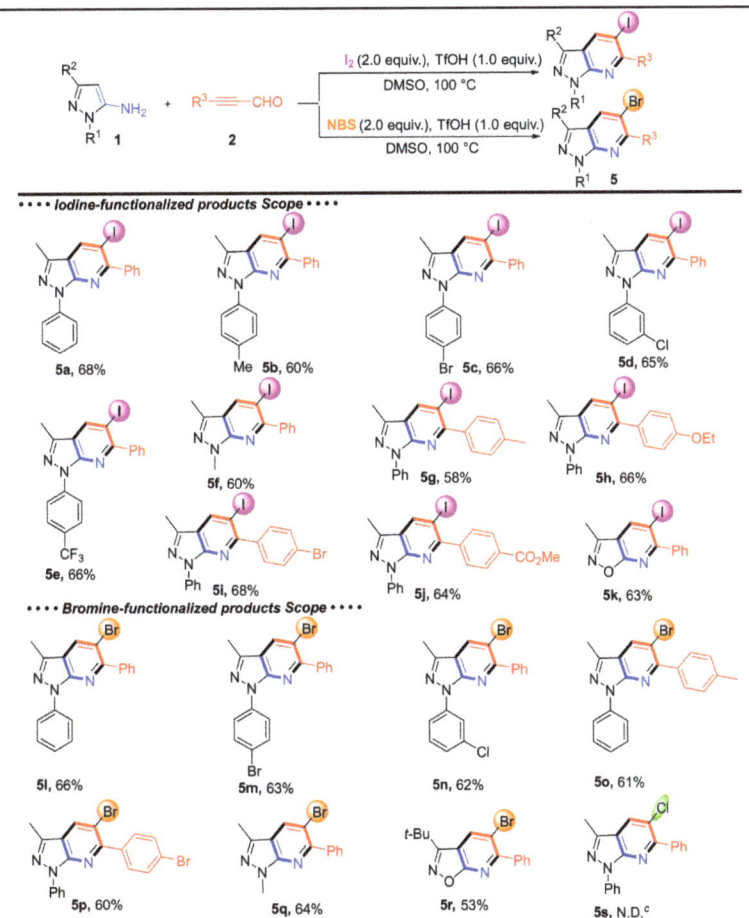

Scheme 4. The substrate scope and Isolated yield of halogen-functionalized products. Reaction conditions: **1** (0.2 mmol), **2** (0.2 mmol), I₂ or NBS (2.0 equiv.), TfOH (1.0 equiv.) in DMSO (2 mL) at 100 °C for 6 h. [c] N.D. = not detected.

To demonstrate the applicability of this method, we modified natural products (Scheme 5a). For example, estrone, tormononetin, and eudistomin Y₁ are all biologically active natural products, and these compounds have phenolic hydroxyl groups that undergo conversion into trifluoromethane sulfonates. Those sulfonates undergo Sonogashira coupling and deketalization to afford alkynaldehyde intermediates (**6**, **8**, and **10**). By using these alkynaldehyde intermediates as substrates for this protocol, we successfully obtained three natural product functionalized pyrazolo[3,4-*b*]pyridines in moderate to good yields (**7**, **9**, and **11**). Because heteroaryl iodides are highly useful functional structures in synthetic organic chemistry, additional applications of iodine-functionalized products were conducted (Scheme 5b) [45]. A series of coupling reactions were examined to form iodine-functionalized products, including Suzuki, Sonogashira, and Heck couplings that yielded the expected products **13–15** in good yields. Furthermore, selenization of iodine-functionalized products afforded **16** in very good yield (78%).

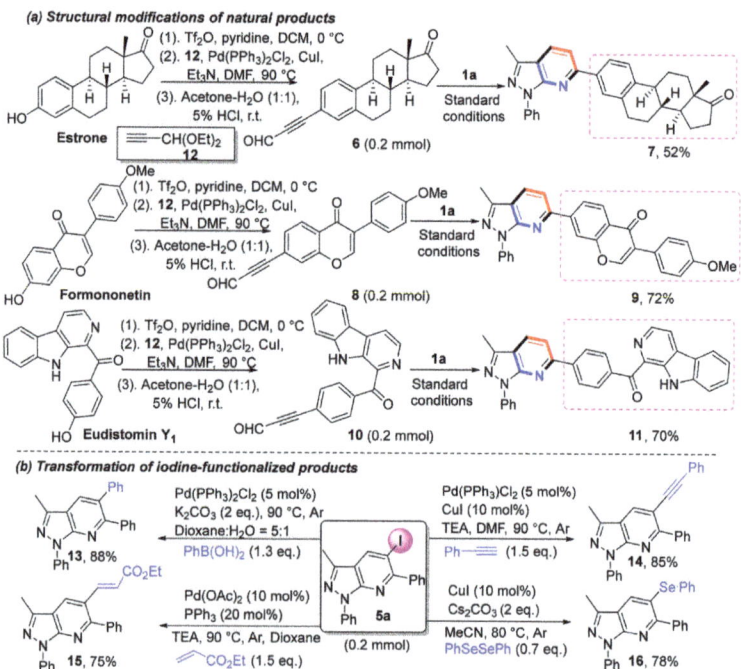

Scheme 5. Strategies to synthesize diverse molecules via C≡C bond activation.

The reaction of **1a** and **2a** was scaled up to 5 mmol to illustrate the potential applications of this method; **3a** and **5a** formed 71% and 53% yields, respectively (Scheme 6). This promising result lays a good foundation for large-scale syntheses.

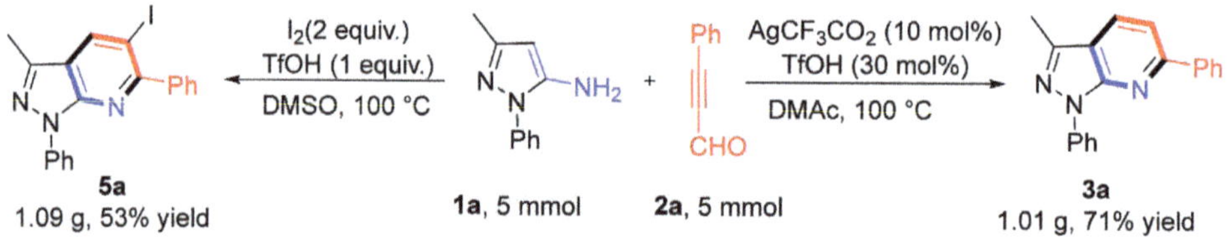

Scheme 6. Scale-up reactions.

Relevant control experiments were conducted to probe the reaction mechanism for the formation of pyrazolo[3,4-b]pyridine frameworks. When the reaction of **1a** with **2a** was conducted without acid for 2 h, it afforded **3a** and **3a′** in 40% and 46% yields, respectively, (Scheme 7a). Intermediate **3a′** was confirmed by TLC-MS(APCI), LC-HRMS, and NMR (the details can be seen in Supplementary Materials). In addition, intermediate **3a′** transform to **3a** in 88% yield under standard conditions (Scheme 7b). These results suggested that **3a′** may serve as the intermediate in this reaction. To illustrate regioselectivity, we chose cinnamaldehyde (**2a′**) as a substrate to react with **1a** under standard conditions. Compared with a standard simple, 3-methyl-1,4-diphenyl-1H-pyrazolo[3,4-b]pyridine (**17**) was obtained in 45% yield, no product **3a** was observed (Scheme 7c). These results confirmed the regioselectivity of this method, as it only afforded the C6 substituted pyrazolo[3,4-b]pyridine for alkynyl aldehydes substrates. Furthermore, when 1 eq. of iodine was added, non-iodinated and iodized products **3a** and **5a** were detected in 46% and 30% yields, respectively (Scheme 7d).

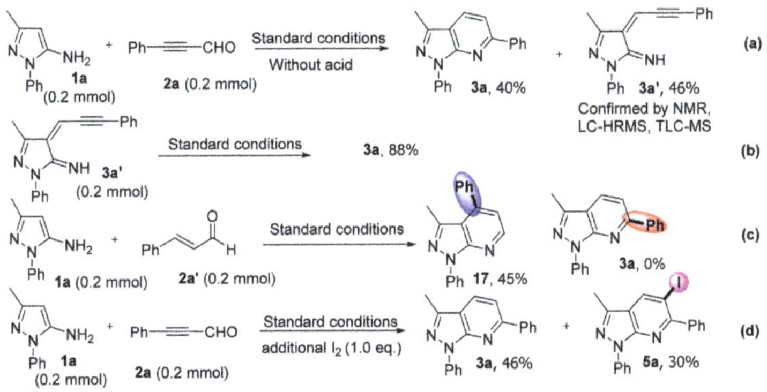

Scheme 7. Control experiments. (**a**) Form intermediate **3a'**. (**b**) Intermediate **3a'** transform to **3a** under standard conditions. (**c**) Validation of regioselectivity experiments. (**d**) Add one equivalent of iodine under standard conditions.

Considering the aforementioned control experiments and earlier works [46–48], a reaction mechanism is shown in Scheme 8 using **1a** with **2a** as a typical reaction. Initially, 3-methyl-1-phenyl-1*H*-pyrazol-5-amine (**1a**) undergoes condensation with 3-phenylpropiolaldehyde (**2a**) to form intermediate **3a'**. Next, the silver salt coordinates to the alkyne of **3a'** to form intermediate **A**; this undergoes 6-*endo*-*dig* cyclization to form **B**. Finally, **B** undergoes demetallation to afford product **3a** (Scheme 8, pathway A). Similarly, I₂ or NBS adds to a triple bond that leads to intermediate **A'**, which undergoes 6-*endo*-*dig* cyclization to form **B'**, followed by acid loss from **B'** to obtain **5a** or **5l** (Scheme 8, pathway B).

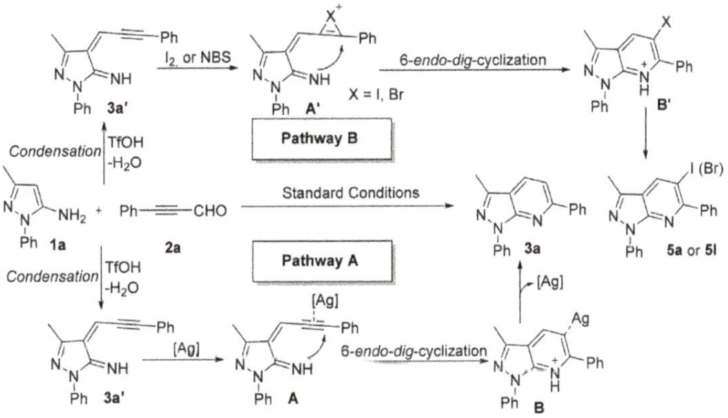

Scheme 8. Plausible mechanistic pathway.

3. Materials and Methods

3.1. General Information

Aminopyrazoles and NBS were purchased from Shanghai Shaoyuan Co. Ltd. (Shanghai, China) 3-substituted propiolaldehyde and Ag(CF₃CO₂) were purchased from Leyan. Unless stated otherwise, all solvents and commercially available reagents were obtained from commercial suppliers and used without further purification. In addition, petroleum ether (b.p. 60–90 °C) was distilled prior to use for Column chromatography. Non-commercial starting materials were prepared as described below or according to literature procedures. TLC analysis was performed using pre-coated glass plates. Column chromatography was performed using silica gel (200–300 mesh). Nuclear magnetic resonance (NMR) spectra were recorded on a Bruker Advance 400 MHz spectrometer at ambient temperature using the non or partly deuterated solvent as internal standard (^{1}H: δ 7.26 ppm and ^{13}C{^{1}H}:

δ 77.0 ppm for CDCl$_3$; ^{1}H: δ 2.50 ppm and ^{13}C{^{1}H}: δ 40.0 ppm for DMSO-d_6). Chemical shifts (δ) are reported in ppm, relative to the internal standard of tetramethylsilane (TMS). The coupling constants (*J*) are quoted in hertz (Hz). Resonances are described as s (singlet), d (doublet), t (triplet), q (quartet), m (multiplet), br (broad) or combinations thereof. High resolution mass-spectrometric (HRMS) were obtained on an Apex-Ultra MS equipped with an electrospray source. Melting points were determined using SGW X-4 apparatus and not corrected. The X-ray diffraction data for the crystallized compound were collected on a Bruker Smart APEX CCD area detector diffractometer (graphite monochromator, Mo Kα radiation, λ = 0.71073 Å) at 296(2) K. All the heating procedures were conducted with an oil bath.

3.2. Synthetic Procedures

Typical Procedure (TP 1) for the Synthesis of 3 and 4 Taking 3a as an Example. A 25 mL pressure vial was charged with **1a** (34.6 mg, 0.20 mmol, 1.0 equiv.), **2a** (26 mg, 24.5 uL, 0.20 mmol, 1.0 equiv.), Ag(CF$_3$CO$_2$) (4.4 mg, 0.02 mmol, 10 mol%), TfOH (9 mg, 5.3 uL, 0.06 mmol, 30 mol%) and DMAc (1.5 mL).The vial was sealed and the reaction mixture was stirred at 100 °C for 2 h under air atmosphere (monitored by TLC). After the reaction was completed, and added 50 mL water to the mixture, then extracted with EtOAc 3 times (3 × 50 mL). The solution was dried over anhydrous Na$_2$SO$_4$, concentrated under reduced pressure, and dried under vacuum. The residue was purified by flash column chromatography by using ethyl acetate/petroleum ether mixture to obtain the corresponding product **3a**.

Typical Procedure (TP 2) for the Synthesis of 5 Taking 5a as an Example. A 25 mL pressure vial was charged with **1a** (34.6 mg, 0.20 mmol, 1.0 equiv.), **2a** (26 mg, 24.5 uL, 0.20 mmol, 1.0 equiv.), I$_2$ (101.5 mg, 0.40 mmol, 2.0 equiv.), TfOH (30 mg, 17.6 uL, 0.20 mmol, 1.0 equiv.) and DMSO (2.0 mL). The vial was sealed, and the reaction mixture was stirred at 100 °C for 6 h under air atmosphere (monitored by TLC). After the reaction was completed, and added 50 mL water to the mixture, then extracted with EtOAc 3 times (3 × 50 mL). The extract was washed with 10% Na$_2$S$_2$O$_3$ solution, dried over anhydrous Na$_2$SO$_4$ and concentrated under reduced pressure. The residue was purified by flash column chromatography by using ethyl acetate/petroleum ether mixture to obtain the corresponding product **5a**.

General Procedure for Synthesis of 13. K$_2$CO$_3$ (0.4 mmol, 2.0 equiv.), phenylboronic acid (0.26 mmol, 1.3 equiv.) and PdCl$_2$(PPh$_3$)$_2$ (5 mol%) were added to a solution of **5a** (0.2 mmol, 1.0 equiv.) in a 5:1 solvent mixture of dioxane and water. The reaction mixture was heated to 90 °C and stirred at this temperature until complete consumption of **5a** was observed (monitored by TLC). After cooling to room temperature, the mixture was diluted with a mixture of EA and water and the aqueous layer was extracted with EtOAc (3 × 50 mL). dried over anhydrous Na$_2$SO$_4$ and concentrated under reduced pressure. The residue was purified by flash column chromatography by using ethyl acetate/petroleum ether mixture to obtain the desired product **13** in 88% yield.

General Procedure for Synthesis of 14. **5a** (0.2 mmol, 1 equiv.), PdCl$_2$(PPh$_3$)$_2$ (5 mol%), CuI (10 mol%) and phenylacetylene (0.3 mmol, 1.5 equiv.) were added to a 25 mL Schlenk flask with a stir bar under an Ar atmosphere. Then DMF (2 mL) and TEA (1 mL) were added sequentially. The reaction mixture was then stirred at 90 °C. Afterwards 15 mL of water were added, and the reaction mixture was extracted with EtOAc (3 × 50 mL). The combined organic fractions were washed with brine and dried over Na$_2$SO$_4$. After filtration, the solvent was removed under reduced pressure. The residue was purified by flash column chromatography by using ethyl acetate/petroleum ether mixture to obtain the desired product **14** in 85% yield.

General Procedure for Synthesis of 15. **5a** (0.2 mmol, 1 equiv.), Pd(OAc)$_2$ (10 mol%), PPh$_3$ (20 mol%) and ethyl acrylate (0.3 mmol, 1.5 equiv.) were added to a 25 mL Schlenk flask with a stir bar under an Ar atmosphere. Then dioxane (2 mL) and TEA (1 mL) were added sequentially. The reaction mixture was then stirred at 90 °C. Afterwards 50 mL of water were added, and the reaction mixture was extracted with EtOAc (3 × 50 mL).

The combined organic fractions were washed with brine and dried over Na$_2$SO$_4$. After filtration, the solvent was removed under reduced pressure. The residue was purified by flash column chromatography by using ethyl acetate/petroleum ether mixture to obtain the desired product **15** in 75% yield.

General Procedure for Synthesis of 16. Adapting a literature procedure [49], A 25 mL Schlenk flask with a stir bar was charged with **5a** (0.2 mmol, 1.0 equiv.), diphenyl diselenide (0.14 mmol, 0.7 equiv.) CuI (0.02 mmol, 10 mol%) and Cs$_2$CO$_3$ (0.4 mmol, 2.0 equiv.) in MeCN (2.0 mL). The vial was sealed and the resulting mixture was stirred at 80 °C for 24 h under an Ar atmosphere. After the reaction completed, and added 50 mL water to the mixture, then extracted with EtOAc 3 times (3 × 50 mL). The extract was washed with brine, dried over anhydrous Na$_2$SO$_4$ and concentrated under reduced pressure. The residue was purified by flash column chromatography by using ethyl acetate/petroleum ether mixture to obtain the desired product **16** in 78% yield.

4. Conclusions

In summary, a cascade *6-endo-dig* cyclization reaction was developed for the switchable synthesis of halogen and non-halogen-functionalized pyrazolo[3,4-*b*]pyridines from 5-aminopyrazoles and alkynyl aldehydes. This method afforded diversified pyrazolo[3,4-*b*]pyridine frameworks via C≡C bond activation with silver, iodine, or NBS. The protocol was characterized by a wide substrate scope, good functional group tolerance, and excellent regional selectivity. The structural modification of estrone, formononetin, and eudistomin Y$_1$ provided new ideas for syntheses of drug molecules. Iodine functionalization allowed several additional transformations, including arylation, alkenylation, alkynylation, and selenization to fabricate useful molecules.

5. Patents

A patent (Yantai University, CN 112300157, and 2021 A) has been derived from this manuscript. The patent is entitled Novel pyrazolopyridine compound with antitumor activity and preparation method thereof.

Supplementary Materials: The following supporting information can be downloaded at: https://www.mdpi.com/article/10.3390/molecules27196381/s1, Characterization data for product **3**, **4**, **5**, **6–11**, and **13–17**, include ^{1}H- and ^{13}C-NMR spectroscopies are available online. CCDC 2075351 contain the supplementary crystallographic data for this paper. These data can be obtained free of charge via www.ccdc.cam.ac.uk/data_request/cif, or by emailing data_request@ccdc.cam.ac.uk, or by contacting The Cambridge Crystallographic Data Centre, 12 Union Road, Cambridge CB2 1EZ, UK; Fax: +44 1223 336033. References [50–56] are cited in the Supplementary Materials.

Author Contributions: Conceptualization, Y.-P.Z.; methodology, Y.-P.Z.; investigation, X.-Y.M. and Y.-J.H.; data curation, F.-R.L. and D.S.; writing—original draft preparation, X.-Y.M., Y.-J.H. and Y.-Y.S.; writing—review and editing, Y.-P.Z.; visualization, F.-R.L.; supervision, A.-X.W. and Y.-P.Z.; project administration, Y.-P.Z. All authors have read and agreed to the published version of the manuscript.

Funding: The authors also thank Talent Induction Program for Youth Innovation Teams in Colleges and Universities of Shandong Province. This work was supported by Science and Technology Innovation Development Plan of Yantai (2020MSGY114) and Yantai "Double Hundred Plan".

Institutional Review Board Statement: Not applicable.

Informed Consent Statement: Not applicable.

Data Availability Statement: Not applicable.

Acknowledgments: The authors also thank Talent Induction Program for Youth Innovation Teams in Colleges and Universities of Shandong Province. Xue-Han Li, Le-Yang Guan, Yu-Ting Han, Meng-Jiao Lei and Jia-Xin Chen are thanked for purification of some compounds.

Conflicts of Interest: The authors declare no conflict of interest.

Sample Availability: Samples of the compounds **3**, **4**, and **5** are available from the authors.

References

1. Orlikova, B.; Chaouni, W.; Schumacher, M.; Aadil, M.; Diederich, M.; Kirsch, G. Synthesis and bioactivity of novel aminopyrazolopyridines. *Eur. J. Med. Chem.* **2014**, *85*, 450–457. [CrossRef] [PubMed]
2. Hu, Y.; Kitamura, N.; Musharrafieh, R.; Wang, J. Discovery of Potent and Broad-Spectrum Pyrazolopyridine-Containing Antivirals against Enteroviruses D68, A71, and Coxsackievirus B3 by Targeting the Viral 2C Protein. *J. Med. Chem.* **2021**, *64*, 8755–8774. [CrossRef] [PubMed]
3. Nagender, P.; Naresh Kumar, R.; Malla Reddy, G.; Krishna Swaroop, D.; Poornachandra, Y.; Ganesh Kumar, C.; Narsaiah, B. Synthesis of novel hydrazone and azole functionalized pyrazolo[3,4-b]pyridine derivatives as promising anticancer agents. *Bioorganic Med. Chem. Lett.* **2016**, *26*, 4427–4432. [CrossRef] [PubMed]
4. Czodrowski, P.; Mallinger, A.; Wienke, D.; Esdar, C.; Pöschke, O.; Busch, M.; Rohdich, F.; Eccles, S.A.; Ortiz-Ruiz, M.J.; Schneider, R.; et al. Structure-Based Optimization of Potent, Selective, and Orally Bioavailable CDK8 Inhibitors Discovered by High-Throughput Screening. *J. Med. Chem.* **2016**, *59*, 9337–9349. [CrossRef]
5. Eagon, S.; Hammill, J.T.; Sigal, M.; Ahn, K.J.; Tryhorn, J.E.; Koch, G.; Belanger, B.; Chaplan, C.A.; Loop, L.; Kashtanova, A.S.; et al. Synthesis and Structure–Activity Relationship of Dual-Stage Antimalarial Pyrazolo[3,4-b]pyridines. *J. Med. Chem.* **2020**, *63*, 11902–11919. [CrossRef]
6. Haydl, A.M.; Breit, B.; Liang, T.; Krische, M.J. Alkynes as Electrophilic or Nucleophilic Allylmetal Precursors in Transition-Metal Catalysis. *Angew. Chem. Int. Ed.* **2017**, *56*, 11312–11325. [CrossRef]
7. Ru, G.; Zhang, T.; Zhang, M.; Jiang, X.; Wan, Z.; Zhu, X.; Shen, W.; Gao, G. Recent progress towards the transition-metal-catalyzed Nazarov cyclization of alkynes via metal carbenes. *Org. Biomol. Chem.* **2021**, *19*, 5274–5283. [CrossRef]
8. Yao, T.; Xia, T.; Yan, W.; Xu, H.; Zhang, F.; Xiao, Y.; Zhang, J.; Liu, L. Copper-Catalyzed Chemodivergent Cyclization of N-(ortho-alkynyl)aryl-Pyrrole and Indoles. *Org. Lett.* **2020**, *22*, 4511–4516. [CrossRef]
9. Yao, T.; Zhang, F.; Zhang, J.; Liu, L. Palladium-Catalyzed Intermolecular Heck-Type Dearomative [4 + 2] Annulation of 2H-Isoindole Derivatives with Internal Alkynes. *Org. Lett.* **2020**, *22*, 5063–5067. [CrossRef]
10. Wu, F.; Zhu, S. A Strategy to Obtain o-Naphthoquinone Methides: Ag(I)-Catalyzed Cyclization of Enynones for the Synthesis of Benzo[h]chromenes and Naphthopyryliums. *Org. Lett.* **2019**, *21*, 1488–1492. [CrossRef]
11. Wu, F.; Cheng, T.; Zhu, S. Construction of Partially Protected Nonsymmetrical Biaryldiols via Semipinacol Rearrangement of o-NQM Derived from Enynones. *Org. Lett.* **2021**, *23*, 71–75. [CrossRef]
12. Rong, M.; Qin, T.; Zi, W. Rhenium-Catalyzed Intramolecular Carboalkoxylation and Carboamination of Alkynes for the Synthesis of C3-Substituted Benzofurans and Indoles. *Org. Lett.* **2019**, *21*, 5421–5425. [CrossRef]
13. Rondla, N.R.; Levi, S.M.; Ryss, J.M.; Vanden Berg, R.A.; Douglas, C.J. Palladium-Catalyzed C-CN Activation for Intramolecular Cyanoesterification of Alkynes. *Org. Lett.* **2011**, *13*, 1940–1943. [CrossRef]
14. Zhang, X.; Zhou, Y.; Wang, H.; Guo, D.; Ye, D.; Xu, Y.; Jiang, H.; Liu, H. Silver-catalyzed intramolecular hydroamination of alkynes in aqueous media: Efficient and regioselective synthesis for fused benzimidazoles. *Green. Chem.* **2011**, *13*, 397–405. [CrossRef]
15. Hu, Y.; Zhou, Y.; Gao, J.; Zhang, H.; Yang, K.; Li, J.; Yan, X.; Li, Y.; Zhu, Y. I_2-Mediated [3 + 2] annulation of methyl-azaarenes with alkyl 2-isocyanoacetates or amino acid ester hydrochlorides: Selective synthesis of iodine-functionalized and non-iodine-functionalized fused imidazoles. *Org. Chem. Front.* **2022**, *9*, 1403–1409. [CrossRef]
16. Godoi, B.; Schumacher, R.F.; Zeni, G. Synthesis of Heterocycles via Electrophilic Cyclization of Alkynes Containing Heteroatom. *Chem. Rev.* **2011**, *111*, 2937–2980. [CrossRef]
17. Fang, G.; Bi, X. Silver-catalysed reactions of alkynes: Recent advances. *Chem. Soc. Rev.* **2015**, *44*, 8124–8173. [CrossRef]
18. Dorel, R.; Echavarren, A.M. Gold(I)-Catalyzed Activation of Alkynes for the Construction of Molecular Complexity. *Chem. Rev.* **2015**, *115*, 9028–9072. [CrossRef]
19. Costello, J.P.; Ferreira, E.M. Regioselectivity Influences in Platinum-Catalyzed Intramolecular Alkyne O-H and N-H Additions. *Org. Lett.* **2019**, *21*, 9934–9939. [CrossRef]
20. Chen, L.; Chen, K.; Zhu, S. Transition-Metal-Catalyzed Intramolecular Nucleophilic Addition of Carbonyl Groups to Alkynes. *Chem* **2018**, *4*, 1208–1262. [CrossRef]
21. Li, Y.; Tang, Z.; Zhang, J.; Liu, L. Gold-catalyzed intermolecular [4+1] spiroannulation via site-selective aromatic $C(sp^2)$–H functionalization and dearomatization of phenol derivatives. *Chem. Commun.* **2020**, *56*, 8202–8205. [CrossRef]
22. Ma, J.; Chen, K.; Fu, H.; Zhang, L.; Wu, W.; Jiang, H.; Zhu, S. Dual Catalysis: Proton/Metal-Catalyzed Tandem Benzofuran Annulation/Carbene Transfer Reaction. *Org. Lett.* **2016**, *18*, 1322–1325. [CrossRef]
23. Pan, Y.; Chen, G.; Shen, C.; He, W.; Ye, L. Synthesis of fused isoquinolines via gold-catalyzed tandem alkyne amination/intramolecular O-H insertion. *Org. Chem. Front.* **2016**, *3*, 491–495. [CrossRef]
24. Zhou, M.; Song, R.; Wang, C.; Li, J. Synthesis of Azepine Derivatives by Silver-Catalyzed [5+2] Cycloaddition of γ-Amino Ketones with Alkynes. *Angew. Chem. Int. Ed.* **2013**, *52*, 10805–10808. [CrossRef]
25. Wu, F.; Zhang, L.; Zhu, S. 1,4-Addition of o-naphthoquinone methides induced by silver-catalyzed cyclization of enynones: An approach to unsymmetrical triarylmethanes and benzo[f]chromenes. *Org. Chem. Front.* **2020**, *7*, 3387–3392. [CrossRef]
26. Li, X.; Han, Y.; Xu, D.; Li, M.; Wei, W.; Liang, Y. Silver Trifluoromethanesulfonate-Catalyzed Annulation of Propargylic Alcohols with 3-Methyleneisoindolin-1-one. *J. Org. Chem.* **2020**, *85*, 2626–2634. [CrossRef]
27. Niu, Y.; Yan, Z.; Gao, G.; Wang, H.; Shu, X.; Ji, K.; Liang, Y. Synthesis of Isoquinoline Derivatives via Ag-Catalyzed Cyclization of 2-Alkynyl Benzyl Azides. *J. Org. Chem.* **2009**, *74*, 2893–2896. [CrossRef]

28. Huo, Z.; Gridnev, I.D.; Yamamoto, Y. A Method for the Synthesis of Substituted Quinolines via Electrophilic Cyclization of 1-Azido-2-(2-propynyl)benzene. *J. Org. Chem.* **2010**, *75*, 1266–1270. [CrossRef]
29. Ouyang, H.; Tang, R.; Zhong, P.; Zhang, X.; Li, J. CuI/I$_2$-Promoted Electrophilic Tandem Cyclization of 2-Ethynylbenzaldehydes with *ortho*-Benzenediamines: Synthesis of Iodoisoquinoline-Fused Benzimidazoles. *J. Org. Chem.* **2011**, *76*, 223–228. [CrossRef]
30. Liu, L.; Chen, D.; Yao, J.; Zong, Q.; Wang, J.; Zhou, H. CuX-Activated *N*-Halosuccinimide: Synthesis of 3-Haloquinolines via Electrophilic Cyclization of Alkynyl Imines. *J. Org. Chem.* **2017**, *82*, 4625–4630. [CrossRef]
31. Unoh, Y.; Hirano, K.; Miura, M. Metal-Free Electrophilic Phosphination/Cyclization of Alkynes. *J. Am. Chem. Soc.* **2017**, *139*, 6106–6109. [CrossRef] [PubMed]
32. Mantovani, A.C.; Hernández, J.G.; Bolm, C. Synthesis of 3-Iodobenzofurans by Electrophilic Cyclization under Solventless Conditions in a Ball Mill. *Eur. J. Org. Chem.* **2018**, *2018*, 2458–2461. [CrossRef]
33. Zhou, J.; Li, W.; Zheng, H.; Pei, Y.; Liu, X.; Cao, H. Visible Light-Induced Cascade Cyclization of 3-Aminoindazoles, Ynals, and Chalcogens: Access to Chalcogen-Containing Pyrimido [1,2-*b*]-indazoles. *Org. Lett.* **2021**, *23*, 2754–2759. [CrossRef] [PubMed]
34. Tang, Z.; Zhang, F.; Yao, T.; Liu, X.; Liu, Y.; Liu, L. Dearomative Iodocyclization of *N*-(*o*-Alkynyl)aryl Isoindole. *J. Org. Chem.* **2022**, *87*, 7531–7535. [CrossRef]
35. Cao, H.; Liu, X.; Liao, F.; Huang, J.; Qiu, H.; Chen, Q.; Chen, Y. Transition Metal-Mediated C=O and C=C Bond-Forming Reactions: A Regioselective Strategy for the Synthesis of Imidazo [1,2-*a*]pyridines and Imidazo [1,2-*a*]pyrazines. *J. Org. Chem.* **2014**, *79*, 11209–11214. [CrossRef]
36. Cao, H.; Liu, X.; Zhao, L.; Cen, J.; Lin, J.; Zhu, Q.; Fu, M. One-Pot Regiospecific Synthesis of Imidazo[1,2-*a*]pyridines: A Novel, Metal-Free, Three-Component Reaction for the Formation of C–N, C–O, and C–S Bonds. *Org. Lett.* **2014**, *16*, 146–149. [CrossRef]
37. Yang, D.; Yu, Y.; Wu, Y.; Feng, H.; Li, X.; Cao, H. One-Pot Regiospecific Synthesis of Indolizines: A Solvent-Free, Metal-Free, Three-Component Reaction of 2-(Pyridin-2-yl)acetates, Ynals, and Alcohols or Thiols. *Org. Lett.* **2018**, *20*, 2477–2480. [CrossRef]
38. Tber, Z.; Hiebel, M.A.; El Hakmaoui, A.; Akssira, M.; Guillaumet, G.; Berteina-Raboin, S. Metal Free Formation of Various 3-Iodo-1*H*-pyrrolo [3′,2′:4,5]imidazo-[1,2-*a*]pyridines and [1,2-*b*]Pyridazines and Their Further Functionalization. *J. Org. Chem.* **2015**, *80*, 6564–6573.
39. Luo, X.; Ge, L.; An, X.; Jin, Y.; Wang, Y.; Sun, P.; Deng, W. Regioselective Metal-Free One-Pot Synthesis of Functionalized 2-Aminothiophene Derivatives. *J. Org. Chem.* **2015**, *80*, 4611–4617. [CrossRef]
40. Wang, C.; Lai, J.; Chen, C.; Li, X.; Cao, H. Ag-Catalyzed Tandem Three-Component Reaction toward the Synthesis of Multisubstituted Imidazoles. *J. Org. Chem.* **2017**, *82*, 13740–13745. [CrossRef]
41. Chen, Z.; Liang, P.; Xu, F.; Qiu, R.; Tan, Q.; Long, L.; Ye, M. Lewis Acid-Catalyzed Intermolecular Annulation: Three-Component Reaction toward Imidazo[1,2-*a*]pyridine Thiones. *J. Org. Chem.* **2019**, *84*, 9369–9377. [CrossRef]
42. Li, Z.; Ling, F.; Cheng, D.; Ma, C. Pd-Catalyzed Branching Cyclizations of Enediyne-Imides toward Furo[2,3-*b*]pyridines. *Org. Lett.* **2014**, *16*, 1822–1825. [CrossRef]
43. Li, Y.; Huang, J.; Wang, J.; Song, G.; Tang, D.; Yao, F.; Lin, H.; Yan, W.; Li, H.; Xu, Z.; et al. Diversity-Oriented Synthesis of Imidazo-Dipyridines with Anticancer Activity via the Groebke–Blackburn–Bienaymé and TBAB-Mediated Cascade Reaction in One Pot. *J. Org. Chem.* **2019**, *84*, 12632–12638. [CrossRef]
44. Liu, X.; Zhou, J.; Lin, J.; Zhang, Z.; Wu, S.; He, Q.; Cao, H. Controllable Site-Selective Construction of 2- and 4-Substituted Pyrimido[1,2-*b*]indazole from 3-Aminoindazoles and Ynals. *J. Org. Chem.* **2021**, *86*, 9107–9116. [CrossRef]
45. Rayadurgam, J.; Sana, S.; Sasikumar, M.; Gu, Q. Palladium catalyzed C–C and C–N bond forming reactions: An update on the synthesis of pharmaceuticals from 2015–2020. *Org. Chem. Front.* **2021**, *8*, 384–414. [CrossRef]
46. Ding, Q.; Wu, J. Lewis Acid- and Organocatalyst-Cocatalyzed Multicomponent Reactions of 2-Alkynylbenzaldehydes, Amines, and Ketones. *Org. Lett.* **2007**, *9*, 4959–4962. [CrossRef]
47. Chen, Z.; Wu, J. Efficient Generation of Biologically Active H-Pyrazolo[5,1-*a*]isoquinolines via Multicomponent Reaction. *Org. Lett.* **2010**, *12*, 4856–4859. [CrossRef]
48. Yang, W.; Zhang, J.; Chen, L.; Fu, J.; Zhu, J. Controllable synthesis of 3-iodo-2*H*-quinolizin-2-ones and 1,3-diiodo-2*H*-quinolizin-2-ones via electrophilic cyclization of azacyclic ynones. *Chem Commun.* **2019**, *55*, 12607–12610. [CrossRef]
49. Dandapat, A.; Korupalli, C.; Prasad, D.J.C.; Singh, R.; Sekar, G. An Efficient Copper(I) Iodide Catalyzed Synthesis of Diaryl Selenides through CAr-Se Bond Formation Using Solvent Acetonitrile as Ligand. *Synthesis* **2011**, *2011*, 2297–2302. [CrossRef]
50. Zhou, B.; Wu, Q.; Dong, Z.; Xu, J.; Yang, Z. Rhodium-Catalyzed 1,1-Hydroacylation of Thioacyl Carbenes with Alkynyl Aldehydes and Subsequent Cyclization. *Org. Lett.* **2019**, *21*, 3594–3599. [CrossRef]
51. Zhu, Y.; Liu, M.; Cai, Q.; Jia, F.; Wu, A. A Cascade Coupling Strategy for One-Pot Total Synthesis of β-Carboline and Isoquinoline-Containing Natural Products and Derivatives. *Chem. Eur. J.* **2013**, *19*, 10132–10137. [CrossRef]
52. Zheng, A.; Zhang, W.; Pan, J. One-Pot and Convenient Conversion of 5-Azidopyrazole-4-carboxaldehyde to Pyrazolo[3,4-*b*]pyridines. *Synth. Commun.* **2006**, *36*, 1549–1556. [CrossRef]
53. Shekarrao, K.; Kaishap, P.P.; Saddanapu, V.; Addlagatta, A.; Gogoi, S.; Boruah, R.C. Microwave-assisted palladium mediated efficient synthesis of pyrazolo[3,4-*b*]pyridines, pyrazolo[3,4-*b*]quinolines, pyrazolo[1,5-*a*]pyrimidines and pyrazolo[1,5-*a*]quinazolines. *RSC Adv.* **2014**, *4*, 24001–24006. [CrossRef]
54. Hamama, W.S.; Ibrahim, M.E.; Zoorob, H.H. Synthesis and Biological Evaluation of Some Novel Isoxazole Derivatives. *J. Heterocycl. Chem.* **2017**, *54*, 341–346. [CrossRef]

55. Iaroshenko, V.O.; Mkrtchyan, S.; Gevorgyan, A.; Miliutina, M.; Villinger, A.; Volochnyuk, D.; Sosnovskikh, V.Y.; Langer, P. 2,3-Unsubstituted chromones and their enaminone precursors as versatile reagents for the synthesis of fused pyridines. *Org. Biomol. Chem.* **2012**, *10*, 890–894. [CrossRef]
56. Qiu, R.; Qiao, S.; Peng, B.; Long, J.; Yin, G. A mild method for the synthesis of bis-pyrazolo[3,4-*b*:4′,3′-*e*]pyridine derivatives. *Tetrahedron Lett.* **2018**, *59*, 3884–3888. [CrossRef]

Article

Fabricated Gamma-Alumina-Supported Zinc Ferrite Catalyst for Solvent-Free Aerobic Oxidation of Cyclic Ethers to Lactones

Naaser A. Y. Abduh [1,*], Abdullah A. Al-Kahtani [1], Mabrook S. Amer [1], Tahani Saad Algarni [1] and Abdel-Basit Al-Odayni [2,*]

1. Department of Chemistry, King Saud University, Riyadh 11451, Saudi Arabia; akahtani@ksu.edu.sa (A.A.A.-K.); msamer@ksu.edu.sa (M.S.A.)
2. Restorative Dental Sciences Department, College of Dentistry, King Saud University, Riyadh 11545, Saudi Arabia
* Correspondence: 439106262@student.ksu.edu.sa (N.A.Y.A.); aalodayni@ksu.edu.sa (A.-B.A.-O.)

Abstract: The aim of this work was to fabricate a new heterogeneous catalyst as zinc ferrite (ZF) supported on gamma-alumina (γ-Al_2O_3) for the conversion of cyclic ethers to the corresponding, more valuable lactones, using a solvent-free method and O_2 as an oxidant. Hence, the ZF@γ-Al_2O_3 catalyst was prepared using a deposition–coprecipitation method, then characterized using TEM, SEM, EDS, TGA, FTIR, XRD, ICP, XPS, and BET surface area, and further applied for aerobic oxidation of cyclic ethers. The structural analysis indicated spherical, uniform ZF particles of 24 nm dispersed on the alumina support. Importantly, the incorporation of ZF into the support influenced its texture, i.e., the surface area and pore size were reduced while the pore diameter was increased. The product identification indicated lactone compound as the major product for saturated cyclic ether oxidation. For THF as a model reaction, it was found that the supported catalyst was 3.2 times more potent towards the oxidation of cyclic ethers than the unsupported one. Furthermore, the low reactivity of the six-membered ethers can be tackled by optimizing the oxidant pressure and the reaction time. In the case of unsaturated ethers, deep oxidation and polymerization reactions were competitive oxidations. Furthermore, it was found that the supported catalyst maintained good stability and catalytic activity, even after four cycles.

Keywords: cyclic ethers; tetrahydrofuran; supported catalyst; lactones; aerobic oxidation; oxygen

1. Introduction

Nowadays, environmental changes have triggered scientists to focus on finding solutions to reduce pollution levels from several sources [1]. One example of these is chemical industries, a factor that has an enormous effect on air and water pollution, resource exhaustion, soil contamination, and greenhouse gases [2–4]; there is now a significant increase in interest in searching for remedies that refer to the principles of green chemistry by which pollution is reduced to the minimal levels [5,6]. It is fundamental that catalysts, the atomic economy, harmless and sustainable materials, selective reactions, and other green chemistry principles are put into consideration when searching for unprecedented techniques in the chemical industry to substitute the non-environmental currently approved approaches [7,8].

Lactones are cyclic organic compounds of carboxylic esters that are typically categorized according to the type of rings as having three, four, five, six, and seven members, namely, α-, β-, γ-, δ-, and ω-lactones, respectively [9]. Due to their availability in nature, uses, and stability, the five- and six-membered ones are considered the most significant [10]. The most varied and well-liked gamma-lactones are represented by γ-butyrolactone (GBL) and γ-valerolactone (GVL), while δ-valerolactone (DVL) is the most popular delta-lactone.

They have numerous applications in manufacturing polymers, chemical intermediates, pharmaceuticals, foods, beverages, electrical products, rubber additives, herbicides,

solvents, viscosity modifiers, textiles, agrochemicals, pesticides, and liquid fuels [11–16]. In addition, they are key components of many biological compounds and, as such, play a significant role in biology [17,18]. Hence, recently, the lactone marketing value has grown greatly [19,20].

Until now, lactone production is still based on three methods in general. The first is the oxidative lactonization of corresponding diols, e.g., 1,4 butanediol and 1,5 pentanediol, as raw materials for GBL and DVL production, respectively. The second is reductive lactonization of γ-keto acid(ester)s, e.g., levulinic acid, which is famous for GVL production. The third is the Baeyer–Villiger oxidation of cyclic ketones, which is mostly applied for DVL synthesis [21]. Moreover, the malic anhydride method is a well-known industrial route for GBL production [22]. From the scientific point of view, these methods are environmentally unsafe and economically unsatisfactory [16,23–26]. As a consequence, finding renewable, environmentally friendly, and economical alternative methods for lactone production is of utmost importance [27–29].

Today, biomass-derived products are of high economic and environmental importance as they, from a green chemistry point of view, represent a sustainable raw material [29,30]. It is possible to synthesize lactones from biomaterials using chemical and biological methods, but these methods need to be improved [31–33].

Cyclic ethers such as tetrahydrofuran (THF), tetrahydropyran (THP), and 2-methyl tetrahydrofuran (2-MTHF) can be utilized as excellent sustainable raw materials for GBL, DVL, and GVL production, respectively, for which they can be sourced from biomass-derived material [34–36]. Notably, cyclic ethers have similar backbone as corresponding lactones and could be oxidized in one step to lactones. However, such a process may require an appropriate catalyst that could achieve a satisfactory economic yield and is environmentally unharmful.

Several studies have been conducted to explore new environmentally friendly oxidation processes [37–39]. For example, Higuchi et al. applied Ru porphyrin heteroaromatic N-oxide system for homogenous cyclic ether oxidation [37]. In addition to drawbacks related to homogenous catalysts, such as production waste and economic cost [40], over-oxidation and ring opening reactions also reduce the lactone yield [41]. Tert-butyl hydroperoxide/(diacetoxyiodo) benzene (DIB/TBHP) protocol, depending on free radicals mechanism with no metallic reagent under mild condition, have been used for the oxidative transformation of cyclic ethers to lactones [38]. Although this protocol prevents the over-oxidized product, processing requirements, moisture sensitivity, necessary solvents, and complex product separation make its use on a large industrial scale unfeasible. Bhaumikb et al. [42] studied the catalytic oxidation of cyclic ethers to the corresponding lactones, employing various crystalline microporous metallosilicates as heterogeneous catalysts in the liquid phase. They used hydrogen peroxide (H_2O_2) as one green oxidizing agent for cyclic ether oxidation. It is, however, essential to note that acceptable catalysts should be inexpensive and easy to obtain. In general, hydrogen peroxide is a remarkably effective oxidant due to its oxidative features represented by the extra oxygen atom compared to water's structure [43]. However, handling requirements, high cost, and product separation have limited its utilization [44]. On the other hand, despite its limited ability to oxidize under mild conditions and low selectivity of desired products, molecular oxygen, the ultimate green oxidizing agent, is a low-cost, cheap, and readily available gas [45,46]. Thus, it is desirable to overcome this limitation by finding an appropriate heterogeneous catalyst [47,48].

In this regard, gold (Au) supported by nanorods of CeO_2 has been tested as a heterogeneous catalyst for cyclic ether oxidation; using THF as a model reaction, oxidation was performed under aerobic conditions to produce GBL as a main product and propylformate as a minor product [39]. Nevertheless, it is still required to improve yield and selectivity while avoiding environmental and economic issues.

In our previous article [49], we reported that the use of $ZnFe_2O_4$ catalyst and H_2O_2 as an oxidizing agent for THF conversion into GBL resulted in 47.3% conversion and 87.2%

selectivity. When O_2 was used in place of H_2O_2, a conversion and a GBL selectivity of 18.1% and 63.3%, respectively, were achieved. Indeed, such unsatisfied improvement can be further ameliorated by incorporating certain support. Hence, materials like Al_2O_3, SiO_2, TiO_2, and ZrO_2 can be used as catalysts, composite agents, or supports for enhancing catalyst reactivity [50]. Support could play a synergistic effect to increase catalyst efficiency. γ-Al_2O_3 is an active catalyst and support with many attractive properties such as low cost, surface area, good thermal stability, and acid/base characteristics [51,52].

Herein, $ZnFe_2O_4/\gamma$-Al_2O_3-supported catalyst was successfully fabricated by a simple deposition–coprecipitation method, being further employed as an effective catalyst for cyclic ether oxidation to lactones through a green process, which involves solvent-free aerobic oxidation under mild conditions. The prepared catalyst was extensively characterized using IR, XRD, TGA, EDS, ICP, XPS, BET surface area, SEM, and TEM. The effect of various factors, including reaction temperature, reaction time, oxidant amount, and catalyst dose, were fully investigated for THF as a model reaction of cyclic ether oxidation. Moreover, oxidation of several five- and six-membered saturated and unsaturated cyclic ethers was also explored.

2. Results and Discussion

2.1. Catalyst Characterizations

2.1.1. FTIR Analysis

Figure 1 shows the FTIR spectra of $ZnFe_2O_4$ (ZF), ZF@γ-Al_2O_3, and γ-Al_2O_3. In this figure, peaks of adsorbed water were observed in all spectra around 3460 and 1635 cm^{-1} for stretching and bending vibrations, respectively. The characteristic broad peaks at 838, 757, and 533 cm^{-1} belong to the Al–O–Al vibration of the γ-Al_2O_3 support [53]. Moreover, the spectrum of the ZF spinal catalyst showed two peaks at 550 and 410 cm^{-1}, which are attributed to Zn–O and Fe–O vibrations at tetrahedral and octahedral sites, respectively [54]. Moreover, the shoulder-type peaks at 864 and 718 cm^{-1}, as well as peaks at 543 and 427 cm^{-1}, confirm the formation of the ZF@γ-Al_2O_3-supported catalyst composite.

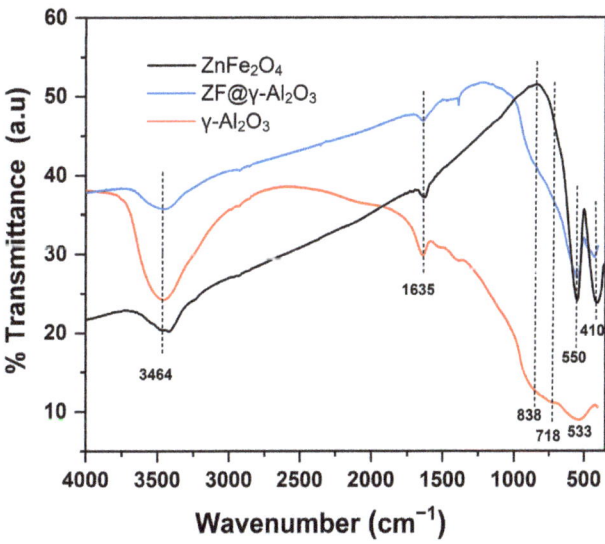

Figure 1. FTIR spectra of $ZnFe_2O_4$ (ZF), ZF@γ-Al_2O_3, and γ-Al_2O_3.

2.1.2. XRD Analysis

The X-ray diffractograms of ZF, ZF@γ-Al_2O_3, and γ-Al_2O_3 are illustrated in Figure 2. The XRD pattern of zinc ferrite with a cubic spinel crystal structure showed a list of 2θ

peaks at 18.45°, 30.42°, 35.65°, 43.53°, 53.87°, 57.61°, 63.19°, 71.75°, 74.79°, and 78.82°, which agreed with the reported JCPDS PDF (card no: 01-077-0011) [55]. The typical bands of γ-Al$_2$O$_3$ were observed at 31.10°, 37.69°, 39.67°, 45.94°, 60.96°, and 67.40° [56], which matched the diffraction planes (220), (311), (222), (400), (511), and (440) in the JCPDS PDF (card no: 01-079-1558), respectively. The XRD pattern of ZF@γ-Al$_2$O$_3$ indicated, besides the peaks of ZF, the presence of Al$_2$O$_3$ prominent peaks at 37.67°, 45.92°, and 67.37°, thus signifying the composite formation. Indeed, the XRD pattern of the catalyst was in good agreement with the literature [57,58].

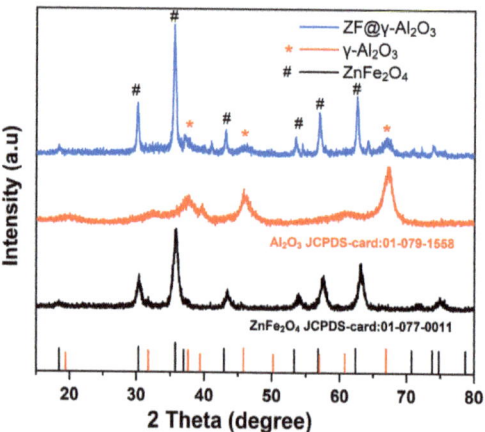

Figure 2. XRD spectra of ZnFe$_2$O$_4$ (ZF), ZF@γ-Al$_2$O$_3$, and γ-Al$_2$O$_3$ with JCPDS card.

The Debye–Scherrer formula was used to calculate the average crystal size of ZF particles in the supported catalyst, which amounted to 18.86 nm. Moreover, the unit constant was also measured, and the result was 8.31, approving the formation of a cubic spinel structure [59].

2.1.3. TGA Analysis

The thermal stabilities of the target substances, the catalyst ZF, the support γ-Al$_2$O$_3$, and the applied γ-Al$_2$O$_3$-supported ZF (ZF@γ-Al$_2$O$_3$) catalyst were monitored using the TGA technique. As can be seen, three degradation steps can be identified in Figure 3. Two weight-loss stages are presented for the ZF catalyst ranging from 25 to 400 °C, with a mass loss of 1.2 and 1.4%, attributed to the loss of physically adsorbed water and lattice water molecule evaporation, respectively [60,61]. However, a mass loss of 0.5% for γ-Al$_2$O$_3$ and 1.43% in the case of ZF@γ-Al$_2$O$_3$ indicated the effect of the amount of γ-Al$_2$O$_3$ in the composite. Moreover, this result confirmed that a stable composite was successfully formed in the calcined catalyst up to 800 °C as there was no significant weight loss.

2.1.4. BET-Surface Area Analysis

The BET-surface area, pore volume, and pore width of the unsupported and supported catalysts were measured using N$_2$ adsorption–desorption isotherms and the Barrett–Joyner–Halenda (BJH) method. According to IUPAC classification, Figure 4 supports the IV-type isotherms for both ZF and γ-Al$_2$O$_3$ and the majority of the mesoporous structure. For hysteresis loop classification, γ-Al$_2$O$_3$ has a type between H1 and H2 [62], whereas ZF has an H1 type. Most importantly, the change in the hysteresis shape of the supported catalyst is further evidence of incorporating ZF particles in the supported matrix. Moreover, the shift in the hysteresis loop at relative pressure (P/P°) from 0.42 for γ-Al$_2$O$_3$ to 0.6 for ZF@γ-Al$_2$O$_3$ indicates increased pore width in the supported catalyst [63]. According to Table 1, upon loading of the ZF to the γ-Al$_2$O$_3$ support, the latter total surface area, external surface area, and average pore volume decreased. The decrease in the surface area of support can be attributed to the dispersed ZF particles on the surface and the blockage of inner pores. In addition, the average pore width was substantially higher for ZF@γ-Al$_2$O$_3$ due to blocking the small pores, which were thus eliminated from pore counting [64].

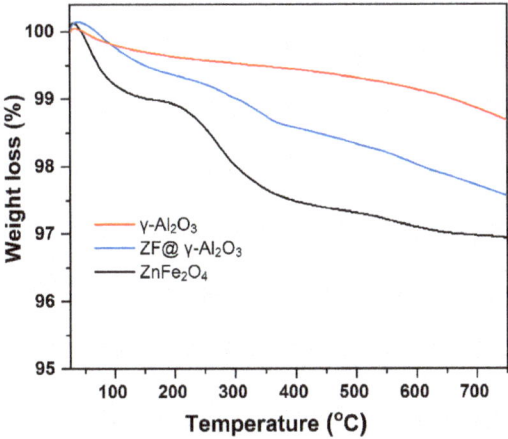

Figure 3. TGA curves of ZnFe$_2$O$_4$ (ZF), γ-Al$_2$O$_3$, and ZF@γ-Al$_2$O$_3$.

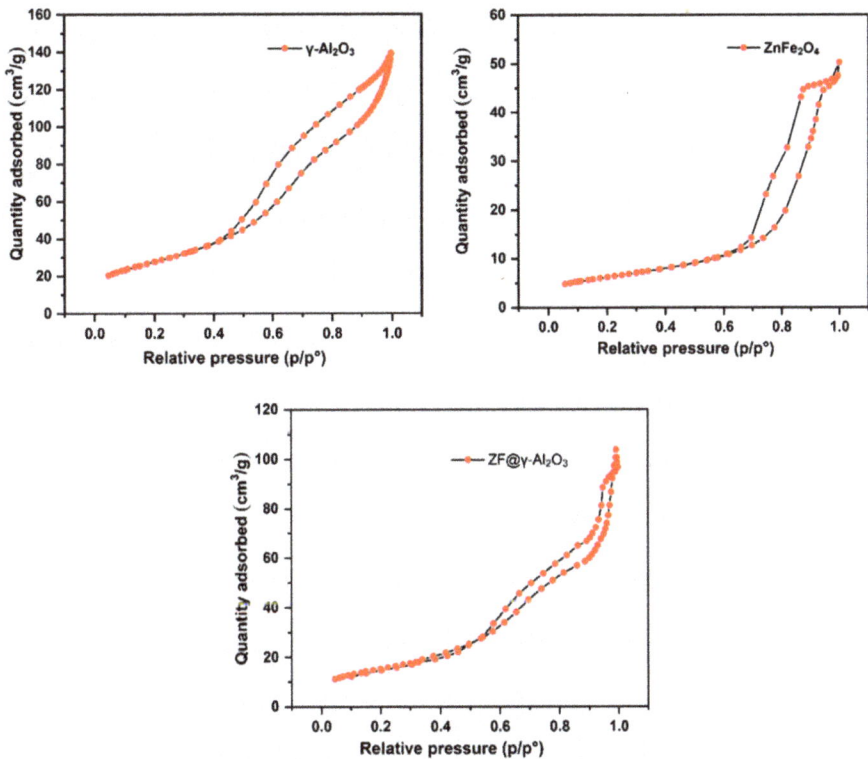

Figure 4. Nitrogen adsorption–desorption isotherms of γ-Al$_2$O$_3$, ZnFe$_2$O$_4$ (ZF), and ZF@γ-Al$_2$O$_3$.

Table 1. BET-surface area and porosity results.

Catalyst	Total Surface Area (m^2/g)	External Surface Area (m^2/g)	Average Pore Volume (cm^3/g)	Average Pore Width (nm)
ZnFe$_2$O$_4$	22.05	14.65	0.08	13.4
γ-Al$_2$O$_3$	100.55	67.35	0.22	78.0
ZF@γ-Al$_2$O$_3$	52.04	36.33	0.16	105.2

2.1.5. TEM Analysis

Surface morphology, shape, and particle size distribution of the obtained supported catalyst were evaluated using a transmission electron microscope (TEM). As shown in Figure 5a, the TEM micrograph confirmed the formation of the cubic ZF structure, which was highly dispersed on the γ-Al$_2$O$_3$ support. Based on the TEM image, particle size and particle distribution were also counted and are presented in Figure 5b. Accordingly, the catalyst particles were of irregular shapes with an average particle size of 24 nm.

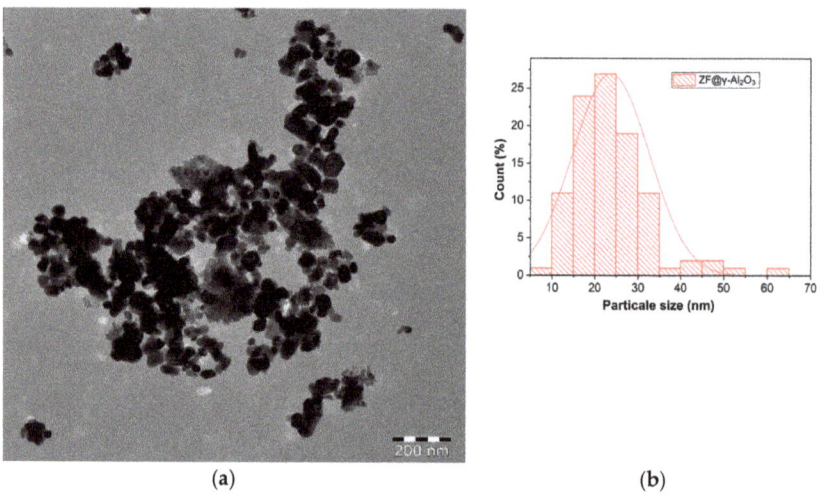

Figure 5. (a) TEM micrograph of ZF@γ-Al$_2$O$_3$, and (b) particle size distribution.

2.1.6. SEM Analysis

The surface morphology of the supported catalyst is illustrated in Figure 6. It is shown that the particles were majorly spherical. The particle size of the dispersed ZF was counted to 26 nm. As can be seen, the obtained particle sizes from both TEM and SEM were almost the same, with further evidence of the catalyst nanostructure and catalyst-support compatibility.

2.1.7. EDS Analysis

The chemical composition of the prepared ZF@γ-Al$_2$O$_3$ catalyst was investigated using the energy-dispersive X-ray spectroscopy (EDS) method, as depicted in Figure 7. The EDS spectrum shows the presence of aluminum (Al), zinc (Zn), iron (Fe), and oxygen (O) peaks, clarifying the composition of the supported catalyst. The mole percentage of the ZF catalyst in the supported catalyst was calculated to be 28%, indicating a homogeneous composition of the supported catalyst.

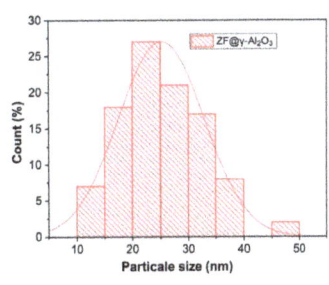

Figure 6. SEM images of ZF@γ-Al$_2$O$_3$.

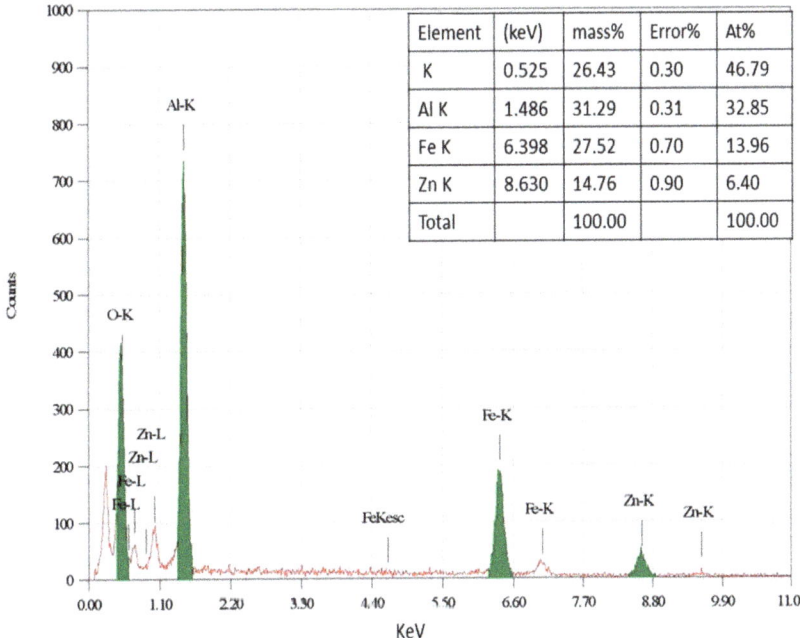

Figure 7. EDS of ZF@γ-Al$_2$O$_3$.

2.1.8. XPS Analysis

The supporting catalyst's atomic composition was identified using X-ray photoelectron spectroscopy (XPS). Zn, Fe, Al, and O elements were detected, as shown in Figure 8. The distinctive peaks of Zn 2p$_{3/2}$ and Zn 2p$_{1/2}$ were found in the spectra of the ZF@γ-Al$_2$O$_3$ at 1022.25 eV and 1045.42 eV, respectively. The peak with characteristic binding energy 531.52 eV and 529.53 eV belongs to the O1s species. In the spectrum of Fe 2p, there were two peaks at 721.40 eV and 725.17 eV, indicating the presence of Fe^{3+}. Furthermore, the peak at 74.22 eV was assigned to Al 2p. Those results agreed with the ZF@γ-Al$_2$O$_3$ reported spectra [57]. In addition, comparing the ZF@Al$_2$O$_3$ spectrum with the ZF XPS spectrum [49] (not shown), a shift in band energies of Zn 2p, Fe 2p, and O 1s was observed, which suggests that the electrical environment in the composite changed [65].

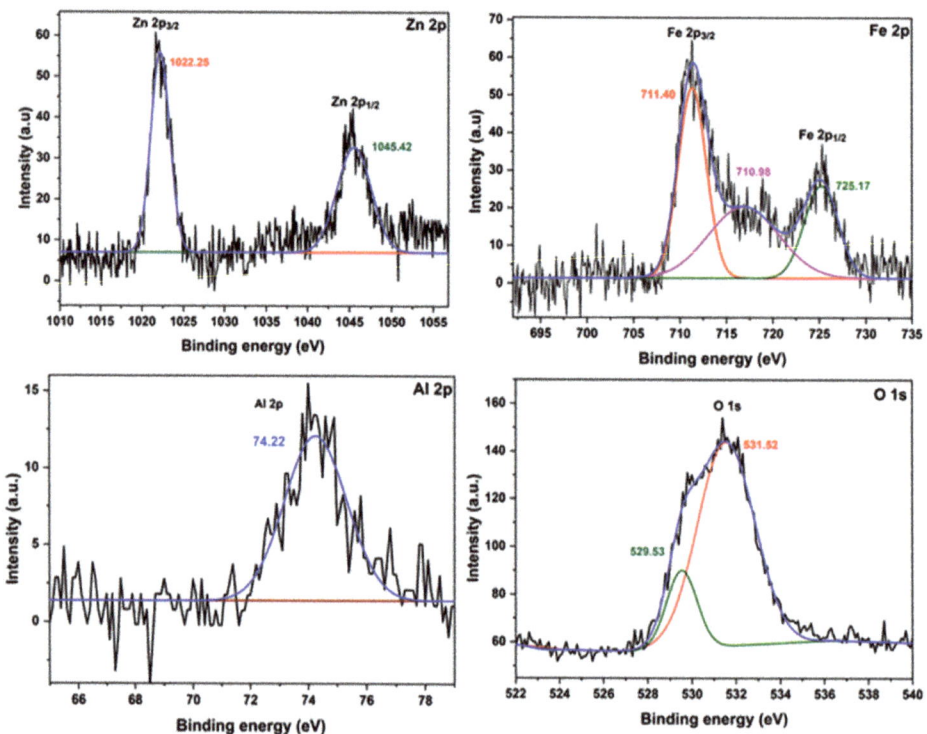

Figure 8. XPS spectra of ZF@γ-Al$_2$O$_3$.

2.1.9. ICP Analysis

Elemental analyses were carried out under inductively coupled plasma spectrometry (ICP) to confirm the chemical composition of the prepared supported catalyst. The results are summarized in Table 2. The molar percentages of zinc, iron, and aluminum revealed by the ICP analysis agreed with those obtained by the EDS analysis (Section 2.1.7).

Table 2. ICP results of ZF@γ-Al$_2$O$_3$.

Catalyst	Al (mol%)	Zn (mol%)	Fe (mol%)
ZF@γ-Al$_2$O$_3$	58.78	12.53	28.69

2.2. Catalytic Activity

The as-prepared supported catalyst was tested for aerobic oxidation of cyclic ethers. The reactions were carried out in a liquid phase without solvent. The identification of the products was achieved by GC-MS. Moreover, the products' conversion and selectivity were determined using gas phase chromatography. THF oxidation was used as a model reaction of cyclic ether oxidation to evaluate the catalyst's performance. The analysis showed that the reaction led to gamma-butyrolactone (GBL) as a mine product, while 2-hydroxy tetrahydrofuran (2-HTHF), 4-hydroxybutyric acid (4-HBA), and 4-hydroxybutaldehyde (4-HBAl) were the minor products (Scheme 1). Furthermore, a blank run experiment was also conducted for comparison, which revealed a negligible conversion, demonstrating that O$_2$ alone cannot oxidize cyclic ethers without a catalyst. According to the preliminary tests, ZF@γ-Al$_2$O$_3$ showed better performance than the individuals γ-Al$_2$O$_3$ and ZnFe$_2$O$_4$ separately. This suggests a synergistic effect of catalyst and support, encouraged by the support's high surface area and physiochemical properties. Large surface areas are advantageous for dispersing ZF as active components, while adequate pore sizes provide reactants additional sites to bind and activate gaseous oxygen species [66]. In essence, the catalytic activity of supporting catalysts depends on their acid–base property. Hence, ether is anticipated to be readily adsorbed and activated on surfaces [67,68].

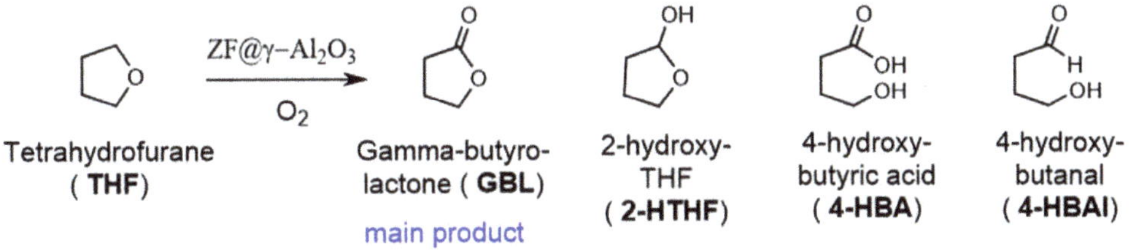

Scheme 1. Reaction products of THF oxidation by O_2 over the ZF@γ-Al_2O_3 catalyst.

The effect of the catalyst-support ratio was investigated for three different catalyst-support mole ratios (Supporting Information, Table S1). According to the results, increasing the support ratio from (1:1) to (1:2) enhanced conversion and lactone selectivity. However, when the catalyst-support ratio was raised to (1:3), more Lewis acid and Brnsted acid sites were present, which were responsible for the ring opening process and subsequent oxidation [69]. Therefore, the ratio (1:2) was found optimal and thus was used for further investigation.

Considering these results, the aerobic oxidation of THF was studied using ZF@γ-Al_2O_3 as the catalyst by varying the following factors: reaction temperature, O_2 pressure, and reaction time. Furthermore, the optimized conditions were applied to other cyclic ether oxidations.

2.2.1. THF Oxidation

The effect of the reaction parameters on the THF oxidation is presented in Figure 9. Hence, the temperature effect was studied in the range from 70 °C to 90 °C and shown in Figure 9a. It is clear that the conversion value of GBL was remarkably dependent on the reaction temperature, as its value approximately doubled when temperature was raised from 70 °C to 85 °C. On the other hand, for the same range, the selectivity value of GBL increased from 74.39% to 83.99%. However, GBL selectivity decreased as temperatures increased to 90 °C with increased acid formation due to additional oxidation events [39]. To investigate the effect of the pressure of oxygen (PO_2) on the rate of THF oxidation, the reaction was performed with varying PO_2 from 1.5 to 7 bars. As seen in Figure 9b, the conversion of THF was significantly improved from 10.5% to 48.1% upon increasing PO_2 from 1 to 3.5 bars. This increase was due to the decrease in the amount of 2-HTHF. As PO_2 increased, GBL selectivity decreased with the formation of oxygenated products such as 4-HBAl and 4-HBA, which can be explained in terms of the fact that GBL underwent deep oxidation in the presence of excessive oxygen. Hence, these results reflect the importance of O_2 pressure control. The reaction time effect is depicted in Figure 9c. The conversion linearly increased from 38.4 to 52.7% as reaction time increased from 5 to 15 h. On the other hand, the selectivity of GBL increased linearly by increasing reaction time up to 12 h. After 12 h, the selectivity of GBL decreased in favor of 4-HBA production. The catalyst dose was also investigated in the range of 0.15 to 1.0 g per 10 mL of substrate, as presented in Figure 9d. When the amount of the catalyst increased from 0.15 to 0.5 g, the conversion increased from 26.3 to 48.1%. This suggests that ZF@γ-Al_2O_3 functioned as active sites for the oxidation and, with the increase in catalyst amount, the number of active sites available for the reaction to progress increased. A gradual increase in GBL selectivity was also noticed with the increase in the catalyst up to 0.5 g, beyond which no significant change was detected. Table 3 summarizes the findings under the optimal reaction condition.

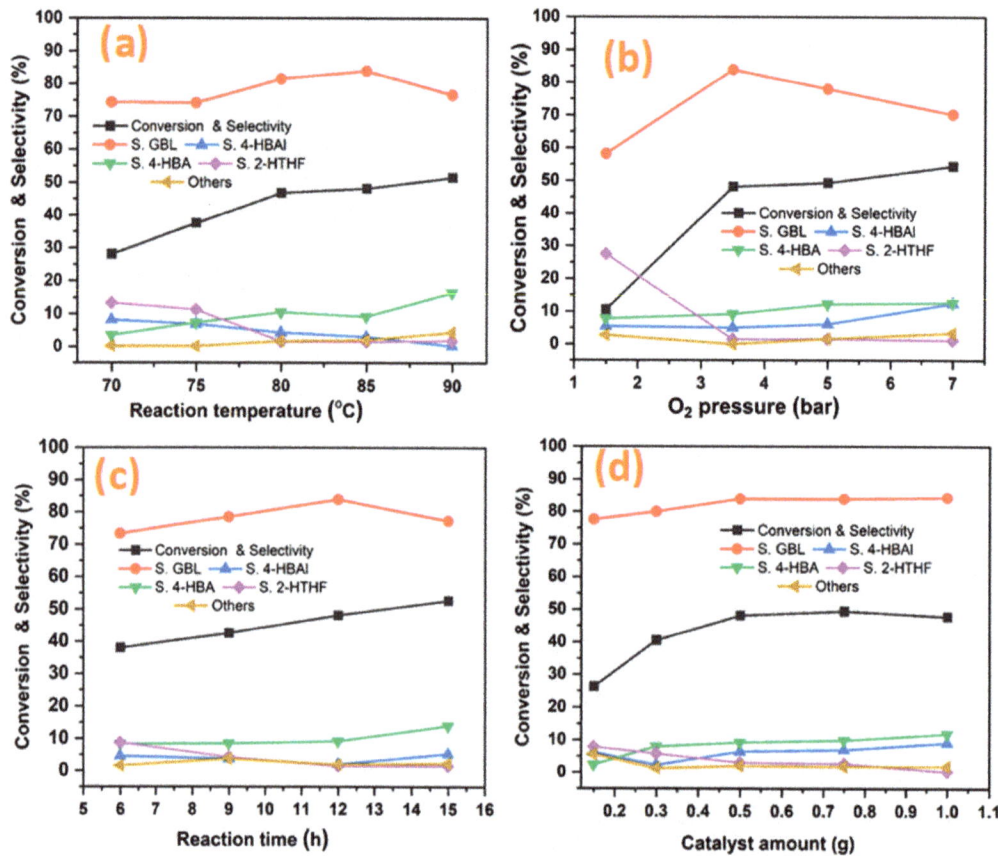

Figure 9. Effect of reaction parameters on the THF oxidation. (**a**) reaction temperature, (**b**) O_2 pressure, (**c**) reaction time, and (**d**) catalyst amount. Reaction conditions (unless otherwise specified): PO_2 = 3.5 bar, THF = 10 mL, 85 °C, 0.5 g catalyst, 12 h.

Table 3. Catalytic results under optimal conditions: PO_2 = 3.5 bar, substrate (THF) = 10 mL, 0.5 g catalyst, 85 °C, 12 h.

Catalyst	Conversion (%)	Selectivity of GBL (%)
ZF@γ-Al$_2$O$_3$	48.1	83.9
ZnFe$_2$O$_4$ (ZF)	18.1	63.3
γ-Al$_2$O$_3$	11.7	32.6
Blank	3.2	0

2.2.2. Cyclic Ethers Oxidation

The oxidative conversion of other five- and six-membered ethers to the corresponding lactones was also investigated and presented in Table 4. According to Entry 1, saturated five-membered cyclic compounds such as 2-methyl tetrahydrofuran (2-MTHF) performed well with 40% conversion and 86% γ-valerolactone (GVL) selectively. Substrates with a six-membered ring showed lower conversion than their five-membered counterparts. This result is in line with Huckel's theory, which states that the six-membered ring is more stable than the five-membered ring (less reactive). For instance, tetrahydropyran (THP) oxidation resulted in a low conversion of 11% but a high selectivity of 89% for δ-valerolactone (DVL) due to the absence of ring opening or excessive oxidation products. Thus, the oxidation was performed under high pressure and a long-time reaction, resulting in both improved conversion and improved selectivity of DVL production. Furthermore, due to their activity, both five- and six-membered unsaturated cyclic ethers demonstrated poor selective outcomes due to deep oxidation and polymerization reactions. Therefore, an experiment was conducted under low pressure and a shorter time, which improved selectivity results towards lactone formation (entries 2, 5 and 6).

Table 4. Cyclic ether oxidation catalyzed by ZF@γ-Al$_2$O$_3$.

Entry	Substrate	Conversion	S. Lactone	Product
1		40	86	
2		76 49 [a]	34 62 [a]	
3		--	--	-
4		11 28 [b]	83 89 [b]	
5		38 49 [a]	61 72 [a]	
6		68	25	
7		11	72	

Reaction condition: substrate = 10 mL, PO$_2$ = 3.5 bar, 85 °C, 0.5 g catalyst, 12 h. [a] Substrate = 10 mL, PO$_2$ = 1.5 bar, 85 °C, 0.5 g catalyst, 6 h. [b] Substrate = 10 mL, PO$_2$ = 7 bar, 85 °C, 0.5 g catalyst, 24 h. For byproducts, see Supporting Information Table S2.

2.3. Reusability

Reusability, along with activity, is considered the most important indicator for the industrial consideration of catalysts. To reduce production costs and minimize waste generation, catalysts must be separated and recycled easily after catalytic reactions [70]. Recycling of the catalyst was performed using centrifugation, washing with aqueous ethanol (25% v/v), and finally drying at 85 °C for 24 h. FTIR and XRD techniques were used to investigate the catalyst's stability, and the results are provided in Figure 10. As can be seen, an acceptable XRD pattern match between the fresh and reused catalysts was detected, indicating high stability of the catalyst system. However, the slight difference in the relative intensities of the peaks before and after the usage of the catalyst was due to the activation process, with the possible occupation of residues that were not removed during the reactivation stage [71]. The traced peak around 2θ of 25° was possibly for carbon residues.

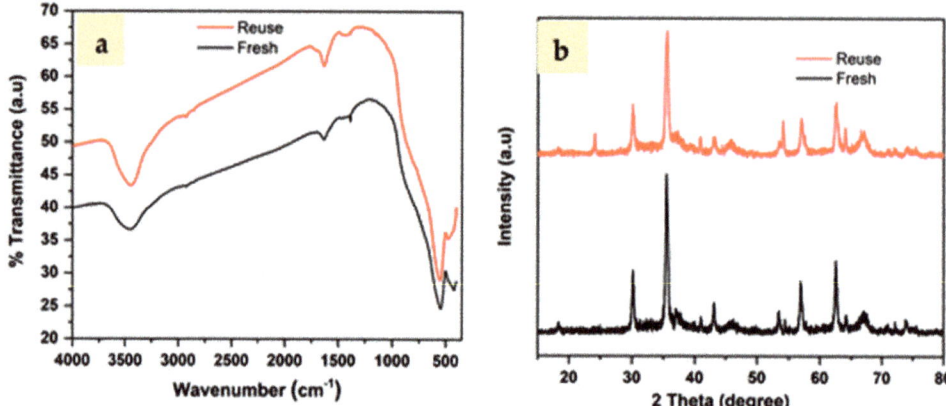

Figure 10. (a) FTIR and (b) XRD for fresh and reused ZF@γ-Al$_2$O$_3$ after the catalytic cycle.

As can be seen in Figure 11, after four runs, a reduction in the THF conversion from 48.1% to 43.8% was observed; however, the GBL selectivity remained almost constant. A controlled leaching experiment was also performed to measure zinc and iron contents. The reaction mixture was filtered and analyzed by ICP. The results showed the absence of ions within the detection capacity of the device of 1 ppb, confirming the stability and reusability of the catalyst. These findings demonstrated the catalyst's stability and ability to oxidize cyclic ethers to the corresponding lactones under an aerobic condition.

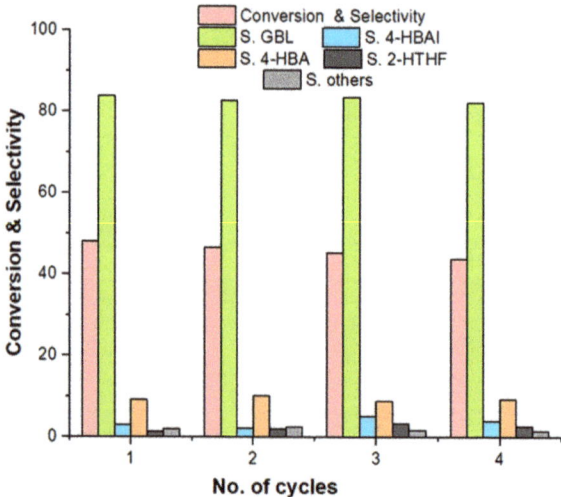

Figure 11. Reusability of ZF@γ-Al$_2$O$_3$. Reaction conditions: PO$_2$ = 3.5 bar, THF = 10 mL, 85 °C, 0.5 g of catalyst, 12 h.

3. Materials and Methods

3.1. Materials

The organic and inorganic chemicals were used as provided without further purification and are as follows: ammonium bicarbonate (NH$_4$HCO$_3$), ferric nitrate nonahydrate (Fe (NO$_3$)$_3$·9H$_2$O), and zinc acetate dihydrate (Zn(CH$_3$COO)$_2$·2H$_2$O) from BDH chemicals, London, UK; THF (99.5%) from Fisher Scientific, Loughborough, UK; THP, 2-MTHF, 3,4-dihydro-2H-pyran, and 2,3-dihydrofuran from Sigma-Aldrich, Burlington, MA, USA;

pyran, furan, and 1.4 dioxane from BDH chemicals, London, UK; and gamma-alumina (γ-Al_2O_3) from Sigma-Aldrich, Burlington, MA, USA.

3.2. Catalyst Preparation

The gamma-alumina-supported zinc ferrite (ZF@γ-Al_2O_3) catalyst was prepared by a deposition–coprecipitation method. Typically, 4.04 g ferric nitrate nonahydrate, 1.1 zinc acetate dihydrate, and 1.1 g gamma-alumina (1:2 mol ratio of ZF/γ-Al_2O_3) were added to a 250 mL round flask containing 30 mL deionized (DI) water. The mixture was stirred at 40 °C for 1 h, and then the appropriate amount of ammonium bicarbonate in 50 mL DI was added drop-wisely. The resulting precipitate was stirred at 75 °C for a further 5 h in a closed flask. After 12 h of aging, the precipitate was filtered, washed, and dried overnight at 75 °C. Calcination at 600 °C was then performed on the catalyst precursor for 5 h.

3.3. Characterization of the Catalysts

Fourier transform infrared (FTIR) spectra were recorded in a PerkinElmer spectrum BX (Perkin Elmer, Waltham, MA, USA) using the KBr disk method on the range of 400 to 4000 cm^{-1}. X-ray diffraction (XRD) measurements were carried out using Rigaku XtaLAB mini II benchtop X-Ray 110 crystallography system (The Woodlands, TX, USA), where Cu K$\alpha\lambda$ = 1.5418 Å was used as the light source; the powdered sample was placed in the sample holder and exposed to X-ray radiation at room temperature, and the diffraction pattern was obtained over the 2θ range of 10° to 80° at a rate of 3°/min. X-ray photoelectron spectroscopy (XPS) was carried out using an ESCALAB 250Xi apparatus and a monochromatic Al Kα X-ray source (1486.6 eV). The specific surface area, pore volume, and average pore diameter of the catalyst were measured in Micromeritics Tristar II 3020 surface area and porosity analyzer (Micromeritics Instrument Corporation, Norcross, GA, USA). To achieve these measurements, catalysts of weights varying from 0.20 g to 0.50 g were used, and the test advancement was monitored by computer and then analyzed by the corresponding software. A JEM-2100F field emission electron microscope (JEOL, Tokyo, Japan) with an 80 kV acceleration voltage was used for transmission electron microscope (TEM) imaging; hence, drops of a prepared sample suspension were deposited on a copper grid made of lacey carbon. Scanning electron microscopy (SEM) was carried out using FESEM (JSM-7600F, JEOL, Tokyo, Japan) with an accelerating voltage of 5 kV, a beam current of 30 A, and image magnification of 30,000× for platinum-coated samples. The energy dispersive X-ray spectroscopy (EDS) profile was obtained using an X-MaxN system from Oxford instruments (Abingdon, UK). Elemental analyses were performed under inductively coupled plasma spectrometry (ICP) measurements using a Nexion 300D Spectrometer from Perkin Elmer (Waltham, MA, USA). For this, a concentrated nitric acid was used to acidify the sample. In the following steps, the acidified samples were digested in an oven at 75 °C for 24 h, cooled, filtered, and diluted as required. On a Mettler Toledo TGA/DSC Star system (Columbus, OH, USA), the thermogravimetric and differential thermal analysis (TGA/DTA) curves of the catalysts were obtained under nitrogen and at a heating rate of 10 °C min^{-1}.

3.4. Experimental Procedure

The reactions were carried out in a 100 mL stainless steel autoclave equipped with a magnetic stirrer, manometer, and vent. Hence, the following experimental conditions were applied unless otherwise specified. A quantity of 10 mL cyclic ether substrate and 0.5 g of the catalyst were charged into the autoclave. The pressure of O_2 was set at 3.5 bar using a regulator and mass flow meter. The reaction temperature was controlled by a heating jacket connected to a water circulator with an accuracy of ± 0.1 °C. The reactor vessel was placed under the desired temperature and continuously stirred for the predefined time. When the target time elapsed, the reactor was cooled to 10 °C and depressurized to atmospheric pressure. The used catalyst was removed by centrifugation, purified, and dried for further use. The resulting solution was analyzed in a gas chromatography (GC) instrument with a flame ionization detector (FID) using a (Rtx-5-length 30 m, ID 0.53 mm) capillary column. The identification of products was achieved occasionally by GC coupled with a mass selective detector (GC-MS). For the separation of the target compounds, the TR-5 MS-SQC capillary column (30 m length × 0.25 mm internal diameter, phase thickness 0.25 µm) was used with helium as the carrier gas (at a flow rate of 1 mL min^{-1}). The total conversion, selectivity, and product yield were calculated using the following equations:

$$Conversion\% = \frac{n_0 - n_r}{n_0} \times 100 \qquad (1)$$

$$Yield\% = \frac{n_i}{n_0} \times 100 \qquad (2)$$

$$Selectivity\% = \frac{yield(i)}{conversion} \times 100 \qquad (3)$$

where n_0 is the number of the initial moles of substrate, n_i is the number of the moles formed from the product i (i = GBL, DVL, . . .), and n_r is the number of the remaining moles from substrate.

4. Conclusions

In this work, the aerobic oxidation of cyclic ethers was examined over the ZF@γ-Al$_2$O$_3$-supported catalyst using O$_2$ as an oxidizing agent under solvent-free conditions. The ZnFe$_2$O$_4$ was successfully incorporated on the γ-Al$_2$O$_3$ support, as indicated by FTIR, XRD, TEM, and BET techniques. The performance of the supported catalyst was better than that of the unsupported one in terms of conversion of the starting model material (THF) and selectivity of the desired product (GBL). The catalytic oxidation condition endorses the production of GBL as the main product. The optimized reaction conditions demonstrated that the catalytic performance was significantly influenced by temperature and amount of oxygen. The catalyst amount also greatly influenced the THF conversion, while GBL selectivity was mostly unaffected. Also, six-membered cyclic ethers were less active than five-membered cyclic ethers in terms of lactone transformation. Due to the over-oxidation and polymerization reactions that occured during the oxidation of unsaturated cyclic ethers, poor lactone selectivity was found. Controlling the reaction time and oxidant pressure are crucial for improving lactone selectivity. Thus, to achieve a high yield of the desired product, the reaction conditions (e.g., reaction time) must be carefully controlled. Hence, it is evidently proven that the applied ZF@γ-Al$_2$O$_3$ is promising for the selective conversion of cyclic ethers to lactones with a high product yield and catalytic activity that remained effective even after four rounds of catalysis.

Supplementary Materials: The following supporting information can be downloaded at https://www.mdpi.com/article/10.3390/molecules28207192/s1, Table S1: Comparison of various catalyst-support mole ratio performances in terms of conversion and selectivity. Condition: PO2 = 3.5 bar, substrate (THF) = 10 mL, 0.5 g catalyst, 85 °C, 12 h. Table S2: Products of cyclic ether oxidation reaction.

Author Contributions: Conceptualization, N.A.Y.A. and A.A.A.-K.; formal analysis, N.A.Y.A. and A.-B.A.-O.; funding acquisition, A.A.A.-K.; investigation, N.A.Y.A. and T.S.A.; methodology, N.A.Y.A. and A.-B.A.-O.; supervision, A.A.A.-K.; validation, M.S.A.; visualization, N.A.Y.A.; writing—original draft, N.A.Y.A. and A.-B.A.-O.; writing—review and editing, A.-B.A.-O. and M.S.A. All authors have read and agreed to the published version of the manuscript.

Funding: This research received no external funding.

Institutional Review Board Statement: Not applicable.

Informed Consent Statement: Not applicable.

Data Availability Statement: The data that support the findings of this study are available on reasonable request from the corresponding authors.

Acknowledgments: The authors extended their appreciation to the Researchers Supporting Project number (RSP2023R266), King Saud University, Riyadh, Saudi Arabia.

Conflicts of Interest: The authors declare no conflict of interest.

Sample Availability: Not applicable.

References

1. Briggs, D. Environmental pollution and the global burden of disease. *Br. Med. Bull.* **2003**, *68*, 1–24. [CrossRef] [PubMed]
2. Cholakov, G.S. Control of pollution in the petroleum industry. *Pollut. Control Technol.* **2009**, *3*, 86–107.
3. Rasul, M.G.; Faisal, I.; Khan, M.M.K. Environmental pollution generated from process industries in Bangladesh. *Int. J. Environ. Pollut.* **2006**, *28*, 144–161. [CrossRef]
4. Makarova, A.S.; Jia, X.; Kruchina, E.B.; Kudryavtseva, E.I.; Kukushkin, I.G. Environmental performance assessment of the chemical industries involved in the Responsible Care®Program: Case study of the Russian Federation. *J. Clean. Prod.* **2019**, *222*, 971–985. [CrossRef]
5. Clark, J.H. Green chemistry: Challenges and opportunities. *Green Chem.* **1999**, *1*, 1–8. [CrossRef]
6. Sharma, S.K.; Chaudhary, A.; Singh, R. Gray chemistry verses green chemistry: Challenges and opportunities. *Rasayan J. Chem.* **2008**, *1*, 68–92.
7. Mulvihill, M.J.; Beach, E.S.; Zimmerman, J.B.; Anastas, P.T. Green chemistry and green engineering: A framework for sustainable technology development. *Annu. Rev. Environ. Resour.* **2011**, *36*, 271–293. [CrossRef]
8. Dunn, P.J. The importance of green chemistry in process research and development. *Chem. Soc. Rev.* **2012**, *41*, 1452–1461. [CrossRef]
9. Syed, N.; Singh, S.; Chaturvedi, S.; Nannaware, A.D.; Khare, S.K.; Rout, P.K. Production of lactones for flavoring and pharmacological purposes from unsaturated lipids: An industrial perspective. *Crit. Rev. Food Sci. Nutr.* **2022**, 1–32. [CrossRef]
10. Kowalczyk, P.; Gawdzik, B.; Trzepizur, D.; Szymczak, M.; Skiba, G.; Raj, S.; Kramkowski, K.; Lizut, R.; Ostaszewski, R. δ-Lactones—A New Class of Compounds That Are Toxic to E. coli K12 and R2–R4 Strains. *Materials* **2021**, *14*, 2956. [CrossRef] [PubMed]
11. Choi, S.; Kim, H.U.; Kim, T.Y.; Kim, W.J.; Lee, M.H.; Lee, S.Y. Production of 4-hydroxybutyric acid by metabolically engineered Mannheimia succiniciproducens and its conversion to γ-butyrolactone by acid treatment. *Metab. Eng.* **2013**, *20*, 73–83. [CrossRef] [PubMed]
12. Bertone, M.E.; Meyer, C.I.; Regenhardt, S.A.; Sebastian, V.; Garetto, T.F.; Marchi, A.J. Highly selective conversion of maleic anhydride to γ-butyrolactone over Ni-supported catalysts prepared by precipitation–deposition method. *Appl. Catal. A Gen.* **2015**, *503*, 135–146. [CrossRef]
13. Díaz-Rodríguez, A.; Borzecka, W.; Lavandera, I.; Gotor, V. Stereodivergent preparation of valuable γ-or δ-hydroxy esters and lactones through one-pot cascade or tandem chemoenzymatic protocols. *ACS Catal.* **2014**, *4*, 386–393. [CrossRef]
14. Tang, X.; Zeng, X.; Li, Z.; Hu, L.; Sun, Y.; Liu, S.; Lei, T.; Lin, L. Production of γ-valerolactone from lignocellulosic biomass for sustainable fuels and chemicals supply. *Renew. Sustain. Energy Rev.* **2014**, *40*, 608–620. [CrossRef]
15. Bińczak, J.; Dziuba, K.; Chrobok, A. Recent developments in lactone monomers and polymer synthesis and application. *Materials* **2021**, *14*, 2881. [CrossRef]
16. Wang, H.; Ding, G.; Li, X.; She, H.; Zhu, Y.; Li, Y. Sustainable production of γ-valerolactone and δ-valerolactone through the coupling of hydrogenation and dehydrogenation. *Sustain. Energy Fuels* **2021**, *5*, 930–934. [CrossRef]
17. Sartori, S.K.; Diaz, M.A.N.; Diaz-Munoz, G. Lactones: Classification, synthesis, biological activities, and industrial applications. *Tetrahedron* **2021**, *84*, 132001. [CrossRef]
18. Janecki, T. *Natural Lactones and Lactams: Synthesis, Occurrence and Biological Activity*; John Wiley & Sons: Hoboken, NJ, USA, 2013.
19. Pazos, D.; Giannasi, P.; Rossy, Q.; Esseiva, P. Combining Internet monitoring processes, packaging and isotopic analyses to determine the market structure: Example of Gamma Butyrolactone. *Forensic Sci. Int.* **2013**, *230*, 29–36. [CrossRef] [PubMed]
20. Kondawar, S.; Rode, C. Ionic liquids for the sustainable transformation of levulinic acid to gamma-valerolactone (GVL). *Curr. Opin. Green Sustain. Chem.* **2022**, *35*, 100607. [CrossRef]
21. Hollmann, F.; Kara, S.; Opperman, D.J.; Wang, Y. Biocatalytic synthesis of lactones and lactams. *Chem. Asian J.* **2018**, *13*, 3601–3610. [CrossRef]
22. Zhu, Y.-L.; Yang, J.; Dong, G.-Q.; Zheng, H.-Y.; Zhang, H.-H.; Xiang, H.-W.; Li, Y.-W. An environmentally benign route to γ-butyrolactone through the coupling of hydrogenation and dehydrogenation. *Appl. Catal. B Environ.* **2005**, *57*, 183–190. [CrossRef]
23. Yu, F.; Chi, Y.; Gao, C.; Chen, R.; Xie, C.; Yu, S. Baeyer-Villiger oxidation of cyclic ketones catalyzed by amino acid ionic liquids. *Chem. Res. Chin. Univ.* **2020**, *36*, 865–869. [CrossRef]
24. Zhang, C.; Chen, L.; Cheng, H.; Zhu, X.; Qi, Z. Atomically dispersed Pd catalysts for the selective hydrogenation of succinic acid to γ-butyrolactone. *Catal. Today* **2016**, *276*, 55–61. [CrossRef]
25. Wettstein, S.G.; Alonso, D.M.; Chong, Y.; Dumesic, J.A. Production of levulinic acid and gamma-valerolactone (GVL) from cellulose using GVL as a solvent in biphasic systems. *Energy Environ. Sci.* **2012**, *5*, 8199–8203. [CrossRef]

26. Winoto, H.P.; Ahn, B.S.; Jae, J. Production of γ-valerolactone from furfural by a single-step process using Sn-Al-Beta zeolites: Optimizing the catalyst acid properties and process conditions. *J. Ind. Eng. Chem.* **2016**, *40*, 62–71. [CrossRef]
27. Zhou, C.; Xiao, Y.; Xu, S.; Li, J.; Hu, C. γ-valerolactone production from furfural residue with formic acid as the sole hydrogen resource via an integrated strategy on Au-Ni/ZrO$_2$. *Ind. Eng. Chem. Res.* **2020**, *59*, 17228–17238. [CrossRef]
28. Tang, D.; Shen, Z.; Lechler, S.; Lu, G.; Yao, L.; Hu, Y.; Huang, X.; Muhler, M.; Zhao, G.; Peng, B. Aerobic oxidative lactonization of diols at room temperature over defective titanium-based oxides in water. *J. Catal.* **2023**, *418*, 237–246. [CrossRef]
29. Budroni, G.; Corma, A. Gold and gold–platinum as active and selective catalyst for biomass conversion: Synthesis of γ-butyrolactone and one-pot synthesis of pyrrolidone. *J. Catal.* **2008**, *257*, 403–408. [CrossRef]
30. Machado, G.; Leon, S.; Santos, F.; Lourega, R.; Dullius, J.; Mollmann, M.E.; Eichler, P. Literature review on furfural production from lignocellulosic biomass. *Nat. Resour.* **2016**, *7*, 115–129. [CrossRef]
31. Alessio, C.; Marco, N.; Maurizio, S.; Alvise, P. Upgrading of Biobased Lactones with Dialkylcarbonates. *ACS Sustain. Chem. Eng.* **2014**, *2*, 2131–2141.
32. Silva, R.; Coelho, E.; Aguiar, T.Q.; Domingues, L. Microbial biosynthesis of lactones: Gaps and opportunities towards sustainable production. *Appl. Sci.* **2021**, *11*, 8500. [CrossRef]
33. Zhang, Z. Synthesis of γ-Valerolactone from Carbohydrates and its Applications. *ChemSusChem* **2016**, *9*, 156–171. [CrossRef] [PubMed]
34. Yan, K.; Wu, G.; Lafleur, T.; Jarvis, C. Production, properties and catalytic hydrogenation of furfural to fuel additives and value-added chemicals. *Renew. Sustain. Energy Rev.* **2014**, *38*, 663–676. [CrossRef]
35. Soszka, E.; Jędrzejczyk, M.; Keller, N.; Ruppert, A.M. High yield production of 2-methyltetrahydrofuran biofuel with reusable Ni-Co catalysts. *Fuel* **2023**, *332*, 126118. [CrossRef]
36. Dastidar, R.G.; Kim, M.S.; Zhou, P.; Luo, Z.; Shi, C.; Barnett, K.J.; McClelland, D.J.; Chen, E.Y.-X.; Van Lehn, R.C.; Huber, G.W. Catalytic production of tetrahydropyran (THP): A biomass-derived, economically competitive solvent with demonstrated use in plastic dissolution. *Green Chem.* **2022**, *24*, 9101–9113. [CrossRef]
37. Kato, N.; Hamaguchi, Y.; Umezawa, N.; Higuchi, T. Efficient oxidation of ethers with pyridine N-oxide catalyzed by ruthenium porphyrins. *J. Porphyr. Phthalocyanines* **2015**, *19*, 411–416. [CrossRef]
38. Zhao, Y.; Ang, J.Q.L.; Ng, A.W.T.; Yeung, Y.-Y. Oxidative transformation of cyclic ethers/amines to lactones/lactams using a DIB/TBHP protocol. *RSC Adv.* **2013**, *3*, 19765–19768. [CrossRef]
39. Liu, S.; Li, S.; Shen, X.; Wang, Y.; Du, J.; Chen, B.; Han, B.; Liu, H. Selective aerobic oxidation of cyclic ethers to lactones over Au/CeO$_2$ without any additives. *Chem. Commun.* **2020**, *56*, 2638–2641. [CrossRef]
40. Ali, M.E.; Rahman, M.M.; Sarkar, S.M.; Hamid, S.B.A. Heterogeneous metal catalysts for oxidation reactions. *J. Nanomater.* **2015**, *2014*, 209. [CrossRef]
41. Prati, L.; Rossi, M. Chemoselective catalytic oxidation of polyols with dioxygen on gold supported catalysts. In *Studies in Surface Science and Catalysis*; Elsevier: Amsterdam, The Netherlands, 1997; Volume 110, pp. 509–516.
42. Sasidharan, M.; Bhaumik, A. Catalytic oxidation of cyclic ethers to lactones over various titanosilicates. *J. Mol. Catal. A Chem.* **2011**, *338*, 105–110. [CrossRef]
43. Jiang, Y.; Ni, P.; Chen, C.; Lu, Y.; Yang, P.; Kong, B.; Fisher, A.; Wang, X. Selective electrochemical H$_2$O$_2$ production through two-electron oxygen electrochemistry. *Adv. Energy Mater.* **2018**, *8*, 1801909. [CrossRef]
44. Dou, J.; Tao, F.F. Selective epoxidation of cyclohexene with molecular oxygen on catalyst of nanoporous Au integrated with MoO3 nanoparticles. *Appl. Catal. A Gen.* **2017**, *529*, 134–142. [CrossRef]
45. Shi, Z.; Zhang, C.; Tang, C.; Jiao, N. Recent advances in transition-metal catalyzed reactions using molecular oxygen as the oxidant. *Chem. Soc. Rev.* **2012**, *41*, 3381–3430. [CrossRef] [PubMed]
46. Sahle-Demessie, E.; Gonzalez, M.A.; Enriquez, J.; Zhao, Q. Selective oxidation in supercritical carbon dioxide using clean oxidants. *Ind. Eng. Chem. Res.* **2000**, *39*, 4858–4864. [CrossRef]
47. Wu, W.; Jiang, H. Palladium-catalyzed oxidation of unsaturated hydrocarbons using molecular oxygen. *Acc. Chem. Res.* **2012**, *45*, 1736–1748. [CrossRef] [PubMed]
48. Zhang, P.; Lu, H.; Zhou, Y.; Zhang, L.; Wu, Z.; Yang, S.; Shi, H.; Zhu, Q.; Chen, Y.; Dai, S. Mesoporous MnCeO x solid solutions for low temperature and selective oxidation of hydrocarbons. *Nat. Commun.* **2015**, *6*, 8446. [CrossRef]
49. Abduh, N.A.; Al-Kahtani, A.; Algarni, T.S.; Al-Odayni, A.-B. Selective Oxidation of Tetrahydrofuran to Gamma-Butyrolactone over Spinel ZnFe$_2$O$_4$ Nanoparticle Catalyst. *Catalysts* **2023**, *13*, 692. [CrossRef]
50. Xie, Z.; Liu, Z.; Wang, Y.; Yang, Q.; Xu, L.; Ding, W. An overview of recent development in composite catalysts from porous materials for various reactions and processes. *Int. J. Mol. Sci.* **2010**, *11*, 2152–2187. [CrossRef]
51. Xue, Z.; Zhong, Z.; Zhang, B.; Xu, C. Performance of catalytic fast pyrolysis using a γ-Al$_2$O$_3$ catalyst with compound modification of ZrO$_2$ and CeO$_2$. *Catalysts* **2019**, *9*, 849. [CrossRef]
52. Firdous, N.; Janjua, N.K. CoPtx/γ-Al$_2$O$_3$ bimetallic nanoalloys as promising catalysts for hydrazine electrooxidation. *Heliyon* **2019**, *5*, e01380. [CrossRef]
53. Atrak, K.; Ramazani, A.; Taghavi Fardood, S. Green synthesis of amorphous and gamma aluminum oxide nanoparticles by tragacanth gel and comparison of their photocatalytic activity for the degradation of organic dyes. *J. Mater. Sci. Mater. Electron.* **2018**, *29*, 8347–8353. [CrossRef]

54. Saafan, S.A.; El-Nimr, M.K.; Hussein, M.M.; Omar, M.K. FTIR, DC, and AC electrical measurements of Mg Zn Nano-ferrites and their composites with Polybenzoxazine. *Appl. Phys. A* **2021**, *127*, 800. [CrossRef]
55. Sarala, E.; Madhukara Naik, M.; Vinuth, M.; Rami Reddy, Y.; Sujatha, H. Green synthesis of Lawsonia inermis-mediated zinc ferrite nanoparticles for magnetic studies and anticancer activity against breast cancer (MCF-7) cell lines. *J. Mater. Sci. Mater. Electron.* **2020**, *31*, 8589–8596. [CrossRef]
56. Ahmed, M.I.; Jahin, H.S.; Dessouki, H.A.; Nassar, M.Y. Synthesis and characterization of γ-Al_2O_3 and α-Al_2O_3 nanoparticles using a facile, inexpensive auto-combustion approach. *Egypt. J. Chem.* **2021**, *64*, 2509–2515. [CrossRef]
57. Jiang, H.; Xu, X.; Zhang, R.; Zhang, Y.; Chen, J.; Yang, F. Nano ferrites (AFe_2O_4, A= Zn, Co, Mn, Cu) as efficient catalysts for catalytic ozonation of toluene. *RSC Adv.* **2020**, *10*, 5116–5128. [CrossRef]
58. Varpe, A.S.; Deshpande, M.D. Study of structural, optical, and dielectric properties of sol–gel derived $ZnFe_2O_4$–Al_2O_3 composite nanoparticles. *J. Sol-Gel Sci. Technol.* **2020**, *96*, 718–727. [CrossRef]
59. Ali, S.Y.; Eid, O.I.; Siddig, M.A. The Influence of Cu on the Dielectric Properties of $NiZnFe_2O_4$ Synthesized by Solid State Reaction Method. *J. Mater. Sci. Chem. Eng.* **2020**, *8*, 14–23.
60. Etemadinia, T.; Allahrasani, A.; Barikbin, B. $ZnFe_2O_4$@SiO_2@Tragacanth gum nanocomposite: Synthesis and its application for the removal of methylene blue dye from aqueous solution. *Polym. Bull.* **2019**, *76*, 6089–6109. [CrossRef]
61. Shafiee, M.; Hafez Ghoran, S.; Bordbar, S.; Gholami, M.; Naderian, M.; Dehghani, F.S.; Amani, A.M. Rutin: A Flavonoid Precursor for Synthesis of $ZnFe_2O_4$ Nanoparticles; Electrochemical Study of Zinc Ferrite-chitosan Nanogel for Doxorubicin Delivery. *J. Nanostructures* **2021**, *11*, 114–124.
62. Urbonavicius, M.; Varnagiris, S.; Pranevicius, L.; Milcius, D. Production of gamma alumina using plasma-treated aluminum and water reaction byproducts. *Materials* **2020**, *13*, 1300. [CrossRef]
63. Zarezadeh-Mehrizi, M.; Afshar Ebrahimi, A.; Rahimi, A. Comparison of γ and δ-Al_2O_3 supported CoMo catalysts in the ydrodesulfurization of straight-run gas oil. *Sci. Iran.* **2019**, *26*, 1555–1565.
64. Dobrosielska, M.; Zieliński, M.; Frydrych, M.; Pietrowski, M.; Marciniak, P.; Martyła, A.; Sztorch, B.; Przekop, R.E. Sol–Gel Approach for Design of Pt/Al_2O_3-TiO_2 System—Synthesis and Catalytic Tests. *Ceramics* **2021**, *4*, 667–680. [CrossRef]
65. Das, K.K.; Patnaik, S.; Nanda, B.; Pradhan, A.C.; Parida, K. $ZnFe_2O_4$-decorated mesoporous Al_2O_3 modified MCM-41: A solar-light-active photocatalyst for the effective removal of phenol and Cr (VI) from water. *ChemistrySelect* **2019**, *4*, 1806–1819. [CrossRef]
66. Khan, S.; Shah, S.S.; Janjua, N.K.; Yurtcan, A.B.; Nazir, M.T.; Katubi, K.M.; Alsaiari, N.S. Alumina supported copper oxide nanoparticles (CuO/Al_2O_3) as high-performance electrocatalysts for hydrazine oxidation reaction. *Chemosphere* **2023**, *315*, 137659. [CrossRef]
67. Winiarska, K.; Klimkiewicz, R.; Tylus, W.; Sobianowska-Turek, A.; Winiarski, J.; Szczygieł, B.; Szczygieł, I. Study of the catalytic activity and surface properties of manganese-zinc ferrite prepared from used batteries. *J. Chem.* **2019**, *2019*, 1–14. [CrossRef]
68. Liu, L.; Li, H.; Tan, Y.; Chen, X.; Lin, R.; Yang, W.; Huang, C.; Wang, S.; Wang, X.; Liu, X.Y. Metal-support synergy of supported gold nanoclusters in selective oxidation of alcohols. *Catalysts* **2020**, *10*, 107. [CrossRef]
69. Zhao, Z.; Yang, C.; Sun, P.; Gao, G.; Liu, Q.; Huang, Z.; Li, F. Synergistic Catalysis for Promoting Ring-Opening Hydrogenation of Biomass-Derived Cyclic Oxygenates. *ACS Catal.* **2023**, *13*, 5170–5193. [CrossRef]
70. Miceli, M.; Frontera, P.; Macario, A.; Malara, A. Recovery/Reuse of Heterogeneous Supported Spent Catalysts. *Catalysts* **2021**, *11*, 591. [CrossRef]
71. Al-Iessa, M.S.; Al-Zaidi, B.Y.; Almukhtar, R.S.; Shakor, Z.M.; Hamawand, I. Optimization of Polypropylene Waste Recycling Products as Alternative Fuels through Non-Catalytic Thermal and Catalytic Hydrocracking Using Fresh and Spent Pt/Al_2O_3 and NiMo/Al_2O_3 Catalysts. *Energies* **2023**, *16*, 4871. [CrossRef]

Disclaimer/Publisher's Note: The statements, opinions and data contained in all publications are solely those of the individual author(s) and contributor(s) and not of MDPI and/or the editor(s). MDPI and/or the editor(s) disclaim responsibility for any injury to people or property resulting from any ideas, methods, instructions or products referred to in the content.

Review

Comparsion of Catalyst Effectiveness in Different Chemical Depolymerization Methods of Poly(ethylene terephthalate)

Marcin Muszyński [1,2], Janusz Nowicki [1], Mateusz Zygadło [3] and Gabiela Dudek [3,*]

[1] Łukasiewicz Research Network, Institute of Heavy Organic Synthesis "Blachownia", Energetyków 9, 47-225 Kędzierzyn-Koźle, Poland; marcin.muszynski@icso.lukasiewicz.gov.pl (M.M.); janusz.nowicki@icso.lukasiewicz.gov.pl (J.N.)

[2] Department of Physical Chemistry and Technology of Polymers, PhD School, Silesian University of Technology, ks. M. Strzody 9, 44-100 Gliwice, Poland

[3] Department of Physical Chemistry and Technology of Polymers, Faculty of Chemistry, Silesian University of Technology, ks. M. Strzody 9, 44-100 Gliwice, Poland; matezyg797@student.polsl.pl

* Correspondence: gabriela.maria.dudek@polsl.pl

Abstract: This paper presents an overview of the chemical recycling methods of polyethylene terephthalate (PET) described in the scientific literature in recent years. The review focused on methods of chemical recycling of PET including hydrolysis and broadly understood alcoholysis of polymer ester bonds including methanolysis, ethanolysis, glycolysis and reactions with higher alcohols. The depolymerization methods used in the literature are described, with particular emphasis on the use of homogeneous and heterogeneous catalysts and ionic liquids, as well as auxiliary substances such as solvents and cosolvents. Important process parameters such as temperature, reaction time, and pressure are compared. Detailed experimental results are presented focusing on reaction yields to allow for easy comparison of applied catalysts and for determination of the most favorable reaction conditions and methods.

Keywords: chemical recycling; depolymerization; catalysis; glycolysis; transesterification; hydrolysis

Citation: Muszyński, M.; Nowicki, J.; Zygadło, M.; Dudek, G. Comparsion of Catalyst Effectiveness in Different Chemical Depolymerization Methods of Poly(ethylene terephthalate). *Molecules* 2023, *28*, 6385. https://doi.org/10.3390/molecules28176385

Academic Editor: Lu Liu

Received: 31 July 2023
Revised: 28 August 2023
Accepted: 29 August 2023
Published: 31 August 2023

Copyright: © 2023 by the authors. Licensee MDPI, Basel, Switzerland. This article is an open access article distributed under the terms and conditions of the Creative Commons Attribution (CC BY) license (https://creativecommons.org/licenses/by/4.0/).

1. Introduction

Plastics are a massively used product in the global economy. It is estimated that about 83,000 million tons of plastics have been produced since 1950 [1]. Plastics are widely used, among others, as construction materials, in the textile industry, as well as in packaging, transport and electronics [2]. The most commonly produced plastics are polyolefins, of which a significant part is poly(ethylene terephthalate) (PET), which is mainly used in the production of packaging and textiles [1]. The mass production of plastics and the lack of proper practices and regulations regarding waste collection and treatment have led to an increasing accumulation of waste in the natural environment. Untreated and poorly deposited waste enters the natural environment, leading to pollution in both land and aquatic environments. The latter has recently been a source of particular interest due to the presence of microplastics in the natural environment, which accumulates in the environment due to poor biodegradability [3,4]. The accumulation of waste and the introduction of legal regulations has resulted in an increase in interest in plastic recycling. Over the years, various methods of recycling plastics have been developed. Among these methods, energy, mechanical and chemical recycling are distinguished. Thermal recycling is used to generate energy and is used for materials that are not suitable for processing using other methods [5]. It is used for degraded plastics that can no longer be recycled, cross-linked polymers and composites whose recycling is not economical. Mechanical recycling is most commonly used for thermoplastics that can be easily recycled and reused. Chemical recycling is the process by which a polymer undergoes a chemical reaction that produces monomers or other valuable chemical compounds. Of the massively used polymers, only some can be chemically recycled. Important factors determining the possibility of chemical recycling of polymers include their susceptibility to depolymerization reaction, availability on the market and cost-effectiveness. The material most often subjected to the depolymerization process is poly(ethylene terephthalate) (PET), although there are also known methods of depolymerization of other materials such as other polyesters [6], polycarbonates [7], and polyurethanes [8]. Poly(ethylene terephthalate) is chemically recycled using such methods as hydrolysis [9], glycolysis [10] and alcoholysis [11].

A significant number of the review papers describe general issues in plastic recycling or focus on the selected polymer and describe available recycling methods in a general way. There are relatively few review publications from the last ten years describing in-depth available methods for the chemical depolymerization of poly(ethylene terephthalate) by transesterification reaction. This review includes the chemical recycling: hydrolysis, methanolysis, glycolysis, alcoholism using alcohols C_2 to C_7, as well as 2-ethylhexanol of PET towards useful products. These compounds can be reused in the synthesis of polymers using recovered compounds, for the synthesis of new products with practical applications or directly to new compounds with increased added value such as dioctyl terephthalate (DOTP) which can be applied as a PVC plasticizer. Furthermore, this review was carried out in terms of the types of catalysts used, divided into homogeneous catalysts, heterogeneous catalysts, as well as non-catalytic systems such as processes carried out in sub-critical and supercritical conditions. The general methodology of the conducted research and selected results obtained by the authors with an emphasis on the efficiency of the processes and the activity of catalytic systems are presented.

2. Chemical Recycling
2.1. Hydrolysis

The hydrolytic decomposition of synthetic polymers that provide starting monomers which can be reused for polymer synthesis is one of the most important methods in the chemical recycling of post-consumer polymeric materials [12]. Not every type of polymer can be processed by this method. However, polyethylene terephthalate is one of the polymers for which this processing path is possible to carry out (Figure 1).

The hydrolytic decomposition of PET for producing starting monomers was the earlier method of waste PET recycling. From a chemical point of view, the hydrolytic decomposition of PET can be carried out in neutral, alkaline and acidic conditions [13,14]. The disadvantage of this method is the relatively high cost of purifying terephthalic acid (TPA) from the post-reaction mixture, which limits the commercial use of this method for the production of high-quality polymer, e.g., intended for contact with food [14].

The neutral hydrolysis method of PET depolymerization uses water or steam without the addition of an acid or basic catalyst, usually at temperatures between 250–300 °C and pressures between 1.5–4.0 MPa. The weight ratio of PET: water is generally from 1:2 to 1:12. The neutral hydrolysis method leads directly to monomers for their subsequent use in polyester synthesis. It can be carried out both in stationary (batch) and continuous modes. In this method, various metal catalysts are often used, which positively affect the efficiency of the process, but the addition of these catalysts has a negative impact on the separation and purification of monomers, especially terephthalic acid. The literature provides a number of examples describing these types of solutions. Stanica-Ezeanu and Matei described a method of waste PET hydrolysis in neutral conditions carried out in a seawater environment [15]. The process was carried out at a temperature of 215 °C and a pressure of 4 MPa. Under these conditions, only 85–87% conversion was achieved and the TPA yield was only 76–84%.

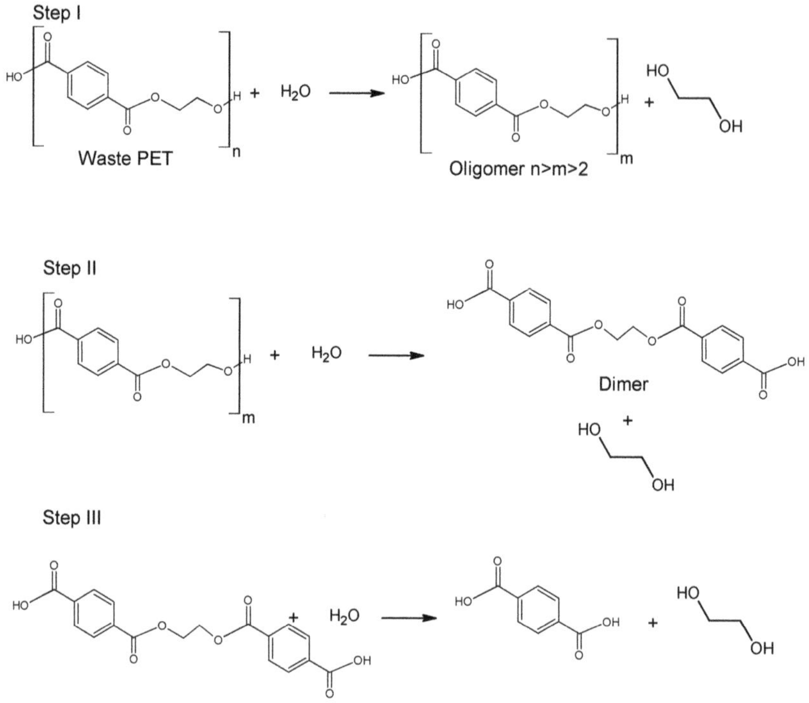

Figure 1. Hydrolysis of poly(ethylene terephthalate).

Under the standard hydrothermal process (250 °C, 39–40 bar, 30 min.), 90–92% PET conversion can be achieved [16,17]. Depolymerization of the colored polymer is slightly less effective. Conversion of PET obtained in this case was only 85% [17]. The process can also be carried out in supercritical conditions (H_2O or CO_2) [17] and also supported by microwave heating [18,19]. Although this method can be considered as effective and more ecological, it requires the use of more drastic conditions of temperature, pressure and reaction time, which can be up to 5 h.

Based on recent literature, methods of PET depolymerization (hydrolysis) carried out in alkaline conditions, most often using NaOH as an alkaline catalyst, are definitely dominant [20–23]. Both the PET conversion and TPA yields obtained were typically >90%. There are also known solutions in which the hydrolysis process is carried out in a mixture containing an additional solvent, such as ethanol [24] or γ-valerolactone used as pre-solvent of waste PET [25]. Specific phase transfer catalysts were also used as catalysts supporting the hydrolysis process [26–28]. They turned out to be very effective in the tested reaction, as evidenced by the very high PET conversion values (99%) obtained. Both classical ammonium salts and highly specific ammonium phosphotungstates were used as the phase transfer catalyst (Table 1).

Table 1. PET hydrolysis in alkaline process conditions.

TPA Yield [%]	Reaction Temperature [°C]	Reaction Time [min]	Pressure [bar]	Catalyst	Solvent	Reference
97	120–200	60	n/a	NaOH	Water	[12]
92	200	25	n/a	NaOH	Water	[20]
95	220	n/a	26	NaOH	Water	[23]
95	80	20	1	NaOH	Water/Ethanol	[24]
99	80	20	1	NaOH	Water	[25]
95	n/a [1]	60	1	NaOH + TBAJ	Water	[26] [1]
99	110	300	n/a	NaOH + [CTA]$_3$PW	Water	[27] [2]
93	145	120	n/a	NaOH + [CTA]$_3$PW	Water	[28] [2]

[1] under microwave irradiation, [2] cetyltrimethylammonium phosphotungstate, n/a—not applicable.

In alkaline hydrolysis, separation of terephthalic acid requires an additional precipitation operation using acid solutions, which generates an additional stream of waste inorganic salts. These problems do not occur in hydrolysis processes carried out under acidic conditions. As catalysts, both strong sulfonic acids, such as H_2SO_4 [28] or *p*-toluenesulphonic acid, were commonly applied [29]. Some heterogeneous catalysts with acidic properties, such as heteropolyacids [28] and "superacid" type SO_4^{2-}/TiO_2 [30] and WO_3/SiO_2 [31] were also used as catalysts in PET hydrolysis process. In the case of sulfuric acid, a PET conversion of 83% was obtained with a TPA yield of about 75%. Similar results were obtained for heteropolyacids; however, superior results were obtained for ptc-type quaternary ammonium phosphotungstates, for which PET conversion was even 100% [28]. Yang et al. described an interesting process of PET hydrolysis catalyzed by easily recyclable terephthalic acid [32]. The advantage of this method is that no compounds are introduced into the process that would require removal from the post-synthesis mixture. PET conversion was close to 100% with a TPA yield of 95.5% (Table 2).

Table 2. PET hydrolysis in acidic process conditions.

TPA Yield [%]	Reaction Temperature [°C]	Reaction Time [min]	Pressure [bar]	Catalyst	Solvent	Reference
75	180	180	n/a	H_2SO_4	Water	[28]
73	180	180	n/a	$H_3PW_{12}O_{40}$	Water	[28]
93	180	180	n/a	[CTA]$_3$PW	Water	[28]
96.2	150	90	n/a	PTSA	Water	[29]
99.2	160	12 h	150	SO_4^{2-}/TiO_2	Water	[30] [1]
99.5	160	15 h	150	WO_3/SiO_2	Water	[31] [1]
95.5	220	180	n/a	TPA	Water	[32]

Note [1] under supercritical CO_2, n/a—not applicable.

2.2. PET Alcoholysis

Alcoholysis of poly(ethylene terephthalate) (Figure 2) is, apart from alkaline hydrolysis, one of the basic methods of chemical depolymerization of waste PET. Alcoholysis involves the degradation of PET in an alcohol environment under conditions of high temperature and pressure. Alcoholysis is considered to be one of the more reliable and effective methods applied for waste PET recycling [33,34]. Alcoholysis processes use a wide group of well-known alcohols, among which the most often used are: methanol, ethanol, butanol and isooctyl alcohol (2-ethylhexanol). Among the described reports in this field, various variants of methanolysis processes definitely dominate, which is related to the key role of dimethyl terephthalate obtained in these processes. As it is known, dimethyl terephthalate is the basic raw material in the synthesis technologies of phthalate polyesters. Alcoholysis using alcohols > C_1 has become less common and esters of terephthalic acid and higher alcohols are often used as raw materials in processes other than the synthesis of polyesters.

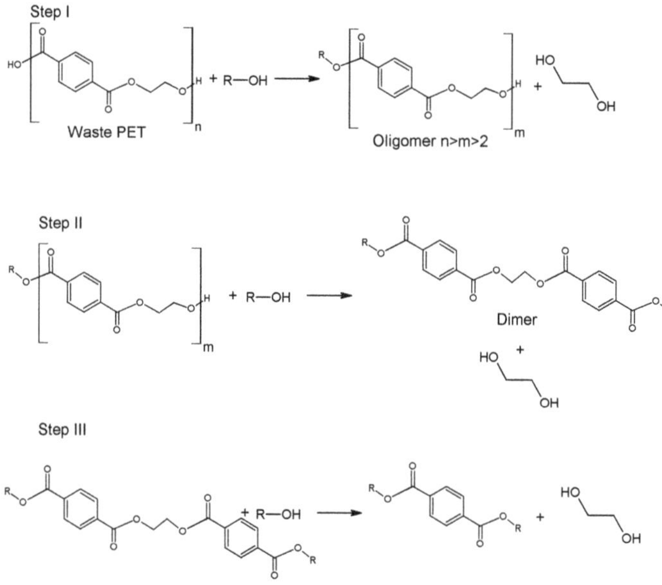

Figure 2. Alcoholysis of poly(ethylene terephthalate).

2.2.1. Methanolysis

Methanolysis process is used to produce dimethylterephthalate (DMT) which then can be applied in poly(ethylene terephthalate). DMT is produced via the reaction of PET with methyl alcohol usually under increased pressure due to the low boiling point of methanol. Reaction is conducted in the presence of various types of catalysts such as Bronsted and Lewis acids, hydroxides, organic bases, oxides and ionic liquids. The process is most often conducted in liquid phase; however, gas-phase methanolysis [35] and processes conducted in supercritical conditions are described in the scientific literature [36]. Separation and repeated usage of catalysts from depolymerization products is an important aspect of designing a viable process that can be implemented as a working chemical technology. Homogenous catalysts can be difficult to remove and to reuse. To remove this obstacle, the use of heterogeneous catalysts in chemical depolymerization of PET is investigated.

One of the catalysts in depolymerization reaction of poly(ethylene terephthalate) is zinc acetate. Its application as a PET depolymerization catalyst was described in numerous scientific publications and is characterized by high activity and its application results in high product yields. Hofmann et al. [37] investigated methanolysis of waste PET using zinc acetate catalyst and waste PET. The process was conducted in the presence of dichloromethane which acted as a solvent. The authors obtained high yields of dimethyl terephthalate which reached 98% after a 20 min time period. However, a large amount of methanol was used in this process. Equivalent methanol-to-PET repeating unit ratio varied from 46.2 to 92.5. Catalyst amount was weight in respect to amount of PET used in the reaction. Interestingly, reaction yield dropped significantly after lowering temperature to 140 °C. In this case, after 20 min reaction yield was under 1%wt and reached 92%wt after 60 min of reaction. Employed catalyst is also susceptible to contaminants in waste PET. When reaction was carried out using dyed bottles, this resulted in lower reaction yield of 38%wt. Investigation of thermodynamics and kinetics of waste PET methanolysis was conducted by Mishra et al. [38] using zinc and lead acetates as catalysts. Influence of PET particle size on reaction kinetics was investigated and reaction was also optimized. Optimal reaction time was determined to be 120 min at temperatures ranging from 130 to 140 °C with PET particle size of 127.5 μm. Zinc acetate was also employed as a catalyst in chemical recycling of mixture of waste PET and polylactic acid (PLA) [37] using methanol as well as a number of other solvents, i.e., ethanol, ethylene glycol, etc. After reaction conducted in 15 h time under boiling point of methanol, there was no discernable effect on PET while all of the PLA was depolymerized. This result is attributed to differences between solubility of tested polymers in methanol. Methanolysis of PET in microwave reactor was investigated in [6] and the process is characterized by a very short reaction time and low amount of catalyst used. Within 10 min of the process with catalyst loading of 0.01 g per 1 g of PET, 88%wt of PET can be converted.

Some inorganic as well as organic catalysts are active at low temperatures in methanolysis of waste poly(ethylene terephthalate). Catalysts like potassium carbonate 1,5,7-triazabicyclo[4.4.0]dec-5-ene (TBD) and potassium methoxide CH_3OK exhibit good catalytic properties in methanolysis reaction conducted in low temperatures [39]. Methanolysis conducted at a temperature of 25 °C over a period of 24 h using K_2CO_3 yielded 93.1%wt dimethyl terephthalate. To obtain such high yields, a large excess of methanol and dichloromethane was used as well as large amounts of catalyst. Implementation of TBD and CH_3OK resulted in lower yields of product 89.3 and 85.5%wt, respectively. Interestingly, K_2CO_3 exhibits highest catalytic activity although it did not completely dissolve in applied solvent. Other catalysts like $KHCO_3$, KOAc, Na_2CO_3, and CaO among others had given significantly lower concentrations of dimethyl terephthalate.

Calcinated sodium silicate (Na_2SiO_3) was used in the methanolysis of PET by Tang et al. [40]. The authors investigated catalyst obtained by calcination under different temperatures as well as influence of catalyst concentration, reaction temperature methanol-to-PET weight ratio and reaction time on reaction yield and PET conversion. Applied silicate exhibits good catalytic properties in relatively small catalyst loadings from 3 to 7%wt reaching up to 63% yield and 74% conversion rate using as much as 5% of the catalyst. The process was tested in temperatures ranging from 160 to 200 °C. Interestingly, the authors used relatively low alcohol-to-PET ratio which ranged from 3 to 7. Process conducted under optimized conditions resulted in obtaining dimethyl terephthalate with 95% yield and 100% conversion. Recycling of catalyst was also investigated. Silicate catalyst was reused four times with some loss in activity attributed to adsorption of water. Magnesium phosphate catalyst obtained in the presence of pectin was used in PET methanolysis [41]. The use of pectin resulted in obtaining catalyst with large BET surface area of 19.51 m^2/g and average pore size of 26.01, which was significantly higher compared to catalyst obtained without the use of pectin. Process was conducted at 180 °C for 150 min with 3%wt of catalyst achieving yield of 74%wt. MgP catalyst is stable when reused as it was reused four times with low loss in PET conversion. However, a methanol-to-PET weight ratio of 200 was used in this study. Such large alcohol excess will have a negative influence on the overall amount of obtained DMT per synthesis.

Heterogeneous catalysts obtained from bio wastes can also be implemented in methanolysis of waste poly(ethylene terephthalate) [42]. The authors used a catalyst obtained by calcination of bamboo leaf at 700 °C in methanolysis of PET waste. The obtained catalyst was composed mainly of SiO_2 and a mixture of various oxides of other metals such as calcium, potassium, iron, manganese, magnesium, etc. Methanolysis reaction using such catalyst allowed for achieving DMT with a yield of 78%wt after two hours in relatively low methanol excess of 7.5 and catalyst loading of 20.8%wt in relation to the mass of PET. Interestingly, increase in catalyst loading, reaction temperature and time resulted in lower yield of DMT. Reusability tests have shown the loss in activity of the catalyst. After four cycles, DMT yield lowered from 78%wt to 67%wt. Nanocatalysts in the form of zinc oxide dispersions are found to be active in methanolysis of poly(ethylene terephthalate) [43]. Depolymerization process conditions were optimized regarding reaction time, methanol-to-PET ratio and catalyst concentration. The tested catalyst exhibits very good activity achieving DMT yield of 97%wt after 15 min at 170 °C and subsequent trials conducted to test the possibility of catalyst reuse have shown a decrease in activity by approximately 20%. Overall, ZnO nanodispersion has proven to be an active catalytic system which allows for obtaining high yields in very short time periods. Heterogeneous hydrotalcite (Mg-Al) has proven to be an effective degradation catalyst when used in conjunction with dimethyl sulfoxide (DMSO) [44]. Degradation process was completed in a 10 min time period obtaining PET oligomer. Obtained product was then reacted with methanol in the presence of sodium hydroxide (NaOH) at 35 °C for 60 min.

Ionic liquids are used as methanolysis catalysts in depolymerization of poly(ethylene terephthalate). Liu et al. [45] tested a series of ionic liquids in methanolysis of various polymers including poly(ethylene terephthalate), polycarbonate, Polyhydroxybutyrate and polylactic acid. [HDBU][Im] and [Bmim]$_2$[CoCl$_4$] were used in PET methanolysis. Reaction was conducted at 170 °C over a period of four hours for [Bmim]$_2$[CoCl$_4$] and 140 °C over a period of three hours for [HDBU][Im]. Reactions yielded 78 and 75%wt, respectively.

Chemical recycling of waste poly(ethylene terephthalate) under supercritical conditions can be employed successfully for PET methanolysis [46]. Process conducted at 298 °C for a duration of 112 min with excess methanol results in DMT yield of 99.79%wt (Table 3).

Table 3. Methanolysis reaction parameters.

Yield [%]	Alcohol/Polymer Ratio	Reaction Temperature [°C]	Reaction Time [min]	Pressure [bar]	Catalyst	Catalyst Amount [%wt.]	Reference
98	46.2	160	20	n/a	Zn(OAc)$_2$	1	[37]
97.76	2.38	130	120	8.5	Zn(OAc)$_2$/Pb(OAc)$_2$	5.1	[38]
0.0	2.16	64.7	900	1	Zn(OAc)$_2$	0.07	[37]
93.1	50	25	1440	1	K$_2$CO$_3$	0.2 [a]	[39]
89.3	50	25	1440	1	TBD	0.2 [a]	[39]
85.5	50	25	1440	1	CH$_3$OK	0.2 [a]	[39]
95	5	200	45	n/a	Na$_2$SiO$_3$	5	[40]
74	200	180	150	n/a	MgP	3	[41]
78	7.5	200	120	n/a	BLA	20.8	[42]
97	6	170	15	n/a	ZnO nanodispersion	3.5	[43]
78	5	170	240	n/a	[Bmim]$_2$[CoCl$_4$]	1.6	[47]
75	5	140	180	n/a	[HDBU][Im]	1.6	[47]
99.79	6	298	112	n/a	-	-	[48]

[a] % of PET degradation; n/a—not applicable.

2.2.2. Glycolysis

Glycolysis is a widely used method for the chemical processing of PET. Most commonly, as the reactive solvent, ethylene glycol (EG) and diethylene glycol (DEG) are used [14]. From a chemical point of view, the glycolysis reaction is the decomposition of PET with glycols, in the presence of transesterification catalysts, where ester linkages break and are replaced with hydroxyl terminals. One of the earliest patents of this process was proposed in 1964 by Ostrysz et al. and concerns the preparation of unsaturated polyester resins (UPR) bearing a structure as presented in Figure 3 [49,50].

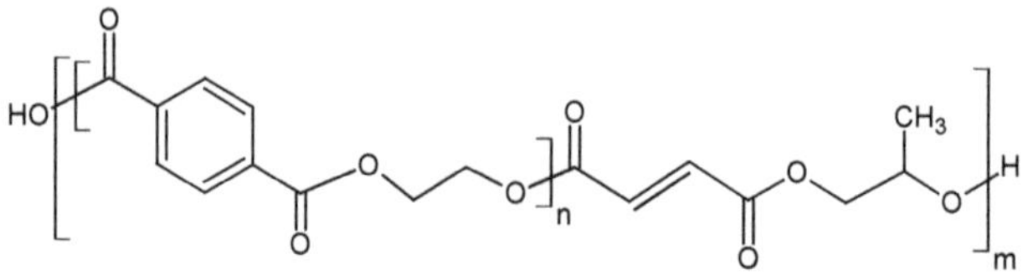

Figure 3. Structure of Thixotropic and Chemoresistant Unsaturated Polyester Resins [3].

Glycolysis proceeds in a several steps as presented in Figure 4. Initially, PET is converted to its dimer form through oligomer molecules at a high rate, and then slow transition takes place. Process is mainly carried out in mild conditions under atmospheric pressure and at a temperature ranging between 180–240 °C. A major advantage of this procedure is the application of less volatile glycols compared to alcohols of similar structure [51,52]. It was also found that specific order of parameters: catalyst concentration > glycolysis temperature > glycolysis time affects the yield of obtained product [53,54]. However, the disadvantage of this process is that the yield is insufficient for industrial application. In order to increase efficiency of the process, catalysts are used [5,55]. The most widely discussed group of catalysts in the recent literature are salts of transition group metals. Zinc (II) Acetate ($Zn(OAc)_2$) had been found to be the most effective catalyst. $Zn(OAc)_2$ was applied for the first time as a catalyst for PET glycolysis in 1989 by Vaidya and Nadkarni; however, other molecules have been topics of interest in recent studies. These are mainly transition metals salts and complexes, ionic liquids and metal oxides [5,56].

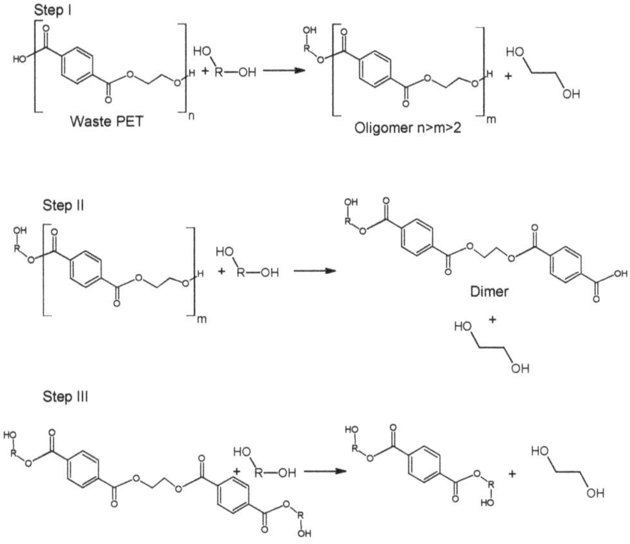

Figure 4. Glycolysis of poly(ethylene terephthalate) [57].

Processed PET in the form of polyhydric alcohols or oligomers with applied glycol on the end-groups have found numerous applications. Depolymerized PET can be used as an additive to poly(vinyl chloride) (PVC) [5]. Another application of oligoesters is in the production of polyurethanes, as polyol groups containing reactant. The depolymerized PET with ethylene glycol can be reused to produce fresh poly(ethylene terephthalate) molecules [57].

Metals salts and complex compounds

$Zn(OAc)_2$ was one of the first catalysts applied in the depolymerization process of poly(ethylene terephthalate). Wang et al. [58] glycolyzed PET fibers with EG at a weight ratio of 1:3 and $Zn(OAc)_2$ as the catalyst. The reaction was carried out at 196 °C under nitrogen atmosphere at standard pressure for 0.5 h for differing catalyst concentrations ranging from 0 to 0.8%wt of PET load. They found that in absence of the catalyst, the reaction almost fails to proceed reaching only 2.79% PET conversion, while with the use of a catalyst at 0.2% concentration, PET conversion exceeded 93%. The highest productivity was obtained at 196 °C with 0.2% of $Zn(OAc)_2$ and PET/EG ratio 1:3 in 2 h, reaching 100% PET conversion and an 81.8% yield of bis(hydroxyethyl) terephthalate (BHET). Song et al. [59] analyzed the process of PET glycolysis with EG applying tropine and tropine-$Zn(OAc)_2$ complex as catalysts. Process was conducted at atmospheric pressure at five temperatures between 150 °C and 190 °C with 10 °C steps. The most effective conditions were established at 170 °C with 5%wt of the catalyst with respect to PET and was achieved over the course of 2 h. The final yields of BHET were then found to be 91% and 182% higher for tropine and tropine-Zn(OAc)$_2$ complex, respectively, than for Zn(OAc)$_2$ reaching only about 15%. Moncada et al. [60] studied progress of PET glycolysis at different temperatures 165 °C, 175 °C and 185 °C, with Zn(OAc)$_2$ catalyst at a concentration of 1%wt and 0.5%wt of initial mass of PET and in presence of EG (2.4 mL along with 1 g of PET). Reaction progress was established by measuring water-insoluble mass fraction of depolymerized PET along 80 min in the oil bath, with 10 min sampling intervals. They found that the highest depolymerization progress is observed after first 30 min at 185 °C reaching about 13% of initial average molecular weight while at 165 °C it was only about 35%. Similar results were found for both catalyst concentrations. PET with decreased molecular weight was then used for block co-polymer building, consisting of PET oligomers and poly(ethylene glycol) [61].

Garcia et al. [62] conducted a series of experiments on metal salts and complexes as catalysts for PET glycolysis with ethylene glycol. Applying different concentration of catalysts and reaction conditions, they attempted to find a substance that would effectively catalyze the PET glycolysis process. The results of their work are presented in Table 4.

Table 4. Results of PET glycolysis [62].

Catalyst	Catalyst Concentration	Reaction Temperature	Reaction Time [h]	PET Conversion [%]	BHET Yield [%]
-	-	-	-	0	0
Ni(NO$_3$)$_2$ × 6H$_2$O	1 [%wt.]	190 °C	3	13	4
FeSO$_4$ × 7H$_2$O	1 [%wt.]	190 °C	3	44	12
CuSO$_4$ × 2H$_2$O	1 [%wt.]	190 °C	3	49	17
CuCl$_2$ × 2H$_2$O	1 [%wt.]	190 °C	3	53	19
Zn(OAc)$_2$ × 2H$_2$O	1 [%wt.]	190 °C	3	55	23
FeCl$_3$ × 6H$_2$O	1 [%wt.]	190 °C	3	57	42
AlCl$_3$	1 [%wt.]	190 °C	3	59	44
ZnCl$_2$	1 [%wt.]	190 °C	3	60	45
MnCl$_2$ × 4H$_2$O	1 [%wt.]	190 °C	3	62	48
NiSO$_4$ × 7H$_2$O	1 [%wt.]	190 °C	3	100	55
NiCl$_2$ × 6H$_2$O	1 [%wt.]	190 °C	3	100	55
ZnSO$_4$ × 7H$_2$O	1 [%wt.]	190 °C	3	100	57
CoCl$_2$ × 6H$_2$O	1 [%wt.]	190 °C	3	100	65
ZrCl$_4$	1 [%wt.]	190 °C	3	100	69
BEt$_3$	1 [%wt.]	190 °C	3	100	56
CoCl$_2$ 1:3 dcype	1.5 [mol % of CoCl$_2$]	190 °C	3	-	10
CoCl$_2$ 1:1 dcype	1.5 [mol % of CoCl$_2$]	190 °C	1	-	58
[Co(dcype)Cl$_2$]	1.5 [mol % of CoCl$_2$]	170 °C	3	-	10
[Co(dcype)Cl$_2$]	1.5 [mol % of CoCl$_2$]	190 °C	1	-	58
[Co(dcype)Cl$_2$]	1.5 [mol % of CoCl$_2$]	190 °C	2	-	75
[Co(dcype)Cl$_2$]	1.5 [mol % of CoCl$_2$]	190 °C	3	-	75
[Co(dcype)Cl$_2$]	3 [mol % of CoCl$_2$]	190 °C	3	-	82
Ni(COD)$_2$	1 [%wt. of Ni(COD)$_2$]	190 °C		100	55

Table 4. *Cont.*

Catalyst	Catalyst Concentration	Reaction Temperature	Reaction Time [h]	PET Conversion [%]	BHET Yield [%]
[Ni(COD)$_2$] 1:4 PPh$_3$	1 [%wt. of Ni(COD)$_2$]	190 °C	3	27	13
[Ni(COD)$_2$] 1:4 P(OiPr)$_3$	1 [%wt. of Ni(COD)$_2$]	190 °C	3	0	0
[Ni(COD)$_2$] 1:2 dcype	1 [%wt. of Ni(COD)$_2$]	190 °C	3	100	59
[Ni(COD)$_2$] 1:2 dppe	1 [%wt. of Ni(COD)$_2$]	190 °C	3	100	67
[Ni(COD)$_2$] 1:2 dppf	1 [%wt. of Ni(COD)$_2$]	190 °C	3	30	15
[(COD)Ni(dppe)]	1 [%wt. of Ni(COD)$_2$]	190 °C	3	-	44
[Ni(COD)$_2$] 1:1 dppe	1 [%wt. of Ni(COD)$_2$]	190 °C	3	-	48
[Ni(COD)$_2$] 1:2 dppe	1 [%wt. of Ni(COD)$_2$]	190 °C	1	-	30
[Ni(COD)$_2$] 1:2 dppe	1 [%wt. of Ni(COD)$_2$]	190 °C	2	-	57
[Ni(COD)$_2$] 1:2 dppe	1 [%wt. of Ni(COD)$_2$]	190 °C	3	-	67
[Ni(COD)$_2$] 1:2 dppe	1 [%wt. of Ni(COD)$_2$]	190 °C	5	-	59
[Ni(COD)$_2$] 1:2 dppe	1 [%wt. of Ni(COD)$_2$]	200 °C	3	-	71

Note: Pressure = 1 [atm]; Glycol:PET ratio = 10 mL:1 g.

As it can be seen, the most efficient catalyst for the PET glycolysis process is a complex compound of cobalt dichloride with 1,2-Bis(dicyclohexylphosphino)ethane ([Co(dcype)Cl$_2$]). They also pointed out a reverse correlation between the amount of ligand added and reaction yield. After increasing the number of dcype, the reaction yield decreased. Furthermore, the efficiency of the process depends on the applied ligands and for catalytic purposes the bidentate ligand should be used, while monodentate inhibits the process of PET glycolysis [62].

Liu et al. [63] chose another approach for reaction yield improvement. They decided to conduct the glycolysis reaction in a homogenous environment. For this purpose, they applied dimethyl sulfoxide (DMSO) as a solvent for PET. They conducted the glycolysis reaction in two variants. First, 5 g of PET powder was mixed with 30 g of EG and heated to temperatures ranging from 160 °C to 190 °C for 5 to 300 min. Second, 5 g of PET powder was mixed with 10 g of EG and 20 g of solvent, then heated in temperature range from 155 °C to 190 °C for 1 to 20 min. For both cases, the catalyst was added when the assumed temperature was reached. Results of their experiments from HPLC analysis are presented in Table 5.

The obtained results directly indicate a high positive correlation between solvent application and BHET yield. With each catalyst, the yield increased by a significant amount. Moreover, the reaction time is extremely small. In most cases, lengthy time periods in the order of hours are required to achieve a BHET yield of about 80%, while the addition of solvent reduces this time to minutes. This has a great impact on energy efficiency of the process. However, the process of applying organic solvents is not environmentally friendly and should be avoided in order to reduce the impact on living organisms. It is for this reason that such a process might be considered economically suitable, but environmentally harmful [63].

Table 5. Results of PET glycolysis with EG and DMSO with 0.25 [g] of a catalyst [63].

Catalyst	Time [min]	Yield of BHET with Solvent [%]	BHET Yield without Solvent [%]
$Zn(OAc)_2 \times 2H_2O$	5	83.88 ± 1.12	42.98 ± 1.76
$Zn(OAc)_2 \times 2H_2O$	1	82.97 ± 2.44	20.11 ± 2.01
$Zn(NO_3)_2 \times 6H_2O$	5	78.64 ± 2.60	25.69 ± 2.37
$Zn(NO_3)_2 \times 2H_2O$	1	53.48 ± 2.46	7.32 ± 2.33
$ZnSO_4 \times 7H_2O$	5	57.21 ± 2.14	1.27 ± 0.29
$Co(OAc)_2 \times 4H_2O$	5	78.69 ± 1.87	24.73 ± 1.88
$Ni(OAc)_2 \times 4H_2O$	5	21.65 ± 1.82	1.22 ± 0.20
$Cu(OAc)_2 \times H_2O$	5	14.68 ± 1.01	1.16 ± 0.32
$Mn(OAc)_2 \times 4H_2O$	5	80.77 ± 1.70	42.15 ± 2.37
$Urea/Zn(OAc)_2$	5	77.90 ± 2.28	48.17 ± 1.51
$K_6SiW_{11}ZnO_{39}(H_2O)$	5	65.94 ± 2.86	38.77 ± 2.62
$[Bmim] Zn(OAc)_3$	5	72.44 ± 2.29	38.40 ± 2.53

Note: Pressure = 1 [atm]; Glycol:PET wt. ratio = 6:1 without solvent; Glycol:PET wt. ratio = 2:1 with solvent; Temperature = 190 °C.

Lei et al. [60] applied tin (II) chloride ($SnCl_2$) as a catalyst in glycolysis process. They conducted studies on increasing glycolysis efficiency by stepwise PET addition process. Process was carried out in diethylene glycol (DEG) at 200 °C, 210 °C, and 220 °C. Weight ratios of PET-to-DEG were 1:2, 1:1.5, and 1:1 accompanied with tin (II) chloride ($SnCl_2$) 0.3%wt of loaded PET for up to 150 min. In the one-step process, all reactants were mixed together and heated to the desired temperature. For the stepwise process, DEG was mixed with $SnCl_2$ and heated, then the PET had been added in small portions with regular intervals. Time measuring was started when the last portion was added. The reaction yield (WSM%) was established as an amount of water soluble monomer (WSM) in ratio with a theoretical amount of water soluble monomer. The highest efficiency was obtained at 220 °C after 90 min of the reaction. Presented results exhibit high improvement on the reaction yield for stepwise process, reaching over twice as much WSM% as the one-step addition. Moreover, higher WSM% was obtained for PET waste bottles and purchased particles than for PET fibers. Authors explained that phenomenon with higher crystallinity of PET fibers, that impedes the penetration of the degrading agent. They also conclude that the reaction takes place mostly at the solid/liquid interface of PET and DEG, though stepwise PET inserting allows for more rapid swelling and dissolving of solids. This has a significant influence on the reaction efficiency and explains the difference between single and multiple step procedures.

Metal oxides

In a study by Son et al. [64] on PET glycolysis, the authors applied exfoliated nanosheets of manganese dioxide (e-MON) as an improvement of the process conducted with typically used MnO_2 forms. e-MON was obtained from δ–MnO_2 in a fluid dynamic reactor in Taylor–Couette flow, and then separated with a centrifuge. With this method, they obtained MnO_2 sheets with a few nanometers thickness and relatively large surface area, what has a crucial influence on the catalyst efficiency. Glycolysis was carried out with an EG in mass ratios EG/PET in range from 6.7 up to 55.5. Process temperatures were controlled from 150 °C to 200 °C. An amount of the e-MON was kept constant 0.01%wt of PET. Comparison experiments with bulk MnO_2 were also conducted. The results clearly show that e-MON is a much more efficient catalyst than bulk MnO_2, reaching 100% yield of BHET synthesis in 30 min at 200 °C, while reaction with bulk MnO_2 reached only about 80%. The highest BHET yields were obtained for EG/PET mass ratio starting from 18.5 and higher reaching 100%. After the process, e-MON was recovered by washing with deionized water and filtration. Recovered e-MON was then used four-more times without losing its catalytic properties still reaching 100% BHET yield synthesis.

Wang et al. [65] used as a PET glycolysis catalysts $CoFe_2O_4$ magnetic nanoparticles (MNP) modified with ionic liquid as a surfactant. Modified NMP's were synthesized by precipitation from aqueous solution of iron and cobalt salts with addition of certain ionic liquid. Modified NMP's were crushed out of the solution through the addition of acetone and separated using magnets. Size distribution of modified MNP's depends on the applied ionic liquid ranging from 4 nm to below 20 nm, while not "coated" MNP has a size distribution range from 20 nm up to 35 nm. For the glycolysis reaction, they used 5 g of PET and 0.1 g of the catalyst. The most efficient amount of EG was distinguished as 25 g within the range 10–30 g, while the reaction time with the highest BHET synthesis yield was found as 2.5 h within the 1–3.5 h range. The results of their work are presented in Table 6.

Table 6. Results of the PET glycolysis in the most efficient parameters [65].

Catalyst	Temperature [°C]	PET Conversion [%]	BHET Yield [%]
$CoFe_2O_4$	195	91.2	72.6
$CoFe_2O_4$/[C_6COOHbim]Br	195	95.3	79.7
$CoFe_2O_4$/[C_{10}COOHbim]Br	195	98.4	88.03
$CoFe_2O_4$/[C_{10}COOHbim]NTf_2	195	99.6	86.6
$CoFe_2O_4$/[C_{10}COOHbim]OAc	195	100	95.4
[C_6COOHbim]Br	205	19.8	8.5
[C_{10}COOHbim]Br	205	22.9	9.5
[C_{10}COOHbim]NTf_2	205	20.4	7.5
[C_{10}COOHbim]OAc	205	60.5	32.1

Note: $CoFe_2O_4$—MNP; bim—butyl imidazole; NTf_2—bis[(trifluoromethyl) sulfonyl] imide; OAc—CH_3COO^-; C_6COOH—hexanoic acid; C_{10}COOH—decanoic acid; Pressure = 1 [atm]; Glycol:PET wt. ratio = 5:1; Duration 2.5 h.

The results in Table 6 clearly indicate that from the tested catalysts, the most efficient is MNP coated with ionic liquid with a formula [C_{10}COOHbim]OAc reaching a BHET synthesis yield of 95.4%. A huge advantage of using modified MNP's is the possibility to easily separate them from the post-reaction mixture by applying a magnetic field. Separating this way, $CoFe_2O_4$/[C_{10}COOHbim]OAc was used over six glycolysis cycles without impacting PET conversion and BHET yield. From seventh to tenth attempts, BHET yields slightly decreased to about 95% [63].

Guo et al. [66] introduced micro iron oxide particles coated with magnesium aluminum oxide (Mg-Al-O@Fe_3O_4) to the PET glycolysis reaction. Mg-Al-O@Fe_3O_4 was prepared from two solutions. Solution A contains sodium hydroxide and sodium carbonate and solution B consists of iron oxide suspension, magnesium nitrate and aluminum nitrate. Solution A was dropwise added into the solution B until pH of the mixture reached 10, then it was left for aging. After 24 h, the mixture was freeze dried and annealed at 1000 °C. Glycolysis reaction was conducted using 5 g of PET along with 25 g of EG. Analyzed amounts of the catalyst were 0.1, 0.25, 0.5, 1.0 and 2.0%wt of PET. Reaction was carried out at different temperatures 200 °C, 220 °C, 240 °C and 250 °C. The reaction time ranged from 20 to 150 min.

They found that the most efficient glycolysis parameters are 240 °C for 90 min with the catalyst concentration 0.5%wt of PET. Under such conditions, BHET synthesis yield reaches 82%mol. The great advantage of such catalyst is its easy recovery through application of an external magnetic field. The catalyst can be used in a cycle. For the second reaction, BHET yield dropped only slightly; however, the third run reached only about a half of the initial yield. For the fourth cycle, the catalyst becomes fully deactivated obtaining less than 10%mol of BHET yield. Fortunately, catalyst regeneration by annealing at 1000 °C recovers most of the catalyst efficiency reaching nearly the same yield as in the second run.

Fang et al. [67] studied the catalytic properties of sandwich-structure polyoxometalates (POMs) for PET glycolysis. They analyzed molecules with the formula $Na_{12}[WZnM_2(H_2O)_2(ZnW_9O_{34})_2]$ where M = Zn^{2+}, Mn^{2+}, Co^{2+}, Cu^{2+} and Ni^{2+}. For the reaction, 5 g of PET was mixed with 20 g of EG in a reactor. PET-to-catalyst molar ratio was 0.018%mol, the temperature was kept at 190 °C and the time duration ranged from 40 min to 110 min. Experimental results are given in Table 7.

Table 7. Results of PET glycolysis reactions with POMs catalysts [67].

Catalyst	Reaction Time [min]	PTE Conversion [%]	BHET Yield [%]
$Na_{12}[WZn_3(H_2O)_2(ZnW_9O_{34})_2]$	40	100.00	84.48
$Na_{12}[WZnCo_2(H_2O)_2(ZnW_9O_{34})_2]$	50	100.00	83.95
$Na_{12}[WZnCo_2(H_2O)_2(ZnW_9O_{34})_2]$	40	99.83	84.61
$Na_{12}[WZnNi_2(H_2O)_2(ZnW_9O_{34})_2]$	60	100.00	83.96
$Na_{12}[WZnNi_2(H_2O)_2(ZnW_9O_{34})_2]$	40	98.13	77.48
$Na_{12}[WZnCu_2(H_2O)_2(ZnW_9O_{34})_2]$	65	100.00	83.61
$Na_{12}[WZnCu_2(H_2O)_2(ZnW_9O_{34})_2]$	40	99.18	78.42
$Na_{12}[WZnMn_2(H_2O)_2(ZnW_9O_{34})_2]$	44	100.00	84.11
$Na_{12}[WZnMn_2(H_2O)_2(ZnW_9O_{34})_2]$	40	99.90	82.88
$Na_{12}[WZn_3(H_2O)_2(CoW_9O_{34})_2]$	110	100.00	82.42
$Na_{12}[WZn_3(H_2O)_2(CoW_9O_{34})_2]$	40	66.90	69.88

Note: Pressure = 1 [atm]; Glycol:PET wt. ratio = 4:1; Temperature = 190 °C.

They found that the most efficient POM was $Na_{12}[WZnCo_2(H_2O)_2(ZnW_9O_{34})_2]$ reaching a BHET yield of 84.61% in only 40 min. The authors also state that POMs are relatively cheap and with low toxicity catalysts. They also recover the remaining catalyst from the post-reaction mixture and use it in a cycle. Within four cycles, the catalyst kept nearly the same BHET yield confirming its stability as declared by the authors.

Putisompon et al. [68] in their research found that calcium oxide (CaO) has applications as a catalyst for transesterification in biodiesel production and decided to test its potential to catalyze PET glycolysis process. For this purpose, as a source of CaO ostrich eggshells, chicken eggshells, mussel shells, geloina shells, and oyster shells from Thailand local flesh market were chosen and $Zn(OAc)_2$ as a reference catalyst. Shells were washed, ground, and calcined in variety of temperatures from 600 °C to 1000 °C with 100 °C step. Calcined shells were then ground to a fine powder. To the reactor, 5 g of PET was loaded along with 1%wt of a catalyst. Weight ratio of PET to EG 1:15 was applied. The reaction was carried out at 192 °C for 2 h. Then, obtained BHET was extracted with water, filtered out, dried, and weight to establish the final yield as a percent of BHET.

The most efficient calcination temperature of 1000 °C was established, reaching 72 and 76% of BHET for a chicken eggshell and an ostrich eggshell, respectively. These results increased from barely several percent at a calcination temperature equal to 600 °C. When the decomposition of calcium carbonate to CaO starts at 700 °C, temperature increase also reduces organic impurities present in the shell structures and increase crystallinity of obtained CaO [65].

Zangana et al. [69] covered the topic of microwave irradiation support for the catalytic glycolysis of PET. They conducted the process using different molar ratios of PET:EG such as 1:4, 1:6, 1:10 and 1:20 with reaction times up to 5 min. As a catalyst, CaO was applied in concentrations 3%wt, 5%wt and 10%wt of the reaction mixture (PET and EG). Reaction was carried out at EG boiling temperature 197 °C, supported by microwave irradiation with power 800 W. For optimized conditions, supported CaO on activated carbon (CaO/AC) as a catalyst was applied. For the CaO, the most optimal reaction conditions are 10%wt of the catalyst, PET:EG ratio 1:10 and 4 min of microwave irradiation reaching about 75% BHET yield, while for CaO/AC 15%wt of catalyst along 3.5 min of microwave irradiation with the same PET:EG ratio similar yield was obtained. They also analyzed catalyst and EG recovery as well as their re-use over 3 cycles. For each run, CaO concentration was kept constant; however, the reaction time was increased for each cycle reaching 16 min on third run. Catalyst efficiency decreased on each run, but what is interesting is that CaO/AC shows much smaller differences than CaO. Recovered CaO and EG behaves at the second run as recovered CaO [66]. It is worth mentioning that Zangana et al. [69] and

Putisompon et al. [68] obtained similar BHET yields; however, Zananga et al. [69] reached the final concentration about 30-times faster due to microwave support.

For the glycolysis process, Lalhmangaihzuala et al. [70] applied orange peels ash (OPA) as the catalyst. Orange peels were initially washed with distilled water and dried at atmospheric conditions or in an oven. In the next step, dry peels were burned. Glycolysis reaction was conducted with 480 mg of PET with certain amount of EG and OPA. Reactor was placed in an oil bath and heated to 190 °C until complete disappearance of PET. Reaction parameters and results are presented in Table 8.

Table 8. Results of PET glycolysis reactions with OPA catalyst [70].

Catalyst Amount [mg]	EG Amount [g]	Time [h]	PET Conversion [%]	BHET Yield [%]
20	2	1	100	63
30	2	1	100	71
50	2	1	100	75
70	2	1	100	69
100	2	1	100	71
50	1	1	100	61
50	1.5	1	100	73
50	2.5	1	100	78
50	3	1	100	76
50	2.5	0.5	80	54
50	2.5	1.5	100	79
50	2.5	2	100	76
50	2.5	2.5	100	71

Note: Pressure = 1 [atm]; PET amount = 480 mg; Temperature = 190 °C.

As it can be seen, the most efficient parameters of PET (480 mg) glycolysis with OPA at 190 °C were a reaction time set as 1.5 h, 2.5 g mass of EG and 50 mg of OPA, which produces 521%wt (of PET) EG and 10.42%wt (of PET) OPA [67]. In conclusion, it is possible to apply waste orange peels as an efficient catalyst for the glycolysis reaction of a waste PET.

Ionic liquids

Shuangjun et al. [71] investigated Lewis acidic ionic liquids (LAIL) as a possible active catalytic system for the glycolysis process of the post consumption PET. For this purpose, LAIL's consist of 1-hexyl-3-methylimidazolium (Hmim) and metal halides (MH) (zinc, cobalt, iron and copper chlorides) were prepared in different molar ratios of Hmim and MH, and mixtures of two LAIL's were also analyzed. To conduct glycolysis reaction, 2 g of PET was loaded into the reactor along with 22 g of EG and 0.1%wt (of PET) of the catalyst. The reaction was carried out for 2 h at 190 °C. Results of these processes are presented in Table 9.

Among all the catalytic compositions, the most efficient composition reached 87.1 BHET yield along with 100% PET conversion, which was $[Hmim]ZnCl_3:[Hmim]CoCl_3$ at a 1:1 molar ratio. What is also interesting is that pure components [Hmim]Cl and MH exhibit lower BHET yields than when mixed forming LAIL's. Moreover, systems with pure $[Hmim]CuCl_2$ and mixture of $[Hmim]ZnCl_2$ and $[Hmim]CuCl_2$ in molar ratio 1:3, respectively have very low PET conversion along with no BHET detected. Basing on their results, authors purposed a decreasing order of metal anions activity: $[ZnCl_3]^- > [CoCl_3]^- > [FeCl_4]^- > [CuCl_3]^-$ [71].

Cano et al. [72] applied MNP consists of the core made of Fe_3O_4 coated with SiO_2 and ionic liquid (IL) as a catalyst. General procedure for PET glycolysis was done by mixing 100 mg of PET and 1 mL of EG in the reactor with 15 mg of MNP. The reaction was preceded for 24 h at 160 °C or 180 °C. They also recovered MNP using an external magnetic field, then applied the same catalyst portion in over a dozen times. At 160 °C reaction yields of BHET were rather low reaching less than 60% with a maximal PET conversion about 64% dropping into poor yield about 20% after fifteenth cycle. In contrast, reaction conducted at 180 °C reached 100% yield after first cycle, while for the next 10 repetitions yield was kept above 90%. In the last twelfth step yield dropped to the still significant yield 84%. All twelve cycles carried out at 180 °C exhibit 100% PET conversion.

Table 9. Results of PET glycolysis reactions with LAIL's catalysts [71].

Molar Ratios [Hmim]Cl:ZnCl$_2$:CoCl$_2$:FeCl$_3$:CuCl$_2$	PET Conversion [%]	BHET Yield [%]
1:1:0:0:0	100	76.8
1:0:1:0:0	85.8	38.1
1:0.25:0.75:0:0	100	75.2
1:0.5:0.5:0:0	100	87.1
1:0.75:0.25:0:0	100	79.6
1:0:0:1:0	99.5	51.5
1:0.25:0:0.75:0	100	60.8
1:0.5:0:0.5:0	98	55.1
1:0.75:0:0.25:0	99.8	37.5
1:0:0:0:1	4.8	–
1:0.25:0:0:0.75	14.5	–
1:0.5:0:0:0.5	99	73.0
1:0.75:0:0:0.25	100	76.8

Note: [Hmim]Cl:ZnCl$_2$—1:1 = [Hmim]ZnCl$_3$; [Hmim]Cl:ZnCl$_2$:CoCl$_2$—1:0.5:0.5 = [Hmim]ZnCl$_3$:[Hmim]CoCl$_3$—1:1; Pressure = 1 [atm]; Glycol:PET wt. ratio = 11:1; Duration = 2 h; Temperature = 190 °C.

Najafi-Shoa et al. [73] in their studies on PET glycolysis applied IL grafted on graphene. For that purpose 1-(triethoxysilyl) propyl-3-methylimidazolium chloride ([TESPMI]Cl) is attached to the graphene oxide (GO), which is then reduced with hydrazine forming rGO/[TESPMI]Cl. The final catalyst is prepared by dispersion of rGO/[TESPMI]Cl at CoCl$_2$ solution, what leads to formation of rGO/[TESPMI]$_2$CoCl$_4$. They have done the glycolysis reaction by mixing 1 g of PET with 8 to 14 g of EG along with 0.05 up to 0.25 g of catalyst. The reaction was carried out from 1 to 4 h at the temperature in range from 150 °C up to 190 °C. What they found is the most efficient conditions are for the process with 14 g of EG proceed for 3 h at 190 °C with 0.15 g of a catalyst. That allows to reach 95.22% BHET yield with 100% PET conversion. Authors also verified the reusability of the catalyst. What they obtained, is that after fifth cycle BHET yield drops to about 90%, while the PET conversion remains unchanged at 100% level.

Others catalytic systems

Hong Le et al. [74] noticed that most of the glycolysis processes require high energy consumption, which is mostly allocated to heating of the reaction system. They have decided to apply anisole as co-solvent for reduction of an energy requirements of the process. Glycolysis was carried out for 10 g of PET with 38.75 g of EG and 22.51 g of anisole. Reaction had been performed for two hours. The variety of catalyst had been tested, such as Zn(OAc)$_2$, NaOAc, KOAc, Na$_2$CO$_3$, K$_2$CO$_3$, NaHCO$_3$, KHCO$_3$, MgCO$_3$, potassium methoxide (CH$_3$OK), 1,1-DMU, 1,3-DMU and 1,5,7-triazabicyclo [4.4.0]dec-5-ene (TBD). Comparing results for the reaction done at 197 °C without co-solvent and at 153 °C with anisole with molar ratios EG:anisole:catalyst:PET equal to 12:4:0.04:1 KOAc was depicted as the most efficient catalyst reaching (at 153 °C with anisole) about 87% BHET yield with 100% PET conversion, while reaction at 197 °C without co-solvent with KOAc reached about 84% BHET yield with the same PET conversion. Nevertheless, reaction carried out at 153 °C with the molar ratios of EG:anisole:catalyst:PET equal to 10:3:0.02:1 only TBD got BHET yield above 80% and 100% PET conversion. Other co-solvents were also tested, however like in the previous sentence, only for anisole 80% BHET yield boundary was crossed reaching also 100% PET conversion.

Veregue et al. [75] in their research work on PET glycolysis involved cobalt nanoparticles (CoNP) too check its activity as a catalytic system. For that purpose CoNP were synthesized through the reduction reaction of cobalt (II) chloride with sodium borohydride. The depolymerization reaction was achieved for 4 g of PET with 100 mL of EG and 60 mg of

CoNP. The reaction was carried out at 180 °C for 2, 3 or 4 h. They have reached maximum BHET yield with the time of the process equal to 3 h reaching 77%. Catalyst was then recovered from the postreaction mixture and used four more times to check its activity. They found that CoNP retained its catalytic properties for all five cycles.

In contrary to the most of research done on the PET glycolysis, Wang et al. [76] have decided to check organocatalyst cyanamide efficiency for this process and find optimal reaction conditions. In their procedure, they mixed 2 g of PET with 20 g of EG. They have analyzed reaction yield in terms of temperature in a range between 160 °C to 200 °C, time in a range 0.5 h up to 3 h and catalyst weight percent with respect to PET in a range from 2.5% to 15%. They obtained the most efficient reaction conditions at the temperature 190 °C in 2.5 h with 5%wt of cyanamide. With these parameters they have reached 100% PET conversion along with about 95% BHET yield. What remains interesting, further increase of mentioned parameters was inversely related with BHET yield, which started to decrease, however PET conversion was kept at 100% level.

Wang et al. [77] checked catalytic activity of graphite carbon nitride colloid (GCNC) in the PET glycolysis process. Moreover, they attempt to establish the most efficient parameters for this reaction. For that purpose they have mixed 2 g of PET with 20 g of EG. Then following reaction conditions were investigated: temperature in a range from 160 °C to 196 °C; time in a range from 5 to 120 min; GCNC mass in a range from 0.01 g to 0.15 g. They found optimal parameters for PET glycolysis with GCNC as 196 °C for 0.5 h along with 0.5 g of the catalyst. In such conditions 80.3% of BHET yield had been co-established with 100% PET conversion.

2.2.3. Ethanolysis

Ethanolysis of poly(ethylene terephthalate) is widely regarded as an environmentally friendly alternative to methanolysis, which has been well known for many years. This is mainly due to the possibility of replacing harmful methanol with safer ethanol [78]. Diethyl terephthalate formed in the depolymerization (hydrolysis) reaction can be reused for the synthesis of poly(ethylene terephthalate) (Figure 5).

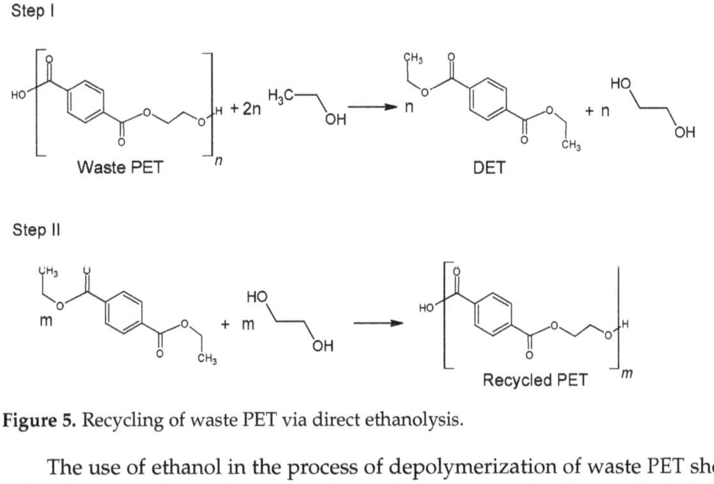

Figure 5. Recycling of waste PET via direct ethanolysis.

The use of ethanol in the process of depolymerization of waste PET should be considered in a much broader aspect than only as a process leading to diethyl terephthalate. Speaking about the process of ethanolysis of poly(ethylene terephthalate), one should also take into account the process whose final product is terephthalic acid and it is a process similar to the process of alkaline PET hydrolysis described earlier. Three reactions take place under alkaline catalysis (NaOH) in a water-ethanol medium: depolymerization of PET to a mixture of diethyl terephthalate (Step I), its hydrolysis to TPA disodium salt (Step II) and precipitation of free TPA (Step III, Figure 6). Ethanol is recycled and can be reused (Closed-loop recycling process), making it very competitive with conventional alkaline hydrolysis processes [24,79,80].

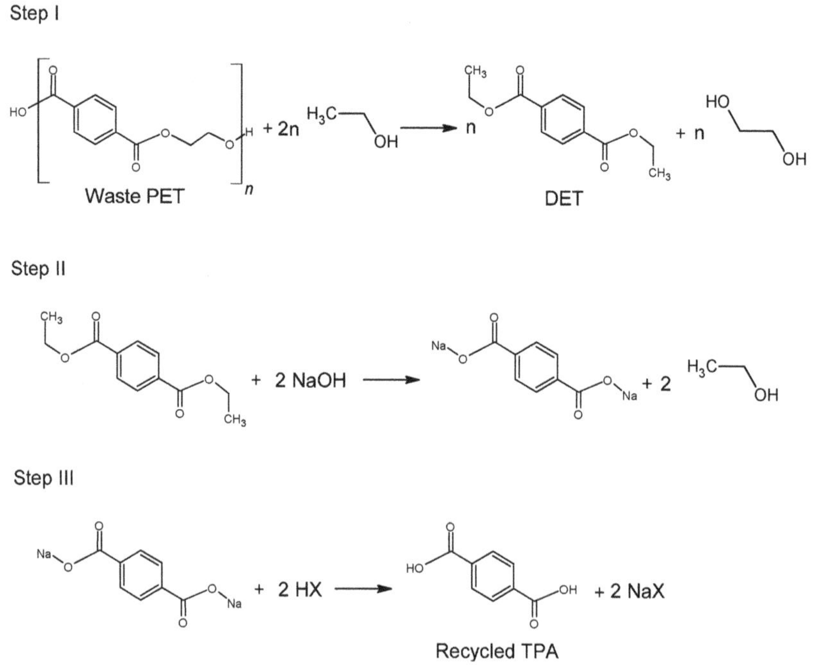

Step I

Step II

Step III

Figure 6. Three-step recycling process of waste PET into terephthalic acid.

The process can be carried out under relatively mild conditions (reaction temperature 50–80 °C, the proportion of ethanol to water 20–100 vol%, the amount of NaOH 5–15%wt) [81]. Under optimal conditions, TPA efficiency reaches 95%. In the method presented in Figure 6, diethyl terephthalate formed in Step I is not separated. Under process conditions, it hydrolyzes to terephthalic acid.

Diethyl terephthalate is obtained by direct ethanolysis of PET (Figure 5). The direct PET ethanolysis process requires temperatures above 200 °C and sometimes the addition of a catalyst. Li et al. [82] described a method of direct PET ethanolysis carried out under pressure and at a temperature of 180–200 °C without the addition of a catalyst. The author reported that under the assumed reaction conditions, nearly 100% PET conversion can be obtained and the DET yield reached 97%. Zinc acetate is an effective catalyst for PET ethanolysis, which allows DET yields of 96–97% to be achieved [83,84]. From a technological point of view, the method is very convenient. Due to differences in solubility, the ethylene glycol formed in the reaction is the lower, immiscible phase, which can be separated from the upper phase containing diethyl terephthalate. Compounds containing acidic sulfonic groups, such as, sulfonic ionic liquids, can also act as catalysts for the PET ethanolysis reaction [85]. By carrying out the process for 14 h at 80 °C in the presence of sulfobutylammonium ionic liquid, diethyl terephthalate can be obtained with a yield of 96%. However, using the techniques of depolymerization of waste PET leading to the production of useful phthalate monomers (terephthalic acid esters), solvolysis processes carried out in supercritical conditions prevail. The basics of the process were presented in the last century in Japan, where a method of PET depolymerization using supercritical water and methanol was developed [78]. Supercritical solvents are very attractive media for conducting many chemical processes mainly because the solvent and transport properties of a single solution can be appreciably and continuously varied with relatively minor changes in either temperature or pressure. Variation in the supercritical fluid density also influences the chemical potential of solutes, reaction rate and equilibrium constant [86]. Depolymerization of waste PET in supercritical methanol is the subject of extensive research [87]. In recent years, however, there have been increasing reports on solvolysis of waste polyester polymers in supercritical ethanol [78,88]. As described by Castro et al. [78], the ethanolysis process was carried out in a supercritical ethanol environment at a temperature of 255 °C, a pressure of 115 or 165 bar and reaction times between 5.0 and 6.5 h without the addition of a catalyst. Under these process conditions, PET was practically completely depolymerized and the main product of the reaction, apart from ethylene glycol and ethanol, was diethyl terephthalate. Later studies included the addition of catalysts, such as metal oxides [89,90] and ionic liquids [91,92]. As the research has shown, the addition of catalysts to the PET ethanolysis process does not have a major impact on the conversion rates, but conclusively shortens the reaction time. Later studies showed that increasing the temperature to 275–350 °C removed the need for the addition of catalysts while maintaining high PET conversion rates and DET efficiency [93,94]. Recent studies have also shown the high efficiency of the PET ethanolysis process in supercritical conditions and the possibility of using this method on a technical scale (Table 10).

Table 10. PET alcoholysis with ethanol.

DET Yield [%]	Reaction Temperature [°C]	Reaction Time [min]	Pressure [bar]	Catalyst	Solvent	Reference
97.3	200	210	n/a	Zn(OAc)$_2$	Ethanol	[82]
92	220	120	n/a	Zn(OAc)$_2$	Ethanol	[83]
97	180	60	n/a	Zn(OAc)$_2$ + Al(OH)$_3$	Ethanol	[84]
95.8	80	20	1	Sulphonic IL	Ethanol	[85]
98.5	240	300	115–165 [1]	-	Ethanol	[94]
>99	255	90	115 [1]	Co$_3$O$_4$ or NiO	Ethanol	[89]
92.2	270	60	n/a [1]	ZnO/Al$_2$O$_3$	Ethanol	[90]
98	255	45	115	[bmim]BF$_4$	Ethanol	[91,92]
95	275–300	90	n/a [1]	-	Ethanol	[93]
98	310	60	n/a [1]	-	Ethanol	[94]

[1] under supercritical conditions, n/a—not applicable.

2.2.4. Alcoholysis with Alcohols > C$_2$

Diethyl terephthalate is used as a safer alternative to dimethyl terephthalate in the synthesis of polyesters. Terephthalates of higher alcohols are no longer used in this specific application. However, they are equally valuable due to their plasticizing properties. This also applies to dibutyl terephthalate, which is produced on an industrial scale. Dibutyl terephthalate has fast melting and low migration properties and provides greater flexibility of finished products. These properties mean that it is used, for example, in the production of flexible floor coverings based on PVC, adhesives and sealants, and in the production of printing inks [95,96].

Dibutyl terephthalate is most often produced by esterification of terephthalic acid, but the use of waste for its synthesis seems to be a more ecological alternative. Depolymerization of waste PET by butanol alcoholysis requires a completely different process than in the case of alcohols such as methanol or ethanol. It also requires the addition of a catalyst. According to the description given in the patent CN102603532, the PET depolymerization process in the presence of 1-butanol and sulfonic ionic liquid allows dibutyl terephthalate to be obtained with a yield of 93.9%. According to the description, the alcoholysis process was carried out at 190 °C for 9 h [85]. In turn, the patent description WO2022112715 provides a general method for the synthesis of terephthalic acid esters, including dibutyl ester [97]. In this method, waste PET is reacted with excess butanol at a temperature between 50–70 °C for 1.5–3 h in the presence of a catalyst being a suitably selected mixture of cyclic guanidine (TBD) or amidine (DBU) derivatives and sodium methoxide. In this method, dibutyl terephthalate can be obtained with a yield between 80–85%. The original method of depolymerization of waste polyesters, including PET, was described in [48]. The essence of the method is the use of specific catalysts based on DBU and various imidazole derivatives (Figure 7).

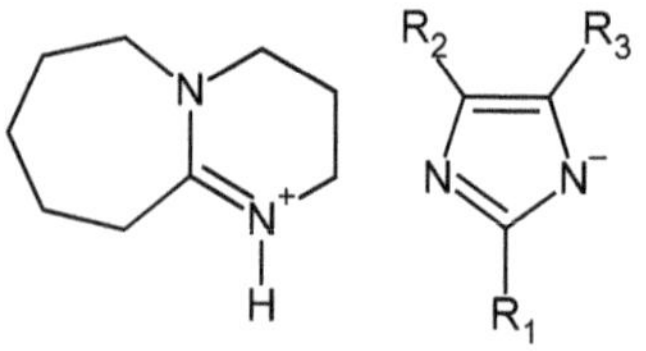

Figure 7. DBU ionic liquid as catalyst for alcoholysis of waste polyesters (PC, PLA, PHB, and PET) [48].

The alcoholysis process was carried out in relatively mild conditions (70 °C) for 2 h. In the case of butanol, a PET conversion of 91% was obtained with a DBT yield of 73%, but for alcohols C5–C6 the yield of the corresponding terephthalic acid ester decreases.

For higher alcohols, clearly higher rates of yield of the appropriate diesters of terephthalic acid can only be obtained in processes carried out in more conventional parameters, i.e., higher temperature and pressure. According to the patent description CN102234227, the PET depolymerization process in the presence of higher alcohols can be carried out at a temperature of 180–200 °C for 4–9 h [98]. Zinc acetate is an effective catalyst for the reaction, but the described method does not provide detailed information on the conversion of PET and the yield of the corresponding terephthalic acid esters. Two other patents CN105503605 and CN106986326 describe the method of PET depolymerization with monohydric alcohols C4-C8 catalyzed by tetrabutyl titanate [99,100]. In addition, this method requires a slightly higher temperature range between 200–230 °C. However, the authors of both solutions also do not provide PET conversion rates and yields of the corresponding esters (Table 11).

Table 11. PET alcoholysis with alcohols C3–C7.

DET Yield [%]	Alcohol	Reaction Temperature [°C]	Reaction Time [min]	Catalyst	Reference
92.3	1-Butanol	190	300	Sulphonic IL	[85]
95.3	1-Hexanol	190	600	Sulphonic IL	[85]
80–85	1-Butanol	50–70	90–180	TBD(DBU) + CH_3ONa	[97]
78	1-Propanol	70	120	[HDBU][Im]	[48]
73	1-Butanol	70	120	[HDBU][Im]	[48]
n/a	C4–C8	180–200	240–540	$Zn(OAc)_2$	[98]
n/a	C4–C8	200–230	120–600	TBT	[99]
n/a	C4–C8	200–230	120–600	TBT	[100]

2.2.5. Alcoholysis with C8 Alcohols

Depolymerization of poly(ethylene terephthalate) can by conducted using higher alcohols, i.e., 2-ethylhexanole, isononyl and isodecyl alcohol. Products obtained in this manner are valuable chemicals that can be used as PVC plasticizers. This process can be catalyzed by organometallic catalysts, ionic liquids and superbases. One of the earliest works on alcoholysis of PET was presented by Gupta et al. [101]. Authors investigated the application of organotin catalyst in depolymerization of waste PET from different sources. The authors conducted depolymerization of waste PET from beverage and food bottles, packaging film, fabrics, car parts, photographic and X-ray films. Additionally, the catalyst was tested in depolymerization reaction of crystalized PET and glycol modified

PET as well as PBT. Obtained DOTP samples were tested with PVC to determine their plasticizing properties. It was concluded that all obtained samples exhibit similar properties to commercial DOTP obtained from terephthalic acid. The most important difference was the difference in color of the product which is heavily influenced by the color of used substrate. Hardness, brittle point, tensile strength and elongation at break were at similar values regardless of the origin of applied plasticizer.

Ionic liquids

Ionic liquids are used as catalysts in alcoholysis of waste PET. A series of (3–sulfonic acid) propyltriethylammonium chloroironinate [$HO_3S–(CH_2)_3–NEt_3$]Cl–$FeCl_3$ ionic liquids obtained with different molar fractions of $FeCl_3$ as well as zinc and copper chloride derivatives were tested in depolymerization of waste polyethylene terephthalate with 2-ethylhexanol and their activity was compared to common transesterification catalysts like zinc acetate, zinc chloride and tetrabuthyl titaniate [102]. Reaction was carried out at temperature of 210 °C for the duration of 8 h using 2-ethylhexanol excess of 3.39. Analysis of postreaction mixture has shown that the highest yield of dioctyl terephthalate was obtained while using ionic liquids containing 0.67 and 0.75 molar fraction of $FeCl_3$ and 0.67 molar fraction of $ZnCl_2$. [$HO_3S–(CH_2)_3–NEt_3$]Cl–$FeCl_3$ ionic liquid was reused seven times with only a small decrease in dioctyl terephthalate yield. [$HO_3S–(CH_2)_3–NEt_3$]Cl–$ZnCl_2$ achieved also very good yields of dioctyl terephthalate (95.7%wt) while [$HO_3S–(CH_2)_3–NEt_3$]Cl–$CuCl_2$ yielded only 37.3%wt of DOTP. Titanium butoxide, which is commonly used as a catalyst in synthesis of DOTP from terephthalic acid and 2-ethylhexanol, gave DOTP yield of 89.0%wt which is lower when compared to active ionic liquid catalysts. Similar results were obtained when $ZnCl_2$ and $Zn(CH_3COO)_2$ were applied. Sulfuric acid has shown poor activity in depolymerization reaction with DOTP yield of 67.5 and PET conversion of 70.1%wt. This result is to be expected since PET depolymerization is a transesterification reaction and Bronsted acids are usually less active compounds in this reaction. Tested ionic liquid catalysts have largely shown good activity; however, a high catalyst concentration of catalyst was used in the reaction. Ionic liquids also plays a role as a cosolvent in PET alcoholysis [103]. A series of methylimidazolium ionic liquids containing different anions (Cl, Br, NO3, etc.) were tested in PET degradation trials in the presence of 2-ethylhexanole. Propylmethylimidazolium ([Amim]Cl) and butylmethylimidazolium ([Bmim]Cl) chlorides have proven to be the most effective cosolvents achieving PET degradation rates of 56.1 and 57.3%wt and DOTP yields of 42.4 and 43.2%wt, respectively. The process was conducted using 2:2:2 IL:2-Eh:PET weight ratios. Catalyst activity of titanium butoxide (TnBt) and zinc acetate was tested using [Bmim]Cl as cosolvent. [Bmim]Cl was reused four times with only marginal changes in PET degradation and DOTP yield. Application of ionic liquid cosolvent has significant impact on degradation of PET as well as on obtained yields of DOTP.

Deep eutectic solvents

Choline chloride deep eutectic solvents (DES) are also used in PET alcoholysis. Zhou et al. [104] investigated the use of DES comprised of choline chloride and various metal salts in PET depolymerization in the presence of 2-ethylhexanole (Table 12). Their activity was compared with other transesterification catalysts such as zinc, manganese and cobalt acetates among others. DES that were applied in depolymerization reaction showed high catalytic activity which resulted in high conversion rates of PET and good DOTP yields. ChCl/Za(Ac)$_2$ has shown the best results achieving conversion close to 100%wt and 84.2%wt yield of DOTP. The depolymerization process was conducted at 185 °C for 60 min with 2-Eh/PET ratio of 3.4 and DES concentration of 5%wt regarding initial PET mass. Choline chloride deep eutectic solvents have shown very good catalytic activity when compared to standard transesterification catalysts such as titanium butoxide or zinc acetate, which under the same reaction conditions yielded lower conversion rates of 78.5%wt and 46.4%wt, respectively, and DOTP yields (63.8%wt and 40.6%wt).

Table 12. Results of PET alcoholysis with 2-ethylhexanole.

Catalyst	Catalyst Concentration [%wt.]	Reaction Temperature [°C]	Reaction Time [min]	Conversion [%]	Yield [%]	Ref.
$ZnCl_2$	20	210	480	93.2	87.6	[102]
$Zn(CH_3COO)_2$	20	210	480	92.5	88.2	[102]
$(CH_3CH_2CH_2CH_2O)Ti$	20	210	480	93.2	89.0	[102]
H_2SO_4	20	210	480	70.1	67.5	[102]
$[HO_3S-(CH_2)_3-NEt_3]Cl$	20	210	480	40.1	35.4	[102]
$[C_4H_6N_2(CH_2)_3SO_3H]_3PW_{12}O_{40}$	20	210	480	97.5	94.7	[102]
$[HO_3S-(CH_2)_3-NEt_3]-FeCl_3$ (x = 0.50)	20	210	480	42.4	38.6	[102]
$[HO_3S-(CH_2)_3-NEt_3]Cl-FeCl_3$ (x = 0.60)	20	210	480	86.0	84.2	[102]
$[HO_3S-(CH_2)_3-NEt_3]Cl-FeCl_3$ (x = 0.64)	20	210	480	94.2	88.6	[102]
$[HO_3S-(CH_2)_3-NEt_3]Cl-FeCl_3$ (x = 0.67)	20	210	480	100	97.6	[102]
$[HO_3S-(CH_2)_3-NEt_3]Cl-FeCl_3$ (x = 0.75)	20	210	480	100	97.9	[102]
$[HO_3S-(CH_2)_3-NEt_3]Cl-ZnCl_2$ (x = 0.67)	20	210	480	100	95.7	[102]
$[HO_3S-(CH_2)_3-NEt_3]Cl-FeCl_2$ (x = 0.67)	20	210	480	87.7	78.9	[102]
$[HO_3S-(CH_2)_3-NEt_3]Cl-CuCl_2$ (x = 0.67)	20	210	480	40.3	37.3	[102]
$[C_4mim]Cl-FeCl_3$ (x = 0.67)	20	210	480	92.1	91.4	[102]
-	-	190	240	1.7	1.2	[103]
[Amim]Cl [b]	-	190	240	56.1 [a]	42.4	[103]
[Bmim]Cl [b]	-	190	240	57.3 [a]	43.4	[103]
[Bmim]Br [b]	-	190	240	46.2 [a]	37.3	[103]
[Bmim]NO_3 [b]	-	190	240	10.5 [a]	6.5	[103]
[Hmim]CF_3SO_3 [b]	-	190	240	5.2 [a]	3.7	[103]
[Bmim]HSO_4 [b]	-	190	240	28.6 [a]	20.5	[103]
[Bmim]BF_4 [b]	-	190	240	4.3 [a]	3.1	[103]
B[mim]PF_6 [b]	-	190	240	3.7 [a]	2.2	[103]
$Ti(OC_4H_9)_4$ [c]	1.2	190	240	98.1	86.7	[103]
$Zn(CH_3COO)_2$ [c]	1.2	190	240	97.5	85.9	[103]
ChCl	5	185	60	3.4	-	[104]
ChCl/$Zn(Ac)_2$ (1:1)	5	185	60	100	84.2	[104]
ChCl/$Mn(Ac)_2$ (1:1)	5	185	60	96.5	80.6	[104]
ChCl/$Co(Ac)_2$ (1:1)	5	185	60	90.4	74.1	[104]
ChCl/$Cu(Ac)_2$ (1:1)	5	185	60	91.7	76.2	[104]
ChCl/$FeCl_3$ (1:1)	5	185	60	94.5	78.3	[104]
$Zn(Ac)_2$	5	185	60	46.4	40.6	[104]
$Mn(Ac)_2$	5	185	60	44.2	36.6	[104]
$Co(Ac)_2$	5	185	60	32.9	27.1	[104]
$Cu(Ac)_2$	5	185	60	33.8	27.6	[104]
$FeCl_3$	5	185	60	43.8	35.7	[104]
$CoCl_2$	5	185	60	31.6	26.5	[104]
H_2SO_4	5	185	60	41.4	34.5	[104]
$Ti(OC_4H_9)_4$	5	185	60	78.5		[104]

Note. [a] % of PET degradation, [b] cosolvent, [c] [Bmim]Cl as cosolvent.

3. Conclusions

The significant amount of waste generated by poly(ethylene terephthalate) requires the development of a recycling process chain in which chemical recycling plays an important role. On the one hand, it allows the depolymerization of degraded plastics that do not meet the quality requirements to be used in mechanical recycling, and on the other hand, provides an opportunity to process cheap waste and obtain products with greater added value. It can be widely used in the recycling of both packaging plastics and textiles, or other waste generated with PET.

Chemical depolymerization processes of poly(ethylene terephthalate) can be carried out in several ways, using the susceptibility of the ester group to select chemical reactions such as hydrolysis and transesterification. The products obtained as a result can be divided into two groups, the first of which includes compounds obtained by hydrolysis, methanolysis, ethanolysis and glycolysis used in the synthesis of polymers. The second group includes terephthalic esters with plasticizing properties of plastics obtained by depolymerization with higher monohydric alcohols.

The use of hydrolysis allows terephthalic acid to be obtained, which after purification can be reused in the polymerization reaction to obtain plastics. This process is relatively widely described in the scientific literature and has found practical application in industry. It requires the use of high temperatures and pressures and, in the case of alkaline hydrolysis, a significant number of bases. Inorganic hydroxides, inorganic acids, organic acids and heteropolyacids, as well as "superacids" and quaternary ammonium salts are used as catalysts in the process. However, it allows for high PET conversion rates and yields in excess of 90%. An important problem in the use of this method is the purification of crude terephthalic acid to the purity required by polymer manufacturers.

The alcoholysis process includes PET depolymerization using monohydric alcohols such as methanol, ethanol, butanol and higher alcohols such as butanol or 2-ethylhexanol. The methanolysis reaction is another chemical recycling process that allows raw materials to be obtained that can be directly used in the synthesis of a new polymer. The process usually requires high temperatures, which due to the low boiling point of methanol necessitates the use of a high-pressure environment. Typical catalysts used in this process are metal acetates, with zinc acetate exhibiting the highest activity. Other salts, "superbases" and ionic liquids are also used. The process can also be carried out in supercritical conditions, which ensures high efficiency without the need for a catalyst. Polyethylene terephthalate ethanolysis is seen as an alternative to the methanolysis process to eliminate harmful methyl alcohol. The process is carried out analogously to methanolysis at elevated temperature and pressure using analogous catalysts. The process can also be carried out under supercritical conditions. Glycolysis is one of the most widely described PET depolymerization processes in the scientific literature. The presented results cover a wide range of catalysts used from typical compounds such as metal salts, metal oxides, metal complexes, polyoxometalates to ionic liquids.

Higher alcohols such as butanol or 2-ethylhexanol can be used in the synthesis of terephthalic plasticizers using catalysts such as superbases, ionic liquids, Lewis acids, acetates and deep eutectic solvents. Due to the high boiling point of the alcohols used, depolymerization processes do not require the use of high pressures. The obtained products have plasticizing properties similar to plasticizers obtained by traditional methods. They are potential substitutes for plasticizers obtained by traditional methods.

The re-use of waste poly(ethylene terephthalate) is an increasingly common practice, mainly due to legal regulations requiring producers to introduce more waste raw material into production stream. Recycling of waste PET by introducing waste substrate in to production leads to gradual degradation of the processed material, which forces the development of effective methods of its chemical recycling. Chemical recycling of PET can be divided into processes that result in compounds used in the production of fresh poly(ethylene terephthalate) and upcycling processes, which result in products used in other fields, with higher added value. Due to their high potential for practical application, it is expected that their future development will focus on developing methods to achieve high conversions and yields in order to obtain a high-quality product. In particular, the future development of PET chemical recycling methods will focus on the development of catalysts that will achieve high product performance with low-mass feedstock ratios. The use of superacids in hydrolysis and acetates, nanodispersion and processes under supercritical conditions in methanolysis and acetates in the glycolysis process seems to be a future direction. Chemical recycling towards value-added products represents an interesting path for waste PET treatment. From the group of products that can be obtained by PET alcoholization, the most promising are the terephthalates of alcohols C_8 to C_{10}. These compounds are widely used as PVC plasticizers, which are traditionally obtained from terephthalic acid and can be seen as a viable alternative for traditional process.

By characterizing the depolymerization methods and catalysts described in the scientific literature, this review will contribute to a better understanding of the state of knowledge about the chemical recycling of poly(ethylene terephthalate). It will also undoubtedly be of help to scientists planning to start research on this topic.

Author Contributions: Conceptualization, M.M., G.D., M.Z. and J.N.; methodology, M.M., G.D., M.Z. and J.N.; writing—original draft preparation, M.M., M.Z., J.N. and G.D.; writing—review and editing, G.D. and M.M.; visualiza-tion, M.M., M.Z. and J.N.; supervision, G.D.; project administration, G.D., M.M. and M.Z.; funding acquisition, G.D, M.M. and M.Z. All authors have read and agreed to the published version of the manuscript.

Funding: M.M., M.Z and G.D. would like to thank the Ministry of Education and Science for funding under the project No SKN/SP/535370/2022. M.M. would like to thank for the financial support the Ministry of Education and Science under project No DWD/4/21/2020. M.Z. and G.D. would like to thank the Silesian University of Technology for providing partial financial support under projects No 31/010/SDU20/0006-10.

Institutional Review Board Statement: Not applicable.

Informed Consent Statement: Not applicable.

Data Availability Statement: No new data was created.

Conflicts of Interest: The authors declare no conflict of interest.

References

1. Geyer, R.; Jambeck, J.R.; Law, K.L. Production, use, and fate of all plastics ever made. *Sci. Adv.* **2017**, *3*, e1700782. [CrossRef] [PubMed]
2. Namazi, H. Polymers in our daily life. *Bioimpacts* **2017**, *7*, 73–74. [CrossRef] [PubMed]
3. Smith, R.; Oliver, C.; Williams, D. The enzymatic degradation of polymers in vitro. *J. Biomed. Mater. Res.* **1987**, *21*, 991–1003. [CrossRef] [PubMed]
4. Zhang, J.; Wamg, X.; Gong, J.; Gu, Z. A study on the biodegradability of polyethylene terephthalate fiber and diethylene glycol terephthalate. *J. Appl. Polym. Sci.* **2004**, *93*, 1089–1096. [CrossRef]
5. Geyer, B.; Lorenz, G.; Kandelbauer, A. Recycling of poly(ethylene terephthalate)—A review focusing on chemical methods. *EXPRESS Polym. Lett.* **2016**, *10*, 559–586. [CrossRef]
6. Hong, M.; Chen, E. Chemically recyclable polymers: A circular economy approach to sustainability. *Green Chem.* **2017**, *19*, 3692–3706. [CrossRef]
7. Antonakou, E.V.; Achilias, D.S. Recent Advances in Polycarbonate Recycling: A Review of Degradation Methods and Their Mechanisms. *Waste Biomass Valor.* **2013**, *4*, 9–21. [CrossRef]
8. Zia, K.M.; Bhatti, H.N.; Bhatti, I.A. Methods for polyurethane and polyurethane composites, recycling and recovery: A review. *React. Funct. Polym.* **2007**, *67*, 675–692. [CrossRef]
9. Yoshioka, T.; Motoki, T.; Okuwaki, A. Kinetics of Hydrolysis of Poly(ethylene terephthalate) Powder in Sulfuric Acid by a Modified Shrinking-Core Model. *Ind. Eng. Chem. Res.* **2001**, *40*, 75–79. [CrossRef]
10. Ghaemy, M.; Mossaddegh, K. Depolymerisation of poly(ethylene terephthalate) fibre wastes using ethylene glycol. *Polym. Degrad. Stab.* **2005**, *90*, 570–576. [CrossRef]
11. Liu, S.; Wang, Z.; Li, J.; Yu, S.; Xie, C.; Liu, F. Butanol alcoholysis reaction of polyethylene terephthalate using acidic ionic liquid as catalyst. *J. Appl. Polym. Sci.* **2013**, *130*, 1840–1844. [CrossRef]
12. Zhou, J.; Hsu, T.; Wang, J. Mechanochemical Degradation and Recycling of Synthetic Polymers. *Angew. Chem. Int. Ed.* **2023**, *62*, e202300768. [CrossRef]
13. Cao, F.; Wang, L.; Zheng, R.; Guo, L.; Chena, Y.; Qian, X. Research and progress of chemical depolymerization of waste PET and high-value application of its depolymerization products. *RSC Adv.* **2022**, *12*, 31564–31576. [CrossRef] [PubMed]
14. Karayannidis, G.P.; Achilias, D.S. Chemical recycling of poly(ethylene terephthalate). *Macromol. Mater. Eng.* **2007**, *292*, 128–146. [CrossRef]
15. Stanica-Ezeanu, D.; Matei, D. Natural depolymerization of waste poly(ethylene terephthalate) by neutral hydrolysis in marine water. *Sci. Rep.* **2021**, *11*, 4431. [CrossRef] [PubMed]
16. Valh, J.V.; Voncina, B.; Lobnik, A.; Zemljic, L.F.; Skodic, L.; Vajnhandl, S. Conversion of polyethylene terephthalate to high-quality terephthalic acid by hydrothermal hydrolysis: The study of process parameters. *Textile Res. J.* **2020**, *90*, 1446–1461. [CrossRef]
17. Colnik, M.; Knez, Z.; Skerget, M. Sub- and supercritical water for chemical recycling of polyethylene terephthalate waste. *Chem. Eng. Sci.* **2021**, *233*, 116389. [CrossRef]
18. Kazutoshi, I.; Takahiro, I.; Katsuki, K. Hydrolysis of PET by Combining Direct Microwave Heating with High Pressure. *Procedia Eng.* **2016**, *148*, 314–318. [CrossRef]
19. Siddiqui, M.N.; Achilias, D.S.; Redhwi, H.H.; Bikiaris, D.N.; Katsogiannis, K.A.G.; Karayannidis, G.P. Hydrolytic depolymerization of PET in a microwave reactor. *Macromol. Mater. Eng.* **2010**, *295*, 575–584. [CrossRef]
20. Singh, S.; Sharma, S.; Umar, A.; Kumar Mehta, S.; Bhatti, M.S.; Kansal, S.K. Recycling of waste poly(ethylene terephthalate) bottles by alkaline hydrolysis and recovery of pure nanospindle-shaped terephthalic acid. *J. Nano. Nanotechnol.* **2018**, *18*, 5804–5809. [CrossRef]

21. Štrukil, V. Highly Efficient Solid-State Hydrolysis of Waste Polyethylene Terephthalate by Mechanochemical Milling and Vapor-Assisted Aging. *ChemSusChem* **2021**, *14*, 330–338. [CrossRef] [PubMed]
22. Bhogle, C.S.; Pandit, A.B. Ultrasound-Assisted Alkaline Hydrolysis of Waste Poly(Ethylene Terephthalate) in Aqueous and Nonaqueous Media at Low Temperature. *Indian Chem. Eng.* **2018**, *60*, 122–140. [CrossRef]
23. Zope, V. Hydrolysis of poly-ethylene terephthalate waste using high pressure autoclave: A chemical recycling. *Int. J. Basic Appl. Chem. Sci.* **2022**, *12*, 1–5.
24. Ugduler, S.; Van Geem, K.M.; Denolf, R.; Roosen, M.; Mys, N.; Ragaert, K.; De Meester, S. Closed-loop recycling of multilayer and colored PET plastic waste by alkaline hydrolysis. *Green Chem.* **2020**, *22*, 5376–5394. [CrossRef]
25. Chen, W.; Yang, Y.; Lan, X.; Zhang, B.; Zhang, X.; Mu, T. Biomass-derived γ-valerolactone: Efficient dissolution and accelerated alkaline hydrolysis of polyethylene terephthalate. *Green Chem.* **2021**, *23*, 4065–4073. [CrossRef]
26. Khalaf, H.I.; Hasan, O.A. Effect of quaternary ammonium salt as a phase transfer catalyst for the microwave depolymerization of polyethylene terephthalate waste bottles. *Chem. Eng. J.* **2012**, *192*, 45–48. [CrossRef]
27. Wang, Y.; Wang, H.; Chen, H.; Liu, H. Towards recycling purpose: Converting PET plastic waste back to terephthalic acid using pH responsive phase transfer catalyst. *Chin. J. Chem. Eng.* **2022**, *51*, 53–60. [CrossRef]
28. Zhang, L.; Gao, J.; Zou, J.; Yi, F. Hydrolysis of poly(ethylene terephthalate) waste bottles in the presence of dual functional phase transfer catalysts. *J. Appl. Polym. Sci.* **2013**, *130*, 2790–2795. [CrossRef]
29. Yang, W.; Wang, J.; Jiao, L.; Song, Y.; Li, C.; Hu, C. Easily recoverable and reusable p-toluenesulfonic acid for faster hydrolysis of waste polyethylene terephthalate. *Green Chem.* **2022**, *24*, 1362–1372. [CrossRef]
30. Li, X.; Lu, H.; Guo, W.; Cao, G.; Liu, H.; Shi, Y. Reaction Kinetics and Mechanism of Catalyzed Hydrolysis of Waste PET Using Solid Acid Catalyst in Supercritical CO_2. *AIChE J.* **2015**, *61*, 200–214. [CrossRef]
31. Guo, W.; Lu, H.; Li, X.; Cao, G. Tungsten-promoted titania as solid acid for catalytic hydrolysis of waste bottle PET in supercritical CO_2. *RSC Adv.* **2016**, *6*, 43171–43184. [CrossRef]
32. Yang, W.; Liu, C.R.; Chang, L.; Song, Y.; Hu, C. Hydrolysis of waste polyethylene terephthalate catalyzed by easily recyclable terephthalic acid. *Waste Manag.* **2021**, *135*, 267–274. [CrossRef]
33. Scremin, D.M.; Miyazaki, D.Y.; Lunelli, C.E.; Silva, S.A.; Zawadzki, S.F. PET recycling by Alcoholysis using a new heterogeneous catalyst: Study and its use in polyurethane adhesives preparation. *Macromol. Symp.* **2019**, *383*, 1800027. [CrossRef]
34. Shojaei, B.; Abtahi, M.; Najafi, M. Chemical recycling of PET: A stepping-stone toward sustainability. *Polym. Adv. Technol.* **2020**, *31*, 2912–2938. [CrossRef]
35. Brivio, L.; Tollini, F. Chapter Six—PET recycling: Review of the current available technologies and industrial perspectives. *Adv. Chem. Eng.* **2022**, *60*, 215–267. [CrossRef]
36. Kim, B.K.; Hwang, G.C.; Bae, S.Y.; Yi, S.C.; Kumazawa, H. Depolymerization of polyethyleneterephthalate in supercritical methanol. *J. Appl. Polym. Sci.* **2001**, *81*, 2102–2108. [CrossRef]
37. Hofmann, M.; Sundermeier, J.; Alberti, C.; Enthaler, S. Zinc(II) acetate Catalyzed Depolymerization of Poly(ethylene terephthalate). *ChemistrySelect* **2020**, *5*, 10010–11001. [CrossRef]
38. Liu, Y.; Wang, M.; Pan, Z. Catalytic depolymerization of polyethylene terephthalate in hot compressed water. *J. Supercrit. Fluids* **2012**, *62*, 226–231. [CrossRef]
39. Mishra, S.; Goje, A.S. Kinetic and thermodynamic study of methanolysis of poly(ethylene terephthalate)waste powder. *Polym. Int.* **2003**, *52*, 337–342. [CrossRef]
40. Sanchez, A.; Collinson, S. The selective recycling of mixed plastic waste of polylactic acid and polyethylene terephthalate by control of process conditions. *Eur. Polym. J.* **2011**, *47*, 1970–1976. [CrossRef]
41. Siddiqui, M.N.; Redhwi, H.H.; Achilias, D.S. Recycling of poly(ethylene terephthalate) waste through methanolic pyrolysis in a microwave reactor. *J. Anal. Appl. Pyrolysis* **2012**, *98*, 214–220. [CrossRef]
42. Pham, D.D.; Cho, J. Low-energy catalytic methanolysis of poly(ethyleneterephthalate). *Green Chem.* **2021**, *23*, 511–525. [CrossRef]
43. Tang, S.; Li, F.; Liu, J.; Guo, B.; Tian, Z.; Lv, J. Calcined sodium silicate as solid base catalyst for alcoholysis of poly(ethylene terephthalate). *J. Chem. Technol. Biotechnol.* **2022**, *97*, 1305–1314. [CrossRef]
44. Gangotena, P.; Ponce, S.; Gallo-Córdova, A.; Streitwieser, D.; Mora, J. Highly Active MgP Catalyst for Biodiesel Production and Polyethylene Terephthalate Depolymerization. *ChemistrySelect* **2022**, *7*, e202103765. [CrossRef]
45. Laldinpuii, Z.T.; Khiangte, V.; Lalhmangaihzuala, S.; Lalmuanpuia, C.; Pachuau, Z.; Lalhriatpuia, C.; Vanlaldinpuiac, K. Methanolysis of PET Waste Using Heterogeneous Catalyst of Bio-waste Origin. *J. Polym. Environ.* **2022**, *30*, 1600–1614. [CrossRef]
46. Du, J.T.; Sun, Q.; Zeng, X.F.; Wang, D.; Wang, J.X.; Chen, J.F. ZnO nanodispersion as pseudohomogeneous catalyst for alcoholysis of polyethylene terephthalate. *Chem. Eng. Sci.* **2020**, *220*, 115642. [CrossRef]
47. Sharma, V.; Parashar, P.; Srivastava, P.; Kumar, S.; Agarwal, D.; Richharia, N. Recycling of waste PET-bottles using dimethyl sulfoxide and hydrotalcite catalyst. *J. Appl. Polym. Sci.* **2013**, *129*, 1513–1519. [CrossRef]
48. Liu, M.; Guo, J.; Gu, Y.; Gao, J.; Liu, F. Versatile Imidazole-Anion-Derived Ionic Liquids with Unparalleled Activity for Alcoholysis of Polyester Wastes under Mild and Green Conditions. *ACS Sustain. Chem. Eng.* **2018**, *6*, 15127–15134. [CrossRef]
49. Paszun, D.; Spychaj, T. Chemical Recycling of Poly(Ethylene Terephthalate). *Ind. Eng. Chem. Res.* **1997**, *36*, 1373–1383. [CrossRef]
50. Ostrysz, R.; Kłosowska, Z.; Jankowska, F. Method of preparation of thixotropic and chemoresistant unsaturated polyester resins. G.B. Patent 1158561, 16 July 1969.

51. Xin, J.; Zhang, Q.; Huang, J.; Huang, R.; Jaffery, Q.Z.; Yan, D.; Zhou, Q.; Xu, J.; Lu, X. Progress in the Catalytic Glycolysis of Polyethylene Terephthalate. *J. Environ. Manag.* **2021**, *296*, 113267. [CrossRef]
52. Sinha, V.; Patel, M.R.; Patel, J.V. Pet Waste Management by Chemical Recycling: A Review. *J Polym Environ.* **2010**, *18*, 8–25. [CrossRef]
53. Chen, C.-H.; Chen, C.-Y.; Lo, Y.-W.; Mao, C.-F.; Liao, W.-T. Studies of Glycolysis of Poly(Ethylene Terephthalate) Recycled from Postconsumer Soft-Drink Bottles. I. Influences of Glycolysis Conditions. *J. Appl. Polym. Sci.* **2001**, *80*, 943–948. [CrossRef]
54. Chen, C.-H.; Chen, C.-Y.; Lo, Y.-W.; Mao, C.-F.; Liao, W.-T. Studies of Glycolysis of Poly(Ethylene Terephthalate) Recycled from Postconsumer Soft-Drink Bottles. II. Factorial Experimental Design. *J. Appl. Polym. Sci.* **2001**, *80*, 956–962. [CrossRef]
55. Ghosal, K.; Nayak, C. Recent Advances in Chemical Recycling of Polyethylene Terephthalate Waste into Value Added Products for Sustainable Coating Solutions—Hope vs. Hype. *Mater. Adv.* **2022**, *3*, 1974–1992. [CrossRef]
56. López-Fonseca, R.; Duque-Ingunza, I.; de Rivas, B.; Arnaiz, S.; Gutiérrez-Ortiz, J.I. Chemical Recycling of Post-Consumer PET Wastes by Glycolysis in the Presence of Metal Salts. *Polym. Degrad. Stab.* **2010**, *95*, 1022–1028. [CrossRef]
57. Nikles, D.E.; Farahat, M.S. New Motivation for the Depolymerization Products Derived from Poly(Ethylene Terephthalate) (PET) Waste: A Review. *Macromol. Mater. Eng.* **2005**, *290*, 13–30. [CrossRef]
58. Hu, Y.; Wang, Y.; Zhang, X.; Qian, J.; Xing, X.; Wang, X. Synthesis of Poly(Ethylene Terephthalate) Based on Glycolysis of Waste PET Fiber. *J. Macromol. Sci. A* **2020**, *57*, 430–438. [CrossRef]
59. Deng, L.; Li, R.; Chen, Y.; Wang, J.; Song, H. New Effective Catalysts for Glycolysis of Polyethylene Terephthalate Waste: Tropine and Tropine-Zinc Acetate Complex. *J. Mol. Liq.* **2021**, *334*, 116419. [CrossRef]
60. Lei, D.; Sun, X.L.; Hu, S.; Cheng, H.; Chen, Q.; Qian, Q.; Xiao, Q.; Cao, C.; Xiao, L.; Huang, B. Rapid Glycolysis of Waste Polyethylene Terephthalate Fibers via a Stepwise Feeding Process. *Ind. Eng. Chem. Res.* **2022**, *61*, 4794–4802. [CrossRef]
61. Moncada, J.; Dadmun, M.D. The Structural Evolution of Poly(Ethylene Terephthalate) Oligomers Produced via Glycolysis Depolymerization. *J. Mater. Chem. A* **2023**, *11*, 4679–4690. [CrossRef]
62. Esquer, R.; García, J.J. Metal-Catalysed Poly(Ethylene) Terephthalate and Polyurethane Degradations by Glycolysis. *J. Organomet. Chem.* **2019**, *902*, 120972. [CrossRef]
63. Liu, B.; Lu, X.; Ju, Z.; Sun, P.; Xin, J.; Yao, X.; Zhou, Q.; Zhang, S. Ultrafast Homogeneous Glycolysis of Waste Polyethylene Terephthalate via a Dissolution-Degradation Strategy. *Ind. Eng. Chem. Res.* **2018**, *57*, 16239–16245. [CrossRef]
64. Son, S.G.; Jin, S.B.; Kim, S.J.; Park, H.J.; Shin, J.; Ryu, T.; Jeong, J.M.; Choi, B.G. Exfoliated Manganese Oxide Nanosheets as Highly Active Catalysts for Glycolysis of Polyethylene Terephthalate. *FlatChem* **2022**, *36*, 100430. [CrossRef]
65. Wang, T.; Zheng, Y.; Yu, G.; Chen, X. Glycolysis of Polyethylene Terephthalate: Magnetic Nanoparticle CoFe2O4 Catalyst Modified Using Ionic Liquid as Surfactant. *Eur. Polym. J.* **2021**, *155*, 110590. [CrossRef]
66. Guo, Z.; Adolfsson, E.; Tam, P.L. Nanostructured Micro Particles as a Low-Cost and Sustainable Catalyst in the Recycling of PET Fiber Waste by the Glycolysis Method. *Waste Manag.* **2021**, *126*, 559–566. [CrossRef]
67. Fang, P.; Liu, B.; Xu, J.; Zhou, Q.; Zhang, S.; Ma, J.; Lu, X. High-Efficiency Glycolysis of Poly(Ethylene Terephthalate) by Sandwich-Structure Polyoxometalate Catalyst with Two Active Sites. *Polym. Degrad. Stab.* **2018**, *156*, 22–31. [CrossRef]
68. Putisompon, S.; Yunita, I.; Sugiyarto, K.H.; Somsook, E. Low-Cost Catalyst for Glycolysis of Polyethylene Terephthalate (PET). *Key Eng. Mater.* **2019**, *824*, 225–230. [CrossRef]
69. Zangana, K.H.; Fernandez, A.; Holmes, J.D. Simplified, Fast, and Efficient Microwave Assisted Chemical Recycling of Poly (Ethylene Terephthalate) Waste. *Mater. Today Commun.* **2022**, *33*, 104588. [CrossRef]
70. Lalhmangaihzuala, S.; Laldinpuii, Z.; Lalmuanpuia, C.; Vanlaldinpuia, K. Glycolysis of Poly(Ethylene Terephthalate) Using Biomass-Waste Derived Recyclable Heterogeneous Catalyst. *Polymers* **2020**, *13*, 37. [CrossRef]
71. Shuangjun, C.; Weihe, S.; Haidong, C.; Hao, Z.; Zhenwei, Z.; Chaonan, F. Glycolysis of Poly(Ethylene Terephthalate) Waste Catalyzed by Mixed Lewis Acidic Ionic Liquids. *J. Therm. Anal. Calorim.* **2021**, *143*, 3489–3497. [CrossRef]
72. Cano, I.; Martin, C.; Fernandes, J.A.; Lodge, R.W.; Dupont, J.; Casado-Carmona, F.A.; Lucena, R.; Cardenas, S.; Sans, V.; de Pedro, I. Paramagnetic Ionic Liquid-Coated $SiO_2@Fe_3O_4$ Nanoparticles—The next Generation of Magnetically Recoverable Nanocatalysts Applied in the Glycolysis of PET. *Appl. Catal. B* **2020**, *260*, 118110. [CrossRef]
73. Najafi-Shoa, S.; Barikani, M.; Ehsani, M.; Ghaffari, M. Cobalt-Based Ionic Liquid Grafted on Graphene as a Heterogeneous Catalyst for Poly (Ethylene Terephthalate) Glycolysis. *Polym. Degrad. Stab.* **2021**, *192*, 109691. [CrossRef]
74. Le, N.H.; Ngoc Van, T.T.; Shong, B.; Cho, J. Low-Temperature Glycolysis of Polyethylene Terephthalate. *ACS Sustainable Chem. Eng.* **2022**, *10*, 17261–17273. [CrossRef]
75. Veregue, F.R.; Pereira da Silva, C.T.; Moisés, M.P.; Meneguin, J.G.; Guilherme, M.R.; Arroyo, P.A.; Favaro, S.L.; Radovanovic, E.; Girotto, E.M.; Rinaldi, A.W. Ultrasmall Cobalt Nanoparticles as a Catalyst for PET Glycolysis: A Green Protocol for Pure Hydroxyethyl Terephthalate Precipitation without Water. *ACS Sustain. Chem. Eng.* **2018**, *6*, 12017–12024. [CrossRef]
76. Wang, Z.; Jin, Y.; Wang, Y.; Tang, Z.; Wang, S.; Xiao, G.; Su, H. Cyanamide as a Highly Efficient Organocatalyst for the Glycolysis Recycling of PET. *ACS Sustain. Chem. Eng.* **2022**, *10*, 7965–7973. [CrossRef]
77. Wang, Z.; Wang, Y.; Xu, S.; Jin, Y.; Tang, Z.; Xiao, G.; Su, H. A Pseudo-Homogeneous System for PET Glycolysis Using a Colloidal Catalyst of Graphite Carbon Nitride in Ethylene Glycol. *Polym. Degrad. Stab.* **2021**, *190*, 109638. [CrossRef]
78. De Castro, R.E.N.; Vidotti, G.J.; Rubira, A.F.; Muniz, E.C. Depolymerization of Poly(ethylene terephthalate) Wastes Using Ethanol and Ethanol/Water in Supercritical Conditions. *J. Appl. Polym. Sci.* **2006**, *101*, 2009–2016. [CrossRef]

79. Sanda, O.; Taiwo, E.A.; Osinkolu, G.A. Alkaline Solvolysis of Poly(Ethylene Terephthalate) in Butan–1–ol Media: Kinetics and Optimization Studies. *British. J. Appl. Sci. Technol.* **2017**, *21*, 1–12. [CrossRef]
80. Wang, X.; An, W.; Du, R.; Tian, F.; Yang, Y.; Zhao, X.; Xu, S.; Wang, Y. Rapid hydrolysis of PET in high-concentration alcohol aqueous solution by pore formation and spontaneous separation of terephthalate. *J. Environ. Chem. Eng.* **2023**, *11*, 109434. [CrossRef]
81. Fregoso-Infante, A.; Vega-Rangel, R.; Figueroa, G. Chemical Method for Recycling Polyethylene Terephtalate (pet) Wastes. SP Patent WO2005082826, 9 September 2005.
82. Tang, X.H.; Li, N.; Li, G.; Wang, A.; Cong, Y.; Xu, G.; Wang, X.; Zhang, T. Synthesis of gasoline and jet fuel range cycloalkanes and aromatics with poly(ethyleneterephthalate) wastes. *Green Chem.* **2019**, *21*, 2709–2719. [CrossRef]
83. Anderson, R.L.; Sikkenga, D.L. Ethanolysis of PET to form DET and Oxidation DET to form Carboxylic Acids. FR Patent WO2007076384, 5 July 2007.
84. Azimova, M.; Nubel, P.; Bergeron, R. Improved Catalyst Performance for Polyester Recycling. Patent WO2021257920, 23 December 2021.
85. Shiwei, L.; Shitao, Y.; Congxia, X.; Fusheng, L.; Lu, L.; Minfang, Z. Method for Preparing Plasticizer Terephthalate by Alcoholysis of Waste Polyester PET. CN Patent CN102603532, 29 October 2014.
86. Hutchenson, K.W. Supercritical Fluid Technology in Material Science and Engineering. Sung, Y.P., Ed.; Marcel Dekker: New York, NY, USA, 2002; p. 87.
87. Yang, Y.; Lu, Y.; Xiang, H.; Xu, Y.; Li, Y. Study on methanolytic depolymerization of PET with supercritical methanol for chemical recycling. *Polym. Degrad. Stab.* **2002**, *75*, 185–191. [CrossRef]
88. Fávaro, S.L.; Freitas, A.R.; Ganzerli, T.A.; Pereira, A.G.B.; Cardozo, A.L.; Baron, O.; Muniz, E.C.; Girotto, E.M.; Radovanovic, E. PET and Aluminum Recycling from Multilayer Food Packaging Using Supercritical Ethanol. *J. Supercrit. Fluids* **2013**, *75*, 138–143. [CrossRef]
89. Fernandes, J.R.; Amaro, L.P.; Muniz, E.C.; Favaro, S.L.; Radovanovic, E. PET Depolimerization in Supercritical Ethanol Conditions Catalysed by Nanoparticles of Metal Oxides. *J. Supercrit. Fluids* **2020**, *158*, e104715. [CrossRef]
90. Yang, Y.; Chen, F.; Shen, T.; Pariatamby, A.; Wen, X.; Yan, M.; Kanchanatip, E. Catalytic depolymerization of waste polyethylene terephthalate plastic in supercritical ethanol by ZnO/γ-Al$_2$O$_3$ catalyst. *Process Saf. Environ. Protect.* **2023**, *173*, 881–892. [CrossRef]
91. Nunes, C.; Vieira da Silva, M.J.; Cristina da Silva, D.; dos Reis Freitas, A.; Rosa, F.A.; Rubira, A.F.; Muniz, E.C. PET Depolimerisation in Supercritical Ethanol Catalysed by [Bmim][BF$_4$]. *RSC Adv.* **2014**, *4*, 20308–20316. [CrossRef]
92. Nunes, C.; Souza, P.; Freitas, A.; Silva, M.; Rosa, F.; Muniz, E. Poisoning Effects of Water and Dyes on the [Bmim][BF$_4$] Catalysis of Poly(Ethylene Terephthalate) (PET) Depolymerization under Supercritical Ethanol. *Catalysts* **2017**, *7*, e43. [CrossRef]
93. Lozano-Martinez, P.; Torres-Zapata, T.; Martin-Sanchez, N. Directing Depolymerization of PET with Subcritical and Supercritical Ethanol to Different Monomers through Changes in Operation Conditions. *ACS Sustain. Chem. Eng.* **2021**, *9*, 9846–9853. [CrossRef]
94. Yan, M.; Yang, Y.; Shen, T.; Grisdanurak, N.; Pariatamby, A.; Khalid, M.; Hantoko, D.; Wibowo, H. Effect of operating parameters on monomer production from depolymerization of waste polyethylene terephthalate in supercritical ethanol. *Process Saf. Environ. Protect.* **2023**, *169*, 212–219. [CrossRef]
95. Hillert, G.; Schäfer, M. Dispersion adhesive and use of a dispersion adhesive. Patent Germany DE102009020497A1, 3 April 2010.
96. Castle, L.; Mayo, A.; Gilbert, J. Migration of plasticizers from printing inks into foods. *Food Addit. Contam.* **1989**, *6*, 437–444. [CrossRef]
97. Medimagh, R. Improved Method for Recycling Pet by Alcoholysis. WO2022112715, 2 June 2022.
98. Ye, D. Method for Preparing Polyvinyl Chloride (PVC) Plasticizer through Ester Exchange Reaction of Waste Polyethylene Terephthalate (PET) and Monohydric Alcohol. CN Patent CN102234227, 30 April 2010.
99. Hu, H.; Wang, L. Method for Preparing Dimethyl Terephthalate from Polybutylene Terephthalate Waste Material. CN Patent CN105503605A, 14 December 2015.
100. Fang, C.; Zhou, X.; Yang, R.; Pan, S.; Lei, W. Carbon Nanotube Prepared from Waste PET, Preparation Method and Application in Conductive Material, Super Capacitor and Catalyst Support. CN Patent CN106986326, 8 May 2017.
101. Dupont, L.A.; Gupta, V.P. Degradative transesterification of terephthalate polyesters to obtain DOTP plasticizer for flexible PVC. *J. Vinyl Addit. Technol.* **1993**, *15*, 100–104. [CrossRef]
102. Liu, S.; Zhou, L.; Li, L.; Yu, L.; Liu, F.; Xie, C.; Song, Z. Isooctanol alcoholysis of waste polyethylene terephthalate in acidic ionic liquid. *J. Polym. Res.* **2013**, *20*, 310. [CrossRef]
103. Chen, J.; Lv, J.; Ji, Y.; Ding, J.; Yang, X.; Zou, M.; Xing, L. Alcoholysis of PET to produce dioctyl terephthalate by isooctyl alcohol with ionic liquid as cosolvent. *Polym. Degrad. Stab.* **2014**, *107*, 178–183. [CrossRef]
104. Zhou, L.; Lu, X.; Ju, Z.; Liu, B.; Yao, H.; Xu, J.; Zhou, Q.; Hu, Y.; Zhang, S. Alcoholysis of Polyethylene Terephthalate for Dioctyl Terephthalate Using Choline Chloride-Based Deep Eutectic Solvents as Efficient Catalysts. *Green Chem.* **2019**, *4*, 897–906. [CrossRef]

Disclaimer/Publisher's Note: The statements, opinions and data contained in all publications are solely those of the individual author(s) and contributor(s) and not of MDPI and/or the editor(s). MDPI and/or the editor(s) disclaim responsibility for any injury to people or property resulting from any ideas, methods, instructions or products referred to in the content.

 *molecules*

Article

A Sustainable Green Enzymatic Method for Amide Bond Formation

György Orsy [1], Sayeh Shahmohammadi [1,2] and Enikő Forró [1,*]

[1] Institute of Pharmaceutical Chemistry, University of Szeged, Eötvös u. 6, H-6720 Szeged, Hungary; orsy.gyorgy@szte.hu (G.O.); sayeh.s@pharm.u-szeged.hu (S.S.)
[2] Stereochemistry Research Group, Eötvös Loránd Research Network, University of Szeged, Eötvös u. 6, H-6720 Szeged, Hungary
* Correspondence: forro.eniko@szte.hu; Tel.: +36-62-544964

Abstract: A sustainable enzymatic strategy for the preparation of amides by using *Candida antarctica* lipase B as the biocatalyst and cyclopentyl methyl ether as a green and safe solvent was devised. The method is simple and efficient and it produces amides with excellent conversions and yields without the need for intensive purification steps. The scope of the reaction was extended to the preparation of 28 diverse amides using four different free carboxylic acids and seven primary and secondary amines, including cyclic amines. This enzymatic methodology has the potential to become a green and industrially reliable process for direct amide synthesis.

Keywords: sustainable enzymatic strategy; direct amide synthesis; green solvent; CALB; carboxylic acid; amine

1. Introduction

The amide bond is a fundamental linkage in nature. It is the main chemical bond that links amino acid building blocks together to give peptides and proteins, which occur worldwide [1–3]. Furthermore, as an important moiety of pharmaceutically active compounds, it can be found in a significant array of commercial drugs worldwide. For example, Acetaminophen, a common pain reliever and an antipyretic agent, is used to treat various conditions such as headache, muscle aches, and arthritis [4]. Amide-based local anesthetics are applied to numb a specific area of the body, before a medical procedure or surgery [5,6]. β-Lactam antibiotics are a group of antibiotics that are used to treat bacterial infections, including pneumonia, bronchitis, and urinary tract infections [7–9]. Celecoxib is a nonsteroidal anti-inflammatory drug (NSAID) utilized to treat pain and inflammation associated with conditions, such as arthritis, menstrual cramps, and sport injuries [10,11]. These are just a few examples of amide drugs, and there are many others used in the treatment of various medical conditions.

A large number of synthetic methods resulting in the formation of amide bonds have been devised in the last decade [12–19]. However, there are only a limited number of strategies that are both efficient and environmentally benign [20–22]. The most common processes utilize coupling reagents or activating agents with larger stoichiometric ratio to couple a free carboxylic acid with an amine. However, these are generally hazardous/poisonous reagents and, consequently, they put a heavy burden on the environment. Furthermore, the purification of the crude products is problematic, since it requires a large quantity of organic solvents, due to the formation of large quantities of by-products [23,24]. Therefore, there is a great demand to develop simple amide-bond-forming reactions to access amides from free carboxylic acids and amines in a green and efficient way.

The green chemistry concept has 12 principles aimed at the design of chemical products and processes, that reduce or eliminate the application and generation of hazardous

substances, which are harmful for human health or the environment [25,26]. Using enzymes in synthetic chemistry has always been a hot topic due to the ability of enzymes to catalyze chemical transformations with high catalytic efficiency and specificity [27–31]. For instance, the reactions mentioned above carried out under harsh conditions can be induced to proceed faster under mild conditions (lower temperatures and pressures, neutral pH), with fewer work-up steps and higher yields. All of these result in an improvement in efficiency and save energy. Enzymes have emerged as preferred tools in green chemistry by replacing hazardous/poisonous reagents, generating the formation of fewer by-products [32–35].

In the case of amide bond formation, enzymes, particularly the members of the lipase family, are powerful and effective biocatalysts for esterification reactions [36–38]. Several lipases have been reported to exhibit high catalytic activity and stability in organic solvents [39,40]. In particular, *Candida antarctica* lipase B (CALB; Novozyme 435) prefers anhydrous conditions and it has been widely applied in esterification and hydrolysis studies [41–46]. CALB could also catalyze amidation reactions [47–52] when the amine is used as a nucleophile in anhydrous organic media. In these reactions, amine amidation with a free carboxylic acid takes place, resulting in the amide product. A previous study reported by the Manova group [53] showed that CALB could be a simple and convenient biocatalyst for the efficient, direct amidation of free carboxylic acids with amines applied for a wide range of substrates, including lipoic acid.

Thus, the CALB enzymatic approach could offer the possibility of accomplishing direct amide coupling in an efficient and sustainable way without any additives in green organic solvents, providing amides with high yields and excellent purity.

Herein, we report a sustainable amidation strategy through CALB-catalyzed coupling of free acyclic carboxylic acids with different primary and secondary amines in a prominent green solvent (Scheme 1). In order to follow the progress of the enzymatic reactions, we also develop an adequate gas chromatography–mass spectrometry (GC-MS) analytic method.

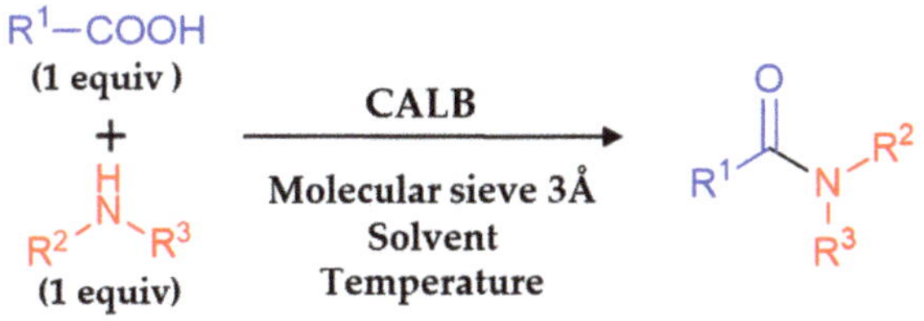

Scheme 1. CALB-catalyzed synthesis of amides.

2. Results and Discussion

In view of the results on the enzymatic amidation of free carboxylic acids with amines [49], CALB-catalyzed amidation of octanoic acid (**1**) with benzylamine (**5**) in the presence of a molecular sieve in toluene at 60 °C was performed (Figure 1a curve I). An excellent conversion of >99% after 30 min was observed.

In order to increase the reaction rate, further preliminary experiments were performed at different temperatures (Figure 1a curves II–IV). When the amidation was performed at 25 °C, the desired product with 78% conversion was obtained after 30 min (curve IV). By applying a higher temperature of 50 °C, the conversion of the reaction improved remarkably, reaching a >99% conversion in 60 min (curve III). However, at temperatures higher than 60 °C, the conversion decreased slightly, because of the thermal denaturation process of CALB protein chains (curve II). The optimal temperature of 60 °C was chosen for further reactions.

Next, we screened organic solvents with different types of polarity such as acetonitrile and N,N-dimethylformamide (DMF). In addition, we focused on the application of greener alternative solvents, such as propylene carbonate (PC), 2-methyltetrahydrofuran (2-MeTHF), diisopropyl ether (DIPE), and cyclopentyl methyl ether (CPME). While the reaction rate in acetonitrile was relatively low, CPME and PC were the most promising green solvents with conversions reaching >99% in 30 min (Figure 1b).

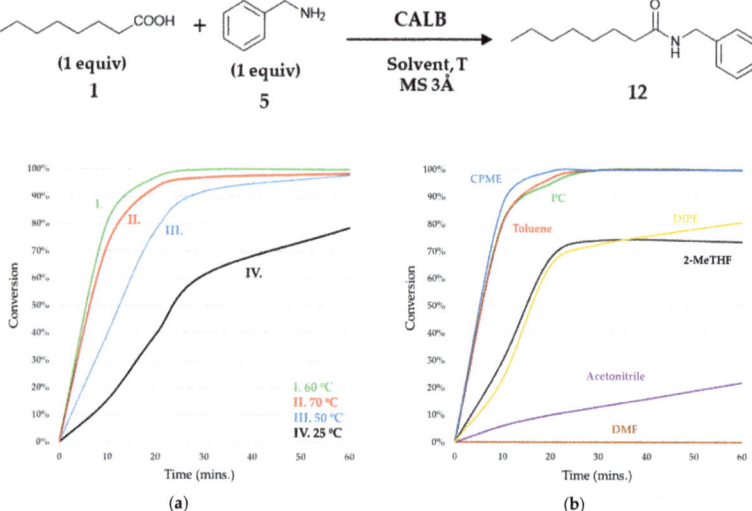

Figure 1. The effect of temperature (**a**) and solvents (**b**) on the reaction conversion catalyzed by CALB.

In an attempt to increase the efficiency of the present method, the initial concentration of the substrate of 46 mM was increased to 92, 460, and 920 mM (Figure 2a,b). Reactions were performed in CPME and PC solvents with a constant enzyme concentration of 50 mg mL^{-1}. Slightly lower reaction rates with 92 mM concentration (curves II(**a**) and (**b**)) were observed. Despite the much lower reaction rates found at concentrations of 460 and 920 mM, the amide formation was still significant after 60 min (curves III(**a**) and (**b**) and IV(**a**) and (**b**)). Such robust behavior of commercially available CALB might be an important parameter not only for laboratory-scale but also industrial-scale reactions [54,55].

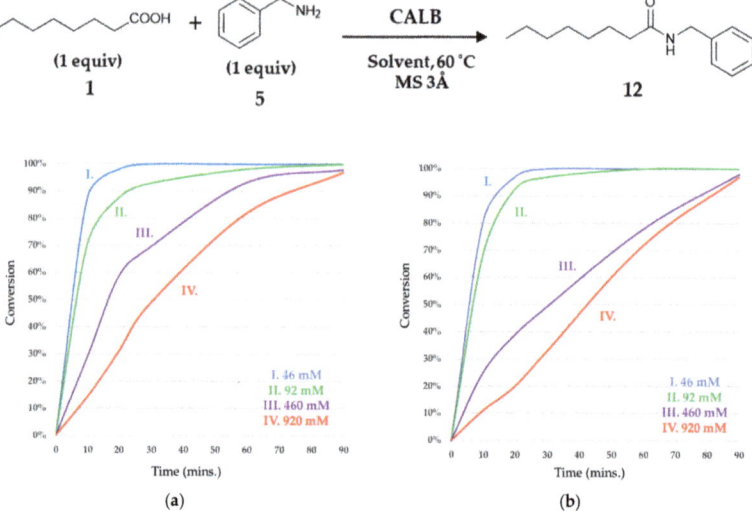

Figure 2. The efficiency of amide formation catalyzed by CALB in CPME (**a**) and PC (**b**).

Having in hand the optimal conditions (CALB, molecular sieve 3 Å, CPME solvent, 60 °C, substrate concentration 920 mM), we performed further amidation reactions and obtained 28 different amides (**12–39**) with excellent conversions (>92%) and yields (>90%) in 90 min (Table 1). The study involved the use of four different free carboxylic acids and seven amines including primary, secondary, and secondary cyclic amines. According to GC-MS analysis, all reactions were completed in 90 min. The product molecules might be used as potential intermediates or building blocks in the synthesis of biologically active compounds [56,57].

Table 1. Substrate scope of amide formation in CPME solvent with conversion data.

$$R^1\text{-COOH} + R^2\text{-NH-}R^3 \xrightarrow[\text{CPME, 60°C, MS 3Å, 90 min}]{\text{CALB}} R^1\text{-C(O)-N}R^2R^3$$

(1 equiv, c = 920 mM) + (1 equiv, c = 920 mM)

1–4 + **5–11** → **12–39**

Substrates	Octanoic Acid (1)	Hexanoic Acid (2)	Butyric Acid (3)	4-Phenylbutyric Acid (4)
benzylamine (**5**)	**12**, 99%	**13**, 98%	**14**, 98%	**15**, 99%
allylamine (**6**)	**16**, 96%	**17**, 96%	**18**, 94%	**19**, 98%
propargylamine (**7**)	**20**, 97%	**21**, 98%	**22**, 92%	**23**, 97%
piperidine (**8**)	**24**, 95%	**25**, 93%	**26**, 92%	**27**, 97%
morpholine (**9**)	**28**, 99%	**29**, 98%	**30**, 94%	**31**, 96%
N^1,N^1-dimethylethane-1,2-diamine (**10**)	**32**, 96%	**33**, 98%	**34**, 93%	**35**, 97%
N^1,N^1-dimethylpropane-1,3-diamine (**11**)	**36**, 99%	**37**, 98%	**38**, 92%	**39**, 99%

3. Materials and Methods

3.1. General

All solvents and reagents were of analytical grade and used directly without further purification. The lipase acrylic resin (CALB, ≥5000 U/g, recombinant, expressed in Aspergillus niger, quality level 200, Catalogue no. L4777), cyclopentyl methyl ether (inhibitor-free, anhydrous, ≥99.9%), propylene carbonate (ReagentPlus®, 99%), toluene (anhydrous, 99.8%), 2-methyltetrahydrofuran (BioRenewable, anhydrous, ≥99%, inhibitor-free), acetonitrile (suitable for HPLC, gradient grade, ≥99.9%), N,N-dimethylformamide (suitable for HPLC, ≥99.9%), and 3 Å molecular sieve (beads, 8–12 mesh) used in this study were purchased from Merck Life Science Kft., an affiliate of Merck KGaA, Darmstadt, Germany (Budapest, Hungary). All reactions were carried out in an Eppendorf™ Innova™ 40R Incubator Shaker. Melting points were determined on a Kofler apparatus. ^{1}H NMR and

^{13}C NMR spectra were recorded on a Bruker Avance NEO 500.1 spectrometer, in CDCl$_3$ as solvent, with tetramethylsilane as an internal standard at 500.1 and 125 MHz, respectively. GC-MS analyses were performed on a Thermo Scientific Trace 1310 Gas Chromatograph coupled with a Thermo Scientific ISQ QD Single-Quadrupole Mass Spectrometer using a Thermo Scientific TG-SQC column (15 m × 0.25 mm ID × 0.25 µm film). Measurement parameters were as follows: column oven temperature: from 50 to 300 °C at 15 °C min^{-1}; injection temperature: 240 °C; ion source temperature: 200 °C; electrospray ionization: 70 eV; carrier gas: He at 1.5 mL min^{-1}; injection volume: 5 µL; split ratio: 1:50; and mass range: 25–500 m/z.

3.2. General Procedure for CALB-Catalyzed Amidation

In preliminary amidation experiments, the carboxylic acid and amine substrates (1:1 equiv) were dissolved in an organic solvent (1 mL) to provide a solution with a given concentration (46, 92, 460, or 920 mM). CALB (50 mg), molecular sieve (50 mg 3 Å size) to avoid the reversible hydrolysis reaction, and *n*-heptadecane (2 µL) as an internal standard were added to the above solution. The mixture was shaken at a selected temperature (25, 50, 60 or 70 °C) in an incubator shaker. The progress of the reaction was followed by taking samples from the reaction mixture at intervals and analyzing them by GC-MS measurements. Conversion of the starting materials (moles of the converted molecules/moles of the initial starting materials × 100) was calculated by *n*-heptadecane as an internal standard, in percent yield of the desired amides (actual yield/theoretical yield × 100) (see Supplementary Information). In order to obtain reliable yields, we filtered the CPME samples through a silica gel plug followed by vacuum evaporation of the solvent. All samples were analyzed by ^{1}H and ^{13}C NMR spectroscopy without any prior purification. Amidations, to form amides **12–39**, were performed with the above protocol under the optimized conditions (50 mg CALB, 920 mM substrate concentration, 1 mL solvent CPME, 50 mg molecular sieve, 2 µL *n*-heptadecane, 60 °C).

3.2.1. N-Benzyloctanamide (**12**)

CAS number: 70659-87-9, white solid, mp = 65.1–66.3 °C [58]. ^{1}H-NMR (CDCl$_3$, 500.1 MHz): δ = 7.25–7.34 (m, 5H, Ar), 5.69 (s, 1H, NH), 4.44 (d, *J* = 5.69 Hz, 2H, CH2), 2.20 (t, *J* = 7.86 Hz, 2H, CH2CO), 1.62–1.68 (m, 2H, CH2), 1.27–1.33 (m, 8H, CH2CH2CH2CH2), 0.87 (t, *J* = 7.16 Hz, 3H, CH3), ^{13}C-NMR (CDCl$_3$, 125 MHz): δ = 14.05, 22.59, 25.77, 28.99, 29.27, 29.69, 31.68, 36.84, 43.60, 127.50, 127.84, 128.71, 138.45, 172.93.

3.2.2. N-Benzylhexanamide (**13**)

CAS number: 6283-98-3, yellowish white solid, mp = 55.1–55.5 °C [59]. ^{1}H-NMR (CDCl$_3$, 500.1 MHz): δ = 7.25–7.33 (m, 5H, Ar), 5.84 (s, 1H, NH), 4.42 (d, *J* = 5.73 Hz, 2H, CH2), 2.19 (t, *J* = 7.87 Hz, 2H, CH2CO), 1.62–1.68 (m, 2H, CH2), 1.27–1.34 (m, 4H, CH2CH2), 0.88 (t, *J* = 7.04 Hz, 3H, CH3), ^{13}C-NMR (CDCl$_3$, 125 MHz): δ = 13.92, 22.39, 25.45, 29.69, 31.48, 36.75, 43.56, 53.42, 127.46, 127,80, 128.68, 138.47, 173.03.

3.2.3. N-Enzylbutyramide (**14**)

CAS number: 10264-14-9, yellowish solid, mp = 43.0–43.6 °C [60]. ^{1}H-NMR (CDCl$_3$, 500.1 MHz): δ = 7.23–7.31 (m, 5H, Ar), 6.14 (s, 1H, NH), 4.39 (d, *J* = 5.75 Hz, 2H, CH2), 2.16 (t, *J* = 7.64 Hz, 2H, CH2CO), 1.62–1.69 (m, 2H, CH2), 0.93 (t, *J* = 7.42 Hz, 3H, CH3), ^{13}C-NMR (CDCl$_3$, 125 MHz): δ = 13.78, 19.18, 29.69, 38.58, 43.46, 53.45, 127.38, 127.73, 128.63, 138.54, 173.00.

3.2.4. N-Benzyl-4-phenylbutanamide (**15**)

CAS number: 179923-27-4, yellowish solid, mp = 79.2–80.2 °C [61]. ^{1}H-NMR (CDCl$_3$, 500.1 MHz): δ = 7.13–7.32 (m, 10H, Ar), 5.84 (s, 1H, NH), 4.39 (d, *J* = 5.73 Hz, 2H, CH2), 2.63 (t, *J* = 7.54 Hz, 2H, CH2), 2.18 (t, *J* = 7.78 Hz, 2H, CH2CO), 1.94–2.00 (m, 2H, CH2),

^{13}C-NMR (CDCl$_3$, 125 MHz): δ = 27.13, 29.72, 35.22, 35.85, 43.59, 46.42, 70.52, 72.51, 125.99, 127.51, 127.83, 128.41, 128.51, 128.71, 138.41, 141.48, 172, 57.

3.2.5. N-Allyloctanamide (16)

CAS number: 70659-85-7, yellow solid, mp = 27.6–28.3 °C [62]. ^{1}H-NMR (CDCl$_3$, 500.1 MHz): δ = 5.84 (ddt, J = 15.97 Hz, 10.29 Hz, 5.09 Hz, 1H, CH), 5.53 (s, 1H, NH), 5.11–5.19 (m, 2H, CH2), 3.88 (t, J = 5.73 Hz, 2H, CH2), 2.19 (t, J = 7.79 Hz, 2H, CH2CO), 1.61–1.66 (m, 2H, CH2), 1,25–1.31 (m, 8H, CH2CH2CH2CH2), 0.87 (t, J = 7.03 Hz, 3H, CH3), ^{13}C-NMR (CDCl$_3$, 125 MHz): δ = 14.04, 22.59, 25.77, 29.00, 29.27, 31.67, 36.81, 41.86, 116.27, 134.42, 172.94.

3.2.6. N-Allylhexanamide (17)

CAS number: 128007-44-3, yellowish oil. ^{1}H-NMR (CDCl$_3$, 500.1 MHz): δ = 5.84 (ddt, J = 15.96 Hz, 10.27 Hz, 5.04 Hz, 1H, CH), 5.55 (s, 1H, NH), 5.11–5.20 (m, 2H, CH2), 3.88 (t, J = 5.74 Hz, 2H, CH2), 2.19 (t, J = 7.81 Hz, 2H, CH2CO), 1.61–1.67 (m, 2H, CH2), 1.29–1.35 (m, 4H, CH2CH2), 0.89 (t, J = 7.08 Hz, 3H, CH3), ^{13}C-NMR (CDCl$_3$, 125 MHz): δ = 13.91, 22.39, 25.45, 31.47, 36.77, 41.86, 116.29, 134.41, 172.97.

3.2.7. N-Allylbutyramide (18)

CAS number: 2978-29-2, yellowish oil. ^{1}H-NMR (CDCl$_3$, 500.1 MHz): δ = 5.84 (ddt, J = 16.00 Hz, 10.32 Hz, 5.06 Hz, 1H, CH), 5.46 (s, 1H, NH), 5.12–5.20 (m, 2H, CH2), 3.89 (t, J = 5.74 Hz, 2H, CH2), 2.17 (t, J = 7.68 Hz, 2H, CH2CO), 1.64–1.71 (m, 2H, CH2), 0.96 (t, J = 7.36 Hz, 3H, CH3), ^{13}C-NMR (CDCl$_3$, 125 MHz): δ = 13.77, 19.16, 29.69, 38.72, 41.86, 116.33, 134.40.

3.2.8. N-Allyl-4-phenylbutanamide (19)

CAS number: 430450-20-7, yellow oil. ^{1}H-NMR (CDCl$_3$, 500.1 MHz): δ = 7.17–7.29 (m, 5H, Ar), 5.83 (ddt, J = 16.00 Hz, 10.28 Hz, 5.06 Hz, 1H, CH), 5.40 (s, 1H, NH), 5.11–5.19 (m, 2H, CH2), 3.87 (t, J = 5.77 Hz, 2H, CH2), 2.66 (t, J = 7.53 Hz, 2H, CH2), 2.19 (t, J = 7.84 Hz, 2H, CH2CO), 1.94–2.02 (m, 2H, CH2), ^{13}C-NMR (CDCl$_3$, 125 MHz): δ = 27.08, 29.70, 35.19, 35.86, 41.91, 116.44, 125.98, 128.40, 128.49, 134.30, 141.46, 172.42.

3.2.9. N-(Prop-2-yn-1-yl)octanamide (20)

CAS number: 422284-34-2, white solid, mp = 72.4–73.4 °C [63]. ^{1}H-NMR (CDCl$_3$, 500.1 MHz): δ = 5.60 (s, 1H, NH), 4.06 (dd, J = 5.18 Hz, J = 2.53 Hz 1H, CH2), 4.04 (dd, J = 5.18 Hz, J = 2.53 Hz, 1H, CH2), 2.22 (t, J = 2.55 Hz, 1H, CH), 2.19 (t, J = 7.77 Hz, 2H, CH2CO), 1.62–1.66 (m, 2H, CH2), 1.25–1.30 (m, 8H, CH2CH2CH2CH2), 0.87 (t, J = 7.05 Hz, 3H, CH3), ^{13}C-NMR (CDCl$_3$, 125 MHz): δ = 14.04, 22.58, 25.54, 28.97, 29.14, 29.20, 31.65, 36.49, 71.53, 79.67, 172.70.

3.2.10. N-(Prop-2-yn-1-yl)hexanamide (21)

CAS number: 62899-12-1, white solid, mp = 47.3–48.0 °C [64]. ^{1}H-NMR (CDCl$_3$, 500.1 MHz): δ = 5.74 (s, 1H, NH), 4.05 (dd, J = 5.21 Hz, J = 2.54 Hz 1H, CH2), 4.04 (dd, J = 5.21 Hz, J = 2.54 Hz, 1H, CH2), 2.22 (t, J = 2.53 Hz, 1H, CH), 2.19 (t, J = 7.84 Hz, 2H, CH2CO), 1.61–1.67 (m, 2H, CH2), 1.29–1.35 (m, 4H, CH2CH2), 0.89 (t, J = 6.98 Hz, 3H, CH3), ^{13}C-NMR (CDCl$_3$, 125 MHz): δ = 13.89, 22.36, 25.23, 29.13, 31.40, 36.42, 71.49, 79.68, 172.80.

3.2.11. N-(Prop-2-yn-1-yl)butyramide (22)

CAS number: 2978-28-1, yellowish solid, mp = 26.1–26.6 °C [65]. ^{1}H-NMR (CDCl$_3$, 500.1 MHz): δ = 5.56 (s, 1H, NH), 4.06 (dd, J = 5.21 Hz, J = 2.54 Hz 1H, CH2), 4.05 (dd, J = 5.19 Hz, J = 2.52 Hz, 1H, CH2), 2.22 (t, J = 2.55 Hz, 1H, CH), 2.17 (t, J = 7.66 Hz, 2H, CH2CO), 1.64–1.71 (m, 2H, CH2), 0.95 (t, J = 7.42 Hz, 3H, CH3), ^{13}C-NMR (CDCl$_3$, 125 MHz): δ = 13.71, 18.97, 29.14, 29.69, 38.37, 71.55.

3.2.12. 4-Phenyl-N-(prop-2-yn-1-yl)butanamide (**23**)

CAS number: 1250568-47-8, colorless oil. ^{1}H-NMR (CDCl$_3$, 500.1 MHz): δ = 7.16–7.29 (m, 5H, Ar), 5.54 (s, 1H, NH), 4.04 (dd, J = 5.24 Hz, J = 2.56 Hz 1H, CH2), 4.03 (dd, J = 5.19 Hz, J = 2.51 Hz, 1H, CH2), 2.66 (t, J = 7.53 Hz, 2H, CH2), 2.22 (t, J = 2.58 Hz, 1H, CH), 2.19 (t, J = 7.83 Hz, 2H, CH2CO), 1.95–2.01 (m, 2H, CH2), ^{13}C-NMR (CDCl$_3$, 125 MHz): δ = 26.84, 29.15, 29.17, 35.08, 35.49, 71.59, 79.58, 126.02, 128.42, 128.50, 141.33, 172.22.

3.2.13. 1-(Piperidin-1-yl)octan-1-one (**24**)

CAS number: 20299-83-6, colorless oil. ^{1}H-NMR (CDCl$_3$, 500.1 MHz): δ = 3.54 (t, J = 5.47 Hz, 2H, CH2N), 3.39 (t, J = 5.36 Hz, 2H, CH2N), 2.30 (t, J = 7.91 Hz, 2H, CH2CO), 1.50–1.66 (m, 8H, CH2CH2CH2CH2), 1.25–1.32 (m, 8H, CH2CH2CH2CH2), 0.87 (t, J = 7.01 Hz, 3H, CH3), ^{13}C-NMR (CDCl$_3$, 125 MHz): δ = 14.05, 22.60, 24.60, 25.50, 25.59, 26.58, 29.10, 29.50, 31.72, 33.49, 42.58, 46.72, 171.52.

3.2.14. 1-(Piperidin-1-yl)hexan-1-one (**25**)

CAS number: 15770-38-4, yellowish oil. ^{1}H-NMR (CDCl$_3$, 500.1 MHz): δ = 3.54 (t, J = 5.47 Hz, 2H, CH2N), 3.39 (t, J = 5.47 Hz, 2H, CH2N), 2.30 (t, J = 7.77 Hz, 2H, CH2CO), 1.51–1.65 (m, 8H, CH2CH2CH2CH2), 1.30–1.35 (m, 4H, CH2CH2), 0.90 (t, J = 6.91 Hz, 3H, CH3), ^{13}C-NMR (CDCl$_3$, 125 MHz): δ = 13.96, 22.49, 24.61, 25.18, 25.60, 26.59, 31.72, 33.45, 42.59, 46.72, 171.53.

3.2.15. 1-(Piperidin-1-yl)butan-1-one (**26**)

CAS number: 4637-70-1, yellow oil. ^{1}H-NMR (CDCl$_3$, 500.1 MHz): δ = 3.54 (t, J = 5.48 Hz, 2H, CH2N), 3.39 (t, J = 5.34 Hz, 2H, CH2N), 2.29 (t, J = 7.78 Hz, 2H, CH2CO), 1.50–1.69 (m, 8H, CH2CH2CH2CH2), 0.96 (t, J = 7.41 Hz, 3H, CH3), ^{13}C-NMR (CDCl$_3$, 125 MHz): δ = 14.04, 18.88, 24.61, 25.61, 26.58, 35.40, 42.58, 46.71, 171.34.

3.2.16. 4-Phenyl-1-(piperidin-1-yl)butan-1-one (**27**)

CAS number: 41208-51-9, white solid, mp = 153.9–154.4 °C [66]. ^{1}H-NMR (CDCl$_3$, 500.1 MHz): δ = 7.16–7.29 (m, 5H, Ar), 3.54 (t, J = 5.53 Hz, 2H, CH2N), 3.31 (t, J = 5.47 Hz, 2H, CH2N), 2.67 (t, J = 7.66 Hz, 2H, CH2), 2.32 (t, J = 7.84 Hz, 2H, CH2CO), 1.93–1.99 (m, 2H, CH2), 1.49–1.64 (m, 6H, CH2CH2CH2), ^{13}C-NMR (CDCl$_3$, 125 MHz): δ = 24.58, 25.60, 26.54, 26.84, 29.69, 33.54, 35.42, 42.63, 46.61, 125.86, 128.33, 128.49, 141.85, 171.01.

3.2.17. 1-Morpholinooctan-1-one (**28**)

CAS number: 5338-65-8, colorless oil. ^{1}H-NMR (CDCl$_3$, 500.1 MHz): δ = 3.66 (t, J = 5.14 Hz, 4H, CH2OCH2), 3.61 (t, J = 3.75 Hz, 2H, CH2N), 3.46 (t, J = 4.53 Hz, 2H, CH2N), 2.30 (t, J = 7.83 Hz, 2H, CH2CO), 1.59–1.65 (m, 2H, CH2), 1.25–1.34 (m, 8H, CH2CH2CH2CH2), 0.88 (t, J = 7.08 Hz, 3H, CH3), ^{13}C-NMR (CDCl$_3$, 125 MHz): δ = 14.03, 22.58, 25.25, 29.05, 29.40, 31.68, 33.11, 41.85, 46.06, 66.68, 66.95, 171.89.

3.2.18. 1-Morpholinohexan-1-one (**29**)

CAS number: 17598-10-6, yellowish oil. ^{1}H-NMR (CDCl$_3$, 500.1 MHz): δ = 3.66 (t, J = 5.18 Hz, 4H, CH2OCH2), 3.61 (t, J = 3.75 Hz, 2H, CH2N), 3.46 (t, J = 4.56 Hz, 2H, CH2N), 2.30 (t, J = 7.85 Hz, 2H, CH2CO), 1.60–1.66 (m, 2H, CH2), 1.30–1.36 (m, 4H, CH2CH2), 0.90 (t, J = 6.95 Hz, 3H, CH3), ^{13}C-NMR (CDCl$_3$, 125 MHz): δ = 13.92, 22.45, 24.93, 31.62, 33.07, 41.86, 46.06, 66.68, 66.96, 171.90.

3.2.19. 1-Morpholinobutan-1-one (**30**)

CAS number: 5327-51-5, colorless oil. ^{1}H-NMR (CDCl$_3$, 500.1 MHz): δ = 3.66 (t, J = 5.20 Hz, 4H, CH2OCH2), 3.61 (t, J = 4.56 Hz, 2H, CH2N), 3.46 (t, J = 4.56 Hz, 2H, CH2N), 2.29 (t, J = 7.70 Hz, 2H, CH2CO), 1.63–1.70 (m, 2H, CH2), 0.97 (t, J = 7.46 Hz, 3H, CH3), ^{13}C-NMR (CDCl$_3$, 125 MHz): δ = 13.96, 18.66, 35.02, 41.86, 46.05, 66.69, 66.97, 171.75.

3.2.20. 1-Morpholino-4-phenylbutan-1-one (31)

CAS number: 61123-44-2, white solid, mp = 40.7–42.5 °C [67]. ^{1}H-NMR (CDCl$_3$, 500.1 MHz): δ = 7.17–7.29 (m, 5H, Ar), 3.59–3.66 (m, 6H, NCH2CH2OCH2), 3.37 (t, J = 4.68 Hz, 2H, NCH2), 2.68 (t, J = 7.59 Hz, 2H, CH2), 2.30 (t, J = 7.71 Hz, 2H, CH2CO), 1.95–2.01 (m, 2H, CH2), ^{13}C-NMR (CDCl$_3$, 125 MHz): δ = 26.54, 29.70, 32.10, 35.26, 45.89, 45.92, 66.63, 66.95, 125.98, 128.40, 128.48, 141.57, 171.42.

3.2.21. N-(2-(Dimethylamino)ethyl)octanamide (32)

CAS number: 114011-26-6, colorless oil. ^{1}H-NMR (CDCl$_3$, 500.1 MHz): δ = 6.14 (s, 1H, NH), 3.30–3.34 (m, 2H, CONHCH2), 2.41 (t, J = 5.90 Hz, 2H, CH2), 2.23 (s, 6H, CH3CH3), 2.17 (t, J = 7.82 Hz, 2H, CH2CO), 1.59–1.65 (m, 2H, CH2), 1.27–1.30 (m, 8H, CH2CH2CH2CH2), 0.87 (t, J = 7.09 Hz, 3H, CH3), ^{13}C-NMR (CDCl$_3$, 125 MHz): δ = 14.04, 22.58, 25.78, 29.00, 29.26, 29.68, 31.69, 36.64, 36.73, 45.07, 57.94, 173.37.

3.2.22. N-(2-(Dimethylamino)ethyl)hexanamide (33)

CAS number: 114011-25-5, colorless oil. ^{1}H-NMR (CDCl$_3$, 500.1 MHz): δ = 6.07 (s, 1H, NH), 3.31–3.34 (m, 2H, CONHCH2), 2.41 (t, J = 5.90 Hz, 2H, CH2), 2.23 (s, 6H, CH3CH3), 2.17 (t, J = 7.81 Hz, 2H, CH2CO), 1.60–1.66 (m, 2H, CH2), 1.28–1.36 (m, 4H, CH2CH2), 0.89 (t, J = 7.04 Hz, 3H, CH3), ^{13}C-NMR (CDCl$_3$, 125 MHz): δ = 13.94, 22.41, 25.46, 29.69, 31.49, 36.62, 36.72, 45.08, 57.91, 173.30.

3.2.23. N-(2-(Dimethylamino)ethyl)butyramide (34)

CAS number: 63224-16-8, yellowish oil. ^{1}H-NMR (CDCl$_3$, 500.1 MHz): δ = 6.12 (s, 1H, NH), 3.31–3.35 (m, 2H, CONHCH2), 2.42 (t, J = 5.88 Hz, 2H, CH2), 2.24 (s, 6H, CH3CH3), 2.15 (t, J = 7.69 Hz, 2H, CH2CO), 1.62–1.70 (m, 2H, CH2), 0.94 (t, J = 7.36 Hz, 3H, CH3), ^{13}C-NMR (CDCl$_3$, 125 MHz): δ = 13.77, 19.19, 29.69, 36.57, 38.65, 45.03, 57.19, 173.14.

3.2.24. N-(2-(Dimethylamino)ethyl)-4-phenylbutanamide (35)

CAS number: 63224-25-9, yellowish oil. ^{1}H-NMR (CDCl$_3$, 500.1 MHz): δ = 7.17–7.28 (m, 5H, Ar), 3.30–3.33 (m, 2H, CONHCH2), 2.65 (t, J = 7.60 Hz, 2H, CH2), 2.40 (t, J = 5.88 Hz, 2H, CH2), 2.22 (s, 6H, CH3CH3), 2.18 (t, J = 7.79 Hz, 2H, CH2CO), 1.94–2.00 (m, 2H, CH2), ^{13}C-NMR (CDCl$_3$, 125 MHz): δ = 22.68, 27.13, 29.35, 29.69, 35.23, 25.84, 36.66, 45.07, 57.90, 125.91, 128.35, 128.51, 141.61, 172.85.

3.2.25. N-(3-(Dimethylamino)propyl)ctanamide (36)

CAS number: 22890-10-4, colorless oil. ^{1}H-NMR (CDCl$_3$, 500.1 MHz): δ = 6.89 (s, 1H, NH), 3.30–3.34 (m, 2H, CONHCH2), 2.37 (t, J = 6.41 Hz, 2H, CH2), 2.23 (s, 6H, CH3CH3), 2.14 (t, J = 7.80 Hz, 2H, CH2CO), 1.57–1.68 (m, 4H, CH2CH2), 1.25–1.30 (m, 8H, CH2CH2CH2CH2), 0.87 (t, J = 7.04 Hz, 3H, CH3), ^{13}C-NMR (CDCl$_3$, 125 MHz): δ = 14.04, 22.59, 25.75, 26.14, 29.02, 29.26, 29.68, 31.68, 36.96, 39.15, 45.34, 58.50, 173.16.

3.2.26. N-(3-(Dimethylamino)propyl)hexanamide (37)

CAS number: 73603-23-3, yellowish oil. ^{1}H-NMR (CDCl$_3$, 500.1 MHz): δ = 6.92 (s, 1H, NH), 3.30–3.34 (m, 2H, CONHCH2), 2.37 (t, J = 6.45 Hz, 2H, CH2), 2.22 (s, 6H, CH3CH3), 2.14 (t, J = 7.77 Hz, 2H, CH2CO), 1.57–1.67 (m, 4H, CH2CH2), 1.28–1.34 (m, 4H, CH2CH2), 0.89 (t, J = 7.16 Hz, 3H, CH3), ^{13}C-NMR (CDCl$_3$, 125 MHz): δ = 13.90, 22.40, 25.42, 26.17, 29.67, 31.45, 36.89, 39.15, 45.35, 58.49, 173.18.

3.2.27. N-(3-(Dimethylamino)propyl)butyramide (38)

CAS number: 53201-67-5, colorless oil. ^{1}H-NMR (CDCl$_3$, 500.1 MHz): δ = 6.89 (s, 1H, NH), 3.32–3.35 (m, 2H, CONHCH2), 2.43 (t, J = 6.43 Hz, 2H, CH2), 2.27 (s, 6H, CH3CH3), 2.13 (t, J = 7.63 Hz, 2H, CH2CO), 1.60–1.70 (m, 4H, CH2CH2), 0.94 (t, J = 7.39 Hz, 3H, CH3), ^{13}C-NMR (CDCl$_3$, 125 MHz): δ = 13.80, 19.12, 26.01, 29.69, 38.87, 38.98, 45.16, 58.36, 172.97.

3.2.28. N-(3-(Dimethylamino)propyl)-4-phenylbutanamide (39)

CAS number: 885912-19-6, yellowish oil. ^{1}H-NMR (CDCl$_3$, 500.1 MHz): δ = 7.16–7.28 (m, 5H, Ar), 6.88 (s, 1H, NH), 3.30–3.34 (m, 2H, CONHCH2), 2.64 (t, *J* = 7.64 Hz, 2H, CH2), 2.38 (t, *J* = 6.42 Hz, 2H, CH2), 2.22 (s, 6H, CH3CH3), 2.16 (t, *J* = 7.78 Hz, 2H, CH2CO), 1.92–1.98 (m, 2H, CH2), 1.62–1.67 (m, 2H, CH2), ^{13}C-NMR (CDCl$_3$, 125 MHz): δ = 14.11, 26.11, 27.19, 29.69, 35.25, 36.09, 39.17, 45.30, 53.42, 58.52, 125.89, 128.34, 128.46, 141.64, 172.61.

4. Conclusions

We successfully developed a sustainable and green enzymatic strategy for the synthesis of amides from free carboxylic acids and amines by using *Candida antarctica* lipase B as a biocatalyst, using GC-MS analysis to monitor the reaction progress. The green enzymatic amidation is simple and efficient without any additives, with the application of cyclopentyl methyl ether as the solvent, which is a greener and safer solvent alternative in comparison with the usual organic solvents. The scope of the reaction was extended to the preparation of 28 diverse amides, by using four different free carboxylic acids and seven amines, including primary and secondary amines as well as cyclic amines. In every case, excellent conversions and yields were achieved without the need of any intensive purification step. This enzymatic methodology offers a way to synthetize pure amides. That is, this synthetic approach is both eco-friendly and practical for large-scale production.

Supplementary Materials: The following supporting information can be downloaded at https://www.mdpi.com/article/10.3390/molecules28155706/s1, Figures S1–S62: ^{1}H-NMR and ^{13}C-NMR spectra of **12–39** amide products and GC-MS measurement.

Author Contributions: E.F. and G.O. planned and designed the project; G.O. and S.S. performed the syntheses and characterized the synthesized compounds; E.F. and G.O. prepared the manuscript for publication. All authors discussed the results and commented on the manuscript. All authors have read and agreed to the published version of the manuscript.

Funding: The authors' thanks are due to the Hungarian Research Foundation (OTKA No. K-138871), the Ministry of Human Capacities, Hungary (grant TKP-2021-EGA-32).

Institutional Review Board Statement: Not applicable.

Informed Consent Statement: Not applicable.

Data Availability Statement: The data presented in this study are available in the article and Supplementary Materials.

Conflicts of Interest: The authors declare no conflict of interest.

Sample Availability: Samples of compounds **12–39** are available from the authors.

References

1. Mahesh, S.; Tang, K.-C.; Raj, M. Amide Bond Activation of Biological Molecules. *Molecules* **2018**, *23*, 2615. [CrossRef]
2. Pattabiraman, V.R.; Bode, J.W. Rethinking amide bond synthesis. *Nature* **2011**, *480*, 471–479. [CrossRef]
3. Wieland, T.; Bodanszky, M. *The World of Peptides: A Brief History of Peptide Chemistry*; Springer: Berlin/Heidelberg, Germany, 2012; pp. 1–22. ISBN 978-3-642-75850-8.
4. Ohashi, N.; Kohno, T. Analgesic Effect of Acetaminophen: A Review of Known and Novel Mechanisms of Action. *Front. Pharmacol.* **2020**, *11*, 580289. [CrossRef]
5. Day, T.K.; Skarda, R.T. The pharmacology of local anesthetics. *Vet. Clin. N. Am. Large Anim. Pract.* **1991**, *7*, 489–500. [CrossRef] [PubMed]
6. Becker, D.E.; Reed, K.L. Local anesthetics: Review of pharmacological considerations. *Anesth. Prog.* **2012**, *59*, 90–102. [CrossRef] [PubMed]
7. Lima, L.M.; Silva, B.; Barbosa, G.; Barreiro, E.J. β-lactam antibiotics: An overview from a medicinal chemistry perspective. *Eur. J. Med. Chem.* **2020**, *208*, 112829. [CrossRef]
8. Turner, J.; Muraoka, A.; Bedenbaugh, M.; Childress, B.; Pernot, L.; Wiencek, M.; Peterson, Y.K. The Chemical Relationship Among Beta-Lactam Antibiotics and Potential Impacts on Reactivity and Decomposition. *Front. Microbiol.* **2022**, *13*, 807955. [CrossRef]
9. King, D.T.; Sobhanifar, S.; Strynadka, N.C.J. One ring to rule them all: Current trends in combating bacterial resistance to the β-lactams. *Protein Sci.* **2016**, *25*, 787–803. [CrossRef]
10. Saxena, P.; Sharma, P.K.; Purohit, P. A journey of celecoxib from pain to cancer. *Prostaglandins Other Lipid Mediat.* **2020**, *147*, 106379. [CrossRef]

11. Sadée, W.; Bohn, L. How specific are "target-specific" drugs? Celecoxib as a case in point. *Mol. Interv.* **2006**, *6*, 196–198. [CrossRef]
12. Bering, L.; Craven, E.J.; Sowerby Thomas, S.A.; Shepherd, S.A.; Micklefield, J. Merging enzymes with chemocatalysis for amide bond synthesis. *Nat. Commun.* **2022**, *13*, 380. [CrossRef]
13. Allen, C.L.; Williams, J.M.J. Metal-catalysed approaches to amide bond formation. *Chem. Soc. Rev.* **2011**, *40*, 3405–3415. [CrossRef]
14. Massolo, E.; Pirola, M.; Benaglia, M. Amide Bond Formation Strategies: Latest Advances on a Dateless Transformation. *Eur. J. Org. Chem.* **2020**, *2020*, 4641–4651. [CrossRef]
15. Todorovic, M.; Perrin, D.M. Recent developments in catalytic amide bond formation. *Pept. Sci.* **2020**, *112*, e24210. [CrossRef]
16. El-Faham, A.; Albericio, F. Peptide Coupling Reagents, More than a Letter Soup. *Chem. Rev.* **2011**, *111*, 6557–6602. [CrossRef]
17. Valeur, E.; Bradley, M. Amide bond formation: Beyond the myth of coupling reagents. *Chem. Soc. Rev.* **2009**, *38*, 606–631. [CrossRef] [PubMed]
18. Orsy, G.; Fülöp, F.; Mándity, I.M. Direct amide formation in a continuous-flow system mediated by carbon disulfide. *Catal. Sci. Technol.* **2020**, *10*, 7814–7818. [CrossRef]
19. Lundberg, H.; Tinnis, F.; Selander, N.; Adolfsson, H. Catalytic amide formation from non-activated carboxylic acids and amines. *Chem. Soc. Rev.* **2014**, *43*, 2714–2742. [CrossRef]
20. Hazra, S.; Gallou, F.; Handa, S. Water: An Underestimated Solvent for Amide Bond-Forming Reactions. *ACS Sustain. Chem. Eng.* **2022**, *10*, 5299–5306. [CrossRef]
21. Sabatini, M.T.; Boulton, L.T.; Sneddon, H.F.; Sheppard, T.D. A green chemistry perspective on catalytic amide bond formation. *Nat. Catal.* **2019**, *2*, 10–17. [CrossRef]
22. Zhang, R.; Yao, W.-Z.; Qian, L.; Sang, W.; Yuan, Y.; Du, M.-C.; Cheng, H.; Chen, C.; Qin, X. A practical and sustainable protocol for direct amidation of unactivated esters under transition-metal-free and solvent-free conditions. *Green Chem.* **2021**, *23*, 3972–3982. [CrossRef]
23. Dunetz, J.R.; Magano, J.; Weisenburger, G.A. Large-Scale Applications of Amide Coupling Reagents for the Synthesis of Pharmaceuticals. *Org. Process Res. Dev.* **2016**, *20*, 140–177. [CrossRef]
24. Magano, J. Large-Scale Amidations in Process Chemistry: Practical Considerations for Reagent Selection and Reaction Execution Organic. *Org. Process Res. Dev.* **2022**, *26*, 1562–1689. [CrossRef]
25. Ganesh, K.N.; Zhang, D.; Miller, S.J.; Rossen, K.; Chirik, P.J.; Kozlowski, M.C.; Zimmerman, J.B.; Brooks, B.W.; Savage, P.E.; Allen, D.T.; et al. Green Chemistry: A Framework for a Sustainable Future. *Org. Process Res. Dev.* **2021**, *25*, 1455–1459. [CrossRef]
26. Anastas, P.; Eghbali, N. Green Chemistry: Principles and Practice. *Chem. Soc. Rev.* **2010**, *39*, 301–312. [CrossRef]
27. Meyer, H.-P.; Eichhorn, E.; Hanlon, S.; Lütz, S.; Schürmann, M.; Wohlgemuth, R.; Coppolecchia, R. The use of enzymes in organic synthesis and the life sciences: Perspectives from the Swiss Industrial Biocatalysis Consortium (SIBC). *Catal. Sci. Technol.* **2013**, *3*, 29–40. [CrossRef]
28. Silverman, R.B. *Organic Chemistry of Enzyme-Catalyzed Reactions*, 2nd ed.; Academic Press: San Diego, SD, USA, 2002; pp. 1–38. ISBN 978-0-08-051336-2.
29. Strohmeier, G.A.; Pichler, H.; May, O.; Gruber-Khadjawi, M. Application of Designed Enzymes in Organic Synthesis. *Chem. Rev.* **2011**, *111*, 4141–4164. [CrossRef]
30. Winkler, C.K.; Schrittwieser, J.H.; Kroutil, W. Power of Biocatalysis for Organic Synthesis. *ACS Cent. Sci.* **2021**, *7*, 55–71. [CrossRef]
31. Forró, E.; Fülöp, F. Advanced procedure for the enzymatic ring opening of unsaturated alicyclic β-lactams. *Tetrahedron Asymmetry* **2004**, *15*, 2875–2880. [CrossRef]
32. Shoda, S.-i.; Uyama, H.; Kadokawa, J.-i.; Kimura, S.; Kobayashi, S. Enzymes as Green Catalysts for Precision Macromolecular Synthesis. *Chem. Rev.* **2016**, *116*, 2307–2413. [CrossRef]
33. Sheldon, R.A. Engineering a more sustainable world through catalysis and green chemistry. *J. R. Soc. Interface* **2016**, *13*, 20160087. [CrossRef] [PubMed]
34. Augustin, M.M.; Augustin, J.M.; Brock, J.R.; Kutchan, T.M. Enzyme morphinan *N*-demethylase for more sustainable opiate processing. *Nat. Sustain.* **2019**, *2*, 465–474. [CrossRef]
35. Sheldon, R.A.; Brady, D. Green Chemistry, Biocatalysis, and the Chemical Industry of the Future. *ChemSusChem* **2022**, *15*, e202102628. [CrossRef] [PubMed]
36. Stergiou, P.Y.; Foukis, A.; Filippou, M.; Koukouritaki, M.; Parapouli, M.; Theodorou, L.G.; Hatziloukas, E.; Afendra, A.; Pandey, A.; Papamichael, E.M. Advances in lipase-catalyzed esterification reactions. *Biotechnol. Adv.* **2013**, *31*, 1846–1859. [CrossRef] [PubMed]
37. Gamayurova, V.S.; Zinov'eva, M.E.; Shnaider, K.L.; Davletshina, G.A. Lipases in Esterification Reactions: A Review. *Catal. Ind.* **2021**, *13*, 58–72. [CrossRef]
38. Singhania, V.; Cortes-Clerget, M.; Dussart-Gautheret, J.; Akkachairin, B.; Yu, J.; Akporji, N.; Gallou, F.; Lipshutz, B.H. Lipase-catalyzed esterification in water enabled by nanomicelles. Applications to 1-pot multi-step sequences. *Chem. Sci.* **2022**, *13*, 1440–1445. [CrossRef]
39. Kumar, A.; Dhar, K.; Kanwar, S.S.; Arora, P.K. Lipase catalysis in organic solvents: Advantages and applications. *Biol. Proced. Online* **2016**, *18*, 2. [CrossRef] [PubMed]
40. Ingenbosch, K.N.; Vieyto-Nuñez, J.C.; Ruiz-Blanco, Y.B.; Mayer, C.; Hoffmann-Jacobsen, K.; Sanchez-Garcia, E. Effect of Organic Solvents on the Structure and Activity of a Minimal Lipase. *J. Org. Chem.* **2022**, *87*, 1669–1678. [CrossRef]
41. Ortiz, C.; Ferreira, M.L.; Barbosa, O.; dos Santos, J.C.S.; Rodrigues, R.C.; Berenguer-Murcia, Á.; Briand, L.E.; Fernandez-Lafuente, R. Novozym 435: The "perfect" lipase immobilized biocatalyst? *Catal. Sci. Technol.* **2019**, *9*, 2380–2420. [CrossRef]

42. Mangiagalli, M.; Carvalho, H.; Natalello, A.; Ferrario, V.; Pennati, M.L.; Barbiroli, A.; Lotti, M.; Pleiss, J.; Brocca, S. Diverse effects of aqueous polar co-solvents on Candida antarctica lipase B. *Int. J. Biol. Macromol.* **2020**, *150*, 930–940. [CrossRef]
43. Zieniuk, B.; Fabiszewska, A.; Białecka-Florjańczyk, E. Screening of solvents for favoring hydrolytic activity of Candida antarctica Lipase B. *Bioprocess Biosyst. Eng.* **2020**, *43*, 605–613. [CrossRef] [PubMed]
44. Forró, E.; Fülöp, F. Enzymatic Strategies for the Preparation of Pharmaceutically Important Amino Acids through Hydrolysis of Amino Carboxylic Esters and Lactams. *Curr. Med. Chem.* **2022**, *29*, 6218–6227. [CrossRef] [PubMed]
45. Megyesi, R.; Forró, E.; Fülöp, F. Substrate engineering: Effects of different N-protecting groups in the CAL-B-catalysed asymmetric O-acylation of 1-hydroxymethyl-tetrahydro-β-carbolines. *Tetrahedron* **2018**, *74*, 2634–2640. [CrossRef]
46. Shahmohammadi, S.; Faragó, T.; Palkó, M.; Forró, E. Green Strategies for the Preparation of Enantiomeric 5–8-Membered Carbocyclic β-Amino Acid Derivatives through CALB-Catalyzed Hydrolysis. *Molecules* **2022**, *27*, 2600. [CrossRef]
47. Sun, M.; Nie, K.; Wang, F.; Deng, L. Optimization of the Lipase-Catalyzed Selective Amidation of Phenylglycinol. *Front. Bioeng. Biotechnol.* **2020**, *7*, 486. [CrossRef]
48. Pitzer, J.; Steiner, K.; Schmid, C.; Schein, V.K.; Prause, C.; Kniely, C.; Reif, M.; Geier, M.; Pietrich, E.; Reiter, T.; et al. Racemization-free and scalable amidation of l-proline in organic media using ammonia and a biocatalyst only. *Green Chem.* **2022**, *24*, 5171–5180. [CrossRef]
49. Hassan, S.; Tschersich, R.; Müller, T.J.J. Three-component chemoenzymatic synthesis of amide ligated 1,2,3-triazoles. *Tetrahedron Lett.* **2013**, *54*, 4641–4644. [CrossRef]
50. Irimescu, R.; Kato, K. Lipase-catalyzed enantioselective reaction of amines with carboxylic acids under reduced pressure in non-solvent system and in ionic liquids. *Tetrahedron Lett.* **2004**, *45*, 523–525. [CrossRef]
51. Prasad, A.K.; Husain, M.; Singh, B.K.; Gupta, R.K.; Manchanda, V.K.; Olsen, C.E.; Parmar, V.S. Solvent-free biocatalytic amidation of carboxylic acids. *Tetrahedron Lett.* **2005**, *46*, 4511–4514. [CrossRef]
52. Gotor, V. Non-conventional hydrolase chemistry: Amide and carbamate bond formation catalyzed by lipases. *Bioorg. Med. Chem.* **1999**, *7*, 2189–2197. [CrossRef]
53. Manova, D.; Gallier, F.; Tak-Tak, L.; Yotava, L.; Lubin-Germain, N. Lipase-catalyzed amidation of carboxylic acid and amines. *Tetrahedron Lett.* **2018**, *59*, 2086–2090. [CrossRef]
54. Dorr, B.M.; Fuerst, D.E. Enzymatic amidation for industrial applications. *Curr. Opin. Chem. Biol.* **2018**, *43*, 127–133. [CrossRef] [PubMed]
55. Lima, R.N.; dos Anjos, C.S.; Orozco, E.V.M.; Porto, A.L.M. Versatility of Candida antarctica lipase in the amide bond formation applied in organic synthesis and biotechnological processes. *Mol. Catal.* **2019**, *466*, 75–105. [CrossRef]
56. Lo, Y.C.; Lin, C.L.; Fang, W.Y.; Lőrinczi, B.; Szatmári, I.; Chang, W.H.; Fülöp, F.; Wu, S.N. Effective Activation by Kynurenic Acid and Its Aminoalkylated Derivatives on M-Type K$^+$ Current. *Int. J. Mol. Sci.* **2021**, *22*, 1300. [CrossRef]
57. Hollmann, F.; Kara, S.; Opperman, D.J.; Wang, Y. Biocatalytic synthesis of lactones and lactams. *Chem.-Asian J.* **2018**, *13*, 3601–3610. [CrossRef] [PubMed]
58. Kunishima, M.; Watanabe, Y.; Terao, K.; Tani, S. Substrate-Specific Amidation of Carboxylic Acids in a Liquid-Liquid Two-Phase System Using Cyclodextrins as Inverse Phase-Transfer Catalysts. *Eur. J. Org. Chem.* **2004**, *2004*, 4535–4540. [CrossRef]
59. Kim, S.E.; Hahm, H.; Kim, S.; Jang, W.; Jeon, B.; Kim, Y.; Kim, M. A Versatile Cobalt Catalyst for Secondary and Tertiary Amide Synthesis from Various Carboxylic Acid Derivatives. *Asian J. Org. Chem.* **2016**, *5*, 222–231. [CrossRef]
60. Ojeda-Porras, A.; Hernandez-Santana, A.; Gamba-Sanchez, D. Direct amidation of carboxylic acids with amines under microwave irradiation using silica gel as a solid support. *Green Chem.* **2015**, *17*, 3157–3163. [CrossRef]
61. Métro, T.-X.; Bonnamour, J.; Reidon, T.; Duprez, A.; Sarpoulet, J.; Martinez, J.; Lamaty, F. Comprehensive Study of the Organic Solvent-Free CDI-Mediated Acylation of Various Nucleophiles by Mechanochemistry. *Chem. Eur. J.* **2015**, *21*, 12787–12796. [CrossRef]
62. Roe, E.T.; Stutzman, J.M.; Swern, D. Fatty Acid Amides. III.^{2a} N-Alkenyl and N,N-Dialkenyl Amides^{2b}. *J. Am. Chem. Soc.* **1951**, *73*, 3642–3643. [CrossRef]
63. Willand, N.; Desroses, M.; Toto, P.; Dirié, B.; Lens, Z.; Villeret, V.; Rucktooa, P.; Locht, C.; Baulard, A.; Deprez, B. Exploring Drug Target Flexibility Using in Situ Click Chemistry: Application to a Mycobacterial Transcriptional Regulator. *ACS Chem. Biol.* **2010**, *5*, 1007–1013. [CrossRef] [PubMed]
64. Tian, J.; Gao, W.-C.; Zhou, D.-M.; Zhang, C. Recyclable Hypervalent Iodine(III) Reagent Iodosodilactone as an Efficient Coupling Reagent for Direct Esterification, Amidation, and Peptide Coupling. *Org. Lett.* **2012**, *14*, 3020–3023. [CrossRef] [PubMed]
65. Ouerghui, A.; Elamari, H.; Dardouri, M.; Ncib, S.; Meganem, F.; Girard, C. Chemical modifications of poly(vinyl chloride) to poly(vinyl azide) and "clicked" triazole bearing groups for application in metal cation extraction. *Adv. Mater. Res.* **2016**, *100*, 191–197. [CrossRef]
66. Wei, W.; Hu, X.-Y.; Yan, X.-W.; Zhang, Q.; Cheng, M.; Ji, J.-X. Direct use of dioxygen as an oxygen source: Catalytic oxidative synthesis of amides. *Chem. Commun.* **2012**, *48*, 305–307. [CrossRef] [PubMed]
67. Cromwell, N.H.; Creger, P.L.; Cook, K.E. Studies with the Amine Adducts of β-Benzoylacrylic Acid and its Methyl Ester. *J. Am. Chem. Soc.* **1956**, *78*, 4412–4416. [CrossRef]

Disclaimer/Publisher's Note: The statements, opinions and data contained in all publications are solely those of the individual author(s) and contributor(s) and not of MDPI and/or the editor(s). MDPI and/or the editor(s) disclaim responsibility for any injury to people or property resulting from any ideas, methods, instructions or products referred to in the content.

Article

Application of Chitosan/Poly(vinyl alcohol) Stabilized Copper Film Materials for the Borylation of α, β-Unsaturated Ketones, Morita-Baylis-Hillman Alcohols and Esters in Aqueous Phase [†]

Bojie Li [1], Wu Wen [1,2], Wei Wen [1,2], Haifeng Guo [1,2], Chengpeng Fu [1,2], Yaoyao Zhang [1,*] and Lei Zhu [1,2,*]

[1] School of Chemistry and Materials Science, Hubei Key Laboratory of Quality Control of Characteristic Fruits and Vegetables, Hubei Engineering University, Xiaogan 432000, China; ghf02062023@163.com (H.G.)
[2] School of Materials Science and Engineering, Hubei University, Wuhan 430062, China
[*] Correspondence: yaoyaozhang@hbeu.edu.cn (Y.Z.); lei.zhu@hbeu.edu.cn (L.Z.)
[†] Dedicated to the 80th anniversary celebration of Hubei Engineering University.

Abstract: A chitosan/poly(vinyl alcohol)-stabilized copper nanoparticle (CP@Cu NPs) was used as a heterogeneous catalyst for the borylation of α, β-unsaturated ketones, MBH alcohols, and MBH esters in mild conditions. This catalyst not only demonstrated remarkable efficiency in synthesizing organoboron compounds but also still maintained excellent reactivity and stability even after seven recycled uses of the catalyst. This methodology provides a gentle and efficient approach to synthesize the organoboron compounds by efficiently constructing carbon–boron bonds.

Keywords: chitosan/PVA-stabilized copper nanoparticles; copper catalysis; heterogeneous catalyst; aqueous phase; recycle and reuse

Citation: Li, B.; Wen, W.; Wen, W.; Guo, H.; Fu, C.; Zhang, Y.; Zhu, L. Application of Chitosan/Poly(vinyl alcohol) Stabilized Copper Film Materials for the Borylation of α, β-Unsaturated Ketones, Morita-Baylis-Hillman Alcohols and Esters in Aqueous Phase. *Molecules* **2023**, *28*, 5609. https://doi.org/10.3390/molecules28145609

Academic Editor: Lu Liu

Received: 28 June 2023
Revised: 21 July 2023
Accepted: 21 July 2023
Published: 24 July 2023

Copyright: © 2023 by the authors. Licensee MDPI, Basel, Switzerland. This article is an open access article distributed under the terms and conditions of the Creative Commons Attribution (CC BY) license (https://creativecommons.org/licenses/by/4.0/).

1. Introduction

Organoboron compounds are a significant class of organic intermediates that are capable of reacting with various nucleophilic reagents and could be conveniently converted into various chemical bonds, such as C–C, C–O, and C–N bonds, which are very important in organic synthesis [1–3]. For this reason, in the past decades, various transition metals, including rhodium [4], palladium [5,6], platinum [7], cobalt [8], nickel [9], and copper [10–12], have been used as catalysts to synthesize the organoboron compounds. Recently, metal nanoparticles have received much attention because of their large surface area, which could increase catalytic efficiency [13,14]. However, metal nanoparticles sometimes have some drawbacks in the aqueous phase; for example, they are prone to aggregation and precipitation. To overcome these problems, an appropriate carrier was considered to load these metal nanoparticles, which could make these loaded nano-catalytic materials have several good advantages, including easy recovery and uniform dispersion [15,16].

Montmorillonite [17–19], activated carbon [20–22], and chitosan [23–26] are commonly used as carriers for heterogeneous catalytic materials in organic synthesis. Among these materials, chitosan is especially preferred due to its numerous amino and hydroxyl groups. These functional groups could coordinate well with various metals to perform a good catalytic activity. So far, several transition metals have been reported using chitosan as a support, including gold [27], silver [28], ruthenium [29], palladium [30–32], platinum [33], and copper [34–36]. Compared with these precious metals, copper has received more attention for its low price and lower toxicity.

In our previous research, we found that the borylation reactions of chalcone derivatives could be carried out smoothly, and the corresponding target products could also be obtained in good yields when CS@Cu(OH)$_2$ [37], Cell-CuI NPs [38], or CP@Cu NPs [39] were used as catalysts (Scheme 1a). Under oxidative conditions, the resulting organoborons could give rise to the desired β-hydroxy-substituted carbonyl compounds, which are widely found in active molecules [40–42]. Compared with other supports, chitosan has inherent advantages due to its green property, abundance, stability, and ability of chelation [26]. With our continuous efforts in exploring applications of chitosan-supported metal catalysts, we were interested in developing a chitosan composite film of stabilized copper nanoparticles and its application for the synthesis of useful organoboron compounds. In particular, chitosan/poly(vinyl alcohol) composite films loaded with copper nanoparticles (CP@Cu NPs) were found to exhibit high reactivity, as well as excellent reusability and stability. Not only did α, β-unsaturated ketones have good reactivity in borylation reactions, but also α, β-unsaturated esters and amides could react smoothly when CP@Cu NPs was used as a catalyst in the reactions, and even after the catalyst was reused seven times, it still showed very good catalytic activity. However, in our previous work, the CP@Cu NPs were limited to the borylation reactions of 1,2-disubstituted α, β-unsaturated compounds, whereas the borylation reactions of 1,1-disubstituted unsaturated compounds were not explored, including the borylation reactions of MBH alcohols, and esters were also not involved. Morita-Baylis-Hillman alcohols or esters have aroused much attention in organic synthesis as valuable synthons and intermediates for the preparation of many important cyclic and acyclic compounds. Thus, their ready availability and condensed functional groups make them particularly attractive. In recent years, considering the importance of MBH alcohols and esters in organic synthesis, more and more research groups pay attention to their applications, especially as electrophilic reagents in borylation reactions [43–53]. These methods still have some shortcomings; for instance, the precious metal palladium as a catalyst is needed in reactions [43,51–53]. Even with copper as a catalyst, the reaction substrate range is quite limited, and only MBH alcohols or esters are compatible in these methods [44–50]. And more importantly, because all of the above methods are homogeneous reactions, the catalysts in reactions are difficult to be separated and reused after the reactions, which resulted in waste and heavy metal residues. Herein, in this work, we used CP@Cu NPs as a heterogeneous catalyst and 1,1-disubstituted α, β-unsaturated compounds (including α, β-unsaturated ketones, MBH alcohols, and esters) as substrates. The borylation reactions of these compounds could be achieved under mild conditions. Considering that some organoboron compounds are not very stable, the corresponding β-hydroxy-substituted carbonyl compounds were obtained via direct oxidation. Finally, the activity and stability of the catalysts were proved by the recovery experiments (Scheme 1b).

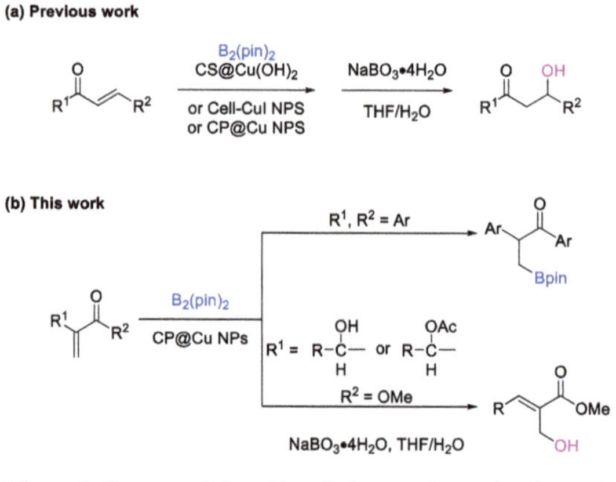

Scheme 1. Copper-catalyzed borylation reactions of α, β-unsaturated compounds.

2. Results and Discussion

2.1. Catalysis of CP@Cu NPs in the Borylation Reaction of α, β-Unsaturated Ketones

The initial experiments commenced with α, β-unsaturated ketone **II-1** (0.2 mmol) as a model substrate. CP@Cu NPs (10.0 mg, 9.0 mol%) was used as a catalyst by using $B_2(pin)_2$ (0.4 mmol, 2.0 equiv) as a boron source in 2.0 mL of solvents. First, various organic solvents were investigated, and no additives were added (Table 1, entries 1–7). When THF and toluene were used as solvents in this reaction, no reaction happened (Table 1, entries 1–2). Surprisingly, when ether was used as a solvent, the reaction occurred, and the target product was obtained in 8% yield (Table 1, entry 3). We continued to explore other solvents, such as MeOH, acetone, and H_2O; the reaction could still happen, but the yields increased not obviously (Table 1, entries 4–6). In our previous work, we found that when we used mixed solvents in the reaction, excellent yields could be gained [37–39]. Considering the role of protons in this reaction, we used MeOH and H_2O as mixed solvents (MeOH/H_2O = 3:1), and the yield was increased to 28% yield (Table 1, entry 7). Next, we intended to examine the effects of additives in the reactions, mainly various organic bases, including 2,2-bipyridine, DMAP, 2-cyanopyridine, 2-chloropyridine, 2-bromopyridine, 2,6-dibromopyridine, and 4-picoline (Table 1, entries 8–14). We found that when we used 4-picoline as an additive in the reaction, the reaction worked very well, and the yield could obviously increase to 60% yield (Table 1, entry 14). Inspired by this result, we considered that the ratio of MeOH and H_2O may have some contribution to these reactions. When we changed the ratio of the mixed solvents, different results were observed (Table 1, entries 15–18). In particular, when the ratio of MeOH to H_2O was 1:1, the best result 92% yield could be obtained (Table 1, entry 16). In the organic synthesis, the reaction time is also one of the important factors that would affect the yield; therefore, we carried out the examination of the reaction time (Table 1, entries 19–21 vs. 16), and it was found that the reaction efficiency was still the highest when the reaction time was 12 h (Table 1, entry 16) and the yield was decreased whether the reaction time was shortened or prolonged. In order to study the effect of the amounts of additives on the reactions, we reduced or increased the amounts of additives and found that it actually had an effect on the reaction, and the yields were reduced to a certain extent (Table 1, entries 22–23 vs. 16). Finally, we investigated the catalyst loading in the reactions. When the catalyst loading was reduced to 4.5 mol%, the yield was still 92% (Table 1, entry 24), but when the amount of catalyst was increased to 13.5 mol%, the yield decreased to 90% (Table 1, entry 25). Therefore, we chose to use 4.5 mol% of catalyst loading to carry out the reactions from the view of economy. Thus, by a series of optimizations of the conditions, the optimal conditions of this research were 4.5 mol% CP@Cu NPs as a catalyst, 2.0 equiv of $B_2(pin)_2$ as a boron source, and 6.0 mol% of 4-picoline as an additive, and the whole reaction was conducted in 2.0 mL of mixed solvents (MeOH/H_2O = 1:1) at room temperature for 12 h (Table 1, entry 24).

With the optimal experimental conditions in hand, we continued to investigate the universality of the reaction, and the results are summarized in Scheme 2. We mainly examined the effects of the substituents on the benzene ring of 1,1-disubstituted α, β-unsaturated ketones on the yields (Scheme 2). We first investigated the para-substituted functional groups on the benzene ring Ar_1; when the substituents were methoxyl and methyl, the yields of the borylation were slightly decreased, and the possible reason was that both methoxyl and methyl were electron-donating groups that had an effect on the electrophilicity of the substrates (**II-2a-II-3a**, 81–85% yields). When the substituents were changed to fluorine and bromine, the yields were very good (**II-4a**, 90% yield; **II-6a**, 93% yield). However, when chlorine was used as a substituent, the yield was decreased to 80% yield (**II-5a**), mainly because it had less electron absorption than fluorine and bromine. Next, we examined the ortho-substituents on the benzene ring Ar_1; the electron-donating group methoxyl had a better yield than the electron-withdrawing group bromine (**II-7a**, 90% yield, vs. **II-8a**, 57% yield). We also investigated the reactivity of meto-position of the benzene ring Ar_1, and the yields of the reactions were still good (**II-9a-II-10a**, 80–92% yields). Last, we found that the substituents on the para-position of the benzene ring Ar_2 had little effect on the yields; neither was it the electron-donating group methyl, nor the electron-withdrawing group fluorine (**II-11a-II-12a**, 81–88% yields, vs. **II-3a**, 81% yield).

Table 1. Optimization of CP@Cu NPs in the borylation reaction of α, β-unsaturated ketones [a].

Entries	CP@Cu NPs	Solvents (2.0 mL)	Additives	Time (h)	NMR Yields (%)
1	(9.0 mol%)	THF	-	12	N.R.
2	(9.0 mol%)	Toluene	-	12	N.R.
3	(9.0 mol%)	Et$_2$O	-	12	8
4	(9.0 mol%)	MeOH	-	12	16
5	(9.0 mol%)	Acetone	-	12	8
6	(9.0 mol%)	H$_2$O	-	12	19
7	(9.0 mol%)	MeOH/H$_2$O = 3:1	-	12	28
8	(9.0 mol%)	MeOH/H$_2$O = 3:1	2,2-bipyridine	12	35
9	(9.0 mol%)	MeOH/H$_2$O = 3:1	DMAP	12	33
10	(9.0 mol%)	MeOH/H$_2$O = 3:1	2-cyanopyridine	12	54
11	(9.0 mol%)	MeOH/H$_2$O = 3:1	2-chloropyridine	12	57
12	(9.0 mol%)	MeOH/H$_2$O = 3:1	2-bromopyridine	12	50
13	(9.0 mol%)	MeOH/H$_2$O = 3:1	2,6-dibromopyridine	12	48
14	(9.0 mol%)	MeOH/H$_2$O = 3:1	4-picoline	12	60
15	(9.0 mol%)	MeOH/H$_2$O = 2:1	4-picoline	12	72
16 [b]	(9.0 mol%)	MeOH/H$_2$O = 1:1	4-picoline	12	92
17	(9.0 mol%)	MeOH/H$_2$O = 1:2	4-picoline	12	89
18	(9.0 mol%)	MeOH/H$_2$O = 1:4	4-picoline	12	80
19	(9.0 mol%)	MeOH/H$_2$O = 1:1	4-picoline	4	79
20	(9.0 mol%)	MeOH/H$_2$O = 1:1	4-picoline	8	87
21 [b]	(9.0 mol%)	MeOH/H$_2$O = 1:1	4-picoline	16	90
22 [b,c]	(9.0 mol%)	MeOH/H$_2$O = 1:1	4-picoline	12	90
23 [b,d]	(9.0 mol%)	MeOH/H$_2$O = 1:1	4-picoline	12	90
24 [b]	(4.5 mol%)	MeOH/H$_2$O = 1:1	4-picoline	12	92
25 [b]	(13.5 mol%)	MeOH/H$_2$O = 1:1	4-picoline	12	90

Reaction conditions: [a] **II-1** (0.2 mmol), B$_2$Pin$_2$ (0.4 mmol), CP@Cu NPS (10.0 mg, 9.0 mol%), additives (6.0 mol%, 0.012 mmol), solvents (2.0 mL) at room temperature. [b] Isolated yield. [c] 4-picoline (5.0 mol%). [d] 4-picoline (7.0 mol%). N.R. = no reaction.

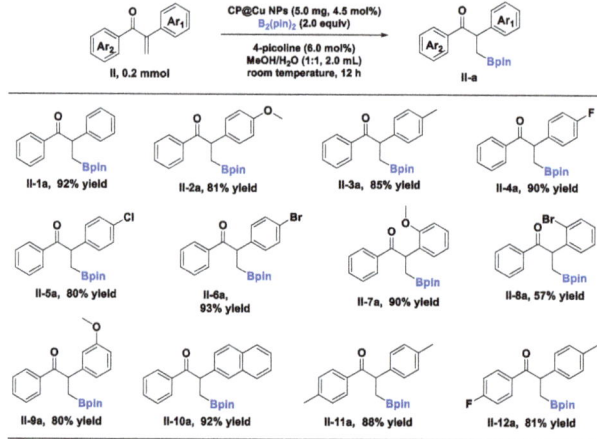

Scheme 2. Screening substrate scope of CP@Cu NPs in the borylation reaction of α, β-unsaturated ketones. Reaction conditions: **II** (0.2 mmol), B$_2$(pin)$_2$ (2.0 equiv), CP@Cu NPs (5.0 mg, 4.5 mol%), 4-picoline (6.0 mol%) in 2.0 mL of mixed solvents (MeOH/H$_2$O =1:1) at room temperature for 12 h.

2.2. Catalysis of CP@Cu NPs in the Borylation Reaction of MBH Alcohols and Esters

MBH alcohols and esters are very important intermediates in organic synthesis, and there is not much research on the borylation reactions of these compounds at present. Therefore, in this work, we planned to use them as reaction substrates for condition optimization. The same as the above condition optimizations, we selected MBH alcohols **III-1** (0.2 mmol) or MBH esters **IV-1** (0.2 mmol) as a model substrate, CP@Cu NPs (5.0 mg, 4.5 mol%) as a catalyst, and $B_2(pin)_2$ (0.4 mmol, 2.0 equiv) as a boron source in 2.0 mL of solvents; the whole reaction was conducted at room temperature for 12 h, and no additives were needed (Table 2). According to the above experimental results we achieved, we believed that proton solvents were beneficial to this reaction, so we just chose methanol and water as the solvents for screening. First, when we used methanol as a solvent, both MBH alcohol **III-1** and **IV-1** could react smoothly, and considering that the intermediates **III-1a** and **IV-1a** were not very stable in these conditions, we directly further oxidized these intermediates to the corresponding β-hydroxy substituted products using excessive $NaBO_3 \cdot 4H_2O$ as an oxidant (entry1, 31% yield; entry 2, 45% yield). And when H_2O was used as a solvent, both reaction yields were not improved (entry 3, 33% yield; entry 4, 20% yield). Then, based on our previous experiment results, the mixed solvents were beneficial to this reaction, and we considered using methanol and water as the mixed solvents for conditional screening. We investigated the ratio of methanol to water in a mixed solvent and found that the highest yield could be obtained while the ratio of methanol to water was 2:1, and the target products could be obtained in 93% (entry 7 for **III-1b**) and 92% (entry 8 for **IV-1b**) isolated yields. Finally, we investigated the catalyst loading and found that the yields of the reactions did not change much. From the economic point of view, 4.5 mol% of catalyst loading was still the best choice in the reactions. Therefore, the optimal conditions for this reaction were MBH alcohols **III-1** (0.2 mmol) or MBH esters **IV-1** (0.2 mmol) as a model substrate, CP@Cu NPs (5.0 mg, 4.5 mol%) as a catalyst, and $B_2(pin)_2$ (0.4 mmol, 2.0 equiv) as a boron source in 2.0 mL of mixed solvent (MeOH/H_2O = 2:1), and the whole reaction was conducted at room temperature for 12 h (entry 7 for **III-1b**, entry 8 for **IV-1b**).

Table 2. Optimization of CP@Cu NPs in the borylation reaction of MBH alcohols and esters [a].

Entries	Substrates	CP@Cu NPs	Solvents (2.0 mL)	NMR Yields (%)
1	III-1	5.0 mg	MeOH	31
2	IV-1	5.0 mg	MeOH	45
3	III-1	5.0 mg	H_2O	33
4	IV-1	5.0 mg	H_2O	20
5	III-1	5.0 mg	MeOH/H_2O =3:1	94
6	IV-1	5.0 mg	MeOH/H_2O =3:1	90
7 [b]	III-1	5.0 mg	MeOH/H_2O =2:1	93
8 [b]	IV-1	5.0 mg	MeOH/H_2O =2:1	92
9	III-1	5.0 mg	MeOH/H_2O =1:1	90
10	IV-1	5.0 mg	MeOH/H_2O =1:1	86
11	III-1	5.0 mg	MeOH/H_2O =1:2	89
12	IV-1	5.0 mg	MeOH/H_2O =1:2	83
13	III-1	5.0 mg	MeOH/H_2O =1:3	88
14	IV-1	5.0 mg	MeOH/H_2O =1:3	85
15	III-1	10.0 mg	MeOH/H_2O =2:1	93
16	IV-1	10.0 mg	MeOH/H_2O =2:1	91

Reaction conditions: [a] **III-1** and **IV-1** (0.2 mmol), $B_2(pin)_2$ (0.4 mmol), CP@Cu (5.0 mg, 4.5 mol%), solvents (2.0 mL) at room temperature for 12 h. [b] Isolated yields.

With the optimized conditions in hand, we investigated the universality of the borylation reactions of MBH alcohols and esters, and the results are summarized in Scheme 3. We first examined the reactions as group R^1 was different; when R^1 was 4-methylphenyl, 4-ethylphenyl, 4-isopropylphenyl, 4-tert-butylphenyl, and 4-methoxyphenyl, the effect of the substituent on the reaction was not significant, whether they were MBH alcohols (**III-2b-III-6b**, 85–98% yields) or esters (**IV-2b-IV-6b**, 80–91% yields). However, when group R^1 was 4-fluorophenyl, 4-chlorophenyl, 4-bromophenyl, and 4-trifluoromethylphenyl, the reaction results were not very good compared with the model reaction. Especially for MBH alcohols, when R^1 was 4-bromophenyl, the target product was not detected (**III-9b**), and when R^1 was 4-trifluoromethylphenyl, the yield was not good because of its strong electron absorption (**III-10b**, 53% yield). When R^1 consisted of 3-substituted benzene rings, the electronic effect of the benzene ring had a great influence on the reactions. When the benzene ring was connected with electron-donating groups, such as 3-methyl and 3-methoxy, the yields were better, but for the electron-deficient group, for example, 3-bromophenyl, the yield had a great influence (**III-13b**, 50% yield; **IV-13b**, 39% yield). However, when R^1 was 2-substituted phenyl, the electronic effect of the aromatic rings had no effect on the reactions, no matter whether they were electron-absorbing substituents or electron-giving substituents (**III-14b-III-16b**, 93–98% yields; **IV-14b-IV-16b**, 87–95% yields). For the disubstituted benzenes of R^1, no matter whether they were electron-absorbing substituents or electron-giving substituents, the reactions could still have good yields (**III-17b-III-20b**, 71–97% yields; **IV-17b-IV-20b**, 67–80% yields). To our delight, when R^1 was the 2-thiophene substituent, the reactions could still take place, and the target products could be obtained in medium yields (**III-21b**, 43% yield; **IV-21b**, 62% yield). Next, we continued to investigate the reactions when R_1 consisted of alkyl groups. From the reaction results, we found that the alkyl substituents could occur smoothly, and the target products could be synthesized in medium to excellent yields (**III-22b-III-27b**, 48–99% yields; **IV-22b-IV-27b**, 40–66% yields). Finally, we investigated the reaction of R^2 and found that when R^2 was ethyl, the reaction activity was still very good, and the corresponding target product could be obtained with good yields (**III-28b**, 86% yield; **IV-28b**, 99% yield).

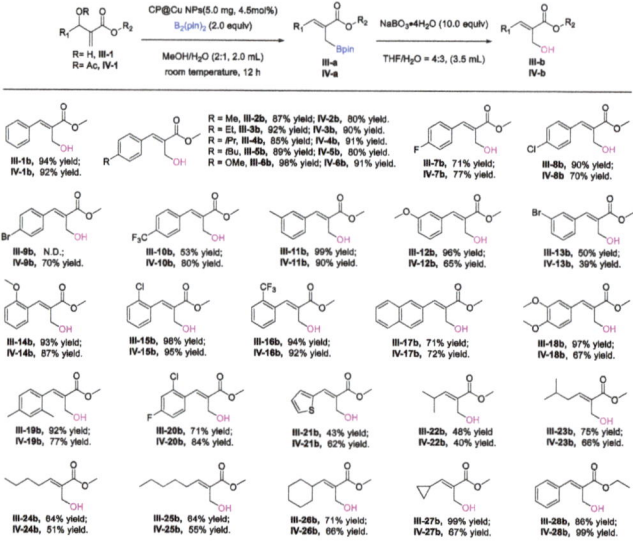

Scheme 3. Screening substrate expansion scope of CP@Cu NPs in the borylation reaction of MBH alcohols and esters. Reaction conditions: **III-1** and **IV-1** (0.2 mmol), B₂Pin₂ (0.4 mmol), CP@Cu (5.0 mg), 2.0 mL of mixed solvents with MeOH/H₂O =3:1 for **III-1** reaction and MeOH/H₂O =2:1 for **IV-1** reaction, at room temperature for 12 h. N.D. = no detection.

2.3. Recycling Experiments of CP@Cu NPs in Borylation Reactions

The main advantage of heterogeneous catalysis in organic synthesis was that the catalyst in the system could be easily recovered and reused. Such a type of operation could not only increase the catalytic efficiency of the catalyst and reduce the cost of the reactions but also avoid the heavy metal residue to the environment. In this work, to assess the reusability and stability of the CP@Cu NPs in borylation reactions, we used MBH alcohols **III-1** as a substrate and CP@Cu NPs as a catalyst. After the completion of the reaction, the catalyst CP@Cu NPs could be recycled with a simple operation. The results showed that the activity of the catalyst stayed very well, and the yield of the product could also still be up to 84% even in the seventh experiment, which confirmed that the catalyst could be recyclable (Figure 1). Notably, the yields of the eighth and ninth cycles were still 83% and 82%, respectively. The slight decrease in the yields that was observed in the recycling experiments was probably due to the formation of a byproduct, which may be absorbed onto the surface of CP@Cu NPs. It must also be mentioned that the catalyst could be reactivated by washing with 10% aq. NaOH solution and dried again after the reaction. By using this process, an average of ~90% yield could be obtained after each cycle. Furthermore, ICP tests of recycled catalyst were carried out, and almost no detectable copper leaching was observed. These results strongly indicated that the CP@Cu NPs was a highly active heterogeneous catalyst for this borylation process.

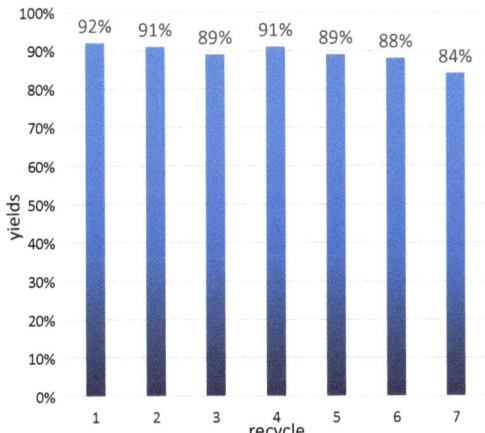

Figure 1. The recycle experiments.

3. Materials and Methods

3.1. Materials

Chitosan/poly(vinyl alcohol) composite film-supported copper nanoparticles (CP@Cu NPs) were prepared, according to the procedures reported [39]. The characterization of CP@Cu NPs was described in the supplementary materials. Bis(pinacolato)diboron ($B_2(pin)_2$, AR), methanol (MeOH, AR), ethanol (EtOH, AR), acetone (AR), tetrahydrofuran (THF, AR), ether (Et_2O, AR), 2,2-bipyridine (AR), 4-dimethylaminopyridine (DMAP, AR), 2-cyanopyridine (AR), 2-chloropyridine (AR), 2-bromopyridine (AR), and 4-picoline (AR) were obtained commercially from Energy Chemical (Shanghai, China).

3.2. Synthesis of α, β-Unsaturated Ketones **II**

In step 1, a mixture of substituted phenylacetonitrile (10 mmol), substituted phenylboronic acid (20 mmol), Pd (OAc)$_2$ (112.3 mg, 0.5 mmol), 2,2'-dipyridine (156.2 mg, 1.0 mmol), TFA (11.4 g, 100 mmol), and H_2O (4 mL) were added into THF (20 mL). Then the mixture was refluxed under nitrogen atmosphere for 2–3 days. The residue was extracted with EtOAc (20 mL) three times. After evaporation of solvent, the crude mixture was purified by flash column chromatograph to afford the intermediate compounds.

In step 2, to the compounds (5 mmol) obtained from step 1, formaldehyde (0.60 g, 20 mmol), piperidine (42.1 mg, 0.5 mmol), AcOH (60.1 mg, 1.0 mmol), and MeOH (5 mL) were added. The mixture was then refluxed for 6 h. After the completion of this reaction, evaporation was carried out to remove MeOH. The residue was washed with CH_2Cl_2 to collect the organic layer, which was washed with brine, dried over Na_2SO_4, and concentrated in vacuo. The desired ketones **II** [54,55] were obtained by further purification by silica gel chromatography.

3.3. Synthesis of MBH Alcohols **III**

Different substituted benzaldehyde (10 mmol), methyl acrylate (1.72 g, 20 mmol) or ethyl acrylate (2.00 g, 20 mmol), and DABCO (1.12 g,10 mmol) were successively added into a 50 mL flask under air. After stirring for 3–7 days at room temperature, the reaction mixture was filtered, and the filtrate was extracted with EtOAc (20 mL) three times. The crude mixture was purified by silica gel chromatography to afford the desired MBH alcohols **III** [56].

3.4. Synthesis of MBH Esters **IV**

Different substituted MBH alcohols **III** (10 mmol), acetic anhydride (1.23 g, 12 mmol), 4-DMAP (122.2 mg, 1 mmol), and DCM (10 mL) were successively added to a 50 mL flask under air. The reaction was monitored by TLC. After completion of reaction, the mixture was filtered, and the filtrate was extracted with EtOAc (20 mL) three times. Then the crude mixture was purified by silica gel chromatography to afford the corresponding MBH esters **IV** [56].

3.5. Analytical Methods

The purification of products was accomplished by using flash column chromatography on silica gel (200–300 mesh) or preparative TLC. Nuclear magnetic resonance (NMR) spectra were recorded on a Bruker Avance III 400 MHz spectrometer (Karlsruhe, Germany) operating at 400 MHz for ^1H and 100 MHz for ^{13}C NMR in $CDCl_3$ unless otherwise noted. $CDCl_3$ served as the internal standard (δ = 7.26 ppm) for ^1H NMR and (δ = 77.0 ppm) for ^{13}C NMR.

3.6. Copper-Catalyzed Borylation Reactions

The reaction procedure is depicted in Scheme 1b. Nano-sized copper loaded onto the membrane material could enable the borylation reaction of α, β-unsaturated ketones, MBH alcohols, and esters under very mild conditions. Because of the difference in reactivity, the borylation of α, β-unsaturated ketones with $B_2(pin)_2$ required the additional bases, whereas the borylation of MBH alcohols and esters could be conducted smoothly without the bases.

3.6.1. Borylation Reactions of α, β-Unsaturated Ketones

At room temperature, 0.2 mmol of α, β-unsaturated ketones **II**, 0.4 mmol of $B_2(pin)_2$, 6 mol% of base, 10.0 mg of CP@Cu NPs as a catalyst, and 2.0 mL of solvent were added in the reaction system. The whole reactions were stirred at room temperature for 12 h, and after completion of the reaction, the mixture was filtered, and the desired products **II-a** were obtained by being purified with column chromatography and characterized by ^1H NMR and ^{13}C NMR (see supplementary materials).

3.6.2. Borylation Reactions of MBH Alcohols and Esters

At room temperature, a reaction mixture containing 0.2 mmol of MBH alcohols or esters (**III** or **IV**), 0.4 mmol of $B_2(pin)_2$, 10.0 mg of CP@Cu NPs, and 2.0 mL of solvent were prepared, and the whole reactions were stirred at room temperature. After the completion of the reactions, the organic phase was separated, and the crude intermediates **III-a** or **IV-a** were added to a mixture of $THF-H_2O$ containing an excess of sodium perborate, and the mixture continued to be stirred for 4 h. When the reaction finished, the desired products

III-b or IV-b were obtained, purified by column chromatography, and characterized by ^{1}H NMR and ^{13}C NMR (see supplementary materials).

3.7. General Procedure for ICP Test of CP@Cu NPs

Chitosan/poly(vinyl alcohol) composite film supported copper nanoparticles (CP@Cu NPs) (~20 mg) were placed in a clean test tube and heated with H_2SO_4 (1 mL) at 200 °C. After 30 min, several drops of concentrated HNO_3 were added carefully, and the tube was shaken occasionally. HNO_3 was continuously added until a clear solution was obtained, and the excess amount of HNO_3 was allowed to evaporate under heating. After the solution was cooled to room temperature, 1 mL of aqua regia was added carefully. The effervescence of gas was observed, and the solution became clearer. The solution was then transferred to a volumetric flask and increased up to 50 mL with water, which was submitted for ICP analysis.

3.8. General Procedure for the Sample Preparation for ICP Analysis to Determine Metal Leaching

After the reaction was finished, the reaction mixture was filtered. The filtrate obtained was concentrated and diluted with 10 mL of THF. Then, 50% v/v of the crude THF solution (5 mL) was then passed through a membrane filter (0.25 or 0.45 μm) into a clean test tube. After the evaporation of the solvent, the solid obtained in the test tube was heated to 200 °C, and 1.0 mL of concentrated H_2SO_4 was added. Following a procedure similar to that described above, concentrated HNO_3 was added at regular intervals until the resulting solution was clear. After the solution was cooled to room temperature, 1 mL of aqua regia was added carefully. The effervescence of the gas was observed, and the solution became clearer. The solution was then transferred to a volumetric flask and increased up to 50 mL with water, which was submitted for ICP analysis.

4. Conclusions

In summary, we reported the preparation of a chitosan-loaded copper catalyst (CP@Cu NPs) and its application in the borylation of α, β-unsaturated ketones, MBH alcohols, and esters with $B_2(pin)_2$ as a boron source. The whole reaction conditions were very mild, and no additives were even needed in the borylation of the MBH alcohols and esters. It demonstrated that the substrate scope of this newly developed method was very broad (more than 40 examples) with very high activity of the catalyst (up to 99% yield). Remarkably, a single heterogeneous catalyst could efficiently catalyze three types of substrates including the borylation of α, β-unsaturated ketones, MBH alcohols, and esters. Moreover, this newly developed strategy could largely solve the recovery of copper catalysts, providing a green and economic way for the efficient synthesis of organoboron compounds in the aqueous phase.

Supplementary Materials: The following supporting information can be downloaded at: https://www.mdpi.com/article/10.3390/molecules28145609/s1.

Author Contributions: Conceptualization, L.Z.; methodology, B.L., W.W. (Wu Wen) and W.W. (Wei Wen); formal analysis, H.G., C.F. and Y.Z.; resources, B.L., Y.Z. and L.Z.; writing—original draft preparation, B.L.; writing—review and editing, Y.Z. and L.Z.; supervision, L.Z.; funding acquisition, Y.Z. and L.Z. All authors have read and agreed to the published version of the manuscript.

Funding: This research was funded by the Key Project of Science and Technology Research of Hubei Provincial Department of Education (D20222701), the National Natural Science Foundation of China (Nos. 22108065, 22278118, 21774029), Key Research and Development Program of Hubei Province (2022BAD083), Central Leading Local Science and Technology Development Special Project of Hubei Province (2022BGE258), Excellent Young and Middle-aged Science and Technology Innovation Team Project of Hubei University (No. T201816).

Conflicts of Interest: The authors declare no conflict of interest.

References

1. Roccaro, A.M.; Vacca, A.; Ribatti, D. Bortezomib in the treatment of cancer. *Recent Pat. Anti-Cancer Drug Discov.* **2006**, *1*, 397. [CrossRef] [PubMed]
2. Coghi, P.S.; Zhu, Y.; Xie, H.; Hosmane, N.S.; Zhang, Y. Organoboron compounds: Effective antibacterial and antiparasitic agents. *Molecules* **2021**, *26*, 3309. [CrossRef] [PubMed]
3. Sacramento, M.; Costa, G.P.; Barcellos, A.M.; Perin, G.; Lenardão, E.J.; Alves, D. Transition-metal-free C−S, C−Se, and C−Te bond formation from organoboron compounds. *Chem. Rec.* **2021**, *21*, 2855. [CrossRef]
4. Kabalka, G.W.; Das, B.C.; Das, S. Rhodium-catalyzed 1,4-addition reactions of diboron reagents to electron deficient olefins. *Tetrahedron Lett.* **2002**, *43*, 2323. [CrossRef]
5. Abu, A.H.; Goldberg, I.; Srebnik, M. Addition reactions of bis(pinacolato)diborane(4) to carbonyl enones and synthesis of (pinacolato)$_2$BCH$_2$B and (pinacolato)$_2$BCH$_2$CH$_2$B by insertion and coupling. *Organometallics* **2001**, *20*, 3962. [CrossRef]
6. Bonet, A.; Gulyas, H.; Koshevoy, I.O.; Estevan, F.; Sanau, M.; Úbeda, M.A.; Fernandez, E. Tandem β-boration/arylation of α,β-unsaturated carbonyl compounds by using a single palladium complex to catalyse both steps. *Chem.–A Eur. J.* **2010**, *16*, 6382. [CrossRef]
7. Lawson, Y.; Norman, N.; Rice, C.; Marder, T. Platinum catalysed 1,4-diboration of α, β-unsaturated ketones. *Chem. Commun.* **1997**, 2051. [CrossRef]
8. Kajiwara, T.; Terabayashi, T.; Yamashita, M.; Nozaki, K. Syntheses, structures, and reactivities of borylcopper and -zinc compounds: 1,4-silaboration of an α, β-unsaturated ketone to form a γ-siloxyallylborane. *Angew. Chem.* **2008**, *120*, 6708. [CrossRef]
9. Hirano, K.; Yorimitsu, H.; Oshima, K. Nickel-catalyzed β-boration of α, β-unsaturated esters and amides with bis(pinacolato)diboron. *Org. Lett.* **2007**, *9*, 5031. [CrossRef] [PubMed]
10. Ito, H.; Yamanaka, H.; Tateiwa, J.; Hosomi, A. Boration of an α, β-enone using a diboron promoted by a copper(I)–phosphine mixture catalyst. *Tetrahedron Lett.* **2000**, *41*, 6821. [CrossRef]
11. Gawande, M.B.; Goswami, A.; Felpin, F.; Asefa, T.; Huang, X.; Silva, R.; Zou, X.; Zboril, R.; Varma, R.S. Cu and Cu-based nanoparticles: Synthesis and applications in catalysis. *Chem. Rev.* **2016**, *116*, 3722. [CrossRef] [PubMed]
12. Kitanosono, T.; Xu, P.; Kobayashi, S. Heterogeneous versus homogeneous copper(II) catalysis in enantioselective conjugate-addition reactions of boron in water. *Chem.–Asian J.* **2014**, *9*, 179. [CrossRef] [PubMed]
13. Yan, N.; Xiao, C.; Kou, Y. Transition metal nanoparticle catalysis in green solvents. *Coord. Chem. Rev.* **2010**, *254*, 1179. [CrossRef]
14. Nørskov, J.K.; Bligaard, T.; Hvolbæk, B.; Abild-Pedersen, F.; Chorkendorff, I.; Christensen, C.H. The nature of the active site in heterogeneous metal catalysis. *Chem. Soc. Rev.* **2008**, *37*, 2163. [CrossRef]
15. Klemm, D.; Heublein, B.; Fink, H.P.; Bohn, A. Cellulose: Fascinating biopolymer and sustainable raw material. *Angew. Chem. Int. Ed.* **2005**, *44*, 3358. [CrossRef]
16. Rezayat, M.; Blundell, R.K.; Camp, J.E.; Walsh, D.A.; Thielemans, W. Green one-step synthesis of catalytically active palladium nanoparticles supported on cellulose nanocrystals. *ACS Sustain. Chem. Eng.* **2014**, *2*, 1241. [CrossRef]
17. Pinnavaia, T.J. Intercalated clay catalysts. *Science* **1983**, *220*, 365. [CrossRef]
18. Kumar, D.D.; Jyoti, B.B.; Pollov, S.P. Recent advances in metal nanoparticles stabilization into nanopores of montmorillonite and their catalytic applications for fine chemicals synthesis. *Catal. Rev.* **2015**, *57*, 257. [CrossRef]
19. Bhat, A.H.; Rangreez, T.A.; Chisti, H. Wastewater treatment and biomedical applications of montmorillonite based nanocomposites: A review. *Curr. Anal. Chem.* **2022**, *18*, 269. [CrossRef]
20. Duan, D.; Chen, D.; Huang, L.; Zhang, Y.; Zhang, Y.; Wang, Q.; Xiao, G.; Zhang, W.; Lei, H.; Ruan, R. Activated carbon from lignocellulosic biomass as catalyst: A review of the applications in fast pyrolysis process. *J. Anal. Appl. Pyrolysis* **2021**, *158*, 105246. [CrossRef]
21. Domínguez, C.M.; Ocón, P.; Quintanilla, A.; Casas, J.A.; Rodriguez, J.J. Highly efficient application of activated carbon as catalyst for wet peroxide oxidation. *Appl. Catal. B Environ.* **2013**, *140*, 663. [CrossRef]
22. Matatov-Meytal, U.; Sheintuch, M. Activated carbon cloth-supported Pd–Cu catalyst: Application for continuous water denitrification. *Catal. Today* **2005**, *102*, 121. [CrossRef]
23. Lee, M.; Chen, B.; Den, W. Chitosan as a natural polymer for heterogeneous catalysts support: A short review on its applications. *Appl. Sci.* **2015**, *5*, 1272. [CrossRef]
24. Sahu, P.K.; Sahu, P.K.; Gupta, S.K.; Agarwal, D.D. Chitosan: An efficient, reusable, and biodegradable catalyst for green synthesis of heterocycles. *Ind. Eng. Chem. Res.* **2014**, *53*, 2085. [CrossRef]
25. Zheng, K.; Xiao, S.; Li, W.; Wang, W.; Chen, H.; Yang, F.; Qin, C. Chitosan-acorn starch-eugenol edible film: Physico-chemical, barrier, antimicrobial, antioxidant and structural properties. *Int. J. Biol. Macromol.* **2019**, *135*, 344. [CrossRef]
26. El, K.A. Chitosan as a sustainable organocatalyst: A concise overview. *ChemSusChem* **2015**, *8*, 217.
27. Mironenko, A.Y.; Sergeev, A.; Nazirov, A.; Modin, E.; Voznesenskiy, S.; Bratskaya, S.Y. H$_2$S optical waveguide gas sensors based on chitosan/Au and chitosan/Ag nanocomposites. *Sens. Actuators B Chem.* **2016**, *225*, 348. [CrossRef]
28. Murugadoss, A.; Chattopadhyay, A. A 'green' chitosan–silver nanoparticle composite as a heterogeneous as well as micro-heterogeneous catalyst. *Nanotechnology* **2007**, *19*, 015603. [CrossRef]
29. Baig, R.N.; Nadagouda, M.N.; Varma, R.S. Ruthenium on chitosan: A recyclable heterogeneous catalyst for aqueous hydration of nitriles to amides. *Green Chem.* **2014**, *16*, 2122. [CrossRef]

30. Rafiee, F. Recent advances in the application of chitosan and chitosan derivatives as bio supported catalyst in the cross coupling reactions. *Curr. Org. Chem.* **2019**, *23*, 390. [CrossRef]
31. Pei, X.; Zheng, X.; Liu, X.; Lei, A.; Zhang, L.; Yin, X. Facile fabrication of highly dispersed Pd catalyst on nanoporous chitosan and its application in environmental catalysis. *Carbohydr. Polym.* **2022**, *286*, 119313. [CrossRef]
32. Zeng, M.; Zhang, X.; Shao, L.; Qi, C.; Zhang, X. Highly porous chitosan microspheres supported palladium catalyst for coupling reactions in organic and aqueous solutions. *J. Organomet. Chem.* **2012**, *704*, 29. [CrossRef]
33. Xie, H.; Yue, H.; Zhang, W.; Hu, W.; Zhou, X.; Prinsen, P.; Luque, R. A chitosan modified Pt/SiO$_2$ catalyst for the synthesis of 3-poly (ethylene glycol) propyl ether-heptamethyltrisiloxane applied as agricultural synergistic agent. *Catal. Commun.* **2018**, *104*, 118. [CrossRef]
34. Shen, C.; Xu, J.; Yu, W.; Zhang, P. A highly active and easily recoverable chitosan@copper catalyst for the C–S coupling and its application in the synthesis of zolimidine. *Green Chem.* **2014**, *16*, 3007. [CrossRef]
35. Mounir, C.; Ahlafi, H.; Aazza, M.; Moussout, H.; Mounir, S. Kinetics and langmuir–hinshelwood mechanism for the catalytic reduction of *para*-nitrophenol over Cu catalysts supported on chitin and chitosan biopolymers. *React. Kinet. Mech. Catal.* **2021**, *134*, 285. [CrossRef]
36. Guo, Y.; Dai, M.; Zhu, Z.; Chen, Y.; He, H.; Qin, T. Chitosan modified Cu$_2$O nanoparticles with high catalytic activity for p-nitrophenol reduction. *Appl. Surf. Sci.* **2019**, *480*, 601. [CrossRef]
37. Li, B.; Wang, L.; Qin, C.; Zhu, L. A green and recyclable chitosan supported catalyst for the borylation of α, β-unsaturated acceptors in water. *Catal. Commun.* **2016**, *86*, 23.
38. Zhou, L.; Han, B.; Zhang, Y.; Li, B.; Wang, L.; Wang, J.; Wang, X.; Zhu, L. Cellulosic CuI nanoparticles as a heterogeneous, recyclable catalyst for the borylation of α, β-unsaturated acceptors in aqueous media. *Catal. Lett.* **2021**, *151*, 3220. [CrossRef]
39. Wen, W.; Han, B.; Yan, F.; Ding, L.; Li, B.; Wang, L.; Zhu, L. Borylation of α, β-unsaturated acceptors by chitosan composite film supported copper nanoparticles. *Nanomaterials* **2018**, *8*, 326. [CrossRef] [PubMed]
40. Chi, W.W.; John, T.G.H.; David, O.H.; Richard, J.R. Tropic acid biosynthesis: The incorporation of (*RS*)-phenyl[2-^{18}O,2-^{2}H]lactate into littorine and hyoscyamine in datura stramonium. *Chem. Commun.* **1998**, 1045.
41. Ushimaru, R.; Ruszczycky, M.W.; Liu, H.-W. Changes in regioselectivity of H atom abstraction during the hydroxylation and cyclization reactions catalyzed by hyoscyamine 6β-Hydroxylase. *J. Am. Chem. Soc.* **2019**, *141*, 1062. [CrossRef] [PubMed]
42. Zhang, K.; Ren, B.-H.; Liu, X.-F.; Wang, L.-L.; Zhang, M.; Ren, W.-M.; Lu, X.-B.; Zhang, W.-Z. Direct and selective electrocarboxylation of styrene oxides with CO$_2$ for accessing β-hydroxy acids. *Angew. Chem. Int. Ed.* **2022**, *61*, e202207660.
43. Guillaume, D.; Nicklas, S.; Kálmán, J.S.; Varinder, K.A. Direct synthesis of functionalized allylic boronic esters from allylic alcohols and inexpensive reagents and catalysts. *Synthesis* **2008**, *14*, 2293.
44. Veeraraghavan, R.; Debarshi, P.; Debanjan, B.; Amit, S.; Venkat, R.R. Novel functionalized trisubstituted allylboronates via hosomi—miyaura borylation of functionalized allyl acetates. *Org. Lett.* **2004**, *6*, 481.
45. Soumya, M.; Srinivas, R.G.; Robert, S.C. Total synthesis of the eupomatilones. *J. Org. Chem.* **2007**, *72*, 8724.
46. Veeraraghavan, R.; Debarshi, P.; Hari, N.G.N.; Matthew, W.; Sadie, S.; Michele, T.Y.; Huangbing, W.; Max, S. Tailored a-methylene-c-butyrolactones and their effects on growth suppression in pancreatic carcinoma cells. *Bioorg. Med. Chem. Lett.* **2010**, *20*, 6620.
47. Veeraraghavan, R.; Hari, N.G.N.; Pravin, G. Syntheses of β- and γ-fluorophenyl cis- and trans-α-methylene-γ-butyrolactones. *Tetrahedron Lett.* **2014**, *55*, 5722.
48. Henry, J.H.; Nader, S.A.; Rusha, P.; Mohamed, N.S.; Veeraraghavan, R. β,γ-Diaryl α-methylene-γ-butyrolactones as potent antibacterials against methicillin-resistant Staphylococcus aureus. *Bioorg. Chem.* **2020**, *104*, 104183.
49. Daniel, P.D.; Chong-Lei, J.; Peng, L.; Kay, M.B. Thiol reactivity of n-aryl α-methylene-γ-lactams: Influence of the guaianolide structure. *J. Org. Chem.* **2022**, *87*, 11204.
50. Xuan, Q.; Wei, Y.; Chen, J.; Song, Q. An expedient E-stereoselective synthesis of multi-substituted functionalized allylic boronates from Morita-Baylis-Hillman alcohols. *Org. Chem. Front.* **2017**, *4*, 1220. [CrossRef]
51. George, W.K.; Venkataiah, B.; Dong, G. Pd-catalyzed cross-coupling of baylis-hillman acetate adducts with bis(pinacolato)diboron: An efficient route to functionalized allyl borates. *J. Org. Chem.* **2004**, *69*, 5807.
52. George, W.K.; Bollu, V. The total synthesis of eupomatilones 2 and 5. *Tetrahedron Lett.* **2005**, *46*, 7385.
53. Feng, J.; Lei, X.; Bao, R.; Li, Y.; Xiao, C.; Hu, L.; Tang, Y. Enantioselective and collective total syntheses of xanthanolides. *Angew. Chem. Int. Ed.* **2017**, *56*, 16323. [CrossRef]
54. Zhang, G.; Fan, Q.; Zhao, Y.; Ding, C. Copper-promoted oxidative intramolecular C–H amination of hydrazones to synthesize ^{1}H-indazoles and ^{1}H-pyrazoles using a cleavable directing group. *Eur. J. Org. Chem.* **2019**, *33*, 5801. [CrossRef]
55. Zhao, P.; Wu, S.; Ke, C.; Liu, X.; Feng, X. Chiral lewis acid-catalyzed enantioselective cyclopropanation and C–H insertion reactions of vinyl ketones with α-diazoesters. *Chem. Commun.* **2018**, *54*, 9837. [CrossRef]
56. Xia, B.; Xu, J.; Xiang, Z.; Cen, Y.; Hu, Y.; Lin, X.; Wu, Q. Stereoselectivity-tailored, metal-free hydrolytic dynamic kinetic resolution of Morita-Baylis-Hillman acetates using an engineered lipase–organic base cocatalyst. *ACS Catal.* **2017**, *7*, 4542. [CrossRef]

Disclaimer/Publisher's Note: The statements, opinions and data contained in all publications are solely those of the individual author(s) and contributor(s) and not of MDPI and/or the editor(s). MDPI and/or the editor(s) disclaim responsibility for any injury to people or property resulting from any ideas, methods, instructions or products referred to in the content.

Review

Research Progress on Propylene Preparation by Propane Dehydrogenation

Cheng Zuo and Qian Su *

College of Chemistry & Chemical and Environmental Engineering, Weifang University, Weifang 261000, China; 17854270427@163.com
* Correspondence: sqian316@wfu.edu.cn

Abstract: At present, the production of propylene falls short of the demand, and, as the global economy grows, the demand for propylene is anticipated to increase even further. As such, there is an urgent requirement to identify a novel method for producing propylene that is both practical and reliable. The primary approaches for preparing propylene are anaerobic and oxidative dehydrogenation, both of which present issues that are challenging to overcome. In contrast, chemical looping oxidative dehydrogenation circumvents the limitations of the aforementioned methods, and the performance of the oxygen carrier cycle in this method is superior and meets the criteria for industrialization. Consequently, there is considerable potential for the development of propylene production by means of chemical looping oxidative dehydrogenation. This paper provides a review of the catalysts and oxygen carriers employed in anaerobic dehydrogenation, oxidative dehydrogenation, and chemical looping oxidative dehydrogenation. Additionally, it outlines current directions and future opportunities for the advancement of oxygen carriers.

Keywords: propylene; propane; oxygen carrier; chemical looping; oxidative dehydrogenation

Citation: Zuo, C.; Su, Q. Research Progress on Propylene Preparation by Propane Dehydrogenation. *Molecules* 2023, *28*, 3594. https://doi.org/10.3390/molecules28083594

Academic Editor: Lu Liu

Received: 7 April 2023
Revised: 18 April 2023
Accepted: 19 April 2023
Published: 20 April 2023

Copyright: © 2023 by the authors. Licensee MDPI, Basel, Switzerland. This article is an open access article distributed under the terms and conditions of the Creative Commons Attribution (CC BY) license (https:// creativecommons.org/licenses/by/ 4.0/).

1. Introduction

Propylene is primarily utilized at room temperature to produce raw materials, such as acrylic acid, acrolein, acrylonitrile, and polypropylene, which are fundamental to the synthesis of plastics, rubber, and fibers. The global demand for propylene is projected to rise in tandem with the development of the social economy. The primary sources of propylene production are steam cracking of refinery gas and catalytic cracking of heavy oils, such as petroleum [1–5]. Nevertheless, the propylene yield in these processes is significantly restricted. The yield ratio of propylene to ethylene in the steam cracking process is 0.4–0.5, while it is only 4.5% in the catalytic cracking process of heavy oil [6–10].

Over the past few years, several technologies have been investigated worldwide to improve propylene yield, such as propane dehydrogenation [11,12], carbon-tetraolefin disproportionation to propylene [13–15], methanol to propylene [16–18], and catalytic cracking of olefins to increase propylene production [19–21]. Among these technologies, propane dehydrogenation has gained considerable attention, and its development potential is vast. Propane dehydrogenation accounts for 4.5% of the total propylene capacity, and is the third-largest source of propylene production globally [22–26]. The primary techniques for propylene preparation are anaerobic dehydrogenation and oxidative dehydrogenation. The former process is associated with high equipment and catalyst costs. In contrast, oxidative dehydrogenation is exothermic, and has lower equipment and catalyst costs compared to conventional anaerobic dehydrogenation. Furthermore, oxidative dehydrogenation has significant potential to address equilibrium conversion limitations and low selectivity. Extensive research has been conducted on the oxidative dehydrogenation of propane, using various gases and solids with the different oxidizing properties as oxidants [27–30]. The primary gas oxidants used in oxidative dehydrogenation are O_2, N_2O, Cl_2, and CO_2.

However, CO$_2$ has garnered more attention from previous researchers, due to its difficulty in oxidizing propylene and propane. Some studies have explored the role of CO$_2$ in the oxidative dehydrogenation of propane, but it is challenging to regulate its impact on the reaction at low temperatures. Moreover, the reaction mechanism is not clear, leading to significant product variability. Anaerobic dehydrogenation typically yields low conversion rates (not exceeding 50%) [31,32], but it has higher dehydrogenation efficiency and propylene selectivity. Therefore, anaerobic dehydrogenation has found practical applications. The aerobic dehydrogenation method entails adding an oxidant to the reaction system. Hydrogen, a byproduct of the decomposition of low-carbon alkanes, reacts with the oxidant to produce H$_2$O, which can be separated from the reaction by condensation, maintaining a positive reaction direction. The oxidative dehydrogenation of low-carbon alkanes is exothermic, resulting in lower reaction temperatures. The catalyst is not deactivated by high temperatures, enhancing its application value. However, oxidative dehydrogenation has problems with process control, necessitating high catalyst selectivity. Hence, the development of catalysts with high selectivity for target olefins is the focus of current research on the aerobic dehydrogenation of low-carbon alkanes.

While anaerobic dehydrogenation is effective in dehydrogenation, it has several problems, such as poor catalyst cycling performance, severe carbon accumulation, and low propane conversion, limiting the development of industrial propane dehydrogenation process technology. As of September 2022, China imported 1,686,800 tons of propylene, whereas only 38,400 tons were exported [33]. The production was much lower than the demand, making it necessary to identify a new pathway for practical and reliable target olefin production [34]. Chemical looping oxidative dehydrogenation utilizes the hydrogen produced by the dehydrogenation of low-carbon alkanes to combine with lattice oxygen provided by metal oxide (MeO) oxygen supports to generate water, which is separated from the reaction system by condensation. This drives the reaction equilibrium in the direction of a positive reaction, increasing propane conversion. Furthermore, the slow release of lattice oxygen effectively controls the rate of propylene production and enhances propylene selectivity. Moreover, the oxygen support used in chemical looping oxidative dehydrogenation typically exhibits better cycling performance and meets industrial production requirements [35–37]. Thus, there is considerable potential for the development of low-carbon olefin production via chemical looping oxidative dehydrogenation.

Although researchers have made many contributions to the propane dehydrogenation, reviews, such as this one, are still necessary to provide direction for future research. In this paper, we will focus on the mechanism and role of catalysts, or oxygen carriers, in propane dehydrogenation reactions.

2. Catalysts for Anaerobic Dehydrogenation Reaction

The catalysts utilized in anaerobic dehydrogenation mainly comprise of Pt-Sn and Cr$_2$O$_3$ catalysts, with other catalysts being less commonly reported. The Pt-Sn catalysts exhibit high catalytic activity, owing to the presence of the noble metal Pt, which also enhances propylene selectivity. Furthermore, the catalysts demonstrate excellent thermal stability and adaptability to a diverse range of reaction conditions, making them industrially viable for over two decades. In contrast, the Cr$_2$O$_3$ catalysts are inexpensive and readily available, but their application is limited, due to the presence of the heavy metal element Cr, which is harmful to the environment, making them less desirable.

2.1. Platinum-Based Catalyst

Platinum (Pt) is a noble metal, often used in the direction of catalyst dehydrogenation. Yu et al. [38] achieved 34.1% propane conversion and 79.2% propylene selectivity using Pt/Al$_2$O$_3$ catalyst at 576 °C. However, due to the excessive acidic bits of Al$_2$O$_3$ support, carbon deposition occurred during the reaction, leading to catalyst deactivation. To address this issue, researchers have improved the activity of Pt/Al$_2$O$_3$ catalysts by adding catalytic agents, or modifying Al$_2$O$_3$. Various studies have demonstrated that adding Sn significantly improves catalyst activity. For instance, Hien et al. [39] investigated the role of Sn in reducing catalyst Pt/γ-Al$_2$O$_3$, and found that Sn addition enabled rapid reductive regeneration of Pt, reducing the occurrence of side reactions. Antolini et al. [40] loaded Pt-Sn onto Al$_2$O$_3$ for propane dehydrogenation and reported that increasing the amount of the active component Sn improved propane conversion and propylene selectivity. The interaction of Sn with Pt produced different types of alloys that modified the defective sites on the catalyst surface, improving Pt dispersion, propane adsorption, and inhibiting the formation of byproducts. Yang et al. [41] investigated the catalytic performance of PtSn catalysts for propane dehydrogenation using first principles calculations and found that the formation of an alloy facilitated the reaction. Vu et al. [42] suggested that the type and stability of PtSn alloy were positively correlated with the activity and stability of the catalytic propane dehydrogenation reaction of this catalyst. Hauser et al. [43] used density function theory (DFT) to study the reaction path of propane dehydrogenation to propylene and found that replacing a Pt atom in the Pt$_4$ cluster with a Sn atom to form a PtSn alloy reduced the activation energy of the rate-controlling step, thereby improving propane conversion and propylene selectivity. Sn transfers electrons to Pt atoms, reducing the desorption energy barrier of propylene and carbon precursors, and hindering the adsorption of propylene on Pt atoms, thus reducing the possibility of side reactions, such as hydrogenolysis and coking.

The size of Pt particles also affects propane dehydrogenation. Kumar et al. [44] prepared Pt/SBA-15 catalysts with varying Pt particle sizes for catalytic propane dehydrogenation and found that Pt particles with a particle size of around 3 nm had higher activity. However, the coking rate and amount of coking were also higher. This is due to smaller Pt particles activating the C-C bond, leading to cleavage reactions.

The results of Nykanen et al. [45] showed that the adsorption energy (0.52 eV) of propylene on the Pt(111) crystal surface is smaller than its energy barrier (0.81 eV) for deep dehydrogenation. While the adsorption energy (0.81 eV) on the Pt(211) crystal surface is larger than its energy barrier (0.54) for deep dehydrogenation. Propylene is prone to deep cracking and coking on the crystalline surface of Pt. Therefore, monometallic Pt-based catalysts have high activity and low selectivity for propylene in the initial stage of the reaction. When the reaction temperature of Pt-based catalysts is high, it is more likely to bring about the sintering problem of Pt nanoparticles. Currently, the Pt-based catalyst activity could be improved by improving the interaction between Pt and the support, in addition to the introduction of metals such as Sn.

2.1.1. Improvement of Support

Different supports can strongly influence catalytic propane dehydrogenation by Pt-based catalysts. Al$_2$O$_3$ and molecular sieves are the main supports currently used for propane dehydrogenation. Kikuchi et al. [46] and Kobayashi et al. [47] mixed Al$_2$O$_3$ with MgO, ZnO, and Fe$_2$O$_3$ to obtain MgO-Al$_2$O$_3$, ZnO-Al$_2$O$_3$, and Fe$_2$O$_3$-Al$_2$O$_3$ supports, respectively, followed by loading Pt and Sn. Experimental results showed that the Pt-Sn/ZnO-Al$_2$O$_3$ catalyst for n-butane dehydrogenation was highly effective. In their catalytic propane dehydrogenation study, Vu et al. [48] employed Pt-Sn catalysts that were loaded with La, Ce, and Y-doped Al$_2$O$_3$ supports. The authors observed that La and Y could form a dispersed phase, whereas Ce formed CeO$_2$, due to an agglomeration phenomenon on the catalyst surface. Notably, PtSn/La-Al$_2$O$_3$ and PtSn/Y-Al$_2$O$_3$ surfaces formed Pt and Sn alloys, respectively. The catalytic activity of these two catalysts was high, due to the low coking amount and the excellent stability of the alloys.

In contrast to metal oxides, molecular sieves used as supports can reduce the adsorption capacity of propylene and minimize side reactions such as product cracking. In a study conducted by Chen et al. [49], a PtSnNaLa/ZSM-5 catalyst was prepared, and was observed to have a lower amount of carbon deposition compared to the PtSnNaLa/γ-Al$_2$O$_3$ catalyst during propane dehydrogenation catalysis. After 880 h of continuous reaction, the propane conversion of the PtSnNaLa/ZSM-5 catalyst remained above 30%, while, after 480 h of continuous reaction with the PtSnNaLa/γ-Al$_2$O$_3$ catalyst, the propane conversion had dropped below 30%. Li et al. [50] prepared Co-doped HZSM-5 catalysts, and it was found that the dehydrogenation reaction rate of propane catalyzed by this catalyst was 12 times higher than that of the HZSM-5 catalyst, and the selectivity of propylene was also high.

2.1.2. Effect of Additives

The addition of metal Sn to the catalyst is also an important factor. Although the Pt-Sn/γ-Al$_2$O$_3$ catalyst showed a significant improvement in catalytic activity, it still suffered from catalyst coking, which shortened its lifetime [51]. To address this issue, Xia et al. [52] used Mg(Al)O-x supports loaded with active components Pt and In to produce Pt-In/Mg(Al)O-x catalysts. The addition of In regulated the acidity of the catalyst surface, improved the dispersion of Pt, and increased the anti-coking ability of the catalyst. Consequently, the Pt-In/Mg(Al)O-4 catalyst reduced carbon accumulation and prolonged the catalyst's lifespan. The initial conversion of propane was 66.4%, and the propane conversion after eight reaction cycles still reached 43.5%. Similarly, Zhang et al. [53] added different levels of La to Al$_2$O$_3$ supports, using the sol-gel method. The best conversion and selectivity of propane were achieved when the mass fraction of La was 1.0%, resulting in 41% propane conversion and 97–98% propylene selectivity. In addition to its effectiveness in the dehydrogenation of propane, the addition of In was also found to be effective in the dehydrogenation of butane. Bocanegra et al. [54] added In to the Pt-Sn system, using MgAl$_2$O$_4$ as a support for the dehydrogenation of butane, which resulted in high selectivity (95–96%) of butene. During anaerobic dehydrogenation, researchers observed that the competitive adsorption of additives decreased the adsorption of low-carbon olefins, but improved the selectivity of target products produced from the dehydrogenation of low-carbon alkanes.

2.2. Cr-Based Catalyst

Cr-based catalysts have gained attention, due to their high catalytic activity and propylene selectivity, as well as their cost-effectiveness compared to noble metal Pt. Also, the better cycling performance of chromium-based catalysts is an important reason for their industrialization. The Cr$_2$O$_3$/Al$_2$O$_3$ catalyst, developed by Cabrera et al. [55], demonstrated propane conversion of up to 47% and propylene selectivity above 90% at a reaction temperature of 600 °C and atmospheric pressure. However, carbon deposition and deactivation of the catalyst remain issues that need to be addressed to improve the conversion of propane and selectivity of propylene. Therefore, modifications to Cr-based catalysts are necessary.

Modification of Supports

Kim et al. [56] examined the impact of varying the ratio of Al$_2$O$_3$ and ZrO$_2$ in Cr$_2$O$_3$ catalyst supports on propylene yield. Their findings showed that the lowest oxygen content of the catalyst was achieved at an Al/Zr ratio of 0.1, resulting in propylene selectivity and yield of 85% and 30%, respectively. The authors speculated that this might be due to the interaction of the active component with the support, leading to the conversion of lattice oxygen to electrophilic oxygen in the catalyst. However, an increase in carbon oxide content (CO$_2$ and CO) was observed, leading to a decrease in propylene selectivity.

It is important to note that Cr is a heavy metal element that poses environmental pollution risks, which greatly limits the widespread industrial use of Cr$_2$O$_3$.

2.3. Introduction of Several Propane Anaerobic Dehydrogenation Industrialization Technologies

Currently, industrial technology for anaerobic propane dehydrogenation mainly consists of the Oleflex process, developed by UP, and the Catofin process, developed by ABB Lummus.

2.3.1. Catofin Process

The Catofin process comprises four stages: propane dehydrogenation to propylene (reaction stage), compression of the reactor discharge (compression stage), and recovery and refining of the product (recovery and refinement stages). The Catofin process employs a CrO_x/Al_2O_3 catalyst, which is cost-effective, and has high cycle times and excellent mechanical properties. The catalyst has a long service life of up to 600 days [21].

The main characteristics of the Catofin process [20] are (1) the use of a low-cost non-precious metal catalyst with excellent mechanical properties and high cycle times, (2) high-pressure reaction, requiring the importation of specialized equipment, and (3) easy separation of products.

Figure 1 shows that process diagram of Catofin dehydrogenation unit.

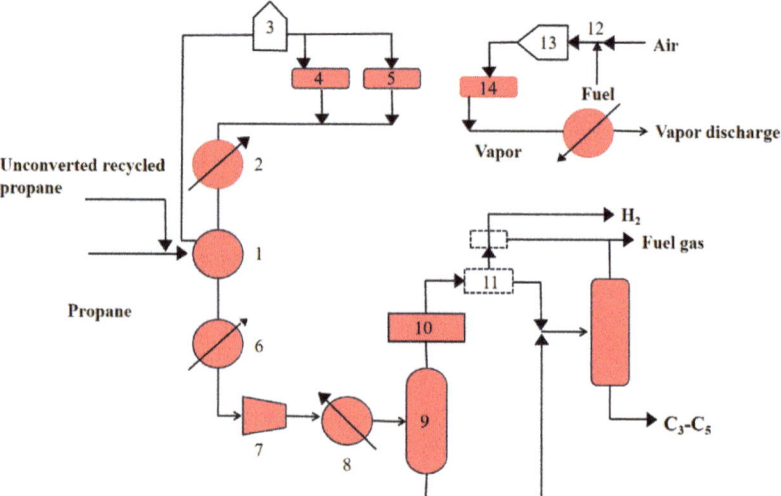

Figure 1. Process diagram of Catofin dehydrogenation unit. (1) Heat exchanger; (2) Steam generator; (3) Heating furnace; (4) Purge section reactor; (5) Process section reactor; (6) Cooler; (7) Compressor; (8) Air cooling; (9) Flash tank; (10) Dryer; (11) Cold box; (12) Gasifier; (13) Heater; (14) Regeneration section reactor.

2.3.2. Oleflex Process

The Oleflex process is divided into three parts: the reaction part, the product separation part, and the catalyst regeneration part. The reaction section uses moving bed reactor. Compared with the Catofin process, the catalyst in the reactor is recycled and has a service life of 2 to 7 days.

The Oleflex process employs the Pt catalyst to carry out the dehydrogenation of propane, and the resulting polymeric grade propylene is obtained by separation and distillation in the presence of the catalyst. This reaction does not require the use of hydrogen or water vapor as diluents, resulting in lower energy consumption and operational costs. The Oleflex process is characterized by (1) high operational safety, a small reaction volume, and easy operation, and (2) a lower one-way conversion and a higher sulfur content limitation (not exceeding 100 ppm) compared to the Catofin process. Table 1 shows the comparison of these two process technologies.

Figure 2 shows that Process diagram of Oleflex dehydrogenation unit. Table 1 shows the comparison of the Catofin and Oleflex propane dehydrogenation processes.

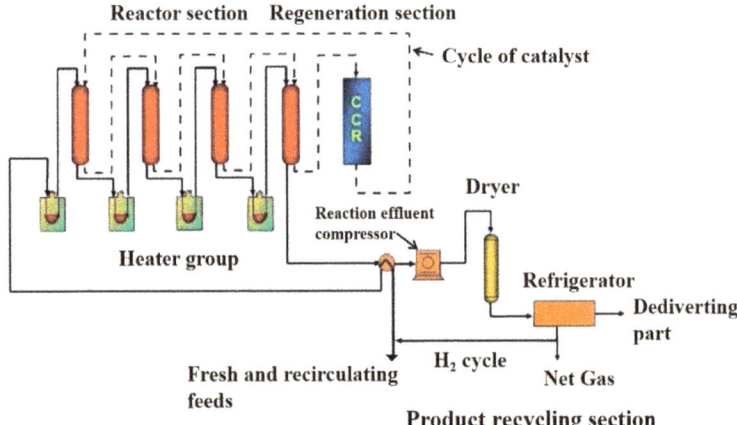

Figure 2. Process diagram of Oleflex dehydrogenation unit.

Table 1. Comparison of Catofin and Oleflex propane dehydrogenation process technologies.

Projects	Process Technology	
	Catofin Process	Oleflex Process
Technology exporter	ABB Lummus	UOP
Reactor type	Fixed Bed	Moving Bed
Total number of reactors	5	3~4
Catalyst	CrO_x/Al_2O_3	$Pt-Sn/Al_2O_3$
Cycle regeneration time	15~30 min	2~7 d
Temperature/°C	600–700	550~620
Pressure/Mpa	0.3~0.5	2~3
Diluent	-	H_2
Propane conversion	48~65	80~88
Propylene selectivity	25	89~91

The reaction is a strong heat absorption reaction, which requires a large amount of external reaction heat supply. Since the dehydrogenation reaction is an equilibrium reaction, increasing the temperature and decreasing the pressure are beneficial for the dehydrogenation reaction to proceed and obtain a high propane conversion. The temperature of industrial propane dehydrogenation reaction is 500–700 °C. However, the high temperature will promote the occurrence of thermal cracking side reactions, which will also produce some heavy hydrocarbons and form a small amount of coking on the catalyst, thus reducing the reactivity. Therefore, the Oleflex process is more selective for propylene than the Catofin process, due to the cyclic regeneration of the catalyst.

3. Catalysts for Oxidative Dehydrogenation Reaction

In contrast to anaerobic dehydrogenation, oxidative dehydrogenation is a highly endothermic reaction that is not limited by thermodynamic equilibrium, thereby increasing propane conversion. However, it is prone to catalyst deactivation, due to carbon deposition. To extend the catalyst lifetime, oxidative dehydrogenation reactions typically require the introduction of a gaseous oxidant. Common oxidants include O_2, N_2O, and CO_2. CO_2, in particular, has been widely studied in the literature, as it does not deeply oxidize propane or propylene. The function of CO_2 as an oxidant can be attributed to two factors [57,58]: (1) the reaction $CO_2 + C \rightarrow 2CO$ can reduce carbon deposition on the catalyst and improve its stability, and (2) it inhibits the adsorption of olefin products on the catalyst surface, thereby improving propylene selectivity.

3.1. Chromium-Based Catalysts

The use of Cr_2O_3/Al_2O_3 catalysts in oxidative dehydrogenation is also common. However, unlike in anaerobic dehydrogenation, the addition of CO_2 as an oxidant does not improve propane conversion and propylene selectivity of the Cr_2O_3/Al_2O_3 catalyst. Its only function is to extend the catalyst's lifetime. Therefore, researchers have explored the use of molecular sieves to modulate the physicochemical properties of Cr_2O_3. Michorczyk et al. [59] loaded Cr_2O_3 onto a MCM-41 molecular sieve and obtained 34.9% propane conversion and 88.5% propylene selectivity at 550 °C. Zhang et al. [60] loaded Cr_2O_3 onto SBA-15, ZrO_2, and ZrO_2/SBA-15 supports and found that the Cr_2O_3/SBA-15 catalyst displayed excellent catalytic activity, with 24.2% propane conversion and 83.9% propylene selectivity at 600 °C.

3.2. Vanadium-Based Catalysts

V_2O_5 is an acidic oxide that exhibits high catalytic activity, but low propylene selectivity. Therefore, V_2O_5 is often loaded onto suitable supports to improve propylene selectivity for propane dehydrogenation. The appropriate support can decrease the deep dehydrogenation capability of V_2O_5 and enhance the selectivity of propylene. The catalytic activity center in vanadium-based catalysts is VO_x [61]. Vanadium oxide with high coordination numbers can deeply oxidize propane, whereas highly dispersed tetrahedral VO_4 provides limited lattice oxygen for propane dehydrogenation. By controlling the release rate of lattice oxygen, selectivity of propylene can be improved. A balanced ratio of acidic and basic sites on the catalyst surface is the key to improving the conversion of propane and propylene selectivity [62]. A more acidic surface activates propane more strongly, improving the propane conversion. On the other hand, the product propylene has a greater electron cloud density compared to propane, and it is more basic. Therefore, a more alkaline surface facilitates propylene desorption and Improves selectivity [63–66]. During the preparation of vanadium-based catalysts, it is crucial to control the vanadium content. Exceeding the theoretical monolayer of vanadium content results in the appearance of octahedral V_2O_5 crystalline phases, with different polymerization deformations on the catalyst surface [67–70]. Therefore, it is essential to disperse tetrahedral VO_4 as much as possible on the catalyst surface to reduce the occurrence of crystalline phase V_2O_5. Hossain et al. prepared vanadium-based CaO-γ-Al_2O_3 supports for the oxidative dehydrogenation of propane [71]. They achieved 25.5% conversion of propane and 94.2% selectivity of propylene at 640 °C, and the most active catalysts were obtained at a mass ratio of CaO to γ-Al_2O_3 of 1:1.

3.3. Gallium-Based Catalysts

Gallium-based catalysts have also been utilized for propane dehydrogenation, along with chromium-based and vanadium-based catalysts. Ga_2O_3 catalysts operate via a heterolysis process, which is distinct from the mechanism of the Cr system. The reaction mechanism is illustrated in Figure 3 [72–76].

Xu et al. [77] and Ren et al. [78] discovered that the impact of CO_2 oxidation on propane dehydrogenation was evident when the reaction rate of (3c) was slow and the reaction rate of (3d) was fast. On the other hand, the presence of CO_2 had little effect on the reaction when the reaction rate of (3c) was fast and the reaction rate of (3d) was slow. However, CO_2 had an inhibitory effect on propane dehydrogenation, as it had to compete with propane for the basic sites on the catalyst surface, which hindered the adsorption of propane on the catalyst surface. When the rate of reaction (3d) was very slow, the conversion of propane and propylene yield decreased with the increase of CO_2 concentration.

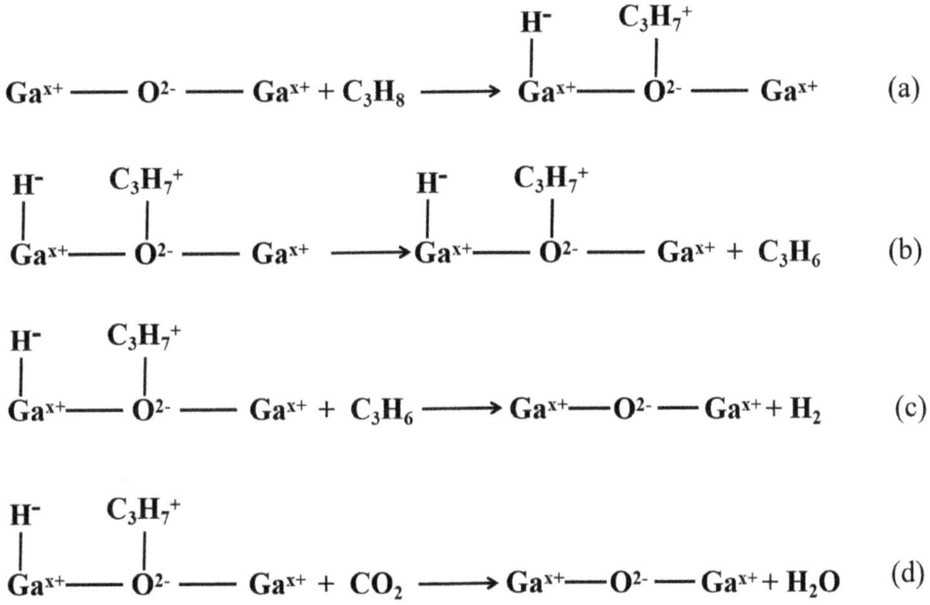

$$Ga^{x+} \longrightarrow O^{2-} \longrightarrow Ga^{x+} + C_3H_8 \longrightarrow \overset{\overset{H^-}{|}}{Ga^{x+}} \longrightarrow \overset{\overset{C_3H_7^+}{|}}{O^{2-}} \longrightarrow Ga^{x+} \quad (a)$$

$$\overset{\overset{H^-}{|}}{Ga^{x+}} \longrightarrow \overset{\overset{C_3H_7^+}{|}}{O^{2-}} \longrightarrow Ga^{x+} \longrightarrow \overset{\overset{H^-}{|}}{Ga^{x+}} \longrightarrow \overset{\overset{C_3H_7^+}{|}}{O^{2-}} \longrightarrow Ga^{x+} + C_3H_6 \quad (b)$$

$$\overset{\overset{H^-}{|}}{Ga^{x+}} \longrightarrow \overset{\overset{C_3H_7^+}{|}}{O^{2-}} \longrightarrow Ga^{x+} + C_3H_6 \longrightarrow Ga^{x+} \longrightarrow O^{2-} \longrightarrow Ga^{x+} + H_2 \quad (c)$$

$$\overset{\overset{H^-}{|}}{Ga^{x+}} \longrightarrow \overset{\overset{C_3H_7^+}{|}}{O^{2-}} \longrightarrow Ga^{x+} + CO_2 \longrightarrow Ga^{x+} \longrightarrow O^{2-} \longrightarrow Ga^{x+} + H_2O \quad (d)$$

Figure 3. Reaction mechanism diagram of Ga_2O_3 catalyst with propane (a–d).

4. The Process of Chemical Looping Oxidative Dehydrogenation

The anaerobic dehydrogenation method has drawbacks, such as the non-recyclability of the catalyst, and being constrained by thermodynamic equilibrium, resulting in low conversion rates. On the other hand, the aerobic dehydrogenation method has issues, such as difficulty controlling the degree of reaction, especially when using CO_2 as the oxidant, leading to varying reaction products. Chemical looping technology uses an oxygen carrier that can be regenerated and slowly releases lattice oxygen to control the degree of reaction, thereby improving the thermodynamic irreversibility of traditional dehydrogenation reactions. Chemical looping oxidative dehydrogenation overcomes the limitations of both anaerobic and oxidative dehydrogenation methods, and has the potential to significantly improve the conversion of low-carbon alkanes and selectivity of low-carbon olefins [79–81].

Chemical looping oxidative dehydrogenation involves oxidation and reduction reactions based on the oxygen carrier's reaction type in two reactors. In the dehydrogenation reactor, the oxygen carrier is used for the dehydrogenation reaction with propane, and is then regenerated with air, releasing heat in the oxidation reactor. During the reaction, the products of low carbon alkanes after dehydrogenation (H_2) combine with the metal oxide oxygen carriers' lattice oxygen to form water, which is removed from the reaction system by condensation, promoting the reaction equilibrium to proceed in the positive reaction direction, thus increasing the conversion rate of low carbon alkanes. The lattice oxygen in the oxygen carrier can be gradually released under specific conditions, controlling the reaction's course, which contributes to enhancing the selectivity of propylene. After the reduction of the oxygen carrier product in the dehydrogenation reactor, it enters the air reactor for oxidation with oxygen to complete the regeneration process. The process flowchart is presented in Figure 4.

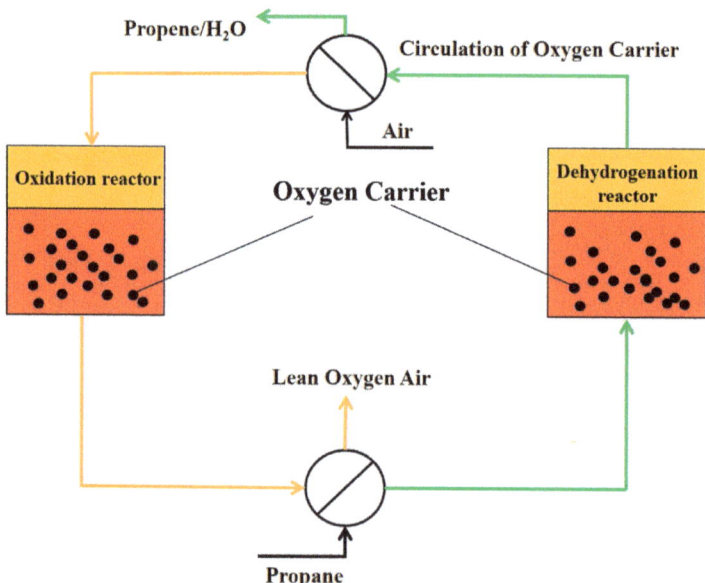

Figure 4. The process flow diagram of chemical looping oxidative dehydrogenation.

4.1. Monometallic Active Oxygen Carriers

Ghamdi et al. [82] investigated catalysts with varying vanadium content (5%, 7%, 10% wt.%), loaded onto γ-Al_2O_3 for chemical looping oxidative dehydrogenation reactions. They achieved a maximum propylene selectivity of 85.94% at a propylene conversion of 11.73%. However, the VO_x/γ-Al_2O_3 catalysts had a limited number of cycles, with a maximum of 10 cycles throughout the reaction. This was likely due to the accumulation of V_2O_5 crystal structures on the catalyst surface as the number of cycles increased, which decreased the propylene yield. In monometallic oxygen carriers, lattice oxygen is released rapidly, resulting in deep propane oxidation during the chemical looping oxidative dehydrogenation of propane. Additional CO_2 is often required to provide an oxygen source for the reaction system. Loading appropriate metal oxides onto catalysts to control the rate of lattice oxygen release and regulate the migration or evolution of the released lattice oxygen from the catalyst surface could reduce the selectivity of CO_x and extend the catalyst's lifespan. Wu et al. [83] compared the activity of Ga-based, Mo-based, and V-based oxygen carriers for propane dehydrogenation at 540 °C, 615 °C, and 650 °C, respectively. The results showed that the V-based oxygen carrier had the best catalytic activity. Meanwhile, the optimum reaction temperature for propane dehydrogenation was also investigated to be 615 °C.

4.2. Bimetallic or Polymetallic Composite Oxygen Carriers

Fukudome et al. [84,85] achieved higher concentrations of isolated VO_x species by incorporating them into the SiO_2 framework using alkoxy exchange between metal alcohol oxygen compounds and polyethylene glycols. It was observed that SiO_2-doped VO_x exhibited higher selectivity for propylene than VO_x loaded onto SiO_2. Gao et al. [86] used polymetallic composite oxygen carriers ($La_xSr_{2-x}FeO_{4-\delta}$) to dehydrogenate ethane into ethylene, with a high yield of 51.6%, and the number of cycles of the oxygen carriers reached up to 30. Thus, there is a need to develop oxygen carriers that can last for more cycles in the future.

Appropriate bimetallic or even polymetallic oxides can release lattice oxygen slowly, which is more effective in controlling the rate of propane dehydrogenation for propylene production than anaerobic dehydrogenation and gaseous oxidants. Moreover, metal oxides can inhibit the conversion of lattice oxygen (O^{2-}) to electrophilic oxygen (O_2^-) and reduce the formation of oxides (e.g., CO_x), thereby improving the conversion of propane and the selectivity of propylene [87–89]. Additionally, the short reaction time of chemical looping oxidative dehydrogenation, ranging from 20 s to 8 min, makes it difficult to study the reaction mechanism of propane dehydrogenation. For industrial promotion, further research is needed to develop oxygen carriers with high oxygen loading capacity and high propane conversion with propylene selectivity.

5. Conclusions and Prospects

(1) The current methods for propylene production are anaerobic and oxidative dehydrogenation. The anaerobic method has been used for many years, but is expensive, due to high equipment and catalyst costs. The oxidative dehydrogenation method is cheaper, but the extent of CO_2 influence on the reaction is difficult to control at certain temperatures, and the reaction mechanism is still unclear, resulting in variable product yields.

(2) In contrast, chemical looping oxidative dehydrogenation resolves the drawbacks of the previous methods. Lattice oxygen release can be controlled by appropriate bimetallic or polymetallic oxides, replacing molecular oxygen. This effectively controls the reaction rate of propane dehydrogenation to produce propylene, and improves the conversion of propane with high selectivity for propylene, compared to oxygen-free dehydrogenation and gas oxidant methods.

(3) The future of chemical looping oxidative dehydrogenation for industrial applications requires the development of multi-component coupled composite oxygen carriers with high oxygen loading, extended cycle life, and high propylene yield.

Author Contributions: C.Z.: Conceptualization, Methodology, Software, Investigation, Writing-original draft. Q.S.: Methodology, Validation, Formal analysis and Visualization. Q.S.: Funding, Acquisition and Supervision. All authors have read and agreed to the published version of the manuscript.

Funding: Support for carrying out this work was provided by Doctoral Research Foundation of Weifang University (2022BS13).

Institutional Review Board Statement: Not applicable.

Informed Consent Statement: Not applicable.

Data Availability Statement: Not applicable.

Conflicts of Interest: The authors declare that they have no known competing financial interests or personal relationships that could have appeared to influence the work reported in this paper.

References

1. Bai, P.T.; Manokaran, V.; Saiprasad, P.S.; Srinath, S. Studies on Heat and Mass Transfer Limitations in Oxidative Dehydrogenation of Ethane Over Cr_2O_3/Al_2O_3 Catalyst. *Procedia Eng.* **2015**, *127*, 1338–1345. [CrossRef]
2. Zeeshan, N. Light alkane dehydrogenation to light olefin technologies: A comprehensive review. *Rev. Chem. Eng.* **2015**, *31*, 413–436.
3. Wang, H.M.; Chen, Y.; Yan, X.; Lang, W.Z.; Guo, Y.J. Cr doped mesoporous silica spheres for propane dehydrogenation in the presence of CO_2: Effect of Cr adding time in sol-gel process. *Micropor. Mesopor. Mat.* **2019**, *284*, 69–77. [CrossRef]
4. Marktus, A.; Fateme, R.; Abbas, J.; Mark, F. Oxidative dehydrogenation of propane to propylene with carbon dioxide. *Appl. Catal. B-Environ.* **2017**, *52*, 429–445.
5. Chen, S.; Chang, X.; Sun, G.D.; Zhang, T.T.; Xu, Y.Y.; Wang, Y.; Pei, C.L.; Gong, J.L. Propane dehydrogenation: Catalyst development, new chemistry, and emerging technologies. *Chem. Soc. Rev.* **2021**, *50*, 3315. [CrossRef]
6. Khadzhiev, S.N.; Usachev, N.Y.; Gerzeliev, I.M.; Belanova, E.P.; Kalinin, V.P.; Kharlamov, V.V.; Kazakov, A.V.; Kanaev, S.A.; Starostina, T.S.; Popov, A.Y. Oxidative dehydrogenation of ethane to ethylene in a system with circulating microspherical metal oxide oxygen carrier: 1. Synthesis and study of the catalytic system. *Pet. Chem.* **2015**, *55*, 651–654. [CrossRef]

7. Darvishi, A.; Davand, R.; Khorasheh, F.; Fattahi, M. Modeling-based optimization of a fixed-bed industrial reactor for oxidative dehydrogenation of propane. *Chin. J. Chem. Eng.* **2016**, *24*, 612–622. [CrossRef]
8. Zhai, Z.; Wang, X.; Licht, R.; Bell, A.T. Selective oxidation and oxidative dehydrogenation of hydrocarbons on bismuth vanadium molybdenum oxide. *J. Catal.* **2015**, *325*, 87–100. [CrossRef]
9. Zea, L.C.H. Oxidative dehydrogenation of propane on Pd-Mo/gamma-Al_2O_3 catalyst: A kinetic study. *Aust. J. Basic Appl. Sci.* **2015**, *9*, 78–83.
10. Zhao, D.; Tian, X.X.; Dmitry, E. In situ formation of ZnOx species for efficientpropane dehydrogenation. *Nature* **2021**, *599*, 234–238. [CrossRef]
11. Zhang, H.J.; Wan, H.; Zhao, Y.; Wang, W.Q. Effect of chlorine elimination from Pt-Sn catalyst on the behavior of hydrocarbon reconstruction in propane dehydrogenation. *Catal. Today* **2019**, *330*, 85–91. [CrossRef]
12. Elbadawi, A.H.; Osman, M.S.; Razzak, S.A.; Hossain, M.M. VO_x-Nb/La-γAl_2O_3 catalysts for oxidative dehydrogenation of ethane to ethylene. *J. Taiwan Inst. Chem. Eng.* **2016**, *61*, 106–116. [CrossRef]
13. Nawaz, Z.; Baksh, F.; Zhu, J.; Wei, F. Dehydrogenation of C3-C4 paraffin's to corresponding olefins over slit-SAPO-34 supported Pt-Sn-based novel catalyst. *J. Ind. Eng. Chem.* **2013**, *19*, 540–546. [CrossRef]
14. Cavani, F.; Trifir ò, F. The oxidative dehydrogenation of ethane and propane as an alternative way for the production of light olefins. *Catal. Today* **1995**, *24*, 307–313. [CrossRef]
15. Sanfilippo, D. Dehydrogenation of Paraffins; Key Technology for Petrochemicals and Fuels. *Cattech* **2000**, *4*, 56–73. [CrossRef]
16. Cavani, F.; Ballarini, N.; Cericola, A. Oxidative dehydrogenation of ethane and propane: How far from commercial implementation? *Catal. Today* **2007**, *127*, 113–131. [CrossRef]
17. Chang, C.D. Methanol Conversion to Light Olefins. *Catal. Rev.* **1984**, *26*, 323–345. [CrossRef]
18. Bricker, J.C. Advanced catalytic dehydrogenation technologies. *Top. Catal.* **2012**, *55*, 1309–1314. [CrossRef]
19. Sattler, J.J.H.B.; Ruiz-Martinez, J.; Santillan-Jimenez, E.; Weckhuysen, B.M. Catalytic dehydrogenation of light alkanes on metals and metal oxides. *Chem. Rev.* **2014**, *114*, 10613–10653. [CrossRef]
20. Chen, J.Q.; Bozzano, A.; Glover, B.; Fuglerud, T.; Kvisle, S. Recent advancements in ethylene and propylene production using the UOP/Hydro MTO process. *Cataly. Today* **2005**, *106*, 103–107. [CrossRef]
21. Vora, B.V. Development of Dehydrogenation Catalysts and Processes. *Top. Catal.* **2012**, *55*, 1297–1308. [CrossRef]
22. Yoshimur, Y.; Kijima, N.; Hayakawa, T.; Murata, K.; Suzuki, K.; Mizukami, F.; Matano, K.; Konishi, T.; Oikawa, T.; Saito, M.; et al. Catalytic cracking of naphtha to light olefins. *Catal. Surv. Jpn.* **2000**, *4*, 157–167. [CrossRef]
23. Hereijgers, B.P.C.; Bleken, F.; Nilsen, M.H.; Svelle, S.; Lillerud, K.P.; Weckhuysen, B.M.; Olsbye, U. Product shape selectivity dominates the Methanol-to-Olefins (MTO) reaction over H-SAPO-34 catalysts. *J. Catal.* **2009**, *264*, 77–87. [CrossRef]
24. Steinfeldt, N.; Muller, D.; Berndt, H. VO_x species on alumina at vanadia loadings and high calcination temperature and their role in the ODP reaction. *Appl. Catal. A-Gen.* **2004**, *272*, 201–213. [CrossRef]
25. Heracleous, E.; Machli, M.; Lemonidou, A.A.; Vasalos, I.A. Oxidative dehydrogenation of ethane and propane over vanadia and molybdena supported catalysts. *J. Mol. Catal. A-Chem.* **2005**, *232*, 29–39. [CrossRef]
26. Skorodumova, N.V.; Simak, S.I.; Lundqvist, B.I.; Abrikosov, I.A.; Johansson, B. Quantum origin of the oxygen storage capability of ceria. *Phys. Rev. Lett.* **2002**, *89*, 166601. [CrossRef]
27. Sadrameli, S.M. Thermal/catalytic cracking of liquid hydrocarbons for the production of olefins: A state-of-the-art review II: Cataltic cracking review. *Fuel* **2016**, *173*, 285–297. [CrossRef]
28. Botavina, M.; Agafonov, Y.A.; Gaidai, N.; Groppo, E.; Corberan, V.C.; Lapidus, A.; Martra, G. Towards efficient catalysts for the oxidative dehydrogenation of propane in the presence of CO_2: Cr/SiO_2 systems prepared by direct hydrothermal synthesis. *Catal. Sci. Technol.* **2016**, *6*, 840–850. [CrossRef]
29. Koirala, R.; Buechel, R.; Pratsinis, S.E.; Baiker, A. Silica is preferred over various single and mixed oxides as support for CO_2-assisted cobalt-catalyzed oxidative dehydrogenation of ethane. *Appl. Catal. A Gen.* **2016**, *527*, 96–108. [CrossRef]
30. Ren, T.; Patel, M.; Blok, K. Olefins from conventional and heavy feedstocks: Energy use in steam cracking and alternative processes. *Energy* **2006**, *31*, 425–451. [CrossRef]
31. Mukherjee, D.; Park, S.-E.; Reddy, B.M. CO_2 as a soft oxidant for oxidative dehydrogenation reaction: An eco benign process for industry. *J. CO2 Util.* **2016**, *16*, 301–312. [CrossRef]
32. Ajayi, B.P.; Rabindran Jermy, B.; Abussaud, B.A.; Al-Khattaf, S. Oxidative dehydrogenation of n-butane over bimetallic mesoporous and microporous zeolites with CO_2 as mild oxidant. *J. Porous. Mater.* **2013**, *20*, 1257–1270. [CrossRef]
33. U.S. Energy Information Administration (EIA). *China Analysis Report*; EIA: Washington, DC, USA, 2022. Available online: http://www.eia.gov/.pdf (accessed on 1 December 2022).
34. Védrine, J. Heterogeneous Partial (amm) oxidation and oxidative dehydrogenation catalysis on mixed metal oxides. *Catalysts* **2016**, *6*, 22. [CrossRef]
35. Elbadawi, A.H.; Ba-Shammakh, M.S.; Al-Ghamdi, S.; Razzak, S.A.; Hossain, M.M. Reduction kinetics and catalytic activity of VO_x/γ-Al_2O_3-ZrO_2 for gas phase oxygen free ODH of ethane. *Chem. Eng. J.* **2016**, *284*, 448–457. [CrossRef]
36. Koc, S.N.; Dayioglu, K.; Ozdemir, H. Oxidative dehydrogenation of propane with K-$MoO_3/MgAl_2O_4$ catalysts. *J. Chem. Sci.* **2016**, *128*, 67–71. [CrossRef]
37. Kong, L.; Li, J.; Zhao, Z.; Liu, Q.; Sun, Q.; Liu, J.; Wei, Y. Oxidative dehydrogenation of ethane to ethylene over Mo-incorporated mesoporous SBA-16 catalysts: The effect of MoO_x dispersion. *Appl. Catal. A-Gen.* **2016**, *510*, 84–97. [CrossRef]

38. Yu, C.; Xu, H.; Chen, X. Preparation etcharacterization, catalytic performance of and PtZn-Sn/SBA-15 catalyst for propane dehydrogenation. *J. Fuel. Chem. Technol.* **2010**, *38*, 308–312. [CrossRef]
39. Hien, N.; Jesper, J.; Bert, M. Role of Sn in the regeneration of Pt/γ-Al$_2$O$_3$ light alkane dehydrogenation catalysts. *Chem. Soc.* **2016**, *6*, 2257–2264.
40. Antolini, E.; Colmati, F.; Gonzalez, E.R. Ethanol oxidation on carbon supported (Pt Sn) alloy/SnO$_2$ and (Pt Sn Pd)alloy/SnO$_2$ catalysts with a fixed Pt/SnO$_2$ atomic ratio: Effect of the alloy phase characteristics. *J. Power. Sources* **2009**, *193*, 555–561. [CrossRef]
41. Yang, M.L.; Zhu, Y.A.; Zhou, X.G. First-principles calculations of propane dehydrogenation over PtSn catalysts. *ACS Catal.* **2012**, *2*, 1247–1258. [CrossRef]
42. Vu, B.K.; Song, M.B.; Ahn, I.Y. Location andstructure of coke generated over Pt-Sn/Al$_2$O$_3$ in propanedehydrogenation. *J. Ind. Eng. Chem.* **2011**, *17*, 71–76. [CrossRef]
43. Hauser, A.W.; Gomes, J.; Bajdich, M. Subnanometer-sized Pt/Sn alloy cluster catalysts for thedehydrogenation of linear alkanes. *Phys. Chem. ChemPhys.* **2013**, *15*, 20727–20734. [CrossRef]
44. Kumar, M.S.; Chen, D.; Walmesley, J.C. Dehydrogenation of propane over Pt-SBA-15: Effect of Pt particle size. *Catal Commun.* **2008**, *9*, 747–750. [CrossRef]
45. Nykanen, L.; Honkala, K. Selectivity in Propene Dehydrogenation on Pt and Pt$_3$Sn Surfaces from First Principles. *ACS Catal.* **2013**, *3*, 3026–3030. [CrossRef]
46. Kikuchi, I.; Haibara, Y.; Ohshima, M. Dehydrogenation of n-butane to butadiene over Pt-Sn/MgO-Al$_2$O$_3$. *J. Jpn. Petrol. Inst.* **2012**, *55*, 33–39. [CrossRef]
47. Kobayashi, S.; Kaneko, S.; Ohshima, M. Effect of iron oxide on isobutene dehydrogenation over Pt/Fe$_2$O$_3$-Al$_2$O$_3$ catalyst. *Appl. Catal. A-Gen.* **2012**, *417–418*, 306–311. [CrossRef]
48. Vu, B.K.; Song, M.B.; Ahn, I.Y. Propane dehydrogenation over Pt-Sn/Rare-earth-doped Al$_2$O$_3$: Influence of La, Ce, or Y on the formation and stabilityof Pt-Sn alloys. *Catal. Today* **2011**, *164*, 214–220. [CrossRef]
49. Chen, F.Q.; Huang, X.L.; Guo, K.Q.; Yang, L.; Sun, H.R.; Xia, W.; Zhang, Z.G.; Yang, Q.W.; Yang, Y.W.; Zhao, D.; et al. Molecular Sieving of Propylene from Propane in Metal-Organic Framework-Derived Ultramicroporous Carbon Adsorbents. *ACS Appl. Mater. Inter.* **2023**, *14*, 30443–30453. [CrossRef]
50. Li, W.; Yu, S.Y.; Meitzner, G.D. Structure andproperties of cobalt-exchanged H-ZSM5 catalysts fordehydrogenation and dehydrocyclization of alkanes. *J. Phys. Chem. B* **2001**, *105*, 1176–1184. [CrossRef]
51. Zhou, J.; Zhang, Y.; Liu, H.; Xiong, C.; Hu, P.; Wang, H.; Chen, S.W.; Ji, H.B. Enhanced performance for propane dehydrogenation through Pt clusters alloying with copper in zeolite. *Nano Res.* **2023**. [CrossRef]
52. Xia, K.; Wan, Z.L.; Pei, P.L. The influences of Mg/Al molar ratio on the properties of PtIn/Mg(Al)O-x catalysts for propane dehydrogenation reaction. *Chem. Eng. J.* **2015**, *9*, 1068–1079. [CrossRef]
53. Zhang, Y.W.; Zhou, Y.M.; Shi, J.J. Propane dehydrogenation over Pt Sn Na/La-doped Al$_2$O$_3$ catalyst: Effect of La content. *Fuel. Process. Technol.* **2013**, *11*, 94–104. [CrossRef]
54. Bocanegra, S.A.; Castro, A.A.; Scelza, O.A.; de Miguel, S.R. Characterization and catalytic behavior in the n-butane dehydrogenation of trimetallic In Pt Sn/MgAl$_2$O$_4$ catalysts. *Appl. Catal. A Gen.* **2007**, *333*, 49–56. [CrossRef]
55. Cabrera, F.; Ardissone, D.; Gorriz, O.F. Dehydrogenation of propane on chromia/alumina catalysts promoted by tin. *Catal. Today* **2007**, *133–135*, 800–804. [CrossRef]
56. Kim, T.H.; Kang, H.H.; Baek, M.S. Dehydrogenation of propane to propylene with lattice oxygen over CrO$_y$/Al$_2$O$_3$-ZrO$_2$ catalysts. *Mol. Catal.* **2017**, *43*, 1–7. [CrossRef]
57. Koirala, R.; Buechel, R.; Krumeich, F.; Pratsinis, S.E.; Baiker, A. Oxidative Dehydrogenation of Ethane with CO$_2$ over Flame-Made Ga-Loaded TiO$_2$. *ACS Catal.* **2015**, *5*, 690–702. [CrossRef]
58. Liu, H.; Zhang, Z.; Li, H.; Huang, Q. Intrinsic kinetics of oxidative dehydrogenation of propane in the presence of CO$_2$ over Cr/MSU-1 catalyst. *J. Nat. Gas Chem.* **2011**, *20*, 311–317. [CrossRef]
59. Michorczyk, P.; Ogonowski, J.; Kustrowski, P. Chromium oxide supported on MUM-41 as a highly active and selective catalyst for dehydrogenation of propane with CO$_2$. *Appl. Catal. A-Gen.* **2008**, *349*, 62–69. [CrossRef]
60. Zhang, X.Z.; Yue, Y.H.; Gao, Z. Chromium oxide supported on mesoporous SBA-15 as propane dehydrogenation and oxidaive dehydrogenation. *Catal. Chem.* **2002**, *83*, 19–25.
61. Frank, B.; Dinse, A.; Ovsitser, O.; Kondratenko, E.V.; Schomäcker, R. Mass and heat transfer effects on the oxidative dehydrogenation of propane (ODP) over a low loaded VO$_x$/Al$_2$O$_3$ catalyst. *Appl. Catal. A Gen.* **2007**, *323*, 66–76. [CrossRef]
62. Elbadawi, A.H.; Ba-Shammakh, M.S.; Al-Ghamdi, S.; Razzak, S.A.; Hossain, M.M.; de Lasa, H.I. A fluidizable VO$_x$/γ-Al$_2$O$_3$-ZrO$_2$ catalyst for the ODH of ethane to ethylene operating in a gas phase oxygen free environment. *Chem. Eng. Sci.* **2016**, *145*, 59–70. [CrossRef]
63. Khan, M.Y.; Al-Ghamdi, S.; Razzak, S.A.; Hossain, M.M.; deLasa, H. Fluidized bed oxidativede hydrogenation of ethane to ethylene over VO$_x$/Ce-γ Al$_2$O$_3$ catalysts: Reduction kinetics and catalyst activity. *Mol. Catal.* **2017**, *443*, 78–91. [CrossRef]
64. Rostom, S.; de Lasa, H.I. Propane Oxidative Dehydrogenation Using Consecutive Feed Injections and Fluidizable VO$_x$/γ Al$_2$O$_3$ and VO$_x$/ZrO$_2$-γ Al$_2$O$_3$ Catalysts. *Ind. Eng. Chem. Res.* **2017**, *56*, 13109–13124. [CrossRef]
65. Elfadly, A.M.; Badawi, A.M.; Yehia, F.Z.; Mohamed, Y.A.; Betiha, M.A.; Rabie, A.M. Selective nano alumina supported vanadium oxide catalysts for oxidative dehydrogenation of ethylbenzene to styrene using CO$_2$ as soft oxidant. *Egypt. J. Pet.* **2013**, *22*, 373–380. [CrossRef]

66. Klose, F.; Wolff, T.; Lorenz, H.; Seidelmorgenstern, A.; Suchorski, Y.; Piorkowska, M.; Weiss, H. Active species on γ-alumina-supported vanadia catalysts: Nature and reducibility. *J. Catal.* **2007**, *247*, 176–193. [CrossRef]
67. Rischard, J.; Antinori, C.; Maier, L.; Deutschmann, O. Oxidative dehydrogenation of n-butane to butadiene with Mo-V-MgO catalysts in a two-zone fluidized bed reactor. *Appl. Catal. A Gen.* **2016**, *511*, 23–30. [CrossRef]
68. Schwarz, O.; Habel, D.; Ovsitser, O.; Kondratenko, E.V.; Hess, C.; Schomäcker, R.; Schubert, H. Impact of preparation method on physico-chemical and catalytic properties of VO_x/γ-Al_2O_3 materials. *J. Mol. Catal. A Chem.* **2008**, *293*, 45–52. [CrossRef]
69. Botková, Š.; Capek, L.; Setnicka, M.; Bulánek, R.; Cicmanec, P.; Kalužov á, A.; Pastva, J.; Zukal, A. VO_x species supported on Al_2O_3-SBA-15 prepared by the grafting of alumina onto SBA-15: Structure and activity in the oxidative dehydrogenation of ethane. *React. Kinet Mech. Catal.* **2016**, *119*, 319–333. [CrossRef]
70. Rubio, O.; Herguido, J.; Men é ndez, M. Oxidative dehydrogenation of n-butane on V/MgO catalysts-kinetic study in anaerobic conditions. *Chem. Eng. Sci.* **2003**, *58*, 4619–4627. [CrossRef]
71. Hossain, M.M. Kinetics of Oxidative Dehydrogenation of Propane to Propylene Using Lattice Oxygen of VO_x/CaO/γAl_2O_3 Catalysts. *Ind. Eng. Chem. Res.* **2017**, *56*, 4309–4318. [CrossRef]
72. Tan, S.; Gil, L.B.; Subramanian, N.; Sholl, D.S.; Nair, S.; Jones, C.W.; Moore, J.S.; Liu, Y.; Dixit, R.S.; Pendergast, J.G. Catalytic propane dehydrogenation over In_2O_3–Ga_2O_3 mixed oxides. *Appl. Catal. A Gen.* **2015**, *498*, 167–175. [CrossRef]
73. Meriaudeau, P.; Naccahe, C. The role of Ga_2O_3 and proton acidlity on the dehydrogenation activity of Ga_2O_3-HZSM-5 catalysts-evidence of a bifunctional mechanism. *J. Mol. A-Chem.* **1990**, *59*, 31–36.
74. Michorczyk, P.; Kuśtrowski, P.; Kolak, A. Ordered mesoporous Ga_2O_3 and Ga_2O_3-Al_2O_3 prepared by nanocasting as effective catalysts for propane dehydrogenation in the presence of CO_2. *Catal. Commun.* **2013**, *35*, 95–100. [CrossRef]
75. Wu, J.L.; Chen, M.; Liu, Y.M.; Cao, Y.; He, H.Y.; Fan, K.N. Sucrose-templated mesoporous β-Ga_2O_3 as a novel efficient catalyst for dehydrogenation of propane in the presence of CO_2. *Catal. Commun.* **2012**, *30*, 61–65. [CrossRef]
76. Xu, B.J.; Zheng, B.; Hua, W. High Si/Al Ratio HZSM-5 Supported Ga_2O_3: A highly stable catalyst for dehydrogenation of propane to propene in the presence of CO_2. *Stud. Surf. Sci. Catal.* **2007**, *170*, 1072–1079.
77. Xu, B.J.; Zheng, B.; Hua, W. Support effect in dehydrogenation of propane in the presence of CO_2 over supported gallium oxide catalysts. *J. Catal.* **2006**, *239*, 470–477. [CrossRef]
78. Ren, Y.J.; Zhang, F.; Hua, W.M. ZnO supported on high silica HZSM-5 as new catalysts for dehydrogenation of propane to propene in the presence of CO_2. *Catal. Today* **2019**, *148*, 316–322. [CrossRef]
79. Blank, J.H.; Beckers, J.; Collignon, P.F.; Rothenberg, G. Redox kinetics of ceria-based mixed oxides in selective hydrogen combustion. *ChemPhysChem* **2007**, *8*, 2490–2497. [CrossRef]
80. Deo, G.; Wachs, I.E. Reactivity of Supported Vanadium Oxide Catalysts: The Partial Oxidation of Methanol. *J. Catal.* **1994**, *146*, 323–334. [CrossRef]
81. Blasco, T.; Lopez-Nieto, J.M. Oxidative dyhydrogenation of short chain alkanes on supported vanadium oxide catalysts. *Appl. Catal. A Gen.* **1997**, *157*, 117–142. [CrossRef]
82. Al-Ghamdi, S.; Moreira, J.; de Lasa, H. Kinetic Modeling of Propane Oxidative Dehydrogenation over VO_x/γ-Al_2O_3 Catalysts in the Chemical Reactor Engineering Center Riser Reactor Simulator. *Ind. Eng. Chem. Res.* **2014**, *53*, 15317–15332. [CrossRef]
83. Wu, T.W.; Yu, Q.B.; Hou, L.M.; Duan, W.J.; Wang, K.; Qin, Q. Selecting suitable oxygen carriers for chemical looping oxidative dehydrogenation of propane by thermodynamic method. *J. Therm. Anal. Calorim.* **2020**, *140*, 1837–1843. [CrossRef]
84. Fukudome, K.; Ikenaga, N.O.; Miyake, T.; Suzuki, T. Oxidative dehydrogenation of alkanes over vanadium oxide prepared with $V(t-BuO)_3O$ and $Si(OEt)_4$ in the presence of polyethyleneglycol. *Catal. Today* **2013**, *203*, 10–16. [CrossRef]
85. Fukudome, K.; Ikenaga, N.; Miyake, T.; Suzuki, T. Oxidative dehydrogenation of propane using lattice oxygen of vanadium oxides on silica. *Catal. Sci. Technol.* **2011**, *1*, 987. [CrossRef]
86. Gao, Y.; Neal, L.M.; Li, F. Li-Promoted $La_xSr_{2-x}FeO_{4-\delta}$ Core—Shell Redox Catalysts for Oxidative Dehydrogenation of Ethane under a Cyclic Redox Scheme. *ACS Catal.* **2016**, *6*, 7293–7302. [CrossRef]
87. Fattahi, M.; Kazemeini, M.; Khorasheh, F.; Rashidi, A. An investigation of the oxidative dehydrogenation of propane kinetics over a vanadium-graphene catalyst aiming at minimizing of the CO_x species. *Chem. Eng. J.* **2014**, *250*, 14–24. [CrossRef]
88. Argyle, M.; Bartholomew, C. Heterogeneous Catalyst Deactivation and Regeneration: A Review. *Catalysts* **2015**, *5*, 145–269. [CrossRef]
89. Gao, Y.F.; Haeri, F.; He, F.; Li, F.X. Alkali Metal-Promoted $La_xSr_{2-x}FeO_{4-\delta}$ redox catalysts for chemical looping oxidative dehydrogenation of ethane. *ACS Catal.* **2018**, *8*, 1757–1766. [CrossRef]

Disclaimer/Publisher's Note: The statements, opinions and data contained in all publications are solely those of the individual author(s) and contributor(s) and not of MDPI and/or the editor(s). MDPI and/or the editor(s) disclaim responsibility for any injury to people or property resulting from any ideas, methods, instructions or products referred to in the content.

Article

Highly Efficient Selective Hydrogenation of Cinnamaldehyde to Cinnamyl Alcohol over CoRe/TiO$_2$ Catalyst

Mengting Chen [1,†], Yun Wang [1,†], Limin Jiang [1], Yuran Cheng [2], Yingxin Liu [1,*] and Zuojun Wei [2,*]

1 College of Pharmaceutical Science, Zhejiang University of Technology, Hangzhou 310014, China
2 Key Laboratory of Biomass Chemical Engineering of the Ministry of Education, College of Chemical and Biological Engineering, Zhejiang University, Hangzhou 310027, China
* Correspondence: yxliu@zjut.edu.cn (Y.L.); weizuojun@zju.edu.cn (Z.W.)
† These authors contributed equally to this work.

Abstract: Allylic alcohols typically produced through selective hydrogenation of $α,β$-unsaturated aldehydes are important intermediates in fine chemical industry, but it is still a challenge to achieve its high selectivity transformation. Herein, we report a series of TiO$_2$-supported CoRe bimetallic catalysts for the selective hydrogenation of cinnamaldehyde (CAL) to cinnamyl alcohol (COL) using formic acid (FA) as a hydrogen donor. The resultant catalyst with the optimized Co/Re ratio of 1:1 can achieve an exceptional COL selectivity of 89% with a CAL conversion of 99% under mild conditions of 140 °C for 4 h, and the catalyst can be reused four times without loss of activity. Meanwhile, the Co$_1$Re$_1$/TiO$_2$/FA system was efficient for the selective hydrogenation of various $α,β$-unsaturated aldehydes to the corresponding $α,β$-unsaturated alcohols. The presence of ReO$_x$ on the Co$_1$Re$_1$/TiO$_2$ catalyst surface was advantageous to the adsorption of C=O, and the ultrafine Co nanoparticles provided abundant hydrogenation active sites for the selective hydrogenation. Moreover, FA as a hydrogen donor improved the selectivity to $α,β$-unsaturated alcohols.

Keywords: cinnamaldehyde; cinnamyl alcohol; CoRe bimetallic catalyst; selective hydrogenation; formic acid

1. Introduction

Selective hydrogenation of $α,β$-unsaturated aldehydes to unsaturated alcohols is an important process used to obtain a great deal of valuable chemicals [1–4]. Cinnamaldehyde (CAL), a typical $α,β$-unsaturated aldehyde, could be selectively hydrogenated to generate cinnamyl alcohol (COL), which is regarded as one of the most promising building blocks in pharmaceutical, agrochemical and fragrance industries [5,6].

Generally, the selective hydrogenation of CAL leads to the reduction of different functional groups, including C=O and C=C, and produces COL, 3-phenylpropionaldehyde (HCAL) and 3-phenylpropanol (HCOL) [7–10]. Owing to a higher binding energy of C=O bonds than C=C bonds (715 vs. 615 KJ·mol^{-1}), the hydrogenation of C=O bonds is more unfavorable in thermodynamics [11–13]. Therefore, it is essential to develop high-performance catalysts to improve the selectivity hydrogenation of C=O bonds and avoid the hydrogenation of C=C bonds.

Supported metal nanoparticle catalysts have been widely used in the industry due to their merits of easy separation and recovery [7,14]. The radial expansion of a metal D-bandwidth and d orbital is related to the selectivity of products (including COL, HCAL and HCOL) [15]. Metal with a small D-bandwidth (such as Ni) is conducive to the formation of HCAL [16], while some noble metals with a relatively large D-bandwidth (such as Ru, Ir, Au and Pt) can be used as catalysts for the synthesis of COL [17–21]. However, the high cost and rarity of noble metal catalysts hinder their industrial application. Non-noble Co has shown potential in the selective hydrogenation of CAL to COL due to its larger D-bandwidth and low price. For example, Zhang et al. [22] investigated the performance of the Co/ZSM-5 catalyst for the hydrogenation of CAL to COL at 90 °C and 20 bar H_2 for 6 h, with a 72.7% conversion of CAL and a 78.5% selectivity to COL. In another work using Co/ZSM-5 as the catalyst, a maximum COL yield of 61.9% was achieved at 100 °C and 20 bar H_2 [23].

However, monometallic Co catalysts have noticeable issues such as poor catalytic activity and high metal dosage. It is widely believed that the introduction of a second metal is an effective way to enhance the catalytic properties of Co nanoparticles, although the catalytic mechanism of the bimetallic catalysts is far from clear [24]. Adjusting the metal–metal/metallic oxide interactions in the catalyst could improve the morphology of dispersed metals and result in electron transfer, thus enhancing the charge density of the active metal and affecting the adsorption/desorption of C=O or C=C bonds on CAL [5,25]. For instance, the $CoGa_3/MgO·Al_2O_3$-LDH catalyst gave 96% COL selectivity in the hydrogenation of CAL at 100 °C and 20 bar H_2 for 8 h, which was significantly higher than the monometallic Co catalyst (42%) [26]. $CoPt/Fe_3O_4$ showed excellent catalytic performance under the conditions of 160 °C and 30 bar H_2, with a CAL conversion of 95% and a COL yield of 84% [27].

Another factor affecting the selectivity of COL in CAL hydrogenation is hydrogen donors. When using molecular hydrogen, a high H_2 pressure is usually needed in the selective hydrogenation of CAL to COL. However, high H_2 pressure requires specialized transportation and handling, which is deemed to be unsafe [28]. In addition, the different phases of H_2 and substrates increase the contact time, thus reducing the reaction efficiency caused by the transport phenomenon [29]. By comparison, the selective hydrogenation by replacing traditional hydrogen with hydrogen donors such as alcohols [30–32], formic acid (FA) [33,34] and silanes [35,36] offers a green, safe, sustainable and atomic economic process. For example, Butt et al. [34] reported that a COL yield of 73% was obtained in the hydrogenation of CAL over a AuNPore catalyst using Et_3SiH as a hydrogen donor at 70 °C for 24 h. When using FA as a hydrogen donor, a COL yield of 97% was achieved over a AuNPore catalyst at 90 °C for 22 h [35]. Herein, we prepared a series of TiO_2-supported CoRe bimetallic catalysts for the hydrogenation of CAL to COL under mild conditions using FA as an effective hydrogen donor, and we further extended the hydrogenation of various α,β-unsaturated aldehydes to α,β-unsaturated alcohols. The choice of Re as the second component is mainly due to the following two considerations. Firstly, high valence Re (+7) can be easily reduced to ReO_x (mainly Re (+4) and Re (+6)), which has many oxygen vacancies, and is conducive to the preferential adsorption of the C=O group on CAL [24,37]. Secondly, ReO_x has a positive effect on the stability of metal nanoparticles [38], thus reducing the aggregation and leakage of Co during the reaction [39,40]. The structure–activity relationship was analyzed by N_2 adsorption–desorption, CO chemisorption, TEM and XPS characterizations. In addition, the effects of the reaction parameters and the stability of the green catalytic system were investigated. Finally, the possible reaction mechanism was proposed. The main purpose here is to develop an efficient methodology for the selective hydrogenation of α,β-unsaturated aldehydes to α,β-unsaturated alcohols.

2. Results and Discussions

2.1. Catalytic Activity Test

The selective hydrogenation of CAL to COL was firstly investigated over the Co/TiO_2, Re/TiO_2 and Co_xRe_y/TiO_2 catalysts (x:y varied from 2:1 to 1:2) using isopropanol as both the solvent and hydrogen donor at 160 °C for 12 h, and the results are shown in Figure 1. It can be seen that the monometallic Co/TiO_2 catalyst showed a low conversion (23%) of CAL, although it gave a high selectivity (83%) to COL. The addition of Re obviously improved the conversion of CAL, which was consistent with the previous report that introducing the hydrophilic metal Re to the catalyst is beneficial for the adsorption of the C=O bond, thus improving the catalytic performance of the catalyst [41]. The maximum CAL conversion of 96% with a COL selectivity of 82% was achieved over the Co_1Re_1/TiO_2 catalyst, better than the Co_1Mo_1/TiO_2, Co_1Ce_1/TiO_2, Co_1Zr_1/TiO_2 catalysts and the Co_1Re_1 catalyst on other supports (SiO$_2$, ZrO$_2$, γ-Al$_2$O$_3$ and ZSM-5, Table S1, in the Supplementary Materials). These phenomena might be attributed to the strong adsorption of TiO$_2$ and ReO$_x$ on exposed the C=O group [42], which improves the diffusion of the substrate and accelerates the hydrogenation of CAL [43].

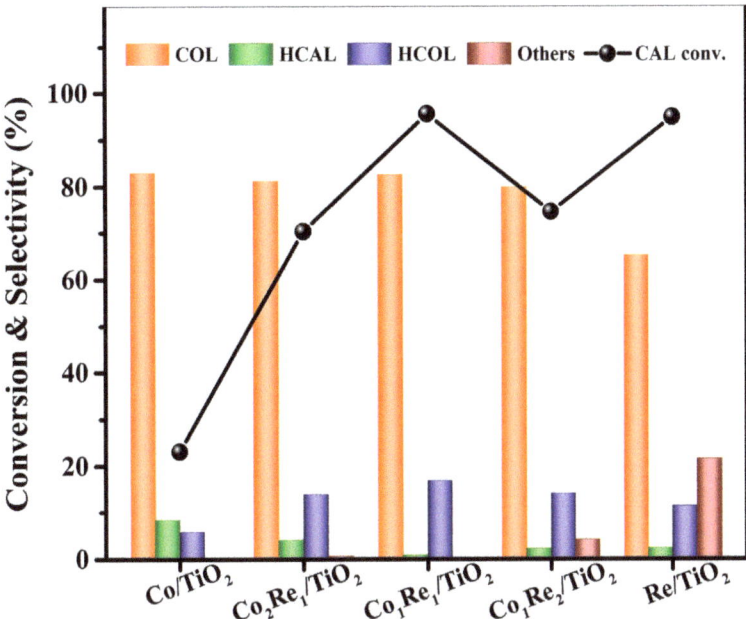

Figure 1. Hydrogenation of CAL to COL over various CoRe/TiO$_2$ catalysts. Reaction conditions: 3 mmol CAL, 80 mg catalyst, 10 mL isopropanol, 160 °C, 12 h. Others: allylbenzene, isopropenylbenzene, 1,1′-(1,5-hexadiene-1,6-diyl)bisbenzene and other unknown by-products.

2.2. Catalyst Characterization

The structural parameters of the samples are summarized in Table 1. Deposition of relatively low Co and Re contents on TiO$_2$ only slightly influenced its specific surface area, and the S$_{BET}$ values of all catalysts are close to TiO$_2$ and remain near 50 m^2·g^{-1}. On the other hand, the average pore diameter D$_p$ of metal-supported catalysts Co/TiO$_2$, Re/TiO$_2$ and Co$_1$Re$_1$/TiO$_2$ were larger than TiO$_2$, and the pore volumes V$_p$ of them were reduced, which could be explained by partial blocking of the narrowest pores by the metallic phase, indicating that the metals were embedded in the carrier pores [44].

Table 1. Textural properties of TiO$_2$, Co/TiO$_2$, Re/TiO$_2$ and Co$_1$Re$_1$/TiO$_2$.

Sample	d$_{TEM}$ (nm)	d$_{Co}$ [a] (nm)	Dispersion of Co [a] (%)	S$_{BET}$ (m^2·g^{-1})	D$_p$ [b] (nm)	V$_p$ [b] (cm^3·g^{-1})
TiO$_2$	/	/	/	52.8	11.3	0.15
Co/TiO$_2$	/	8.1	12.3	52.9	28.9	0.41
Co$_1$Re$_1$/TiO$_2$	1.7	1.8	55.7	48.5	24.5	0.33
Re/TiO$_2$	/	/	/	50.7	27.5	0.37

[a] Determined by CO chemisorption. [b] The pore size and pore volumes were derived from the adsorption branches of isotherms by using the BJH model.

The morphologies of Co/TiO$_2$ and Co$_1$Re$_1$/TiO$_2$ catalysts were characterized by TEM. As illustrated in Figure 2a, no clear Co nanoparticles were observed on the Co/TiO$_2$ catalyst, which is in agreement with Cheng's work [45]. As seen from Figure 2c, no diffraction spot was observed in the fast Fourier transformation (FFT) image of the CoRe particles, which is similar to our earlier work [46], implying its amorphous structure. The size-distribution histogram substantiates that the average size of the CoRe nanoparticles is about 1.7 nm (Figure 2d), which is similar to the size (1.8 nm) measured by CO chemisorption, and is much smaller than monometallic Co (8.1 nm, Table 1). This observation was further demonstrated by the results of Co dispersion measured by CO chemisorption (Table 1). The dispersion of Co on Co_1Re_1/TiO_2 is much higher than that of Co/TiO_2. These phenomena manifest that the introduction of the second metal Re significantly reduced the particle size of the Co nanoparticles and improved Co dispersion, thus providing more hydrogenation active sites for the selective hydrogenation of CAL.

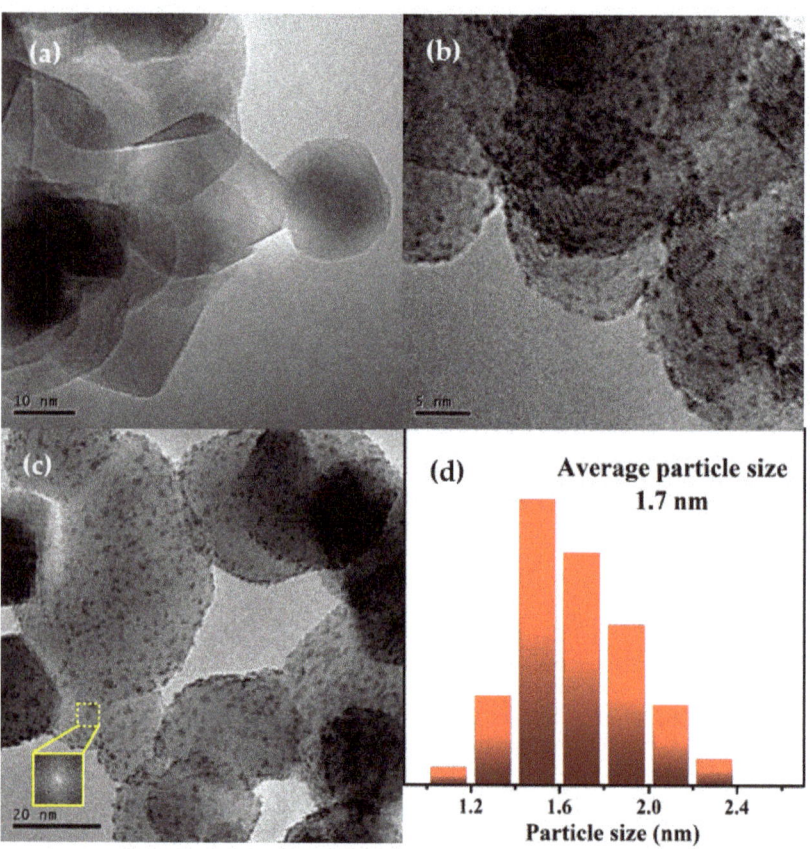

Figure 2. TEM images of (**a**) Co/TiO_2 and (**b**,**c**) Co_1Re_1/TiO_2 catalysts and (**d**) particle size distribution of Co_1Re_1/TiO_2.

The chemical state and surface composition of Co/TiO_2, Re_1/TiO_2 and Co_1Re_1/TiO_2 catalysts were assessed by XPS (Figure 3), and the calculated abundances of different surface Co and Re species are summarized in Table S3. As shown in Figure 3a, the peaks in the Co 2p spectrum of Co/TiO_2 at 778.2 and 793.4 eV are assigned to Co^0, the peaks at 781.0 and 796.8 eV are attributed to CoO, and the peaks at 796.1 and 802.5 eV are satellite peaks [47]. The observation of CoO was due to the oxidation of surface metallic Co nanoparticles in the air. As shown in Figure 3b, the binding energy in the range of 39–50 eV belonged to the Re 4f region, which was deconvoluted into doublet peaks for $4f_{5/2}$ and $4f_{7/2}$ orbits, implying the presence of ReO ($4f_{5/2}$ = 44.2 eV, $4f_{7/2}$ = 41.9 eV), ReO_2 ($4f_{5/2}$ = 46.3 eV, $4f_{7/2}$ = 44.0 eV), and Re_2O_5 ($4f_{5/2}$ = 48.3 eV, $4f_{7/2}$ = 46.0 eV) [48] in Re/TiO_2 with an atomic ratio of 65:16:19

(Table S2), and no metallic Re was detected, which was contributed to the high affinity of Re for oxygen [49]. Compared to their monometallic counterparts, the content of ReO in Co_1Re_1/TiO_2 significantly increased to 71%, which may promote the spillover of hydrogen on its surface and therefore be beneficial to the hydrogenation process [46].

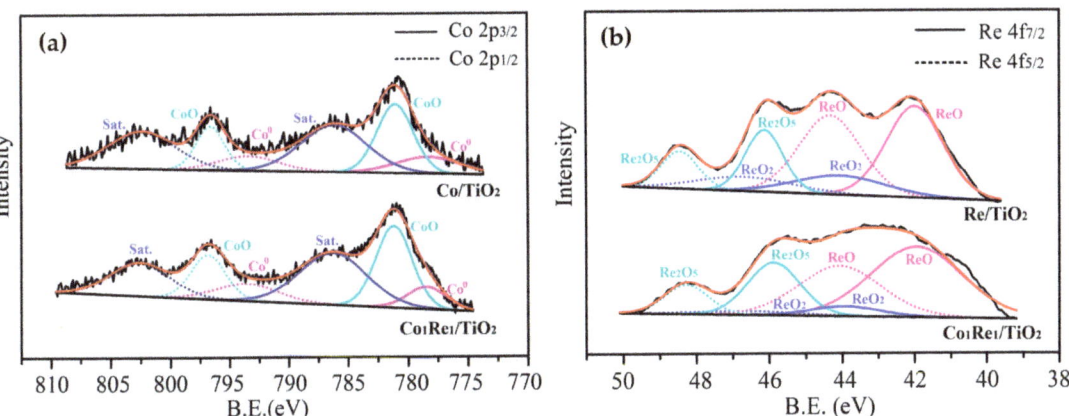

Figure 3. XPS spectra of (**a**) Co 2p and (**b**) Re 4f of the reduced catalysts Co/TiO_2, Re/TiO_2 and Co_1Re_1/TiO_2.

2.3. Effects of Reaction Conditions

As mentioned in Figure 1, using isopropanol as a hydrogen donor over the Co_1Re_1/TiO_2 catalyst can cause a high yield of COL to be obtained. However, a higher temperature (160 °C) and a longer reaction time (12 h) were required, and a great amount of HCOL (17%) was generated, thus decreasing COL selectivity. As shown in Table S3 (entry 4), compared to other common hydrogen donors (isopropanol, triethyl silicane and ammonium formate), when using FA as the hydrogen donor, a high COL yield could be achieved under mild reaction conditions in the inert solvent tetrahydrofuran (THF). Meanwhile, FA is green, sustainable, atom economical and easy to operate. Therefore, it is a good choice of hydrogen donor for the selective hydrogenation of CAL to COL. Thus, we used FA as a hydrogen donor to further study the hydrogenation of CAL to COL. Based on the results in Table 2, the combination of FA and triethylamine (NEt_3) was found to be essential for achieving high activity and selectivity in the hydrogenation of CAL to COL over Co_1Re_1/TiO_2, and a CAL conversion of 99% and a COL selectivity of 89% were obtained at a FA:NEt_3 molar ratio of 1:1 (Table 2, entry 2), which displayed obvious advantages over molecular hydrogen (Table 2, entry 5). The addition of NEt_3 to provide basic sites can substantially facilitate the crucial FA deprotonation process, which appears to be a key factor for achieving high activity of CAL hydrogenation to COL [50]. Too much or too little amounts of NEt_3 in the reaction system, however, led to a decrease in the selectivity of COL (Table 2, entries 3 and 4). In addition, the CAL:FA molar ratio also affected the reactivity of the hydrogenation of CAL to COL, as shown in Table 3. Low amounts of FA led to the deficiency of active hydrogen and to the decrease in the conversion of CAL, while excess FA led to an increase in by-products, and the optimal molar ratio of CAL:FA was 1:2 (Table 3, entry 2).

Table 2. Hydrogenation of CAL to COL using different hydrogen donors.

Entry	Hydrogen Donor	Molar Ratio (FA:NEt$_3$)	Conv. (%)	Sel. (%)			
				COL	HCAL	HCOL	Others
1	FA	/	64	29	4	3	64
2	FA	1:1	99	89	/	10	1
3	FA	2:5	99	72	/	28	/
4	FA	5:2	99	80	/	20	/
5 [a]	Hydrogen	/	57	17	42	25	16

Reaction conditions: 3 mmol CAL, CAL:FA = 1:2, 10 mL THF, 80 mg Co$_1$Re$_1$/TiO$_2$ catalyst, 140 °C, 4 h. [a] 2.5 MPa H$_2$. Others: allylbenzene, isopropenylbenzene, 1,1′-(1,5-hexadiene-1,6-diyl)bisbenzene and other unknown by-products.

Table 3. Effect of CAL:FA molar ratio on the hydrogenation of CAL to COL.

Entry	Molar Ratio (CAL:FA)	Conv. (%)	Sel. (%)			
			COL	HCAL	HCOL	Others
1	1:1.5	92	89	/	11	/
2	1:2	99	89	/	10	1
3	1:3	94	60	/	7	33

Reaction conditions: 3 mmol CAL, FA:NEt$_3$ = 1:1, 10 mL THF, 80 mg Co$_1$Re$_1$/TiO$_2$ catalyst, 140 °C, 4 h. Others: allylbenzene, isopropenylbenzene, 1,1′-(1,5-hexadiene-1,6-diyl)bisbenzene and other unknown by-products.

The effects of the reaction temperature and catalyst dosage for the hydrogenation of CAL to COL were investigated over Co$_1$Re$_1$/TiO$_2$, and the results are shown in Figure 4. Figure 4a shows the results of the hydrogenation of CAL at varied temperatures (from 100 to 160 °C). During the reaction, no HCAL was detected. The catalyst showed low activity for the reaction at a low temperature of 100 °C. Increasing the reaction temperature to 140 °C achieved a remarkable increase in the catalytic performance. However, the excessive reaction temperature (160 °C) led to a significant increase in the excessive hydrogenation of HCOL and other by-products. Thus, 140 °C was the appropriate reaction temperature for CAL-selective hydrogenation to COL. Fixing the reaction temperature at 140 °C, the hydrogenation of CAL was carried out over the Co$_1$Re$_1$/TiO$_2$ catalyst at a dosage in the range from 40 to 100 mg, and the results are shown in Figure 4b. It can be seen that the yield of COL was significantly improved with the increase in catalyst dosage from 40 to 80 mg. However, an excessive catalyst dosage reduced the selectivity of COL increased the excessive hydrogenation by-product HCOL and other by-products. As expected, our rationally designed Co$_1$Re$_1$/TiO$_2$ catalyst exhibited higher activity and selectivity for the hydrogenation of CAL to COL compared with TiO$_2$, Co/TiO$_2$, and Re/TiO$_2$ catalysts (Table S4), and it showed significant advantages compared with the relevant literature (Table S5) [24,27,30,32–35,45,51–61]. The results of the selective hydrogenation of CAL to COL suggest that the synergistic effect among the TiO$_2$ support and Co and Re metals is the main reason for the enhanced performance of the Co$_1$Re$_1$/TiO$_2$ catalyst.

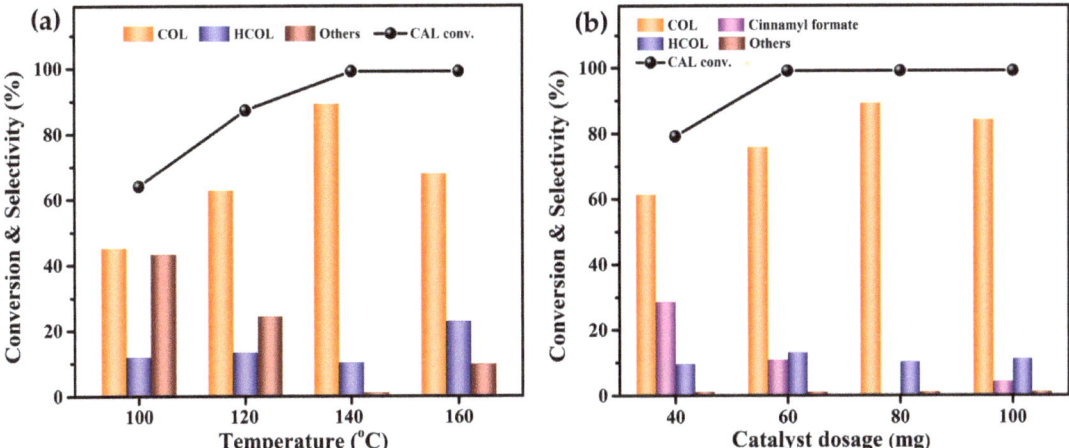

Figure 4. Effect of (**a**) reaction temperature; (**b**) Co_1Re_1/TiO_2 catalyst dosage for the hydrogenation of CAL to COL. Reaction conditions: 3 mmol CAL, CAL:FA:NEt$_3$ = 1:2:2, 10 mL THF, 80 mg Co_1Re_1/TiO_2 catalyst, 140 °C, 4 h. Others: allylbenzene, isopropenylbenzene, 1,1′-(1,5-hexadiene-1,6-diyl)bisbenzene and other unknown by-products.

Subsequently, we investigated the stability of the Co_1Re_1/TiO_2 catalyst for the hydrogenation of CAL to COL. During each cycle, after a complete reaction at 140 °C for 4 h, the catalyst was centrifuged, washed with THF for five times, and reused for the next runs. As shown in Figure 5, the Co_1Re_1/TiO_2 catalyst kept its good performance during recycling, the conversion of CAL was 97%, and the selectivity of COL was 87% after four runs. The XRD pattern of the spent catalyst showed no noticeable morphology changes compared with the fresh one (Figure S1). Moreover, the comparable metal contents in the fresh and spent catalysts determined by ICP-OES indicated no obvious metal leaching (Table S6). These results suggest that the Co_1Re_1/TiO_2 catalyst has excellent stability for the hydrogenation of CAL to COL.

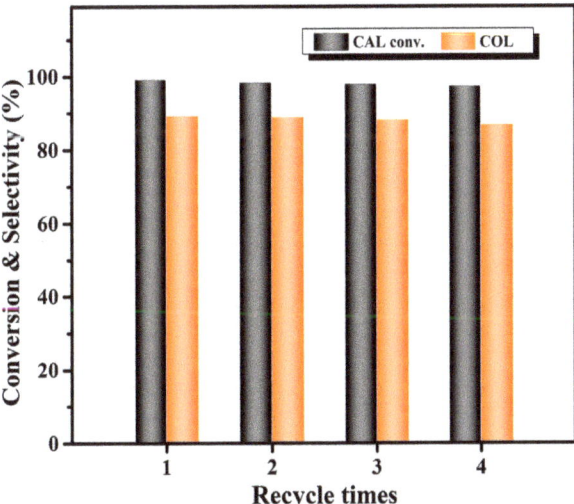

Figure 5. Recyclability of Co_1Re_1/TiO_2 catalyst. Reaction conditions: 3 mmol CAL, CAL:FA:Net$_3$ = 1:2:2, 10 mL THF, 80 mg catalyst, 140 °C, 4 h.

In addition, the applicability of the Co₁Re₁/TiO₂/FA system was tested for the selective hydrogenation of various α,β-unsaturated aldehydes to the corresponding α,β-unsaturated alcohols under the optimized reaction conditions, except for the reaction time, which was adjusted to obtain high yields. The results are listed in Table 4. Similar to COL, α-methyl cinnamaldehyde also achieved high conversion (99%, entry 1) and selectivity (86%), and the remaining 14% selectivity was attributed to simultaneous hydrogenation of C=C and C=O by-product. Moreover, aliphatic crotonaldehyde and citral could be converted well, and the target products were offered with selectivities of 74% and 96%, respectively (entries 2, 3). It is clear that substituent groups greatly affected the performance of the Co₁Re₁/TiO₂ catalyst. In general, the selectivity to α,β-unsaturated alcohols is closely related to the steric prohibition of the substituent groups. The more steric prohibition of C=C by the substitutes at γ-carbon, the higher the C=O selectivity, which is in good agreement with the literature [62]. Furthermore, a high conversion of 99% was obtained after the hydrogenation of cycloaliphatic isophorone (entry 4), with a product selectivity of 84% accompanied by excessive hydrogenation by-product of 16%. These results well verify that the Co₁Re₁/TiO₂/FA system is effective for the selective hydrogenation of α,β-unsaturated aldehydes to α,β-unsaturated alcohols.

Table 4. Hydrogenation of various α,β-unsaturated aldehydes over the Co₁Re₁/TiO₂ catalyst.

Entry	Substrate	Time (h)	Conv. (%)	Sel. (%)	
1	(cinnamaldehyde)	5	99	OH (86)	OH (14)
2	(crotonaldehyde)	8	93	OH (74)	OH (26)
3	(citral)	8	92	OH (96)	OH (5)
4	(isophorone)	12	99	OH (84)	OH (16)

Reaction conditions: 3 mmol substrate, substrate:FA:NEt₃ = 1:2:2, 10 mL THF, 80 mg Co₁Re₁/TiO₂ catalyst, 140 °C.

To speculate the possible reaction pathway, the time course experiments of the hydrogenation of CAL over the Co₁Re₁/TiO₂ catalyst was carried out under optimal reaction conditions (Figure 6a). As shown in Figure 6a, CAL was rapidly converted (73%), and a 42% yield of COL was achieved at the initial 1 h, accompanied by 26% of cinnamyl formate and 5% of HCOL. Under the weak alkaline condition of NEt₃, cinnamyl formate could be easily formed through esterification of COL with formic acid [63], which then smoothly decreased and disappeared within 4 h with the decomposition of formic acid. CAL was almost completely converted at 4 h, and COL and HCOL were offered with yields of 88% and 10%, respectively. It is recognized that the decomposition mechanism of FA in the presence of NEt₃ is as follows [34,64,65] NEt₃ acts as a proton scavenger to facilitate the O–H bond cleavage, thus forming metal–formate species during the initial step of the reaction. Then metal–formate produces molecular hydrogen through a β-elimination pathway. This hydrogen desorption step is irreversible, indicating that it is feasible to use renewable FA as a convenient hydrogen donor instead of molecular H₂ for sustainable and green organic synthesis. The proposed mechanism is shown in Figure 6b. The Co₁Re₁/TiO₂ catalyst and FA produce metal–formate species in the presence of NEt₃, which are then decomposed to generate metal–hydride species and CO₂. Meanwhile, the C=O bond on CAL is adsorbed on the ReOₓ species and is coordinated with metal–hydride. Finally, it is neutralized with $^+$HNEt$_3$ to form COL, thereby realizing the regeneration of the Co$_1$Re$_1$/TiO$_2$ catalyst and NEt$_3$.

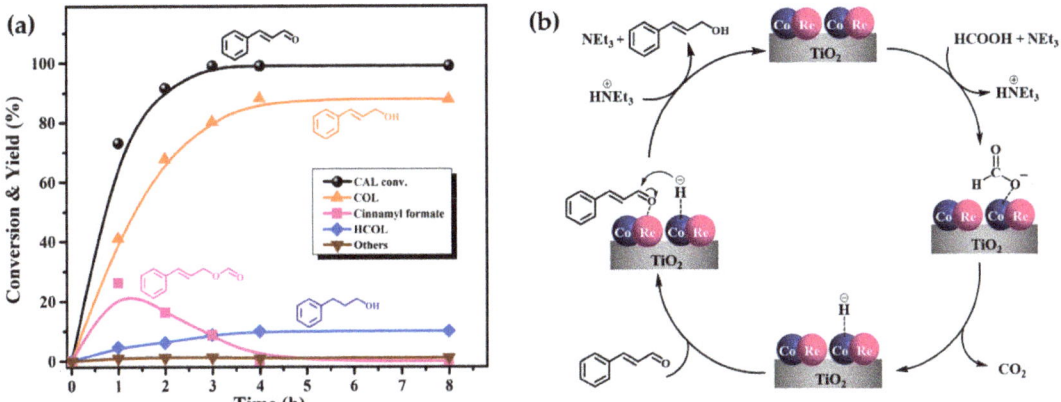

Figure 6. (a) Time courses of the hydrogenation of CAL to COL over Co$_1$Re$_1$/TiO$_2$. Reaction conditions: 3 mmol CAL, CAL:FA:NEt$_3$ = 1:2:2, 10 mL THF, 80 mg Co$_1$Re$_1$/TiO$_2$ catalyst, 140 °C. Others: allylbenzene, isopropenylbenzene, 1,1′-(1,5-hexadiene-1,6-diyl)bisbenzene and other unknown by-products. (b) The proposed mechanism for the hydrogenation of CAL to COL over Co$_1$Re$_1$/TiO$_2$.

3. Experimental Section

3.1. Materials

Co(NO$_3$)$_2$·6H$_2$O (99%) was purchased from Shanghai Jiuling Chemical Co., Ltd. (Shanghai, China). NH$_4$ReO$_4$ (99%) was purchased from Shanghai Macklin Biochemical Co., Ltd. (Shanghai, China). Isopropanol (99.5%), tetrahydrofuran (99.5%) and formic acid (99%) were purchased from Yonghua Chemical Technology Co., Ltd. (Suzhou, China). Triethylamine (99.5%) was purchased from Sinopharm Chemical Reagent Co., Ltd. (Shanghai, China). Other chemicals were purchased from Aladdin Reagent Co., Ltd. (Shanghai, China). All the chemicals used in this work were analytical reagents and were used without further purification.

3.2. Preparation of Various CoRe/TiO$_2$ Catalysts

A series of TiO$_2$-supported CoRe catalysts with variable Co:Re molar ratios (Co:Re = 2:1, 1:1 or 1:2) were prepared by an incipient wetness impregnation method. The loading of Co in the catalysts was kept at 2 wt%. Taking the CoRe/TiO$_2$ catalyst with a Co:Re molar ratio of 1:1 as an example, the preparation method was as follows: firstly, TiO$_2$ was calcined at 500 °C for 4 h to remove the impurities prior to impregnation of the metal precursor. Then, 0.0998 g of Co(NO$_3$)$_2$·6H$_2$O and 0.0920 g of NH$_4$ReO$_4$ were dissolved in ca. 1 mL deionized H$_2$O. An appropriate amount of TiO$_2$ (ca. 0.98 g) was slowly added to the aqueous solution under ultrasound. After being impregnated at room temperature for 24 h, the mixture was dried at 110 °C overnight and finally reduced at 500 °C in a tubular furnace under hydrogen flow for 3 h to obtain the target catalyst, which was denoted as Co$_1$Re$_1$/TiO$_2$. Two monometallic catalysts 2 wt% Co/TiO$_2$ and 2 wt% Re/TiO$_2$ were prepared by using the same method for comparison. The information on the content of Co and Re in the corresponding catalyst is described in Table S7.

3.3. Catalyst Characterization

Brunauer–Emmett–Teller (BET) surface areas and pore structures of the catalyst samples were measured by pulsed N$_2$ adsorption–desorption method at −196 °C using Micromeritics ASAP 2460 analyzer. Before N$_2$ physisorption, the samples were degassed under vacuum at 250 °C for 3 h. Transmission electron microscopy (TEM) images were obtained using a Tecnai G2 F30 S-Twin instrument (FEI Co., Columbia, SC, USA). The particle size distribution of metal nanoparticles in the sample was determined by measuring approximately 100 particles randomly selected from the TEM micrographs. The metal dispersion and particle size were measured by CO chemisorption at 35 °C using a Micromeritics ASAP 2020 system. X-ray diffraction (XRD) was conducted using an X' Pert PRO X-ray diffractometer equipped with Cu Kα radiation at 40 kV and 40 mA (λ = 0.15405 nm). Samples were scanned from 20° to 80° with a scanning rate of $4°\cdot min^{-1}$ and a step size of 0.02°. The content of the metals was measured by inductively coupled plasma-optical emission spectroscopy (ICP-OES) using a Thermo Fisher iCAP PRO. X-ray photoelectron spectroscopy (XPS) spectra were obtained using an Escalab Mark II X-ray spectrometer (VG Co., Manchester, UK) equipped with a magnesium anode (Mg Kα = 1253.6 eV). Energy corrections were performed using a C 1s peak of the pollutant carbon at 284.8 eV.

3.4. Catalytic Performance

The catalytic hydrogenation of α,β-unsaturated aldehydes to α,β-unsaturated alcohols was performed in a 25 mL stainless-steel autoclave equipped with magnetic stirring. In a typical experiment, 3 mmol of α,β-unsaturated aldehyde, 6 mmol of FA, 6 mmol of NEt$_3$, 80 mg of catalyst and 10 mL of THF were added into the reactor. The reactor was sealed and purged with N$_2$ five times. The autoclave was then heated to the required temperature and kept at this temperature for the required time under the continuous stirring speed of 1000 rpm. After the reaction, the autoclave was quickly cooled to room temperature, and the reaction products were separated from the catalyst by centrifugation and quantitatively analyzed with an Agilent 7890 A gas chromatographer equipped with an HP-5 capillary column (30.0 m × 0.32 mm × 0.25 μm) and a flame ionization detector (FID) using *n*-dodecane as an internal standard. The conformation of the products was performed on an Agilent 6890 GC system coupled to a mass spectrometer equipped with an Agilent 5973 quadrupole mass analyzer. CAL conversion and selectivity and the yield of the products were calculated by the following equations.

$$Conversion\ (\%) = \frac{moles\ of\ substrate\ consumed}{moles\ of\ initial\ substrate} \times 100\% \quad (1)$$

$$Selectivity\ (\%) = \frac{moles\ of\ desired\ product\ formed}{moles\ of\ substrate\ consumed} \times 100\% \quad (2)$$

$$Yield\ (\%) = Conversion \times Selectivity \times 100\% \quad (3)$$

4. Conclusions

In summary, CoRe bimetallic catalysts supported on TiO$_2$ were achieved and were first reported in the selective hydrogenation of CAL to COL using FA as a hydrogen donor to replace the traditional molecular hydrogen. Especially, the Co$_1$Re$_1$/TiO$_2$ catalyst performed excellent activity, selectivity and stability, with a 99% conversion of CAL and 89% COL selectivity, and no obvious deactivation was observed after using it four times. Under similar reaction conditions, using α-methyl cinnamaldehyde, crotonaldehyde, citral and isophorone as feedstocks, high conversions and excellent selectivities to allylic alcohols were also achieved. The technique established in this work provides a green, mild and efficient process for the selective hydrogenation of α,β-unsaturated aldehydes to allylic alcohols.

Supplementary Materials: The following supporting information can be downloaded at: https://www.mdpi.com/article/10.3390/molecules28083336/s1, Table S1: Hydrogenation of CAL over different Co-based catalysts; Table S2: Co 2p and Re 4f dispersion on different catalysts; Table S3: Hydrogenation of CAL to COL using different hydrogen donors; Table S4: Hydrogenation of CAL to COL over various catalysts using formic acid as the hydrogen donor; Table S5: Hydrogenation of CAL to COL over various catalytic systems; Table S6: Metal contents in fresh and spent Co$_1$Re$_1$/TiO$_2$

catalyst measured by ICP-OES; Table S7: Labels and calculated metal contents of catalysts; Figure S1: XRD patterns of fresh and spent Co$_1$Re$_1$/TiO$_2$; Figure S2: GC/MS spectra of substrate and products in the hydrogenation of COL; Figure S3: GC/MS spectra of the main products in the hydrogenation of other α,β-unsaturated aldehydes. References [24,27,30,32,33,33–35,45,51–61] are cited in the supplementary materials.

Author Contributions: Writing—original draft preparation, M.C. and Y.W.; writing—review and editing, M.C. and Y.W.; investigation, L.J.; resources, L.J. and Y.C.; supervision, Y.L. and Z.W.; project administration, Y.L. and Z.W. All authors have read and agreed to the published version of the manuscript.

Funding: This work was supported by the Zhejiang Provincial Natural Science Foundation of China (LY23B060006 and LY18B060016).

Institutional Review Board Statement: Not applicable.

Informed Consent Statement: Not applicable.

Data Availability Statement: The data presented in this study are available on request from the corresponding author.

Conflicts of Interest: The authors declare no conflict of interest.

Sample Availability: Not applicable.

References

1. Gallezot, P.; Richard, D. Selective hydrogenation of α,β-unsaturated aldehydes. *Catal. Rev.* **1998**, *40*, 81–126. [CrossRef]
2. Zhao, M.; Yuan, K.; Wang, Y.; Li, G.; Guo, J.; Gu, L.; Hu, W.; Zhao, H.; Tang, Z. Metal-organic frameworks as selectivity regulators for hydrogenation reactions. *Nature* **2016**, *539*, 76–80. [CrossRef]
3. Hao, C.H.; Guo, X.N.; Pan, Y.T.; Chen, S.; Jiao, Z.F.; Yang, H.; Guo, X.Y. Visible-light-driven selective photocatalytic hydrogenation of cinnamaldehyde over Au/SiC catalysts. *J. Am. Chem. Soc.* **2016**, *138*, 9361–9364. [CrossRef] [PubMed]
4. Luneau, M.; Lim, J.S.; Patel, D.A.; Sykes EC, H.; Friend, C.M.; Sautet, P. Guidelines to achieving high selectivity for the hydrogenation of α,β-unsaturated aldehydes with bimetallic and dilute alloy catalysts: A review. *Chem. Rev.* **2020**, *120*, 12834–12872. [CrossRef]
5. Wang, X.; Liang, X.; Geng, P.; Li, Q. Recent advances in selective hydrogenation of cinnamaldehyde over supported metal-based catalysts. *ACS Catal.* **2020**, *10*, 2395–2412. [CrossRef]
6. Liu, Q.; Liu, Q.; Chen, Y.; Li, Y.; Su, H.; Liu, Q.; Li, G. Ir nanoclusters confined within hollow MIL-101(Fe) for selective hydrogenation of α,β-unsaturated aldehyde. *Chin. Chem. Lett.* **2022**, *33*, 374–377. [CrossRef]
7. Kahsar, K.R.; Schwartz, D.K.; Medlin, J.W. Control of metal catalyst selectivity through specific noncovalent molecular interactions. *J. Am. Chem. Soc.* **2014**, *136*, 520–526. [CrossRef]
8. Zhang, R.; Wang, L.; Ren, J.; Hu, C.; Lv, B. Effect of boron nitride overlayers on Co@BNNSs/BN-catalyzed aqueous phase selective hydrogenation of cinnamaldehyde. *J. Colloid Interface Sci.* **2023**, *630*, 549–558. [CrossRef]
9. Zhou, Y.; Chen, C.; Li, Q.; Liu, Y.; Wei, T.; Liu, Y.; Zeng, Z.; Bradshaw, D.; Zhang, B.; Huo, J. Precise control of selective hydrogenation of α,β-unsaturated aldehydes in water mediated by ammonia borane. *Appl. Catal. B Environ.* **2022**, *311*, 121348. [CrossRef]
10. Liu, C.; Zhu, P.; Wang, J.; Liu, H.; Zhang, X. Geometrically embedding dispersive Pt nanoparticles within silicalite-1 framework for highly selective α,β-unsaturated aldehydes hydrogenation via oriented C=O adsorption configuration. *Chem. Eng. J.* **2022**, *446*, 137064. [CrossRef]
11. Chen, H.; Peng, T.; Liang, B.; Zhang, D.; Lian, G.; Yang, C.; Zhang, Y.; Zhao, W. Efficient electrocatalytic hydrogenation of cinnamaldehyde to value-added chemicals. *Green Chem.* **2022**, *24*, 3655–3661. [CrossRef]
12. Xue, Y.; Yao, R.; Li, J.; Wang, G.; Wu, P.; Li, X. Efficient Pt-FeO$_x$/TiO$_2$@SBA-15 catalysts for selective hydrogenation of cinnamaldehyde to cinnamyl alcohol. *Catal. Sci. Technol.* **2017**, *7*, 6112–6123. [CrossRef]
13. Liu, Y.; Wang, X.; Zhang, C.; Xu, Q.; Dang, L.; Zhao, X.; Tan, H.; Li, Y.; Zhao, F. Defect engineering and spilt-over hydrogen in Pt/(WO$_3$–TH$_2$) for selective hydrogenation of C=O bonds. *New J. Chem.* **2022**, *46*, 15950–15958. [CrossRef]
14. Mateen, M.; Akhtar, M.N.; Gao, L.; Cheong, W.-C.M.; Lv, S.; Zhou, Y.; Chen, Z. Engineering electrophilic atomic Ir sites on CeO$_2$ colloidal spheres for selectivity control in hydrogenation of α,β-unsaturated carbonyl compounds. *Nano Res.* **2022**, *15*, 7107–7115. [CrossRef]
15. Johar, P.; McElroy, C.R.; Rylott, E.L.; Matharu, A.S.; Clark, J.H. Biologically bound nickel as a sustainable catalyst for the selective hydrogenation of cinnamaldehyde. *Appl. Catal. B* **2022**, *306*, 121105. [CrossRef]
16. Ning, L.; Zhang, M.; Liao, S.; Zhang, Y.; Jia, D.; Yan, Y.; Gu, W.; Liu, X. Differentiation of Pt–Fe and Pt–Ni$_3$ surface catalytic mechanisms towards contrasting products in chemoselective hydrogenation of α,β-unsaturated aldehydes. *ChemCatChem* **2020**, *13*, 704–711. [CrossRef]

17. Egeberg, A.; Dietrich, C.; Kind, C.; Popescu, R.; Gerthsen, D.; Behrens, S.; Feldmann, C. Bimetallic Nickel-Iridium and Nickel-Osmium alloy nanoparticles and their catalytic performance in hydrogenation reactions. *ChemCatChem* **2017**, *9*, 3534–3543. [CrossRef]
18. Padmanaban, S.; Gunasekar, G.H.; Yoon, S. Direct heterogenization of the Ru-Macho catalyst for the chemoselective hydrogenation of α,β-unsaturated carbonyl compounds. *Inorg. Chem.* **2021**, *60*, 6881–6888. [CrossRef]
19. Tao, R.; Shan, B.Q.; Sun, H.D.; Ding, M.; Xue, Q.S.; Jiang, J.G.; Wu, P.; Zhang, K. Surface molecule manipulated Pt/TiO$_2$ catalysts for selective hydrogenation of cinnamaldehyde. *J. Phys. Chem. C* **2021**, *125*, 13304–13312. [CrossRef]
20. Xin, H.; Li, M.; Chen, L.; Zhao, C.; Wu, P.; Li, X. Lanthanide oxide supported Ni nanoparticles for the selective hydrogenation of cinnamaldehyde. *Catal. Sci. Technol.* **2023**, *13*, 1048–1055. [CrossRef]
21. Tan, Y.; Liu, X.; Zhang, L.; Liu, F.; Wang, A.; Zhang, T. Producing of cinnamyl alcohol from cinnamaldehyde over supported gold nanocatalyst. *Chin. J. Catal.* **2021**, *42*, 470–481. [CrossRef]
22. Zhang, X.B.; Zhang, Y.J.; Chen, F.; Xiang, Y.Z.; Zhang, B.; Xu, L.Y.; Zhang, T.R. Efficient selective hydrogenation of cinnamaldehyde over zeolite supported cobalt catalysts in water. *Reac. Kinet. Mech. Cat.* **2015**, *115*, 283–292. [CrossRef]
23. Zhang, B.; Zhang, X.B.; Xu, L.Y.; Zhang, Y.J.; Qin, Y.H.; Liang, C.F. Selective hydrogenation of cinnamaldehyde over ZSM-5 supported Co catalysts. *Reac. Kinet. Mech. Cat.* **2013**, *110*, 207–214. [CrossRef]
24. Wei, Z.; Zhu, X.; Liu, X.; Xu, H.; Li, X.; Hou, Y.; Liu, Y. Pt-Re/rGO bimetallic catalyst for highly selective hydrogenation of cinnamaldehyde to cinnamylalcohol. *Chin. J. Chem. Eng.* **2019**, *27*, 369–378. [CrossRef]
25. Stucchi, M.; Vasile, F.; Cattaneo, S.; Vomeri, A.; Hungria, A.B.; Prati, L. Pt-WO$_x$/C catalysts for α,β-unsaturated aldehydes hydrogenation: An NMR study of the effect of the reactant adsorption on activity and selectivity. *Eur. J. Org. Chem.* **2022**, *2022*, e202200735. [CrossRef]
26. Yang, Y.; Rao, D.; Chen, Y.; Dong, S.; Wang, B.; Zhang, X.; Wei, M. Selective hydrogenation of cinnamaldehyde over Co-based intermetallic compounds derived from layered double hydroxides. *ACS Catal.* **2018**, *8*, 11749–11760. [CrossRef]
27. Yuan, T.; Liu, D.; Pan, Y.; Pu, X.; Xia, Y.; Wang, J.; Xiong, W. Magnetic anchored CoPt bimetallic nanoparticles as selective hydrogenation catalyst for cinnamaldehyde. *Catal. Lett.* **2018**, *149*, 851–859. [CrossRef]
28. Mabate, T.P.; Meijboom, R.; Bingwa, N. The inorganic perovskite-catalyzed transfer hydrogenation of cinnamaldehyde using glycerol as a hydrogen donor. *Catalysts* **2022**, *12*, 241. [CrossRef]
29. Xiao, P.; Zhu, J.; Zhao, D.; Zhao, Z.; Zaera, F.; Zhu, Y. Porous LaFeO$_3$ prepared by an in situ carbon templating method for catalytic transfer hydrogenation reactions. *ACS Appl. Mater. Interfaces* **2019**, *11*, 15517–15877. [CrossRef]
30. Liu, X.; Cheng, S.; Long, J.; Zhang, W.; Liu, X.; Wei, D. MOFs-derived Co@CN bi-functional catalysts for selective transfer hydrogenation of α,β-unsaturated aldehydes without use of base additives. *Mater. Chem. Front.* **2017**, *1*, 2005–2012. [CrossRef]
31. Wang, H.; Liu, B.; Liu, F.; Wang, Y.; Lan, X.; Wang, S.; Ali, B.; Wang, T. Transfer hydrogenation of cinnamaldehyde catalyzed by Al$_2$O$_3$ using ethanol as a solvent and hydrogen donor. *ACS Sustain. Chem. Eng.* **2020**, *8*, 8195–8205. [CrossRef]
32. Siddiqui, N.; Sarkar, B.; Pendem, C.; Khatun, R.; Sivakumar Konthala, L.N.; Sasaki, T.; Bordoloi, A.; Bal, R. Highly selective transfer hydrogenation of α,β-unsaturated carbonyl compounds using Cu-based nanocatalysts. *Catal. Sci. Technol.* **2017**, *7*, 2828–2837. [CrossRef]
33. Hareesh, H.N.; Minchitha, K.U.; Venkatesh, K.; Nagaraju, N.; Kathyayini, N. Environmentally benign selective hydrogenation of α,β-unsaturated aldehydes and reduction of aromatic nitro compounds using Cu based bimetallic nanoparticles supported on multiwalled carbon nanotubes and mesoporous carbon. *RSC Adv.* **2016**, *6*, 82359–82369. [CrossRef]
34. Butt, M.; Feng, X.; Yamamoto, Y.; Almansour, A.I.; Arumugam, N.; Kumar, R.S.; Bao, M. Unsupported nanoporous Gold-catalyzed chemoselective reduction of α,β-unsaturated aldehydes using formic acid as hydrogen source. *Asian J. Org. Chem.* **2017**, *6*, 867–872. [CrossRef]
35. Takale, B.S.; Wang, S.; Zhang, X.; Feng, X.; Yu, X.; Jin, T.; Bao, M.; Yamamoto, Y. Chemoselective reduction of α,β-unsaturated aldehydes using an unsupported nanoporous gold catalyst. *Chem. Commun.* **2014**, *50*, 14401–14404. [CrossRef]
36. Dhakshinamoorthy, A.; Alvaro, M.; Garcia, H. HKUST-1 catalyzed room temperature hydrogenation of acetophenone by silanes. *Catal. Commun.* **2017**, *97*, 74–78. [CrossRef]
37. Liu, Y.; Liu, K.; Zhang, M.; Zhang, K.; Ma, J.; Xiao, S.; Wei, Z.; Deng, S. Highly efficient selective hydrogenation of levulinic acid to γ-valerolactone over Cu-Re/TiO$_2$ bimetallic catalysts. *RSC Adv.* **2021**, *12*, 602–610. [CrossRef]
38. Zhou, K.; Chen, J.; Cheng, Y.; Chen, Z.; Kang, S.; Cai, Z.; Xu, Y.; Wei, J. Enhanced catalytic transfer hydrogenation of biomass-based furfural into 2-methylfuran over multifunctional Cu–Re bimetallic catalysts. *ACS Sustain. Chem. Eng.* **2020**, *8*, 16624–16636. [CrossRef]
39. Bazin, D.; Borko, L.; Koppany, Z.; Kovacs, I.; Stefler, G.; Sajo, L.I.; Schay, Z.; Guczi, L. Re±Co/NaY and Re±Co/Al$_2$O$_3$ bimetallic catalysts: In situ EXAFS study and catalytic activity. *Catal. Lett.* **2002**, *84*, 169–182. [CrossRef]
40. Liu, C.; He, Y.; Wei, L.; Zhang, Y.; Zhao, Y.; Hong, J.; Chen, S.; Wang, L.; Li, J. Hydrothermal carbon-coated TiO$_2$ as support for Co-based catalyst in Fischer–Tropsch synthesis. *ACS Catal.* **2018**, *8*, 1591–1600. [CrossRef]
41. Tamura, M.; Tokonami, K.; Nakagawa, Y.; Tomishige, K. Rapid synthesis of unsaturated alcohols under mild conditions by highly selective hydrogenation. *Chem. Commun.* **2013**, *49*, 7034–7036. [CrossRef] [PubMed]
42. Yao, M.; Ji, Y.; Wang, H.; Ao, Z.; Li, G.; An, T. Adsorption mechanisms of typical carbonyl-containing volatile organic compounds on anatase TiO$_2$ (001) surface: A DFT investigation. *J. Phys. Chem. C* **2017**, *121*, 13717–13722. [CrossRef]

43. Storsater, S.; Totdal, B.; Walmsley, J.; Tanem, B.; Holmen, A. Characterization of alumina-, silica-, and titania-supported cobalt Fischer–Tropsch catalysts. *J. Catal.* **2005**, *236*, 139–152. [CrossRef]
44. Vít, Z.; Gulková, D.; Kaluža, L.; Kupčík, J. Pd–Pt catalysts on mesoporous SiO_2–Al_2O_3 with superior activity for HDS of 4,6-dimethyldibenzothiophene: Effect of metal loading and support composition. *Appl. Catal. B* **2015**, *179*, 44–53. [CrossRef]
45. Cheng, S.; Lu, S.; Liu, X.; Li, G.; Wang, F. Enhanced activity of alkali-treated ZSM-5 zeolite-supported Pt-Co catalyst for selective hydrogenation of cinnamaldehyde. *Molecules* **2023**, *28*, 1730. [CrossRef] [PubMed]
46. Wei, Z.; Li, Q.; Cheng, Y.; Dong, M.; Zhang, Z.; Zhu, Y.; Liu, Y.; Sun, Y. Low loading of $CoRe/TiO_2$ for efficient hydrodeoxygenation of levulinic acid to γ-valerolactone. *ACS Sustain. Chem. Eng.* **2021**, *9*, 10882–10891. [CrossRef]
47. Huck-Iriart, C.; Soler, L.; Casanovas, A.; Marini, C.; Prat, J.; Llorca, J.; Escudero, C. Unraveling the chemical state of Cobalt in Co-based catalysts during ethanol steam reforming: An in situ study by near ambient pressure XPS and XANES. *ACS Catal.* **2018**, *8*, 9625–9636. [CrossRef]
48. Duke, A.S.; Galhenage, R.P.; Tenney, S.A.; Ammal, S.C.; Heyden, A.; Sutter, P.; Chen, D.A. In situ ambient pressure X-ray photoelectron spectroscopy studies of methanol oxidation on Pt(111) and Pt–Re alloys. *J. Phys. Chem. C* **2015**, *119*, 23082–23093. [CrossRef]
49. Rozmysłowicz, B.; Kirilin, A.; Aho, A.; Manyar, H.; Hardacre, C.; Wärnå, J.; Salmi, T.; Murzin, D.Y. Selective hydrogenation of fatty acids to alcohols over highly dispersed ReO_x/TiO_2 catalyst. *J. Catal.* **2015**, *328*, 197–207. [CrossRef]
50. Fu, Z.; Wang, Z.; Lin, W.; Song, W.; Li, S. High effective conversion of furfural to 2-methylfuran over $Ni-Cu/Al_2O_3$ catalyst with formic acid as a hydrogen donor. *Appl. Catal. A-Gen.* **2017**, *547*, 248–255. [CrossRef]
51. Tang, Y.; Li, H.; Cui, K.; Xia, Y.; Yuan, G.; Feng, J.; Xiong, W. Chemoselective hydrogenation of cinnamaldehyde over amorphous coordination polymer supported Pt-Co bimetallic nanocatalyst. *Chem. Phys. Lett.* **2022**, *801*, 139683. [CrossRef]
52. Rong, Z.; Sun, Z.; Wang, Y.; Lv, J.; Wang, Y. Selective hydrogenation of cinnamaldehyde to cinnamyl alcohol over graphene supported Pt–Co bimetallic catalysts. *Catal. Lett.* **2014**, *144*, 980–986. [CrossRef]
53. Zheng, Q.; Wang, D.; Yuan, F.; Han, Q.; Dong, Y.; Liu, Y.; Zhu, Y. An effective Co-promoted platinum of Co–Pt/SBA-15 catalyst for selective hydrogenation of cinnamaldehyde to cinnamyl alcohol. *Catal. Lett.* **2016**, *146*, 1535–1543. [CrossRef]
54. Li, Y.; Li, Z.G.; Zhou, R.X. Bimetallic Pt-Co catalysis on carbon nanotubes for the selective hydrogenation of cinnamaldehyde to cinnamyl alcohol: Preparation and characterization. *J. Mol. Catal. A Chem.* **2008**, *279*, 140–146. [CrossRef]
55. Lv, Y.; Han, M.; Gong, W.; Wang, D.; Chen, C.; Wang, G.; Zhang, H.; Zhao, H. Fe-Co alloyed nanoparticles catalyzing efficient hydrogenation of cinnamaldehyde to cinnamyl alcohol in water. *Angew. Chem. Int. Ed.* **2020**, *59*, 23521–23526. [CrossRef]
56. Zhao, J.; Malgras, V.; Na, J.; Liang, R.; Cai, Y.; Kang, Y.; Alshehri, A.A.; Alzahrani, K.A.; Alghamdi, Y.G.; Asahi, T.; et al. Magnetically induced synthesis of mesoporous amorphous CoB nanochains for efficient selective hydrogenation of cinnamaldehyde to cinnamyl alcohol. *Chem. Eng. J.* **2020**, *398*, 125564. [CrossRef]
57. Gu, Z.; Chen, L.; Li, X.; Chen, L.; Zhang, Y.; Duan, C. NH_2-MIL-125(Ti)-derived porous cages of titanium oxides to support Pt-Co alloys for chemoselective hydrogenation reactions. *Chem. Sci.* **2019**, *10*, 2111–2117. [CrossRef]
58. Wang, X.; He, Y.; Liu, Y.; Park, J.; Liang, X. Atomic layer deposited Pt-Co bimetallic catalysts for selective hydrogenation of α,β-unsaturated aldehydes to unsaturated alcohols. *J. Catal.* **2018**, *366*, 61–69. [CrossRef]
59. Shi, J.; Nie, R.; Zhang, M.; Zhao, M.; Hou, Z. Microwave-assisted fast fabrication of a nanosized Pt_3Co alloy on reduced graphene oxides. *Chin. J. Catal.* **2014**, *35*, 2029–2037. [CrossRef]
60. Tian, Z.; Liu, C.; Li, Q.; Hou, J.; Li, Y.; Ai, S. Nitrogen- and oxygen-functionalized carbon nanotubes supported Pt-based catalyst for the selective hydrogenation of cinnamaldehyde. *Appl. Catal. A-Gen.* **2015**, *506*, 134–142. [CrossRef]
61. Plessers, E.; De Vos, D.E.; Roeffaers, M.B.J. Chemoselective reduction of α,β-unsaturated carbonyl compounds with UiO-66 materials. *J. Catal.* **2016**, *340*, 136–143. [CrossRef]
62. He, S.; Xie, L.; Che, M.; Chan, H.C.; Yang, L.; Shi, Z.; Tang, Y.; Gao, Q. Chemoselective hydrogenation of α,β-unsaturated aldehydes on hydrogenated MoO_x nanorods supported iridium nanoparticles. *J. Mol. Catal. A Chem.* **2016**, *425*, 248–254. [CrossRef]
63. Perdomo, I.C.; Gianolio, S.; Pinto, A.; Romano, D.; Contente, M.L.; Paradisi, F.; Molinari, F. Efficient enzymatic preparation of flavor esters in water. *J. Agric. Food Chem.* **2019**, *67*, 6517–6522. [CrossRef]
64. Bi, Q.Y.; Du, X.L.; Liu, Y.M.; Cao, Y.; He, H.Y.; Fan, K.N. Efficient subnanometric gold-catalyzed hydrogen generation via formic acid decomposition under ambient conditions. *J. Am. Chem. Soc.* **2012**, *134*, 8926–8933. [CrossRef] [PubMed]
65. Sims, J.J.; Ould Hamou, C.A.; Réocreux, R.; Michel, C.; Giorgi, J.B. Adsorption and decomposition of formic acid on Cobalt(0001). *J. Phys. Chem. C* **2018**, *122*, 20279–20288. [CrossRef]

Disclaimer/Publisher's Note: The statements, opinions and data contained in all publications are solely those of the individual author(s) and contributor(s) and not of MDPI and/or the editor(s). MDPI and/or the editor(s) disclaim responsibility for any injury to people or property resulting from any ideas, methods, instructions or products referred to in the content.

Article

Mesoporous Polymeric Ionic Liquid via Confined Polymerization for Laccase Immobilization towards Efficient Degradation of Phenolic Pollutants

Yu Liang [1], Xinyan Chen [1], Jianli Zeng [2], Junqing Ye [1], Bin He [1], Wenjin Li [1] and Jian Sun [1,3,*]

[1] Key Laboratory of Molecular Medicine and Biotherapy, Ministry of Industry and Information Technology, School of Life Science, Beijing Institute of Technology, Beijing 100081, China

[2] State Key Laboratory of Catalytic Materials and Reaction Engineering, Research Institute of Petroleum Processing, China Petroleum & Chemical Corporation, Beijing 100083, China

[3] Advanced Research Institute of Multidisciplinary Science, Beijing Institute of Technology, Beijing 100081, China

* Correspondence: jiansun@bit.edu.cn

Abstract: Laccase immobilization is a promising method that can be used for the recyclable treatment of refractory phenolic pollutants (e.g., chlorophenols) under mild conditions, but the method is still hindered by the trade-off limits of supports in terms of their high specific surface area and rich functional groups. Herein, confined polymerization was applied to create abundant amino-functionalized polymeric ionic liquids (PILs) featuring a highly specific surface area and mesoporous structure for chemically immobilizing laccase. Benefiting from this strategy, the specific surface area of the as-synthesized PILs was significantly increased by 60-fold, from 5 to 302 m^2/g. Further, a maximum activity recovery of 82% towards laccase was recorded. The tolerance and circulation of the immobilized laccase under harsh operating conditions were significantly improved, and the immobilized laccase retained more than 84% of its initial activity after 15 days. After 10 cycles, the immobilized laccase was still able to maintain 80% of its activity. Compared with the free laccase, the immobilized laccase exhibited enhanced stability in the biodegradation of 2,4-dichlorophenol (2,4-DCP), recording around 80% (seven cycles) efficiency. It is proposed that the synergistic effect between PILs and laccase plays an important role in the enhancement of stability and activity in phenolic pollutant degradation. This work provides a strategy for the development of synthetic methods for PILs and the improvement of immobilized laccase stability.

Keywords: polymeric ionic liquids; confined polymerization; laccase; immobilization; 2,4-DCP; removal

1. Introduction

Chlorophenols, a class of important chemicals that are used in the industrial field, can cause serious pollution due to their aromatic structure and high toxicity [1,2]. Therefore, the degradation of chlorophenol pollution is urgent and has attracted extensive attention. Although a variety of physical, chemical, and biological methods have been used towards this end, most of these have certain limitations, such as high energy consumption, high costs, and low degradation efficiency [3–5], limiting their large-scale application. In recent years, biodegradation has emerged as one of the most promising technologies for treating harmful organic pollutants from the viewpoints of sustainable chemistry and green chemistry, due to its higher environmental friendliness, higher selectivity, and more efficient mineralization compared with other methods [6]. Further, the potential for biodegradation to be used to remove chlorophenols in water has attracted considerable research attention.

Laccases (EC 1.10.3.2), found in fungi, plants, bacteria, and insects, belong to the blue-copper enzyme family. They have broad substrate specificity and eco-friendliness [7]. In addition, laccases can catalyze the one-electron oxidation of various organic and inorganic substrates, including phenols, ketones, phosphates, ascorbates, amines, and lignin. Free laccases extracted from nature have the disadvantages of poor stability, a short lifespan, being difficult to recycle, and high cost, meaning it is generally difficult for them to tolerate harsh industrial processing conditions [8–10]. Researchers have mainly modified enzymes through protein engineering, non-covalent modification, chemical modification, and immobilization [11–13]. However, in practice, protein engineering still faces many challenges. After protein modification, laccases exhibit unstable properties with regard to their preservation and application [14]. Both the covalent modification and non-covalent modification of enzyme systems are unstable, and the activity and stability of enzymes may be reduced after modification. Therefore, enzyme immobilization technology is ideal for compensating for the deficiency of free enzymes [15,16]. One of the significant advantages of enzyme immobilization is that it provides an expected ideal enzyme system that can be reused for a couple of cycles, reducing operation costs [11,17]. Currently, enzyme immobilization strategies mainly focus on supports innovation through the use of metallic compounds [18], carbon-based materials [19], silicon-based nanomaterials [20], metal-organic frameworks or their derivatives [14], polymers [21], etc. Liu and co-workers immobilized laccase into carbon-based mesoporous magnetic composites through adsorption. The system was able to retain above 70% of its initial activity after five cycles [22]. Navarro-Sanchez and colleagues encapsulated laccase in MOF and realized high activity of laccase under high-temperature conditions [23]. Xu and colleagues developed a method of laccase immobilization based on nanofibrous membranes consisting of electrospun chitosan/poly (vinyl alcohol) composites. Compared with free laccase, the combined removal of 2,4-DCP was significantly improved by the immobilized laccase [21]. However, there are still some challenges that need to be resolved prior to enzyme immobilization's industrial application, such as enzyme leakage, toxicity, the low surface area involved, and the supports' high cost. Therefore, it is highly desirable to develop novel supports to achieve the effects of the high load, low leakage, high activity, and high stability of enzymes.

In the past few decades, the application of polymeric ionic liquid (PIL)-based biomaterials has attracted wide attention. PILs are one type of conductive polymer that have an IL structure in the repeating unit, where the anions or cations of the PILs are confined in the matrix of the macromolecules. They have a wide range of applications in polyelectrolytes, flexible materials, stimuli-responsive materials, catalysts, energy materials, and carbon materials [24–27]. To the best of our knowledge, the application of PILs in the immobilization of enzymes is still under development. There are two factors limiting this application. On the one hand, the existing preparation methods for PIL-based porous materials involve low IL contents and a narrow application range [28,29]. On the other hand, although PILs can have high IL contents, they are mostly non-porous or low-specific surface area materials [24,28,29]. Recently, Wang and colleagues introduced hydrophilic polyvinylpyrrolidone chain segments into hydrophobic polydivinylbenzene frameworks through the reinitiation of suspended double bonds, affording a kind of adsorbent with a uniform distribution of functional groups, strong cross-linking structure, and highly specific surface area [30]. Therefore, a higher specific surface area of PILs can be achieved through this kind of confined polymerization, further broadening their application in the biological field. However, research into this subject is still lacking. The higher the specific surface area, the better the enzyme immobilization's performance, owing to there being more available reactive groups [31,32]. Therefore, the strategy of enhancing the specific surface area and mechanical properties of PILs is expected to further broaden their application in the biological field [33,34].

Herein, two amino-functionalized PIL microspheres synthesized by traditional free radical polymerization and confined polymerization, respectively, were compared. They were found to have different specific surface areas, pore structures, and rough surfaces. Laccase was covalently immobilized on the surface of these microspheres via the bridging of glutaraldehyde. A better enzyme immobilization system was selected for the degradation of 2,4-dichlorophenol (2,4-DCP), and its biocatalytic degradation was characterized. The optimum conditions, including the temperature and pH, for the immobilized laccase system were determined using the enzyme relative activity test, and the storage stability and circulation of the same system were further tested. Finally, the biodegradation of 2,4-DCP was systematically evaluated. This work provides a protocol for the design and optimization of the structure of PILs. In addition, the study found that compared with free laccase, the stability and circularity of the PIL-immobilized laccase were greatly improved.

2. Results and Discussion

2.1. Structural Characterization of Polymerized Ionic Liquid Materials

As illustrated in Scheme 1, the amino-functionalized PILs were synthesized by two different polymerization methods (i.e., traditional free radical polymerization and confined polymerization), and the corresponding products were denoted as PIL (1)–NH_2 and PIL (2)–NH_2, respectively. The material synthesized after the immobilization of laccase was recorded as PIL–NH_2–GA–Lac.

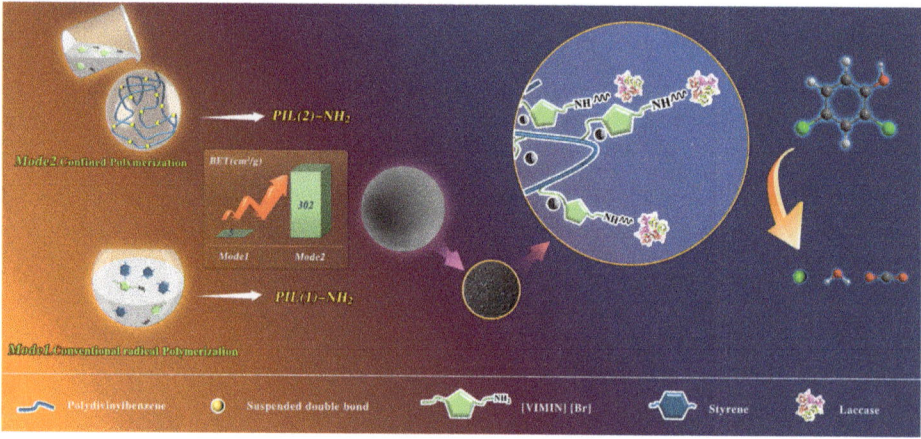

Scheme 1. Preparation of PILs for laccase immobilization and its application in phenol removal.

Figure 1a–c show the morphology of the PIL–NH_2–GA–Lac composites. The PIL (1)–NH_2 composites did not have regular spheroids, nor were they observed to have a uniform surface or pore structure. Figure 1d–f indicate that PIL (2)–NH_2 had a spherical structure with a size distribution between 15 and 100 μm, and the surface of its microsphere was rough, as shown in Figure 1f. In addition, PIL (2)–NH_2–GA–Lac (Figure 1g–i) still retained a spherical structure, while its surface was smoother compared with that of the pristine support (PIL (2)–NH_2). This may have been due to the coating that formed after the immobilization of the laccase.

Surface interactions between the laccase and two supports (i.e., PIL (1)–NH_2 and PIL (2)–NH_2) were investigated by FT-IR (Figure 2a). As expected, in the spectra of the IL monomer and PIL–NH_2, a stretching vibration peak of the N-H bond of a primary amine was observed near 3441 cm^{-1}, and a vibration peak near 1182 cm^{-1} also confirmed the presence of a primary amine. The absorption corresponding to the C-H stretching of the -CH=CH of the ILM was observed at 3026 cm^{-1}, which indicated the aromatic structure of DVB. The peak at 1652 cm^{-1} corresponded to the characteristic peak of the carbon–carbon double bonds of -CH=CH_2, and the peak at 709 cm^{-1} was the C-H characteristic peak in the double bond connecting the benzene ring, which was not affected by the introduction of functional groups [35]. This also verified the successful synthesis of the polymer skeleton. Moreover, the peaks at 1691 cm^{-1} and 1602 cm^{-1} represented the presence of an amide I and an amide II band, respectively, which indicated that laccase was immobilized to the amino-functionalized PIL [36].

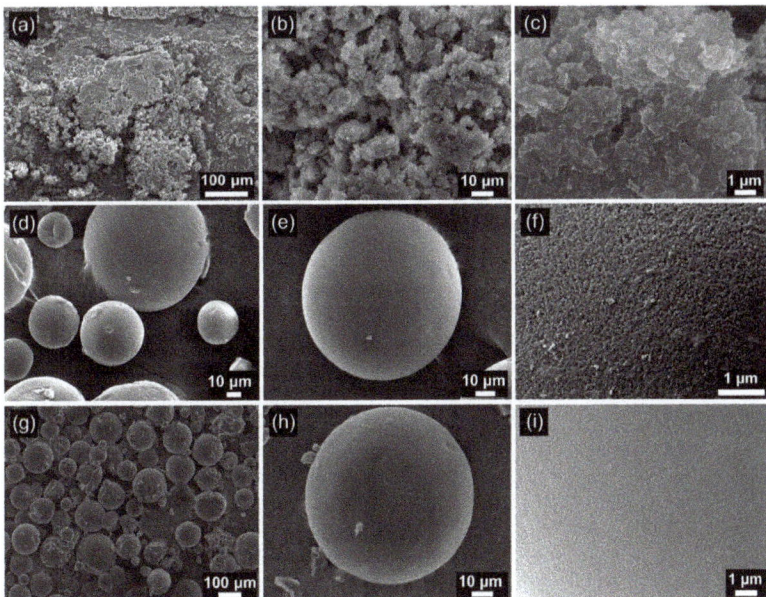

Figure 1. SEM micrographs of PIL (1)–NH$_2$–GA–Lac (**a–c**), PIL (2)–NH$_2$ (**d–f**) and PIL (2)–NH$_2$–GA–Lac (**g–i**).

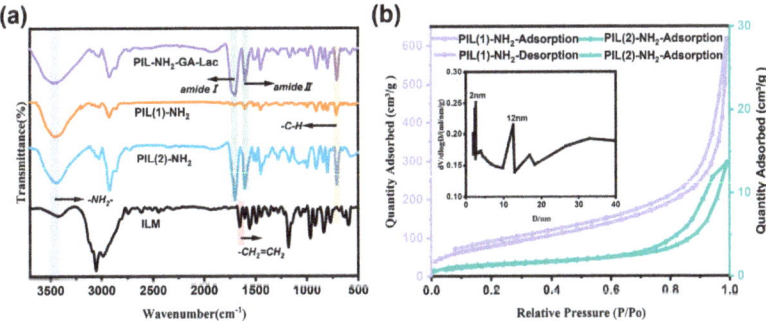

Figure 2. FTIR spectra of PIL–NH$_2$ and PIL–NH$_2$–GA–Lac (**a**). Nitrogen adsorption–desorption isotherms and BJH pore size distributions of PIL–NH$_2$ (**b**).

Figure 2b shows the specific surface area analysis and pore size distribution curves of the two amino-functionalized PIL supports. It can be observed that the adsorption curve of the isotherm was inconsistent with the desorption curve, resulting in hysteresis loops. The N$_2$ isothermal adsorption–desorption curves of PIL (1)–NH$_2$ and PIL (2)–NH$_2$ were IV H3 and H4 hysteresis loops, respectively. It was found that PIL (1)–NH$_2$ had a fairly low specific surface area of 5 cm^2/g. Nevertheless, PIL (2)–NH$_2$ showed a high uptake capacity, having a Brunauer–Emmett–Teller (BET) surface area of 302 cm^2/g. The formation of a high specific surface area can be attributed to the polymerization of the polystyrene skeleton in the early stage of the procedure [30]. The analysis of the pore size distribution in PIL (2)–NH$_2$ showed that it had a mesoporous structure with a pore size of mainly 12 nm. Its pore size distribution between 12 and 50 nm can be interpreted as owing to the diversity of its pore patterns. According to the previous literature, a part of laccase can be immobilized into mesoporous materials, further improving its immobilization efficiency and activity [23,37].

To verify the effect of the specific surface area on the immobilized efficiency, we used the popular crosslinking agent glutaraldehyde (GA), and the amount used (2% v/v) in this work was analogous to that used in other studies (0.5–5% v/v) [26,38]. Glutaraldehyde is a crosslinking agent that has two aldehyde groups during the cross-linking process. In this study, after activation, an aldehyde group was covalently linked to PIL–NH$_2$ and a free formyl of the microspheres. During the immobilization of laccase, free formyl groups on the surface of PIL–NH$_2$ microspheres were covalently linked to laccase macromolecules. However, excessive cross-linking can lead to distortion of the laccase configuration, so a relatively low concentration of glutaraldehyde was used for immobilization to prevent the distortion of the enzyme structure and to reduce the activity [39].

2.2. Immobilization of Laccase

Next, two important immobilized laccase parameters, the initial laccase concentration and the cross-linking time, were investigated. The number of amino-functionalized PIL particles determines the number of laccases. As shown in Figure 3a, with a higher initial concentration, more laccase was able to be immobilized. In addition, the immobilized efficiencies of laccase corresponded to the nature of PILs, such as their specific surface area and pore size. A higher specific surface area and larger pore size can facilitate the entry of enzymes into the support to achieve a higher enzyme immobilization [40]. When the initial concentration of laccase was below 1 mg/mL, both the immobilized laccase and activity recovery increased. It should be noted that when the concentration was 1 mg/mL, the maximum laccase activity recovery was obtained (82%, 0.34 u/mg; the specific activity of free laccase was 0.41 u/mg under the same test conditions). However, the activity recovery did not increase when the concentration was over 1 mg/mL, which might have been due to the fact that an aggregation of the excessive laccase in the pores occurred or that a multilayer structure formed on the surface of the PILs.

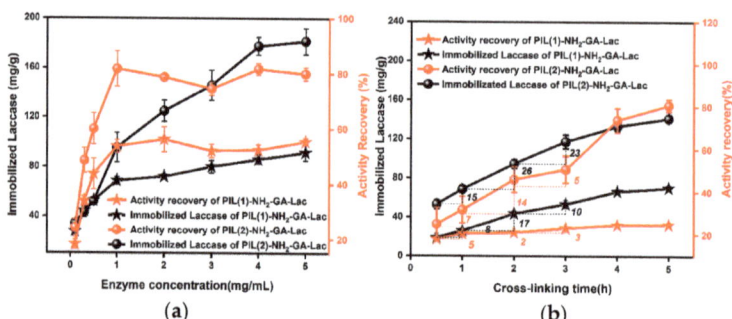

Figure 3. Effect of laccase concentration (**a**) and cross-linking time (**b**) on the activity recovery and the immobilized laccase.

Cross-linking time is another parameter that influences the immobilization efficiency of enzymes. As shown in Figure 3b, the activity recovery of the laccase activity increased from 1 to 5 h due to sufficient cross-linking. However, the immobilization rate gradually slowed as the time extended, as shown by the slope (comparing the data in the figure, it was found that the maximum immobilized rate generally occurred within 2 h) in Figure 3b. For example, when the cross-linking time was 2 h, the immobilized laccase of PIL (1)–NH$_2$–GA–Lac and PIL (2)–NH$_2$–GA–Lac increased by 17 mg/g and 26 mg/g, respectively. In addition, the activity recovery of PIL (2)–NH$_2$–GA–Lac increased by 14% during this period. This was the most efficient period of enzyme immobilization. From the viewpoint of dynamics, a large concentration gradient in the early cross-linking time (0.5–2 h) promoted the immobilization of laccase. A maximum activity recovery was achieved when the cross-linking time was around 5 h (i.e., 82%, 0.31 u/mg). It is worth noting that PIL (2)–NH$_2$–GA–Lac, having a larger specific surface area, had a higher capture ability for laccase than PIL (1)–NH$_2$ under the same enzyme concentration. However, the excessive immobilization of enzymes makes the aperture of the support relatively thin, resulting in the reduced accessibility of the substrate to the active sites [41,42].

2.3. pH and Thermal Stability of Immobilized Laccase

The optimal pH of a laccase immobilization reaction system varies according to the charge in the enzymatic protein and the support used [43]. As shown in Figure 4a, and consistent with previous studies, free laccase showed maximum activity at a pH of 2.5 and rapidly decreased activity as the pH increased, and it was almost completely inactivated under neutral conditions [44]. In general, anionic supports attract protons from a given solution, which results in a higher concentration of H$^+$ in the diffusion layer of an immobilized enzyme than in the external solution. This is why the pH of the external solution should be higher [45]. In the present study, the amino-functionalized PILs were positively charged (+1.7–+5.8 eV); however, the optimal pH value of the immobilized laccase still was higher than that of free laccase. This can be attributed to the imidazolium skeleton in the PILs being positively charged, and the fact that part of the OH$^-$ was enriched in the solution. Therefore, there was more available H$^+$ near the laccase at the other end of the cross-linking arm. Further, this is why the optimum pH of the PIL (2)–NH$_2$–GA–Lac system was increased slightly. Additionally, the activity–pH curve of most of the immobilized enzymes was bell-shaped, and compared with that of the free laccase, the bell-shaped curve of the immobilized laccase, especially PIL (2)–NH$_2$, was flatter. Even in the buffer solution of pH 7, PIL (2)–NH$_2$ was able to retain more than 45% activity.

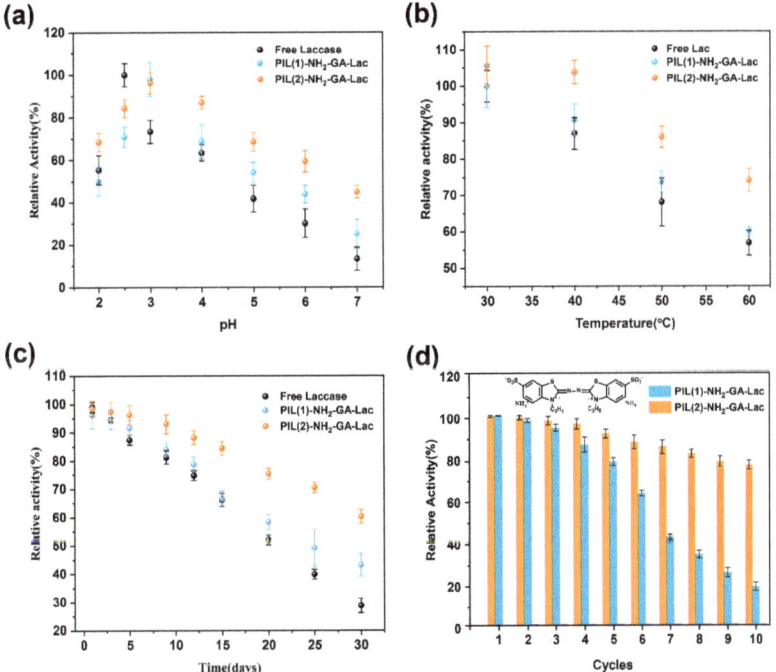

Figure 4. Effect of (**a**) pH and (**b**) temperature on the activity of free laccase and PIL–NH$_2$–GA–Lac. (**c**) Storage stability of free and immobilized laccase and (**d**) reusability of the immobilized laccase. Conditions: ABTS (100 mM), PIL–NH$_2$–GA–Lac (10 mg/mL), pH = 3 (**b**–**d**), and 25 °C (**a**,**c**,**d**).

Figure 4b shows that the relative activities of both the free and immobilized laccase were strongly temperature-dependent. The thermal stability of PIL–NH$_2$–GA–Lac was improved, and its activity decreased more slowly than that of free laccase. In particular, PIL (2)–NH$_2$–GA–Lac maintained more than 70% activity at 60 °C. This may be attributed to the PIL supports' providing covalently attached sites and confined space for the laccase, as well as the inactive active center's being preserved by its binding with the ligand, resulting in the decreased formation of the thermo-unfolded enzyme. Therefore, higher temperatures (above 40 °C) were required to break down the intermolecular interactions and to inactivate the enzyme [46]. In addition, some reports have proved that PILs have a good heat insulation effect, so when the temperature of the external solution rises, the surface and internal temperature of the support will be relatively low, which protects the laccase from being inactivated by high temperatures [47].

2.4. Operational Stability of Immobilized Laccase

Storage stability and reusability are also important characteristics and are prerequisites for the practical application of immobilized enzymes. Storage is particularly difficult because the conformation of an enzyme changes over time and gradually loses its activity in solution. As shown in Figure 4c, it can be observed that with the increase in storage time, the difference between the free and immobilized laccase activities became significant. The relative activity of the free laccase was less than 50% after 15 days, while PIL (2)–NH$_2$–GA–Lac was still able to maintain more than 80%. After 30 days, the relative activity of PIL (2)–NH$_2$–GA–Lac was twice that of free laccase. The storage stability of PIL (1)–NH$_2$–GA–Lac fell between that of PIL (2)–NH$_2$–GA–Lac and free laccase. This shows that immobilization significantly improved the storage stability of the laccase. One possible reason for this is that the laccase was in a confined space provided by PIL–NH$_2$ and that the spatial conformation had little or no tendency to change [48].

The reusability test performed over 10 successive catalytic cycles showed that the immobilized laccase exhibited superior circularity. Despite its loss of activity in cycling, the immobilized laccase (i.e., PIL (2)–NH$_2$–GA–Lac) exhibited good reusability, retaining 80% of its initial activity after 10 cycles (Figure 4d). It should be noted that the amount of immobilized laccase in PIL (2)–NH$_2$–GA–Lac was about two-fold higher than that in PIL (1)–NH$_2$–GA–Lac under theoretical conditions; however, the relative activity of PIL (2)–NH$_2$–GA–Lac was three-fold higher than that of PIL (1)–NH$_2$–GA–Lac after 10 cycles. This can be attributed to the higher specific surface area and stronger binding force of PIL (2)–NH$_2$, by which the laccase activity loss of PIL (2)–NH$_2$–GA–Lac was smaller under the same experimental conditions.

2.5. Analysis of Enzyme Kinetic Parameters

The enzymatic kinetic parameters are shown in Table 1 and Supplementary Materials: Figure S2. The Michaelis–Menten constant reflects the affinity of an enzyme to a substrate. The smaller the K_m, the greater the affinity between the enzyme and the substrate. As shown in Table 1, the K_m of the immobilized laccase was slightly reduced compared with that of the free laccase, indicating its substrate affinity was enhanced. This might be ascribed to the rough surface of the PILs, which is conducive to interactions with the substrates and enzymes. The maximum reaction rate was V_{max}, and K_m/V_{max} was used to evaluate the catalytic efficiency. The reaction rate of the immobilized enzymes was improved compared with that of free laccase. This may have been due to the enrichment of the substrate on the supports, which made it easier for the laccase to contact the substrate (ABTS) and accelerate the reaction.

Table 1. K_m and V_{max} of free laccase and immobilized laccase.

	V_{max} (mM/(mg·min))	K_m (mmol/L)
Free laccase	0.035	1.202
PIL (1)–NH$_2$–GA–Lac	0.043	1.100
PIL (2)–NH$_2$–GA–Lac	0.043	0.998

2.6. Interference Test of Metal Ions

As can be seen from Figure S3, both Fe^{3+} and Cr^{2+} had certain inhibitory effects. This may have been because these metal ions occupy the active centers of laccase. However, Mg^{2+}, Cu^{2+}, and Zn^{2+} were able to promote the activity of the laccase. This is consistent with previous results that have found that Zn^{2+} and Mg^{2+} are catalytic activators of many oxidoreductases [49,50]. The activation effect of Cu^{2+} was consistent with the fact that Cu^{2+} is an active center of laccase and an important component of the laccase molecule [51]. Compared with the free laccase, the support provided some shielding effect for the immobilized laccase and reduced the interference of the metal ions.

2.7. Removal of 2,4-DCP

Next, the catalytic potential of PIL (2)–NH$_2$–GA–Lac was investigated. Many previous studies have shown that various process parameters may significantly affect the removal of harmful pollutants by enzymes [52]. Therefore, after evaluating the stability and reusability of immobilized laccase, it was crucial to determine the effects of various process conditions (e.g., pH, temperature, and substrate concentration) on its efficiency and the optimal conditions for the effective removal of phenolic contaminants. In Figure 5a–c, the bar represents the efficiency of phenol removal over time (12–36 h). The dot plot represents the removal, degradation, and adsorption of 2,4-DCP by the immobilized enzyme within a certain time range. The recycling efficiency of 2,4-DCP removal by the immobilized enzyme was also studied, as shown in Figure 5d.

As shown in Figure 5a, with the increase in the substrate concentration from 1 to 20 mg/L, both the 2,4-DCP removal and degradation increased firstly and then slightly decreased. When the concentration was 15 mg/L, the maximum 2,4-DCP removal and degradation occurred (90% and 55%, respectively). However, the 2,4-DCP removal and degradation decreased slightly when the substrate concentration was 20 mg/L. This can be explained by the fact that at a higher substrate concentration, the phenolic substrate tended to aggregate through intermolecular interactions, which led to a reduction in the substrate adsorption in the immobilized system [53,54]. In addition, it can be seen from the bar that with extension of the reaction time (12–36 h), the removal of 2,4-DCP basically showed a steady rising trend. As one item that contributed to the removal, the 2,4-DCP adsorption increased until it reached around 35%. This also can be understood as the gradual saturation of the immobilized capacity on the support [55].

The removal of 2,4-DCP was investigated over a broad pH range of 2 to 6 at 20 °C (Figure 5b). As can be seen from the bar, when the reaction lasted for 24 h, the 2,4-DCP removal increased substantially, reaching its maximum increment at pH 4. Furthermore, regarding their variation with the pH values, the removal and degradation showed similar trends, while the adsorption did not change significantly, remaining at about 20%. It was also found that the removal changed a little when the pH was 2–4 and reached its maximum value at pH 5 (84%, 36 h). However, when the pH exceeded 5, the degradation was negatively correlated with the pH. It can be seen that the weak acidity of the solution was conducive to the removal and degradation of 2,4-DCP. In a slightly high pH environment, the dissociation equilibrium of 2,4-DCP shifted in the direction of producing more H^+. According to the principle of chemical reaction equilibrium, anions (OH^-) enable and promote the catalytic oxidation reaction [56]. However, in this study, this led to lower catalytic oxidation efficiency when the pH value was further increased. On the other hand, the increase in anions in the dissociation equilibrium brought about the formation of free phenoxy, which gave rise to C-O coupling and further led to the accumulation of 2,4-DCP dimers and a high concentration of oligomer. This may hinder the substrate molecules' ability to enter the laccase's active center, resulting in reduced 2,4-DCP removal [43].

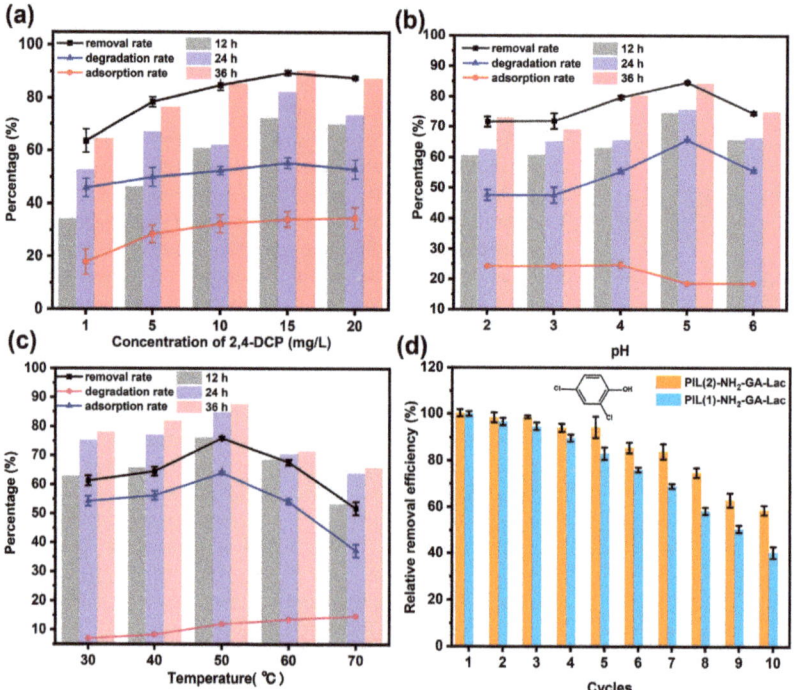

Figure 5. Effect of time and concentration of 2,4-DCP (**a**), pH (**b**), and temperature (**c**) on the efficiency of removal of phenolic compounds and reusability of the immobilized laccase (**d**). Conditions: 2,4-DCP (1 mg/L), PIL–NH_2–GA–Lac (10 mg/mL), pH = 3 (**b**–**d**), 20 °C (**a**,**c**,**d**), and 12 h (**d**).

Figure 5c demonstrates that PIL (2)–NH_2–GA–Lac was characterized by 2, 4-DCP removal over a wide temperature range. As the temperature increased, the removal and degradation gradually increased, with the maximum removal recorded at 50 °C (24 h, 85%). The covalent binding between the laccase and the PIL–NH_2 led to the stronger conformation of the laccase, by which the higher temperature was able to make the laccase-activated groups more active, benefiting the diffusion and electron transfer of the substrate in the material channel. In addition, the synergistic mechanism between the laccase and the support enabled the maximum removal of 2,4-DCP by PIL (2)–NH_2–GA–Lac at 50 °C. However, when the temperature continued to increase, this was not conducive to the adsorption and degradation of 2,4-DCP by PIL (2)–NH_2–GA–Lac. This was because constant high temperatures will cause protein inactivation and greatly reduce the catalytic oxidation capacity of laccase [57]. The trend of the bar also supports this view. The increase in 2,4-DCP removal decreased with time, reaching its minimum value at 60–70 °C.

The cycling properties of the immobilized laccase system are shown in Figure 5d. As expected, the relative removal efficiency of the immobilized laccase system did not decrease significantly as the number of batches increased. After seven cycles, the 2,4-DCP removal was still maintained at more than 80%. Table S1 (in Supplementary Materials) summarizes recent studies on enzyme immobilization and its application in the degradation of phenolic pollutants. Compared with other supports, PILs have higher mechanical properties, more suitable specific surface area and superior circulation performance. They conform with the support selection principle of enzyme immobilization and provide ideas for the further development of enzyme immobilization. Compared with the typical reports listed, the PIL (2)–NH$_2$–GA–Lac system used in this paper showed comparable or better operational stability for the degradation of phenolic pollutants. This can be attributed to the strong binding between the laccase and the support, which helped avoid leakage [58]. The above results illustrate that the reusability of the current immobilized laccase system showed great potential for the system's large-scale application in the future.

3. Experimental Section

3.1. Materials

Laccase was supplied by Chengdu Huaxia Chemical (0.5 U mg^{-1}). Glutaraldehyde (25% in H$_2$O, v/v) was supplied by Sigma Aldrich (St. Louis, MO, USA). Divinyl benzene (DVB), toluene and 2,2′-azo ratio (2-methylpropanitrile) (AIBN, 98%) were provided by Tianjin Damao Chemical Reagent Co., Ltd. (Tianjin, China). Polyvinyl alcohol purchased from Jiangsu Aikang Biomedical Research and Development Co., Ltd. (PVA). N-vinylimidazole and 3-Bromopropylamine hydrobromide were purchased from Meryer. AIBN was used after fresh recrystallization. 2,2′-Azbis (3-ethylbenzothiazoline-6-sulfonic acid) (ABTS), sodium citrate dehydration, FeCl$_3$, AlCl$_3$, CrCl$_3$, ZnCl$_2$, and MgCl$_2$ were purchased from Aladdin Chemical Co., Ltd. Except for AIBN, which needs recrystallization, and DVB, which needs extraction, other solvents and reagents were analytical grade and were able to be used without further purification.

3.2. Characterization of the Support

The PILs were characterized by scanning electron microscopy (SEM, Haitich SU8020, Tokyo, Japan), Fourier-transform infrared spectroscopy (FTIR, Shimadzu IRTracer-100, Kyoto, Japan), specific surface analysis (BET, BSD-PM, Beijing, China), and thermogravimetric analysis (TGA, Setaram Labsys, Lyon, France). The optical density of the sample solution was measured by a 752 UV-Vis Spectrophotometer (Shanghai Spectral Instrument Co., Ltd., Shanghai, China). The concentration of 2,4-DCP was analyzed by high-performance liquid chromatography (HPLC, Agilent 1260, Santa Clara, CA, USA). The content of the laccase in the solution was determined by a microplate reader (INFINITE M PLEX, Männedorf, Switzerland).

3.3. Preparation of Amino PILs

Preparation of amino-ionic liquid monomer (ILM–NH$_2$): N-vinylimidazole (2.8 g, 30 mmol) and acetonitrile (30 mL) were added to a two-necked flask equipped with a magnetic stirrer. The mixture was refluxed at 78 °C under a nitrogen atmosphere. 3-Bromopropylamine (4.4 g, 20 mmol) was added drop by drop into the flask over 12 h. After purification, the ILM–NH$_2$ was obtained.

Preparation of PIL (1)–NH$_2$: ILM–NH$_2$ (0.5 g, 2.68 mmol), DVB (3.2 g, 24.6 mmol), ethanol and deionized water were added to a three-necked flask equipped with a magnetic stirrer. The mixture was refluxed at 70 ± 8 °C over 48 h.

Preparation of PIL (2)–NH$_2$: Two steps were involved. The first step: according to the traditional suspension polymerization, deionized water (810 g) and PVA (90 g, 5%) were mixed in a three-necked flask equipped with a magnetic stirrer. The mixture was refluxed at 40 °C. Then, toluene and DVB were added (mass ratio: 3:1, 300 g). The mixture was refluxed at 70 °C–80 °C over 2–3 h. The heating was stopped after PDVB solidified. After purification, the PIL (2)–NH$_2$ was obtained.

3.4. Laccase Immobilization by PILs

Amino-functionalized PILs (named PIL-NH$_2$) immobilized laccase (Lac) by covalently binding glutaraldehyde (GA). The activated PIL (1)–NH$_2$ and PIL (2)–NH$_2$ (20 mg) were dispersed in a sodium citrate buffer (pH = 3, 10 mL, 100 mM) under ultrasound, and then laccase (1 mg·ml^{-1}) was added. Then, glutaraldehyde (2% v/v) was added drop-wise to the mixture of PIL–NH$_2$ and laccase. The new mixture was cross-linked at room temperature for a certain time (e.g., 1, 2, 3, 4, or 5 h). The immobilization efficiency was estimated using Equation (1). Drawing BSA standard curve [59]: Deionized water was used to prepare BSA solution with a concentration of 1–10 µg/mL. BSA (4 mL) was mixed with Coomassie brilliant blue G-250 protein reagent (1 mL), and the absorbance was measured at 595 nm with a microplate reader after 3 min. Coomassie brilliant blue G-250 staining method was used to determine the ability of amino-functionalized PILs to immobilize laccase. After the enzyme was cross-linked with the support for a certain time, the supernatant was taken and combined with the Coomassie brilliant blue G-250 reagent. The absorbance was determined by colorimetry at 595 nm, and C (final concentrations of protein) was obtained according to the standard curve.

$$M = \frac{(C_0 - C)V}{m} \quad (1)$$

Here, M is the efficiency of immobilized laccase of PIL–NH$_2$ (mg/g); C_0 and C are the initial and final concentrations of proteins (µg/mL), respectively; V is the volume of solution (mL); m is the mass of PIL–NH$_2$ (mg). The laccase immobilized by PIL–NH$_2$–GA–Lac under different initial concentrations of laccase and different cross-linking times was calculated.

3.5. Activity Assays of Free and Immobilized Laccase

The activities of the free laccase and immobilized laccase were determined by an ABTS assay at 420 nm. The reaction mixture consisted of 100 mM ABTS and a suitable amount of free or immobilized laccase. During the process, the increase in the absorbance at 420 nm was measured using a UV-2450 spectrophotometer. One unit (U) of laccase activity was defined as the amount of enzyme needed to oxidize 1 mol of ABTS per minute. The specific activity and activity recovery of immobilized enzyme were calculated by the following formulas [60]:

$$\text{The specific activity of the immobilized enzyme} \left(\frac{IU}{g}\right) = \frac{\Delta A}{M \times m \times t} \quad (2)$$

where ΔA is the change in absorbance in a certain period; M is immobilized laccase (mg/g); m is the mass of PIL–NH$_2$ (mg); t is the reaction time (min).

$$\text{Activity recovery} = \frac{R_i}{R_f} \quad (3)$$

where R_i is the specific activity of the immobilized laccase (U/g); R_f is the specific activity of the free laccase under the same conditions (U/g).

3.6. Effect of pH and Temperature on Laccase Activity

To determine the pH and temperature activity profiles of the free and immobilized laccase, the activity assays were carried out over a pH range of 2.0–7.0 (in sodium citrate buffer (100 mM, pH 2.0–4.0), sodium acetate buffer (100 mM, pH 5.0) and phosphate buffer (100 mM, pH 6.0–7.0) at 25 °C and a temperature range of 30–60 °C. The results were converted to relative activities (percentage of the maximum activity obtained in that series). Each set of experiments was performed in triplicate, and the arithmetic mean values were calculated.

3.7. Storage Stability and Cycle Stability

The free and immobilized laccase were incubated in sodium citrate buffer at 4 °C. To detect the activity, the laccase was taken out of the buffer every few days and tested by ABST assay. The relative activity of freshly immobilized laccase was considered to be 100%.

3.8. Determination of K_m and V_{max}

Determination of K_m and V_{max}: The free and immobilized laccase were assayed at different concentrations of ABTS, ranging from 1 to 5 mM, to determine the laccase kinetics, i.e., K_m and V_{max} values, using Lineweaver–Burk reciprocal plot transformation of Michaelis–Menton equation.

3.9. Effect of Metal Ions

Aqueous solutions of Cr^{2+}, Fe^{3+}, Zn^{2+}, Mg^{2+}, and Al^{3+} ions (e.g., 5 mmol/L) were prepared. Metal salt solution (1 mL), sodium citrate buffer (1 mL, pH = 3.0), free laccase (1 mL) or immobilized laccase (5 mg) and ABTS substrate (2 mL) were mixed in a 50 mL centrifuge tube. The final concentration of metal ions was 1 mmol/L. Reaction took place at 25 °C for 5 min. The laccase activity was measured at 420 nm wavelength by ultraviolet spectrophotometer. Three parallel samples were set in each group. The buffer solution was used instead of the metal salt solution in the blank group.

$$\omega = \frac{L_j}{L_0} \times 100\% \qquad (4)$$

where ω is the metal ions' effect on efficiency; L_j is the laccase activity with metal ions; L_0 is the laccase activity without metal ions.

3.10. Removal of Phenolic Compounds

The ability of the immobilized laccase to remove 2,4-DCP (1–20 mg/L) with different pH (2–6) and temperature (30 °C–70 °C) levels was tested in 10 mL reaction medium (10 mg PIL (2)–NH$_2$–GA–Lac, 12–36 h). Chlorophenols removed by adsorption were also investigated using inactivated PIL–NH$_2$–GA–Lac. The conditions were the same as above, except that PIL–NH$_2$–GA–Lac was inactivated. Each set of experiments was performed in triplicate, and the arithmetic mean values were calculated.

The removal (R), adsorption (R_A) and degradation (R_D) of 2,4-DCP were calculated by the following Equations:

$$R = \frac{C_1 - C_2}{C_1} \times 100\% \qquad (5)$$

$$R_A = \frac{C_1 - C_3}{C_1} \times 100\% \qquad (6)$$

$$R_D = (R - R_A) \qquad (7)$$

where C_0 is the initial concentration of 2,4-DCP, C_1 is the residual concentration of 2,4-DCP removed by immobilized laccase, and C_2 is the residual concentration of 2,4-DCP adsorbed by inactivated PIL–NH$_2$–GA–Lac.

To explore the recyclability of immobilized enzyme, the degradation of 2,4-DCP (1 mg/L) by PIL (2)–NH$_2$–GA–Lac (10 mg) was repeated in 10 cycles. After each batch, the immobilized laccase was isolated by centrifuge, and then an aliquot of fresh reaction medium was added for the next cycle. The relative removal efficiency was correlated with the highest removal percentage (100% represented the highest removal). This was expressed as the following Equation:

$$\text{Relative removal efficiency} = \frac{R_i}{R_{max}} \times 100 \qquad (8)$$

where R_i is the phenol removal of each sample and R_{max} is the highest phenol removal of all samples. In the cycle test, the first phenol removal was generally considered 100%.

The high-performance liquid chromatography: C18 column. The mobile phase was 75% (mass fraction) methanol and 25% (mass fraction) ultrapure water. The flow rate was 1 mL/min. The detection wavelength was 225 nm. The peak time of 2,4-DCP was about 6.5 min at 35 °C.

4. Conclusions

In summary, in this study, confined polymerization was demonstrated to be an efficient strategy for constructing PILs with a high specific surface area (302 m^2/g) and mesoporous structure (~12 nm). (I) The microenvironment provided by PILs may play an important role in shielding external environmental interferences (e.g., pH, temperature). (II) Compared with the free laccase, the immobilized laccase showed improved stability and circulation stability. (III) In the catalytic study of the effects of the immobilized laccase on 2,4-DCP, the interactions between the PILs and 2,4-DCP might have enriched the substrates, enabling the laccase to promote biodegradation. This research demonstrated the versatile potential of PILs in constructing high-efficiency immobilized enzymes, and lends support to the promotion of their application in wastewater treatment and biological technology development.

Supplementary Materials: The following supporting information can be downloaded at: https://www.mdpi.com/article/10.3390/molecules28062569/s1, Figure S1: The H1 NMR of ionic liquids monomer(ILM, [AVIM]Br); Figure S2: Linewaever-Burk plots of free laccase and immobilized laccase; Figure S3: Effect of metal ions on laccase activity. Table S1: Different support enzyme loading performance and phenol compounds removal rate.

Author Contributions: Y.L.: Conceptualization, Methodology, Software, Formal analysis, Writing—original draft. X.C., J.Z., J.Y., B.H., W.L.: Software, Formal analysis, Writing—review & editing. J.S.: Conceptualization, Supervision, Project administration, Writing—review & editing, Funding acquisition. All authors have read and agreed to the published version of the manuscript.

Funding: This work was supported by the State Key Laboratory of Catalytic Materials and Reaction Engineering (RIPP, SINOPEC, 202021641390A) and Startup Foundation of Beijing Institute of Technology, China (3160011182007).

Institutional Review Board Statement: Not applicable.

Informed Consent Statement: Not applicable.

Data Availability Statement: Not applicable.

Acknowledgments: The authors thank the Biological and Medical Engineering Core Facilities and the Analysis & Testing Center, Beijing Institute of Technology, for supporting the experimental tests.

Conflicts of Interest: The authors declare no conflict of interest.

References

1. Zdarta, J.; Jankowska, K.; Bachosz, K.; Degórska, O.; Kaźmierczak, K.; Nguyen, L.N.; Nghiem, L.D.; Jesionowski, T. Enhanced Wastewater Treatment by Immobilized Enzymes. *Curr. Pollut. Rep.* **2021**, *7*, 167–179. [CrossRef]
2. Hu, D.; Song, L.; Yan, R.; Li, Z.; Zhang, Z.; Sun, J.; Bian, J.; Qu, Y.; Jing, L. Valence-mixed iron phthalocyanines/(1 0 0) Bi$_2$MoO$_6$ nanosheet Z-scheme heterojunction catalysts for efficient visible-light degradation of 2-chlorophenol via preferential dechlorination. *Chem. Eng. J.* **2022**, *440*, 135786. [CrossRef]
3. Gupta, V.K.; Nayak, A.; Agarwal, S.; Tyagi, I. Potential of activated carbon from waste rubber tire for the adsorption of phenolics: Effect of pre-treatment conditions. *J. Colloid Interface Sci.* **2014**, *417*, 420–430. [CrossRef]
4. Karimi-Maleh, H.; Fakude, C.T.; Mabuba, N.; Peleyeju, G.M.; Arotiba, O.A. The determination of 2-phenylphenol in the presence of 4-chlorophenol using nano-Fe$_3$O$_4$/ionic liquid paste electrode as an electrochemical sensor. *J. Colloid Interface Sci.* **2019**, *554*, 603–610. [CrossRef]
5. Saravanan, A.; Kumar, P.S.; Karishma, S.; Vo, D.-V.N.; Jeevanantham, S.; Yaashikaa, P.; George, C.S. A review on biosynthesis of metal nanoparticles and its environmental applications. *Chemosphere* **2020**, *264*, 128580. [CrossRef] [PubMed]
6. Chan, S.S.; Khoo, K.S.; Chew, K.W.; Ling, T.C.; Show, P.L. Recent advances biodegradation and biosorption of organic compounds from wastewater: Microalgae-bacteria consortium. *Bioresour. Technol.* **2022**, *344*, 126159. [CrossRef] [PubMed]
7. Fernández-Fernández, M.; Sanromán, M.; Moldes, D. Recent developments and applications of immobilized laccase. *Biotechnol. Adv.* **2013**, *31*, 1808–1825. [CrossRef] [PubMed]
8. Mehra, R.; Muschiol, J.; Meyer, A.S.; Kepp, K.P. A structural-chemical explanation of fungal laccase activity. *Sci. Rep.* **2018**, *23*, 17285. [CrossRef] [PubMed]
9. Mokhtar, A.; Nishioka, T.; Matsumoto, H.; Kitada, S.; Ryuno, N.; Okobira, T. Novel biodegradation system for bisphenol A using laccase-immobilized hollow fiber membranes. *Int. J. Biol. Macromol.* **2019**, *180*, 737–744. [CrossRef] [PubMed]

10. Li, M.; Dai, X.; Li, A.; Qi, Q.; Wang, W.; Cao, J.; Jiang, Z.; Liu, R.; Suo, H.; Xu, L. Preparation and Characterization of Magnetic Metal–Organic Frameworks Functionalized by Ionic Liquid as Supports for Immobilization of Pancreatic Lipase. *Molecules* **2022**, *27*, 6800. [CrossRef]
11. Sheldon, R.A.; Basso, A.; Brady, D. New frontiers in enzyme immobilisation: Robust biocatalysts for a circular bio-based economy. *Chem. Soc. Rev.* **2021**, *50*, 5850–5862. [CrossRef]
12. Wu, H.; Chen, Q.; Zhang, W.; Mu, W. Overview of strategies for developing high thermostability industrial enzymes: Discovery, mechanism, modification and challenges. *Crit. Rev. Food Sci. Nutr.* **2021**, *26*, 1–18. [CrossRef] [PubMed]
13. Mohammadi, M.; As, M.; Salehi, P.; Yousefi, M.; Nazari, M.; Brask, J. Immobilization of laccase on epoxy-functionalized silica and its application in biodegradation of phenolic compounds. *Int. J. Biol. Macromol.* **2018**, *109*, 443–447. [CrossRef] [PubMed]
14. Katyal, P.; Chu, S.; Montclare, J.K. Enhancing organophosphate hydrolase efficacy via protein engineering and immobilization strategies. *Ann. NY Acad. Sci.* **2020**, *1480*, 54–72. [CrossRef] [PubMed]
15. Wu, S.; Snajdrova, R.; Moore, J.C.; Baldenius, K.; Bornscheuer, U.T. Biocatalysis: Enzymatic Synthesis for Industrial Applications, Bornscheuer. *Angew. Chem. Int. Ed.* **2021**, *60*, 88–119. [CrossRef]
16. Dong, Z.; Liu, Z.; Shi, J.; Tang, H.; Xiang, X.; Huang, F.; Zheng, M.-M. Carbon nanoparticle-stabilized pickering emulsion as a sustainable and high-performance interfacial catalysis platform for enzymatic esterification/transesterification. *ACS Sustain. Chem. Eng.* **2019**, *7*, 7619–7629. [CrossRef]
17. Hwang, E.T.; Lee, S. Multienzymatic Cascade Reactions via Enzyme Complex by Immobilization. *ACS Catal.* **2019**, *9*, 4402–4425. [CrossRef]
18. Ariaeenejad, S.; Motamedi, E.; Salekdeh, G.H. Application of the immobilized enzyme on magnetic graphene oxide nano-carrier as a versatile bi-functional tool for efficient removal of dye from water. *Bioresour. Technol.* **2021**, *319*, 124228. [CrossRef]
19. Zdarta, J.; Feliczak-Guzik, A.; Siwińska-Ciesielczyk, K.; Nowak, I.; Jesionowski, T. Materials, Mesostructured cellular foam silica materials for laccase immobilization and tetracycline removal: A comprehensive study. *Microporous Mesoporous Mater.* **2019**, *291*, 109688. [CrossRef]
20. Chen, G.; Huang, S.; Kou, X.; Zhu, F.; Ouyang, G. Embedding Functional Biomacromolecules within Peptide-Directed Metal–Organic Framework (MOF) Nanoarchitectures Enables Activity Enhancement. *Angew. Chem. Int. Ed.* **2020**, *59*, 13947–13954. [CrossRef]
21. Xu, R.; Zhou, Q.; Li, F.; Zhang, B. Laccase immobilization on chitosan/poly(vinyl alcohol) composite nanofibrous membranes for 2,4-dichlorophenol removal. *Chem. Eng. J.* **2013**, *222*, 321–329. [CrossRef]
22. Liu, Y.; Zeng, Z.; Zeng, G.; Tang, L.; Pang, Y.; Li, Z.; Liu, C.; Lei, X.; Wu, M.; Ren, P.; et al. Immobilization of laccase on magnetic bimodal mesoporous carbon and the application in the removal of phenolic compounds. *Bioresour. Technol.* **2012**, *115*, 21–26. [CrossRef]
23. Navarro-Sánchez, J.; Barrios, N.A.; Berlanga, B.L.; Ruiz-Pernía, J.J.; Fonfria, V.A.L.; Tuñón, I.; Martí-Gastaldo, C. Translocation of enzymes into a mesoporous MOF for enhanced catalytic activity under extreme conditions. *Chem. Sci.* **2019**, *10*, 4082–4088. [CrossRef] [PubMed]
24. Aydemir, T.; Güler, S. Characterization and immobilization of *Trametes versicolor* laccase on magnetic chitosan–clay composite beads for phenol removal. *Artif. Cells Nanomed. Biotechnol.* **2015**, *43*, 425–432. [CrossRef] [PubMed]
25. Yuan, J.; Mecerreyes, D.; Antonietti, M. Poly(ionic liquid)s: An update. *Prog. Polym. Sci.* **2013**, *38*, 1009–1036. [CrossRef]
26. Zhang, S.-Y.; Zhuang, Q.; Zhang, M.; Wang, H.; Gao, Z.; Sun, J.-K.; Yuan, J. Poly(ionic liquid) composites. *Chem. Soc. Rev.* **2020**, *49*, 1726–1755. [CrossRef]
27. Qian, W.; Texter, J.; Yan, F. Frontiers in poly(ionic liquid)s: Syntheses and applications. *Chem. Soc. Rev.* **2017**, *46*, 1124–1159. [CrossRef]
28. Song, H.; Wang, Y.; Liu, Y.; Chen, L.; Feng, B.; Jin, X.; Zhou, Y.; Huang, T.; Xiao, M.; Huang, F.; et al. Conferring Poly(ionic liquid)s with High Surface Areas for Enhanced Catalytic Activity. *ACS Sustain. Chem. Eng.* **2021**, *9*, 2115–2128. [CrossRef]
29. Song, H.; Wang, Y.; Xiao, M.; Liu, L.; Liu, Y.; Liu, X.; Gai, H. Design of Novel Poly(ionic liquids) for the Conversion of CO_2 to Cyclic Carbonates under Mild Conditions without Solvent. *ACS Sustain. Chem. Eng.* **2019**, *7*, 9489–9497. [CrossRef]
30. Kou, Z.; Wang, C. Preparation of highly crosslinked polyvinylpyrrolidone–polydivinylbenzene adsorbents based on reinitiation of suspended double bonds to achieve excellent blood compatibility and bilirubin removal. *Mater. Adv.* **2022**, *3*, 4839–4850. [CrossRef]
31. Kujawa, J.; Głodek, M.; Li, G.; Al-Gharabli, S.; Knozowska, K.; Kujawski, W. Highly effective enzymes immobilization on ceramics: Requirements for supports and enzymes. *Sci. Total. Environ.* **2021**, *801*, 149647. [CrossRef]
32. Patel, S.K.S.; Choi, H.; Lee, J.-K. Multimetal-Based Inorganic–Protein Hybrid System for Enzyme Immobilization. *ACS Sustain. Chem. Eng.* **2019**, *7*, 13633–13638. [CrossRef]
33. DiCosimo, R.; McAuliffe, J.; Poulose, A.; Yrookaran; Bohlmann, G. Industrial use of immobilized enzymes. *Chem. Soc. Rev.* **2013**, *42*, 6437–6474. [CrossRef] [PubMed]
34. Barbosa, G.S.D.S.; Oliveira, M.E.P.S.; Dos Santos, A.B.S.; Sánchez, O.C.; Soares, C.M.F.; Fricks, A.T. Immobilization of Low-Cost Alternative Vegetable Peroxidase (*Raphanus sativus* L. peroxidase): Choice of Support/Technique and Characterization. *Molecules* **2020**, *25*, 3668. [CrossRef]
35. Rashid, S.S.; Mustafa, A.H.; Ab Rahim, M.H.; Gunes, B. Magnetic nickel nanostructure as cellulase immobilization surface for the hydrolysis of lignocellulosic biomass. *Int. J. Biol. Macromol.* **2022**, *209*, 1048–1053. [CrossRef] [PubMed]

36. Qiu, X.; Wang, Y.; Xue, Y.; Li, W.; Hu, Y. Laccase immobilized on magnetic nanoparticles modified by amino-functionalized ionic liquid via dialdehyde starch for phenolic compounds biodegradation. *Chem. Eng. J.* **2020**, *391*, 123564. [CrossRef]
37. Liu, J.; Peng, J.; Shen, S.; Jin, Q.; Li, C.; Yang, Q. Enzyme Entrapped in Polymer-Modified Nanopores: The Effects of Macromolecular Crowding and Surface Hydrophobicity. *Chem. Eur. J.* **2013**, *19*, 2711–2719. [CrossRef]
38. Kadam, A.A.; Jang, J.; Jee, S.C.; Sung, J.-S.; Lee, D.S. Chitosan-functionalized supermagnetic halloysite nanotubes for covalent laccase immobilization. *Carbohydr. Polym.* **2018**, *194*, 208–216. [CrossRef] [PubMed]
39. Bilal, M.; Zhao, Y.; Rasheed, T.; Iqbal, H.M.N. Magnetic nanoparticles as versatile carriers for enzymes immobilization: A review. *Int. J. Biol. Macromol.* **2018**, *120*, 2530–2544. [CrossRef]
40. Huang, W.; Zhang, W.; Gan, Y.; Yang, J.; Zhang, S. Laccase immobilization with metal-organic frameworks: Current status, remaining challenges and future perspectives. *Crit. Rev. Environ. Sci. Technol.* **2020**, *52*, 1282–1324. [CrossRef]
41. Zhang, Q.; Kang, J.; Yang, B.; Zhao, L.; Hou, Z.; Tang, B. Immobilized cellulase on Fe_3O_4 nanoparticles as a magnetically recoverable biocatalyst for the decomposition of corncob. *Chin. J. Catal.* **2016**, *37*, 389–397. [CrossRef]
42. Ngin, P.; Cho, K.; Han, O. Immobilization of Soybean Lipoxygenase on Nanoporous Rice Husk Silica by Adsorption: Retention of Enzyme Function and Catalytic Potential. *Molecules* **2021**, *26*, 291. [CrossRef]
43. Es, I.; Vieira, J.D.G.; Amaral, A.C. Principles, techniques, and applications of biocatalyst immobilization for industrial application. *Appl. Microbiol. Biotechnol.* **2015**, *99*, 2065–2082. [CrossRef]
44. Chen, X.; He, B.; Feng, M.; Zhao, D.; Sun, J. Immobilized Laccase on magnetic nanoparticles for enhanced lignin model compounds degradation. *Chin. J. Chem. Eng.* **2020**, *28*, 2152–2159. [CrossRef]
45. Zhang, S.; Gao, S.; Gao, G. Immobilization of beta-galactosidase onto Magnetic Beads. *Appl. Biochem. Biotechnol.* **2010**, *160*, 1386–1393. [CrossRef]
46. Drozd, R.; Rakoczy, R.; Wasak, A.; Junka, A.; Fijałkowski, K. The application of magnetically modified bacterial cellulose for immobilization of laccase. *Int. J. Biol. Macromol.* **2018**, *108*, 462–470. [CrossRef]
47. Xiao, C.; Liang, W.; Chen, L.; He, J.; Liu, F.; Sun, H.; Zhu, Z.; Li, A. Janus Poly(ionic liquid) Monolithic Photothermal Materials with Superior Salt-Rejection for Efficient Solar Steam Generation. *ACS Appl. Energy Mater.* **2019**, *2*, 8862–8870. [CrossRef]
48. Weng, Y.; Song, Z.; Chen, C.-H.; Tan, H. Hybrid hydrogel reactor with metal–organic framework for biomimetic cascade catalysis. *Chem. Eng. J.* **2021**, *425*, 131482. [CrossRef]
49. Xu, X.; Huang, X.; Liu, D.; Lin, J.; Ye, X.; Yang, J. Inhibition of metal ions on Cerrena sp. laccase: Kinetic, decolorization and fluorescence studies. *J. Taiwan Inst. Chem. Eng.* **2018**, *84*, 1–10. [CrossRef]
50. Lv, Y.; Liang, Q.; Li, Y.; Li, X.; Liu, X.; Zhang, D.; Li, J. Effects of metal ions on activity and structure of phenoloxidase in Penaeus vannamei. *Int. J. Biol. Macromol.* **2021**, *174*, 207–215. [CrossRef]
51. Zhou, Y.; You, S.; Zhang, J.; Wu, M.; Yan, X.; Zhang, C.; Liu, Y.; Qi, W.; Su, R.; He, Z. Copper ions binding regulation for the high-efficiency biodegradation of ciprofloxacin and tetracycline-HCl by low-cost permeabilized-cells. *Bioresour. Technol.* **2021**, *344*, 126297. [CrossRef]
52. Chen, Z.; Yao, J.; Ma, B.; Liu, B.; Kim, J.; Li, H.; Zhu, X.; Zhao, C.; Amde, M. A robust biocatalyst based on laccase immobilized superparamagnetic Fe_3O_4–SiO_2–NH_2 nanoparticles and its application for degradation of chlorophenols. *Chemosphere* **2021**, *291*, 132727. [CrossRef] [PubMed]
53. Vineh, M.B.; Saboury, A.A.; Poostchi, A.A.; Rashidi, A.M.; Parivar, K. Stability and activity improvement of horseradish peroxidase by covalent immobilization on functionalized reduced graphene oxide and biodegradation of high phenol concentration. *Int. J. Biol. Macromol.* **2018**, *106*, 1314–1322. [CrossRef]
54. Mahmoodi, N.M.; Saffar-Dastgerdi, M.H.; Hayati, B. Environmentally friendly novel covalently immobilized enzyme bionanocomposite: From synthesis to the destruction of pollutant. *Compos. B Eng.* **2020**, *184*, 107666. [CrossRef]
55. Kamranifar, M.; Allahresani, A.; Naghizadeh, A. Synthesis and characterizations of a novel CoFe2O4@CuS magnetic nanocomposite and investigation of its efficiency for photocatalytic degradation of penicillin G antibiotic in simulated wastewater. *J. Hazard. Mater.* **2019**, *366*, 545–555. [CrossRef] [PubMed]
56. Huang, Y.; Xi, Y.; Yang, Y.; Chen, C.; Yuan, H.; Liu, X. Degradation of 2,4-dichlorophenol catalyzed by the immobilized laccase with the carrier of Fe3O4@MSS–NH_2. *Chin. Sci. Bull.* **2014**, *59*, 509–520. [CrossRef]
57. Hu, Y.; Dai, L.; Liu, D.; Du, W.; Wang, Y. Progress & prospect of metal-organic frameworks (MOFs) for enzyme immobilization (enzyme/MOFs). *Renew. Sustain. Energy Rev.* **2018**, *91*, 793–801. [CrossRef]
58. Zdarta, J.; Meyer, A.S.; Jesionowski, T.; Pinelo, M. Developments in support materials for immobilization of oxidoreductases: A comprehensive review. *Adv. Colloid Interface Sci.* **2018**, *258*, 1–20. [CrossRef]
59. Bradford, M.M. A rapid and sensitive method for the quantitation of microgram quantities of protein utilizing the principle of protein-dye binding. *Anal. Biochem.* **1976**, *72*, 248–254. [CrossRef] [PubMed]
60. Alver, E.; Metin, A. Chitosan based metal-chelated copolymer nanoparticles: Laccase immobilization and phenol degradation studies. *Int. Biodeterior. Biodegradation* **2017**, *125*, 235–242. [CrossRef]

Disclaimer/Publisher's Note: The statements, opinions and data contained in all publications are solely those of the individual author(s) and contributor(s) and not of MDPI and/or the editor(s). MDPI and/or the editor(s) disclaim responsibility for any injury to people or property resulting from any ideas, methods, instructions or products referred to in the content.

Article

Alkyl Levulinates and 2-Methyltetrahydrofuran: Possible Biomass-Based Solvents in Palladium-Catalyzed Aminocarbonylation

Nuray Uzunlu [1], Péter Pongrácz [1], László Kollár [1,2,3] and Attila Takács [1,2,*]

[1] Department of General and Inorganic Chemistry, University of Pécs, Ifjúság Útja 6, H-7624 Pécs, Hungary
[2] János Szentágothai Research Centre, University of Pécs, Ifjúság Útja 20, H-7624 Pécs, Hungary
[3] ELKH-PTE Research Group for Selective Chemical Syntheses, Ifjúság Útja 6, H-7624 Pécs, Hungary
* Correspondence: takacsattila@gamma.ttk.pte.hu; Tel.: +36-72-503-600

Abstract: In this research, ethyl levulinate, methyl levulinate, and 2-methyltetrahydrofuran as bio-derived hemicellulose-based solvents were applied as green alternatives in palladium-catalyzed aminocarbonylation reactions. Iodobenzene and morpholine were used in optimization reactions under different conditions, such as temperatures, pressures, and ligands. It was shown that the XantPhos ligand had a great influence on conversion (98%) and chemoselectivity (100% carboxamide), compared with the monodentate PPh$_3$. Following this study, the optimized conditions were used to extend the scope of substrates with nineteen candidates (various para-, ortho-, and meta-substituted iodobenzene derivatives and iodo-heteroarenes), as well as eight different amine nucleophiles.

Keywords: aminocarbonylation; green solvents; amides; palladium; homogeneous catalysis

1. Introduction

Biomass is the only abundant and concentrated source of non-fossil carbon that is available on Earth, and its conversion into special chemicals and fuels has been the focus of several chemical researches within the past decade [1]. The biomass obtained by plants and other wastes can be used to provide sustainable chemicals. In general, biomass usage will bring benefits, such as a cleaner environment, more security, and projected long-term economic savings [2]. The solvent industry, as one of the biggest and most important global markets, is projected to reach 30.0 billion USD in 2025 (21.8 billion USD in 2020) at a CAGR (compound annual growth rate) of 6.6% during the forecast period and is estimated to reach 34 million metric tons by 2027 [3,4].

The bio-derived solvents obtained from biomass are environmentally benign, biodegradable, and have lower toxicity than conventional organic solvents. Lomba and coworkers found in a detailed study that levulinic acid and its esters showed very low toxicity, with a high biodegrability, which supports their use as green alternatives of traditional chemicals. They investigated the ecotoxicity on *Chlamydomonas reinhardtii*, *Vibrio fischeri*, *Daphnia magna*, and *Eisenia foetida* and ascertained that the toxicities of the levulinates (methyl-, ethyl-, and butyl levulinate) were increased as a function of the length of the alkyl chain [5]. Ventura and her research group evaluated the toxicity of biomass-derived platform molecules by using the Microtox toxicity test. Contrary to the statement of the above-mentioned research group, they found that ethyl levulinate was less toxic (practically harmless, EC$_{50}$ = 694 mg L^{-1}) than levulinic acid (moderately toxic, EC$_{50}$ = 28.4 mg L^{-1}) in the case of *Vibrio fischeri* bacteria [6]. In recent times, 2-MeTHF has been selected over tetrahydrofurane, due to its favorable characteristic properties (e.g., higher boiling point, which allows performing reactions at a relatively higher temperature [7], lower miscibility in water, which favors a cleaner workup [8]). Furthermore, it has been established after detailed toxicological studies that 2-MeTHF is not associated with any genotoxicity and mutagenicity [9].

Because solvents have great influence on reactivity and selectivity, their selection is very crucial in most chemical processes. For this reason, there are solvent selection guides made by researchers in academic and industrial areas (e.g., CHEM21, Sanofi's selection guide) to help choose the appropriate reaction media for a chemical synthesis [10–13]. In these guides, the preferable terms are characterized by favorable environmental, health, and safety (EHS) properties [14], and chemicals are categorized generally as hazardous, problematic, or recommended. On the other hand, these guides are helpful for selecting the best solvent because the "green solvents" deal with constraints that are sometimes contradictory: the chemical reaction efficiency, safety (flash point, resistance, energy decomposition, and peroxydation), health (acute, long-term, and single-target organ toxicity), environment (biodegradability, ecotoxicity, solubility in water, volatility, odor, and life cycle analysis), quality, industrial constraints (boiling and freezing points, density, and recyclability) and cost. Briefly, the best green solvent must meet the specifications by which the concept of "green chemistry" has been defined by the well-known 12 principles [15].

Water is cheap, safe, non-toxic, environmentally benign, and readily available; therefore, the application of it as a suitable reaction medium has received increasing attention during the last decades. However, an efficient homogeneous CC coupling, as well as an aminocarbonylation reaction, are often restricted, due to the limited solubility of reagents (substrates, amine nucleophiles) and amide products or the decomposition of the catalyst. Consequently, many more heterogeneous [16–21] or biphasic [22] aminocarbonylation processes have been described than homogeneous processes [23–25]. Additionally, it has been described by our research group that water can act as an O-nucleophile during aminocarbonylative conditions, resulting in the formation of carboxylic acid when low-reactive nucleophile reaction partners are used [26]. Considering the above-mentioned reasons, our attention turned to using bio-derived solvents instead of water.

Transition metal-catalyzed carbonylation reactions in the presence of nucleophile reagents have become an indispensable tool for the synthesis of several α,β-unsaturated and aromatic carboxylic acid derivatives [27,28]. Although these reactions require low catalyst loading, they generally need a huge amount of solvent as a reaction media. One of these reactions is aminocarbonylation, which has great importance concerning the synthesis of simple building blocks and the functionalization of biologically important skeletons. Carboxamides can be synthesized from easily available starting materials with aminocarbonylation, although the conventional carboxylic acid−carboxylic halide−carboxamide route is difficult to prepare because it has no notable yield in the implementation [29,30]. Moreover, our research group has been investigating this reaction in the conventional solvent DMF for years [29–39], and thanks to the ascertainments of the groups from Skrydstrup [40] and Mika [41–43], our interest has been turned to the investigation of green solvents.

In this study, we plan to investigate some appropriate green solvents as reaction mediums for Pd-catalyzed aminocarbonylation, which has many applications in both industrial and fine chemistry in organic synthesis, including the introduction of amides with a variety of N-substituents [44,45]. Amides are one of the most important classes of organic compounds, especially for the pharmaceutical industry, because most of the drugs contain amide functionality. For example, most of the top 15 best-selling drugs in 2017 contained amide moiety [46].

Levulinic acid (LA), which is one of the platform compounds derived from biomass, can be produced from lignocellulose biomass via two different ways. One is the direct hydrolysis of biomass based on cellulose through 5-(hydroxymethyl)furfural (5-HMF) intermediate, while the other is known as the furfuralcohol catalytic hydrolysis way, producing LA via the hemicellulose–xylose(C_5 unit)–furfural–furfuralcohol pathway (Figure 1). Therefore, levulinic acid can be considered a valuable platform molecule that can be converted into several important chemicals, such as levulinate esters, GVL, 2-MeTHF, etc. In this work, ethyl levulinate was chosen as a biomass-derived solvent that can be synthesized via direct esterification of the platform molecule, levulinic acid, in ethanol (Figure 1). This bio-based solvent is the viable additive for gasoline and diesel transportation fuels, and it can also either be used in the flavoring and fragrance industries or as substrates for various kinds of condensation and addition reactions at the ester and keto groups in organic chemistry [47,48]. Lei et al. investigated the Suzuki–Miyaura coupling of amides with this solvent and obtained 63% conversion during this transformation [49]. Another chosen levulinate ester was methyl levulinate, which is also a certified viable additive for gasoline and diesel transportation fuels, similar to the other levulinate esters [50]. Homogeneous acid catalysts or mixtures of Lewis and Brønsted acids have also been generally employed to produce a high yield of methyl levulinate from cellulose [51].

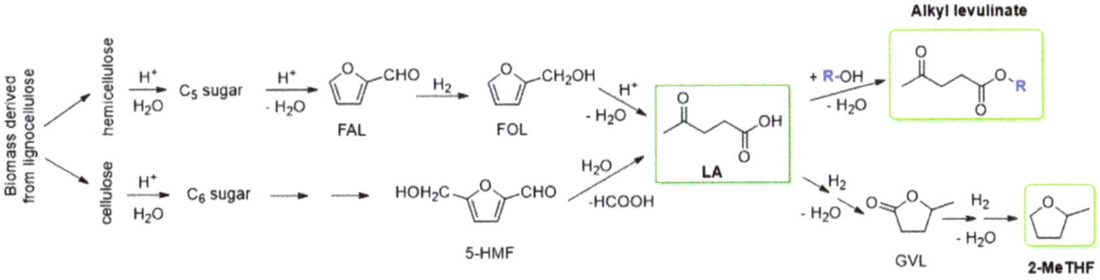

Figure 1. Converting lignocellulose through levulinic acid to alkyl levulinates and 2-MeTHF [52,53].

According to the computational study of Leal Silva's group, the conversion of furfural derived from hemicellulose to 2-methyltetrahydrofuran (2-MeTHF) could be more profitable than ethyl levulinate synthesis [52], which moved our attention to the advantages of 2-MeTHF as a solvent in the aminocarbonylation reaction. It is highly flammable and mostly used as a fuel additive and an alternative solvent of tetrahydrofuran (THF) [47,54]. 2-MeTHF is a promising solvent for transition metal-catalyzed reactions, and it has been justified with quite a large number of articles published in literature [28,54–63]. In this research, considering the green properties of the above-mentioned alkyl levulinates and 2-MeTHF, we investigated their applicability in the palladium-catalyzed aminocarbonylation of iodobenzene and its substituted derivatives, as well as iodo(hetero)arenes, in the presence of various N-nucleophiles.

2. Results and Discussion
2.1. Optimization Study

Iodobenzene and morpholine as the nucleophilic reaction partners were chosen to find the optimized conditions in our aminocarbonylation model reaction performed in green solvents (Scheme 1). Pressure, temperature, and ligand were selected as the variable parameters in the optimization study (Figure 2). First, the reaction was performed in the presence of a $Pd(OAc)_2/2\ PPh_3$ catalyst at 50 °C under 1 or 40 bar CO, conditions which have been generally used and well-studied by our research group [64].

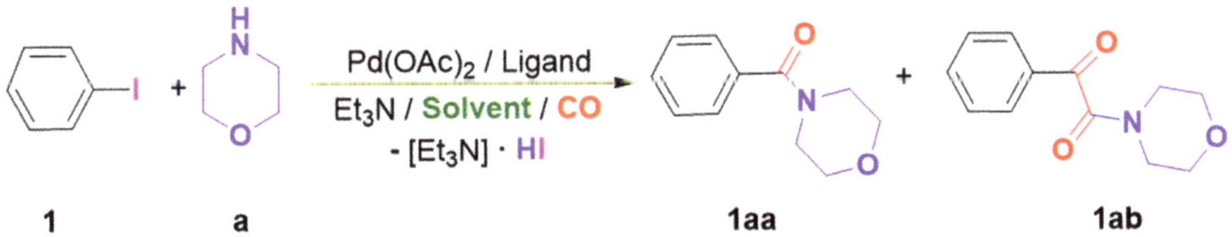

Scheme 1. Palladium-catalyzed aminocarbonylation reaction of iodobenzene with morpholine.

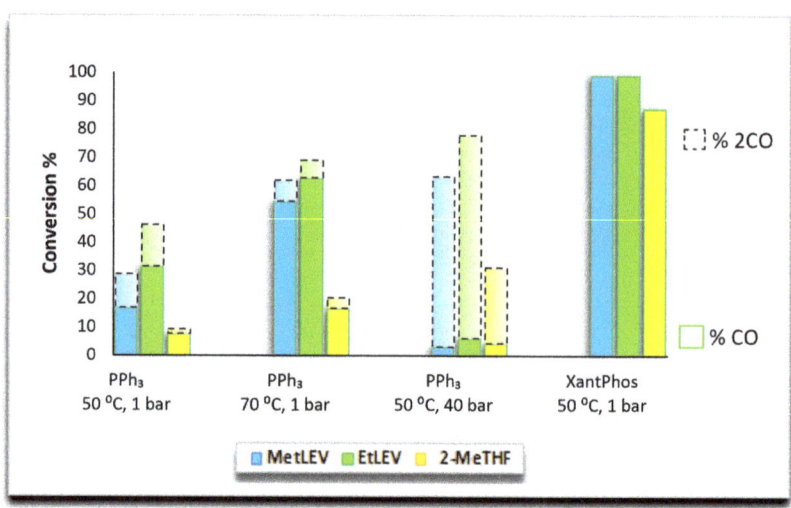

Figure 2. Optimization study of the aminocarbonylation of iodobenzene (**1**) with morpholine (**a**). (Reaction conditions: 0.5 mmol of iodobenzene, 0.75 mmol of morpholine, 0.0125 mmol of Pd(OAc)$_2$, 0.025 mmol of PPh$_3$ or 0.0125 mmol of XantPhos, 0.25 mL of triethylamine, and 5 mL of solvent under a CO atmosphere for 6 h. The conversion was determined by GC using dodecane as an internal standard.)

Under atmospheric carbon monoxide pressure, the reaction showed moderate conversion (29–46%) in the alkyl levulinate-type solvents (MetLEV and EtLEV) and 9% conversion in 2-MeTHF after 6 h. Under these conditions, the reaction was not chemoselective, due to the formation of the corresponding 2-ketocarboxamide-type product (16–41%) caused by the double carbon monoxide insertion. After carrying out the reactions under elevated temperature (70 °C) in the presence of the previously used catalyst, the following considerations are worth mentioning: (i) the chemoselectivity was favorable toward the amide (**1aa**) product, and (ii) the conversions were increased in the case of all solvents after a 6 h reaction time. The CO pressure had a great influence on the product selectivity. After carrying out the reaction at 40 bar of carbon monoxide pressure in the presence of the previously used catalyst at 50 °C, the 1-morpholino-2-phenylethane-1,2-dione (**1ab**) formed with high chemoselectivity (>85%) in all cases. It must be noted that the iodobenzene (**1**) was converted by 79% by using the alkyl levulinate-type solvents, while in 2-MeTHF, the conversion was only 35% after a 24 h reaction. Despite the promising chemoselectivity results at the elevated CO pressure, the above-mentioned low conversions inspired us to find other reaction conditions that could increase the synthetic importance of this reaction in green solvents. The aminocarbonylation of iodobenzene and *tert*-butylamine has been investigated by Marosvölgyi-Haskó and coworkers in DMF [65] and GVL using the Pd(OAc)$_2$/PPh$_3$ catalyst [41]. Although the amines were different, we described similar behavior in our former research (reactivity and selectivity) under the same reaction conditions [66–68]. After carrying out the reactions with *tert*-butylamine in the conventional organic solvent, the following considerations can be stated after a 6 h reaction time: *(i)* the conversion (50%) was higher under atmospheric conditions than in our green solvents (9–46%); *(ii)* the chemoselectivity was better towards the ketoamide-type product in DMF (amide:ketoamide = 22:78) at 1 bar of CO, while the amide formation was more favorable in alkyl levulinates and 2-MeTHF (59–84%); *(iii)* under elevated carbon monoxide pressure, higher conversions were detected in alkyl levulinates (63–78%) than in DMF (53%); *(iv)* excellent chemoselectivity towards the ketoamide was observed in the conventional solvents, as well as in our green solvents. It also has to be mentioned that the conversions and the chemoselectivity values, observed in GVL by Marosvölgyi-Haskó et. al., are similar to our results detected in alkyl levulinates.

Based on our former research [69,70], in which XantPhos was used successfully to convert the substrate selectively to the target product in a short reaction time, we decided to apply this rigid bidentate phosphine ligand with a large bite angle to increase the efficiency (e.g., conversion, selectivity) of the model reaction. After changing the triphenylphosphine to XantPhos and carrying out the reaction under ambient conditions (50 °C and 1 bar of CO pressure), complete conversion was detected in the alkyl levulinate solvents, while the reaction was also almost complete in 2-MeTHF (87%) after 6 h.

To justify the applicability of the bio-derived candidates, we performed our model reaction in DMF under the same reaction conditions as in the green solvents, and the results are summarized in Table 1.

Table 1. Comparison of the aminocarbonylation reaction of iodobenzene (1) and morpholine (a) in conventional and green solvents [a].

Entry	Solvent	Ligand	Temp. [°C]	p_{CO} [bar]	Conv.[b] [%]	Ratio of Products [b]	
						Amide	Ketoamide
1	DMF	PPh$_3$	50	1	71	28	72
2	DMF	PPh$_3$	70	1	74	67	33
3	DMF	PPh$_3$	50	40	79	3	97
4	DMF	XantPhos	50	1	100	88	12
5	MetLev	PPh$_3$	50	1	29	59	41
6	MetLev	PPh$_3$	70	1	62	88	12
7	MetLev	PPh$_3$	50	40	63	5	95
8	MetLev	XantPhos	50	1	99	100	0
9	EtLev	PPh$_3$	50	1	46	68	32
10	EtLev	PPh$_3$	70	1	69	91	9
11	EtLev	PPh$_3$	50	40	78	8	92
12	EtLev	XantPhos	50	1	98	100	0
13	2-MeTHF	PPh$_3$	50	1	9	84	16
14	2-MeTHF	PPh$_3$	70	1	20	82	18
15	2-MeTHF	PPh$_3$	50	40	31	14	86
16	2-MeTHF	XantPhos	50	1	87	100	0

[a] Reaction conditions: 0.5 mmol of iodobenzene, 0.75 mmol of morpholine, 0.0125 mmol of Pd(OAc)$_2$, 0.025 mmol of PPh$_3$ or 0.0125 mmol of XantPhos, 0.25 mL of triethylamine, and 5 mL of solvent under a CO atmosphere after a 6 h reaction time. [b] The conversion and the ratio of the carbonylated products were determined by GC using dodecane as an internal standard.

It can be seen that the conversion in DMF was higher than in our solvents under atmospheric conditions and in the presence of the Pd(OAc)$_2$/2 PPh$_3$ catalyst system. Furthermore, the formation of the **1ab** product was more favorable in DMF, while the amide (**1aa**) formation was preferable in the green solvents (entries 1, 5, 9, and 13). The amide (**1aa**) formation was much more expressed by carrying out the reactions at 70 °C (entries 2, 6, 10, and 14). The conversion and the selectivity values were almost the same in conventional and green solvents (except in 2-MeTHF) under high pressure conditions (entries 3, 7, 11, and 15). After changing the triphenylphosphine to the bidentate XantPhos, the reaction rate was extremely increased, and the amide (**1aa**) was formed with high selectivity (88%) (entries 4, 8, 12, and 16).

Additionally, the morpholino(phenyl)methanone (**1aa**) compounds were isolated in the following yields after performing the aminocarbonylation under optimized conditions: 61% (MetLev), 66% (EtLev), and 65% (2-MeTHF). After comparing these values with the isolated yields (>81%) reached by using conventional solvents [71–73], it can be stated that removing the solvent under reduced pressure followed by column chromatography is an applicable process to isolate the carboxamide products synthesized in bio-derived solvents. In this way, we have appropriate conditions to use in further reactions, in which we can extend the scope of the amine nucleophiles, as well as the substrates.

2.2. Extending the Scope of Amine Nucleophiles

With the optimized conditions on hand (XantPhos, 50 °C, atmospheric carbon monoxide pressure), we extended the scope of the amine nucleophiles in the aminocarbonylation of iodobenzene (Figure 3).

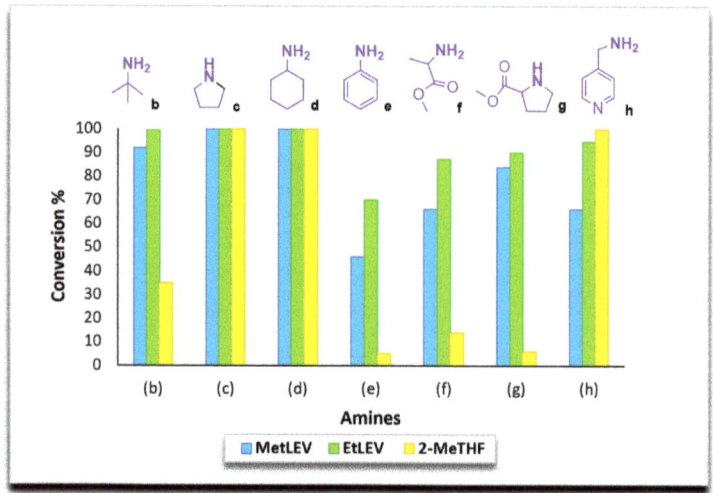

Figure 3. Palladium-catalyzed aminocarbonylation of iodobenzene (**1**) with different amines (**b–h**). (Reaction conditions: 0.5 mmol of iodobenzene, amine nucleophile (1.5 mmol of tert-butylamine, 0.75 mmol of pyrrolidine, 0.75 mmol of cyclohexylamine, 1.0 mmol of aniline, 0.55 mmol of amino acid methyl esters (AlaOMe, ProOMe), and 0.75 mmol of 4-picolylamine), 0.0125 mmol of Pd(OAc)$_2$, 0.0125 mmol of XantPhos, 0.25 mL of triethylamine, and 5 mL of solvent at 50 °C under 1 bar of CO in 6 h. The conversion was determined by GC using dodecane as an internal standard.)

It can be seen, by using simple primary (**b,d**) and secondary amines, that they showed complete conversion in all green solvents. The tert-butylamine (**b**) provided similar reactivity to amine (**c**) and (**d**) alkyl levulinate solvents, while in 2-MeTHF, strikingly lower conversion was detected after 6 h. The lowest basicity among our nucleophiles was in the presence of the aromatic aniline (**e**), with 46 and 70% conversions detected in methyl and ethyl levulinate, respectively. The N-phenylbenzamide (**2ea**) was identified in traces in 2-MeTHF. Alanine methyl ester (**f**) and proline methyl ester (**g**) showed a slightly lower reactivity in methyl levulinate than in ethyl levulinate, but the difference was not significant. Furthermore, the two amino acid methyl esters provided the lowest conversion in 2-MeTHF, which can be explained by their low solubility in this solvent. While the 4-picolylamine (**h**) showed similar reactivity to the amino acid methyl esters in alkyl levulinates, surprisingly, it was completely converted to the corresponding carboxamide (**2ha**) in 2-MeTHF after a 6 h reaction.

2.3. Extending the Scope of Substrates

In the next step, iodobenzene derivatives (**2–9**) possessing various para-substituents were reacted with morpholine (**a**) in the chosen green solvents under carbonylative conditions (Scheme 2).

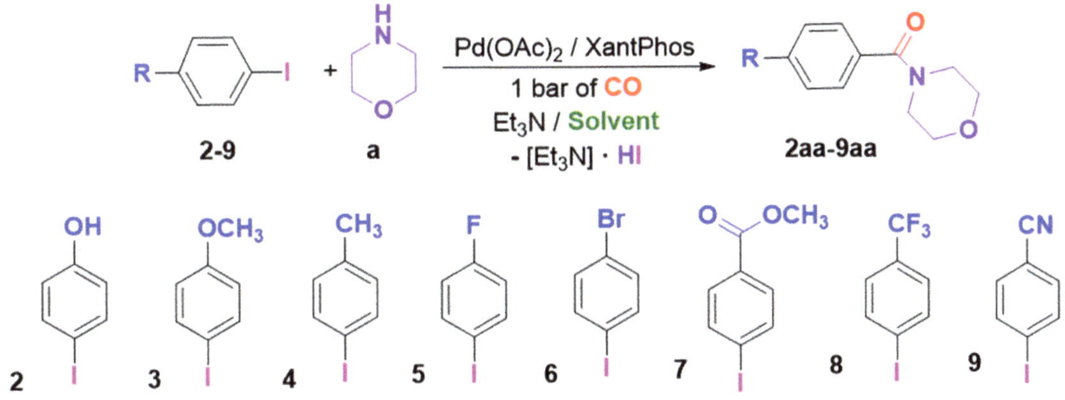

Scheme 2. Palladium-catalyzed aminocarbonylation reaction with para-substituted iodobenzenes (**2–9**) and morpholine (**a**).

Although a strong relation between the reactivity and the Hammett para constant of the substrates was not observed, the substrates (**8**, **9**), having electron-withdrawing substituents, provided lower conversions (55–67%) than the first three para-substituted iodobenzene derivatives (**2–4**) that bore electron-donating groups (Figure 4). Furthermore, substrates **2–4** reacted quite quickly with morpholine and gave high conversion values (76–96%), especially in methyl levulinate, after a 2 h reaction. It can also be seen that 1-fluoro-4-iodobenzene (**5**) and 1-bromo-4-iodobenzene (**6**) showed the lowest reactivity (46–59% conversions) in the alkyl levulinate solvents. After analyzing the reaction mixtures after 6 h, complete conversions were detected in almost all cases except 4-iodoanisole **3** (92%). Each substrate showed quite a high reactivity in alkyl levulinate-type solvents (>46%) in a 2 h reaction. It can also be seen that the behaviors of the para-substituted model compounds are almost completely the same. Only substrate **3** provided a much lower reactivity in the EtLEV (46%) than in the MetLEV (76%/) solvent. Consequently, there was no significant difference between the behaviors of our para-substituted substrates (**2–9**) in the alkyl levulinate-type solvents. The reactions, performed in 2-MeTHF under the same conditions mentioned above, showed strikingly different reactivity than those in the alkyl levulinate solvents. Most of the substrates showed very low conversions (<22%) in 2 h, except 4-iodotoluene (**4**) and methyl 4-iodobenzoate (**7**), in which cases 68% and 45% conversions were observed, respectively. While substrates **4** and **7** provided complete conversion, the others (**2**, **3**, **5**, **6**, **8**, and **9**) were converted by only 24–68% after a 6 h reaction. More than 92% of the target compounds were detected in the reaction mixture after 24 h; only the 4-iodobenzotrifluoride (**8**) provided the corresponding carboxamide in a quantity of 78%.

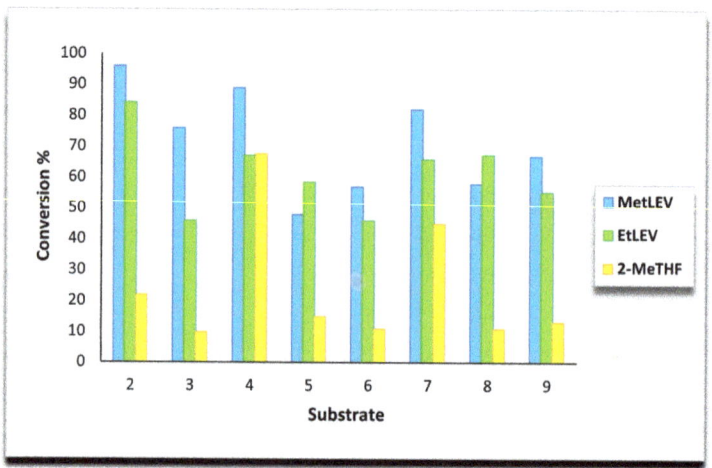

Figure 4. Palladium-catalyzed aminocarbonylation of para-substituted iodobenzene derivatives (4-iodo-phenole, 4-iodoanisole, 4-iodotoluene, 1-fluoro-4-iodobenzene, 1-bromo-4-iodobenzene, methyl 4-iodo-benzoate, 4-iodobenzotrifluoride, and 4-iodobenzonitrile). (Reaction conditions: 0.5 mmol of para-substituted iodobenzene, 0.75 mmol of morpholine, 0.0125 mmol of Pd(OAc)$_2$, 0.0125 mmol of XantPhos, 0.25 mL of triethylamine, and 5 mL of solvent under a CO atmosphere for 2 h. The conversion was determined by GC using dodecane as an internal standard.)

In the continuation, ortho/meta-mono-substituted (**10–12**) and di- and tri-substituted (**13, 14**) iodobenzene derivatives were examined. Compared to para-substituted substrates, the ortho-, di-, and tri-substituted compounds (**10, 13, 14**) provided lower conversions, likely due to steric reasons (Figure 5).

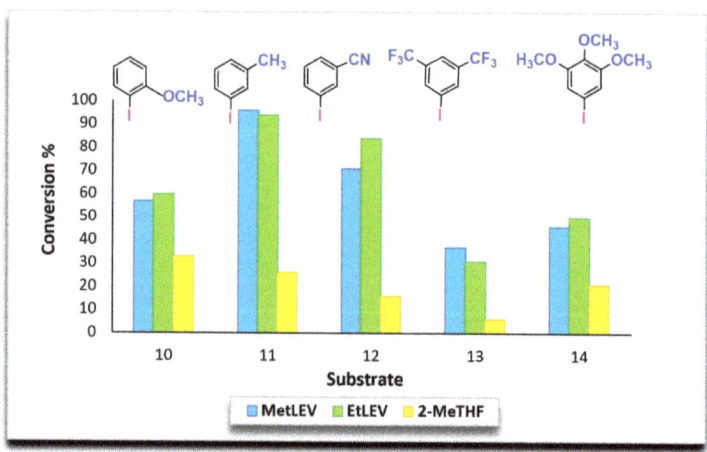

Figure 5. Pd-catalyzed aminocarbonylation of substituted iodobenzenes (**10–14**) with morpholine (**a**). (Reaction conditions: 0.5 mmol of substrate, 0.75 mmol of morpholine, 0.0125 mmol of Pd(OAc)$_2$, 0.0125 mmol of XantPhos, 0.25 mL of triethylamine, and 5 mL of solvent under CO atmosphere in 6 h. The conversion was determined by GC using dodecane as an internal standard.)

The 3-iodotoluene (**11**) (96 and 94% conversion) and 3-iodobenzonitrile (**12**) (71 and 84% conversion) showed high activity in both MetLEV and EtLEV solvents after 6 h reactions, while 2-iodoanisole (**10**) was converted in moderate yields (57–60%), due to the steric hindrance of the methoxy moiety in the ortho position. Substrate **13**, which had trifluoromethyl substituents in ortho- and para-positions, showed less reactivity (32–37%) in alkyl levulinate solvents, while the tri-substituted 5-iodo-1,2,3-trimethoxybenzene (**14**) was slightly more reactive than 1-iodo-3,5-bis(trifluoromethyl)benzene (**13**) during the 6 h reaction. Furthermore, it was demonstrated that substrates **11**, **12**, and **14** were converted completely, in contrast with substrates **10** and **13**, which gave 80–93% conversions in alkyl levulinates after a 24 h reaction. Considering our result in the aminocarbonylation reaction with substrates **11–14** and morpholine in 2-MeTHF, it can be easily seen that each starting material showed very poor reactivity (<33%) after 6 h. Nonetheless, substrate **11** had a complete conversion in 24 h, while substrates **10**, **12**, and **14** were converted by 58–65% in one-day reactions. Only substrate **13** was converted slowly, resulting in the (2,4 bis(trifluoromethyl)phenyl) (morpholino)methanone (**13aa**) product being detectable in traces (16%) after 24 h. Consequently, we can underline that alkyl levulinate-type solvents could be more appropriate solvents than 2-MeTHF in the aminocarbonylation reaction in the presence of ortho/meta-mono-substituted (**10–12**) and di/tri-substituted (**13**, **14**) iodobenzene derivatives.

In the last part of the study, some iodo-heteroaromatic substrates (**15–19**) were reacted with morpholine under optimized conditions. Then, their reactivities were investigated and compared in the green solvents above (Figure 6).

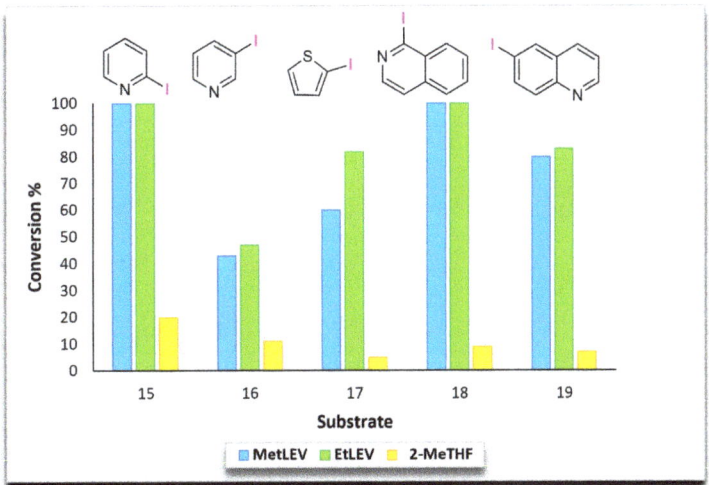

Figure 6. Pd-catalyzed aminocarbonylation of iodoheteroaromatic substrates (**15–19**) with morpholine (**a**). (Reaction conditions: 0.5 mmol of substrate, 0.75 mmol of morpholine, 0.0125 mmol of Pd(OAc)$_2$, 0.0125 mmol of XantPhos, 0.25 mL of triethylamine, and 5 mL of solvent under CO atmosphere in 2 h. The conversion was determined by GC using dodecane as an internal standard.)

2 Iodopyridine (**15**) and 1-iodoisoquinoline (**18**), in which the iodo-functionality is adjacent to the N-atom in the ring, were completely converted into the target carboxamide (**15aa**, **18aa**) in methyl- and ethyl levulinate after a 2 h reaction. 2-Iodothiophene (**17**) and 6-iodoquinoline (**19**) also showed good conversions (60–83%), while 3-iodopyridine (**16**) was the least reactive (43–47%) among them. While substrates **17** and **19** were converted completely in 6 h, 3-iodopyridine (**16**) showed 88% and 92% conversions in MetLEV and EtLEV, respectively. It is clearly seen in Figure 6 that the substrates unambiguously showed much higher reactivity in the alkyl levulinate-type solvents than in 2-MeTHF after 2 h of reaction time; conversion values lower than 20% were detected by GC analysis. The most reactive 2-iodopyridine (**15**) provided the target morpholino(pyridin-2-yl)methanone (**15aa**)

compound with 82% conversion while in the reaction mixture, even though the conversions were just 7–36%, in the case of substrates **16–19**, after 6 h in 2-MeTHF.

Because the conversions, especially in alkyl levulinates, are comparable to the reactions performed in conventional solvents, and the isolated yields in the case of compound **1aa** are similar to the results found in literature, the elementary steps of the catalytic cycle should be the same as in conventional solvents. Considering this concept, the following mechanism could be proposed. The 'starting' palladium(0) complex (Pd(CO)L_n, **A**) formed by reduction, reacts with the organic halide substrate in an oxidative addition, affording the palladium(II)-aryl intermediate (**B**). It is followed by the insertion of carbon monoxide into the palladium–carbon (Pd-Ar) bond, resulting in the corresponding palladium(II)-acyl complex (**C**). The next step is the nucleophile (NuH) attack on the species **C**, giving the catalytic intermediate **D**. It is followed by the HI elimination in the presence of the base (Et_3N) from the complex **D**, providing the amido-acyl-palladium(II) species (**E**). The last step is the reductive elimination, in which the carboxamide (**F**) is formed, and the palladium(0) species is regenerated (Scheme 3).

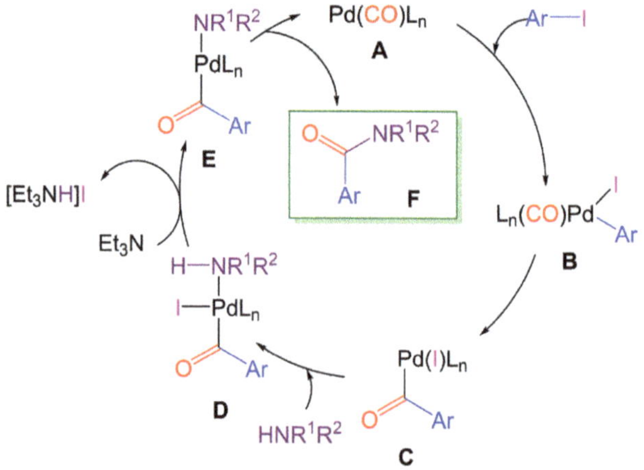

Scheme 3. Proposed mechanism of palladium-catalyzed aminocarbonylation in bio-derived solvents.

3. Materials and Methods

3.1. Compounds and Solvents

Solvents (ethyl levulinate, methyl levulinate, 2-MeTHF, and dichloromethane), substrates (**1–19**), nucleophiles (**a–h**), and triethylamine (Et_3N) were purchased from Sigma-Aldrich (St. Louis, MO, USA), and they were used without any further purification. The Pd(OAc)$_2$ and the ligands (PPh$_3$ and XantPhos) were also purchased from Sigma-Aldrich. TLC plates (silica gel on TLC Al foils with fluorescence indicators of 254 nm) and the silica gel (high-purity grade, average pore size 60 Å (52–73 Å), 70–230 mesh) were purchased from Sigma-Aldrich (St. Louis, MO, USA).

3.2. Aminocarbonylation Reaction under Atmospheric Pressure of CO

Pd(OAc)$_2$ (2.8 mg, 0.0125 mmol), PPh$_3$ (6.5 mg, 0.025 mmol), or XantPhos (7.2 mg, 0.0125 mmol) were measured in a 100 mL three-necked flask equipped with a reflux condenser connected to a balloon filled with argon. An amount of 0.5 mmol of substrate, one of the amine nucleophiles (0.75 mmol of 4-picolylamine, cyclohexylamine, pyrrolidine, and morpholine; 0.55 mmol of ProOMe·HCl and AlaOMe·HCl; 1.0 mmol of aniline; and 1.5 mmol of ^{t}BuNH$_2$), 0.25 mL of triethylamine, 0.25 mmol of internal standard (dodecane), and 5 mL of the green solvent were added under argon. Then the balloon was vacuumed and filled with carbon monoxide. The reaction mixture was stirred at 600 rpm in an oil bath at 50 °C for 24 h. The conversion was determined by GC measurements after 2, 6, and

24 h, and then the reaction was stopped and filtered. All the carboxamides, synthesized in the aminocarbonylation reactions, have already been described and characterized in previous literature. Due to this reason and environmental considerations, the isolation of the products was not performed. We performed the workup and isolation processes only in the case of our model aminocarbonylation reaction, which was performed in the presence of XantPhos; the solvents were removed under reduced pressure, and the crud was purified using a column chromatography with the carefully chosen $CHCl_3$: EtOAc = 8:2 eluent.

3.3. Aminocarbonylation Reaction under High Pressure of CO

$Pd(OAc)_2$ (2.8 mg, 0.0125 mmol), PPh_3 (6.5 mg, 0.025 mmol), or XantPhos (7.2 mg, 0.0125 mmol) were measured into a 100 mL stainless steel autoclave. The reagents and solvents, which were mentioned in Chapter 2.2, were transferred under argon. The reaction vessel was pressurized up to 40 bar total pressure with carbon monoxide, and the reaction mixture was stirred at 600 rpm in an oil bath at 50 °C for 24 h. After the given time, the reaction was stopped, and the autoclave was carefully depressurized in a well-ventilated hood. Then, the reaction mixture was filtered and analyzed by GC measurements after 2, 6, and 24 h.

3.4. Analytical Measurements

The reaction mixtures were analyzed by gas chromatography (Shimadzu Nexis GC-2030, Tokyo, Japan; Agilent J&W GC Column, DB-1MS stationary phase with automatic injection or DB-5MS stationary phase for ethyl levulinate and 1-iodobenzene with manual injection) using the following parameters: injector temperature: 250 °C; oven initial temperature: 50 °C (holding time: 1 min.); heating rate: 15 °C/min; final temperature: 320 °C (holding time: 11 min.); detector temperature: 280 °C; and carrier gas: helium (1 mL/min). The conversion and selectivity of the reactions were determined by GC. Unless otherwise stated, the conversion was checked with the internal standard method using dodecane. Mass spectrometry data were recorded using a GC–MS-QP2020 system (Shimadzu, Tokyo, Japan) with electron spray ionization (ESI) to identify the amides formed in the reactions (See Supplementary materials). The data are given as mass unit per charge (m/z), and the intensities are given in brackets. These data sets can be found in the Supplementary file.

4. Conclusions

In conclusion, we focused our attention on selecting environmentally friendly and greener solvents, which are commercially available and quite new, to facilitate the development of palladium-catalyzed homogeneous aminocarbonylation reactions. After the detailed optimization study, it has been shown that the $Pd(OAc)_2$/XantPhos catalyst system has great activity in the aminocarbonylation reaction of iodobenzene and morpholine, and the substrate converted completely and selectively towards the target carboxamide. By performing our model reaction in a conventional solvent, we demonstrated that the results observed in DMF and in the bio-derived reaction media are similar, justifying the applicability of our green solvents in aminocarbonylation. The appropriate conditions were chosen to extend the scope of amine nucleophiles with eight candidates, as well as nineteen different substrates (various *para*-, *ortho*-, and *meta*-substituted iodobenzene derivatives and iodo-heteroarenes). Considering our results, it can be unambiguously stated that the methyl- and ethyl levulinate are much more effective than the 2-MeTHF under similar conditions. Consequently, the 'alkyl levulinate'-type solvents could be used as alternative solvents for palladium-catalyzed aminocarbonylation reactions, opening a greener procedure for this synthetically relevant transformation.

Supplementary Materials: The following supporting information can be downloaded at: https://www.mdpi.com/article/10.3390/molecules28010442/s1, Table S1: Optimization study of the aminocarbonylation of iodobenzene (1) with morpholine (a); Table S2: Palladium-catalyzed aminocarbonylation of iodobenzene (1) with different primary and secondary amines (b–h); Table S3: Palladium-catalyzed aminocarbonylation of para-substituted iodobenzene derivatives (2–9) with morpholine (a);

Table S4: Palladium-catalyzed aminocarbonylation of substituted iodobenzenes (**10–14**) with morpholine (a); Table S5: Palladium-catalyzed aminocarbonylation of iodoheteroaromatic substrates (**15–19**) with morpholine (a); Table S6: Palladium-catalyzed aminocarbonylation of iodobenzene (**1**) with different primary and secondary amines (**a–h**); Table S7: Palladium-catalyzed aminocarbonylation of para-substituted iodobenzene derivatives (**2–9**) with morpholine (a); Table S8: Palladium-catalyzed aminocarbonylation of substituted iodobenzenes (**10–14**) with morpholine (a); Table S9: Palladium-catalyzed aminocarbonylation of iodoheteroaromatic substrates (**15–19**) with morpholine (a).

Author Contributions: Conceptualization, L.K. and A.T.; methodology, N.U. and A.T.; data curation, N.U. and A.T.; writing—original draft preparation, A.T. and N.U.; writing—review and editing, P.P., L.K., N.U. and A.T. All authors have read and agreed to the published version of the manuscript.

Funding: The research in Hungary was funded by NKFIH within the framework of the project TKP2021-EGA-17. This research was financed by the National Research, Development, and Innovation Office (Grant Number PD-132403). Project no. TKP2020-4.1.1 has been implemented with the support provided from the National Research, Development, and Innovation Fund of Hungary, financed under the Thematic Excellence Program 2020 National Excellence Sub-program.

Institutional Review Board Statement: Not applicable.

Informed Consent Statement: Not applicable.

Data Availability Statement: The data presented in this study are available in the Supplementary Material.

Acknowledgments: The authors are grateful to Gábor Mikle for the GC-MS measurements.

Conflicts of Interest: The authors declare no conflict of interest.

Sample Availability: Samples of the compounds are not available from the authors.

References

1. Clark, J.H. Green and Sustainable Chemistry: An Introduction. In *Green and Sustainable Medicinal Chemistry: Methods, Tools and Strategies for the 21st Century Pharmaceutical Industry*; Summerton, L., Sneddon, H.F., Jones, L.C., Clark, J.H., Eds.; The Royal Society of Chemistry: London, UK, 2016; p. 5. Available online: www.rsc.org (accessed on 1 October 2016).
2. Clark, J.H.; Farmer, T.J.; Hunt, A.J.; Sherwood, J. Opportunities for Bio-Based Solvents Created as Petrochemical and Fuel Products Transition towards Renewable Resources. *Int. J. Mol. Sci.* **2015**, *16*, 17101–17159. [CrossRef] [PubMed]
3. Global Solvents Industry. Available online: https://www.globenewswire.com/news-release/2020/07/10/2060660/0/en/Global-Solvents-Industry.html (accessed on 28 December 2022).
4. Solvents Market by Type (Alcohols, Ketones, Esters, Glycol Ethers, Aromatic, Aliphatic), Application (Paints Coatings, Polymer Manufacturing, Printing Inks), and Region-Global Forecast to 2025. Available online: https://www.marketsandmarkets.com/Market-Reports/solvent-market-1325.html (accessed on 28 December 2022).
5. Lomba, L.; Muñiz, S.; Pino, M.R.; Navarro, E.; Giner, B. Ecotoxicity studies of the levulinate ester series. *Ecotoxicology* **2014**, *23*, 1484–1493. [CrossRef] [PubMed]
6. Ventura, S.P.M.; de Morais, P.; Coelho, J.A.S.; Sintra, T.; Coutinho, J.A.P.; Afonso, C.A.M. Evaluating the toxicity of biomass derived platform chemicals. *Green Chem.* **2016**, *18*, 4733–4742. [CrossRef]
7. Jordan, A.; Hall, C.G.J.; Thorp, L.R.; Sneddon, H.F. Replacement of Less-Preferred Dipolar Aprotic and Ethereal Solvents in Synthetic Organic Chemistry with More Sustainable Alternatives. *Chem. Rev.* **2022**, *122*, 6749–6794. [CrossRef] [PubMed]
8. Pace, V.; Hoyos, P.; Castoldi, L.; De María, P.D.; Alcántara, A.R. 2-Methyltetrahydrofuran (2-MeTHF): A Biomass-Derived Solvent with Broad Application in Organic Chemistry. *Chemsuschem* **2012**, *5*, 1369–1379. [CrossRef]
9. Bijoy, R.; Agarwala, P.; Roy, L.; Thorat, B.N. Unconventional Ethereal Solvents in Organic Chemistry: A Perspective on Applications of 2-Methyltetrahydrofuran, Cyclopentyl Methyl Ether, and 4-Methyltetrahydropyran. *Org. Process. Res. Dev.* **2021**, *26*, 480–492. [CrossRef]
10. Alfonsi, K.; Colberg, J.; Dunn, P.J.; Fevig, T.; Jennings, S.; Johnson, T.A.; Kleine, H.P.; Knight, C.; Nagy, M.A.; Perry, D.A.; et al. Green chemistry tools to influence a medicinal chemistry and research chemistry based organisation. *Green Chem.* **2007**, *10*, 31–36. [CrossRef]
11. Byrne, F.P.; Jin, S.; Paggiola, G.; Petchey, T.H.M.; Clark, J.H.; Farmer, T.J.; Hunt, A.J.; McElroy, C.R.; Sherwood, J. Tools and techniques for solvent selection: Green solvent selection guides. *Sustain. Chem. Process.* **2016**, *4*, 1. [CrossRef]
12. Prat, D.; Pardigon, O.; Flemming, H.-W.; Letestu, S.; Ducandas, V.; Isnard, P.; Guntrum, E.; Senac, T.; Ruisseau, S.; Cruciani, P.; et al. Sanofi's Solvent Selection Guide: A Step Toward More Sustainable Processes. *Org. Process. Res. Dev.* **2013**, *17*, 1517–1525. [CrossRef]

13. Prat, D.; Wells, A.; Hayler, J.; Sneddon, H.; McElroy, C.R.; Abou-Shehada, S.; Dunn, P.J. CHEM21 Selection Guide of Classical- and Less Classical-solvents. *Green Chem.* **2016**, *18*, 288–296. [CrossRef]
14. Capello, C.; Fischer, U.; Hungerbühler, K. What is a green solvent? A comprehensive framework for the environmental assessment of solvents. *Green Chem.* **2007**, *9*, 927–934. [CrossRef]
15. Anastas, P.T.; Warner, J.C. Green Chemistry: Theory and Practice. Available online: https://www.acs.org/content/acs/en/greenchemistry/principles/12-principles-of-green-chemistry.html (accessed on 10 October 2022).
16. Lei, Y.; Wan, Y.; Li, G.; Zhou, X.-Y.; Gu, Y.; Feng, J.; Wang, R. Palladium supported on an amphiphilic porous organic polymer: A highly efficient catalyst for aminocarbonylation reactions in water. *Mater. Chem. Front.* **2017**, *1*, 1541–1549. [CrossRef]
17. Qureshi, Z.S.; Revankar, S.A.; Khedkar, M.V.; Bhanage, B. Aminocarbonylation of aryl iodides with primary and secondary amines in aqueous medium using polymer supported palladium-N-heterocyclic carbene complex as an efficient and heterogeneous recyclable catalyst. *Catal. Today* **2012**, *198*, 148–153. [CrossRef]
18. Khedkar, M.V.; Sasaki, T.; Bhanage, B.M. Immobilized Palladium Metal-Containing Ionic Liquid-Catalyzed Alkoxycarbonylation, Phenoxycarbonylation, and Aminocarbonylation Reactions. *ACS Catal.* **2013**, *3*, 287–293. [CrossRef]
19. Wójcik, P.; Rosar, V.; Gniewek, A.; Milani, B.; Trzeciak, A. In situ generated Pd(0) nanoparticles stabilized by bis(aryl)acenaphthenequinone diimines as catalysts for aminocarbonylation reactions in water. *J. Mol. Catal. A Chem.* **2016**, *425*, 322–331. [CrossRef]
20. Suzuka, T.; Sueyoshi, H.; Ogihara, K. Recyclable Polymer-Supported Terpyridine–Palladium Complex for the Tandem Aminocarbonylation of Aryl Iodides to Primary Amides in Water Using NaN3 as Ammonia Equivalent. *Catalysts* **2017**, *7*, 107. [CrossRef]
21. Mart, M.; Tylus, W.; Trzeciak, A. Pd/DNA as a highly active and recyclable catalyst for aminocarbonylation and hydroxycarbonylation in water: The effect of Mo(CO)6 on the reaction course. *Mol. Catal.* **2019**, *462*, 28–36. [CrossRef]
22. Pace, V.; Castoldi, L.; Alcántara, A.R.; Holzer, W. Highly efficient and environmentally benign preparation of Weinreb amides in the biphasic system 2-MeTHF/water. *RSC Adv.* **2013**, *3*, 10158–10162. [CrossRef]
23. Wu, X.; Ekegren, J.K.; Larhed, M. Microwave-Promoted Aminocarbonylation of Aryl Iodides, Aryl Bromides, and Aryl Chlorides in Water. *Organometallics* **2006**, *25*, 1434–1439. [CrossRef]
24. Bhanage, B.M.; Tambade, P.J.; Patil, Y.P.; Bhanushali, M.J. Pd(OAc)2-Catalyzed Aminocarbonylation of Aryl Iodides with Aromatic or Aliphatic Amines in Water. *Synthesis* **2008**, *2008*, 2347–2352. [CrossRef]
25. Tambade, P.J.; Patil, Y.P.; Qureshi, Z.S.; Dhake, K.P.; Bhanage, B.M. Pd(OAc)$_2$-Catalyzed Carbonylative Coupling of Aryl Iodide with Ortho-Haloamines in Water. *Synth. Commun.* **2011**, *42*, 176–185. [CrossRef]
26. Ács, P.; Takács, A.; Szilágyi, A.; Wölfling, J.; Schneider, G.; Kollár, L. The synthesis of 13α-androsta-5,16-diene derivatives with carboxylic acid, ester and carboxamido functionalities at position-17 via palladium-catalyzed carbonylation. *Steroids* **2009**, *74*, 419–423. [CrossRef]
27. Beller, M.; Wu, X.F. *Transition Metal Catalyzed Carbonylation Reactions*; Springer: Berlin/Heidelberg, Germany, 2013; p. 1.
28. Santoro, S.; Ferlin, F.; Luciani, L.; Ackermann, L.; Vaccaro, L. Biomass-derived solvents as effective media for cross-coupling reactions and C–H functionalization processes. *Green Chem.* **2017**, *19*, 1601–1612. [CrossRef]
29. Takács, A.; Farkas, R.; Kollár, L. High-yielding synthesis of 2-arylacrylamides via homogeneous catalytic aminocarbonylation of α-iodostyrene and α,α'-diiodo-1,4-divinylbenzene. *Tetrahedron* **2008**, *64*, 61–66. [CrossRef]
30. Takács, A.; Farkas, R.; Petz, A.; Kollár, L. Synthesis of 2-naphthylacrylamides and 2-naphthylacrylates via homogeneous catalytic carbonylation of 1-iodo-1-naphthylethene derivatives. *Tetrahedron* **2009**, *65*, 4795–4800. [CrossRef]
31. Gergely, M.; Farkas, R.; Takács, A.; Petz, A.; Kollár, L. Synthesis of N-picolylcarboxamides via palladium-catalysed aminocarbonylation of iodobenzene and iodoalkenes. *Tetrahedron* **2014**, *70*, 218–224. [CrossRef]
32. Ács, P.; Takács, A.; Kiss, M.; Pálinkás, N.; Mahó, S.; Kollár, L. Systematic investigation on the synthesis of androstane-based 3-, 11- and 17-carboxamides via palladium-catalyzed aminocarbonylation. *Steroids* **2011**, *76*, 280–290. [CrossRef]
33. Takács, A.; Ács, P.; Farkas, R.; Kokotos, G.; Kollár, L. Homogeneous catalytic aminocarbonylation of 1-iodo-1-dodecene. The facile synthesis of odd-number carboxamides via palladium-catalysed aminocarbonylation. *Tetrahedron* **2008**, *64*, 9874–9878. [CrossRef]
34. Takács, A.; Ács, P.; Berente, Z.; Wölfling, J.; Schneider, G.; Kollár, L. Novel 13β- and 13α-d-homo steroids: 17a-carboxamido-d-homoestra-1,3,5(10),17-tetraene derivatives via palladium-catalyzed aminocarbonylations. *Steroids* **2010**, *75*, 1075–1081. [CrossRef]
35. Kégl, T.R.; Mika, L.T.; Kégl, T. 27 Years of Catalytic Carbonylative Coupling Reactions in Hungary (1994–2021). *Molecules* **2022**, *27*, 460. [CrossRef]
36. Mikle, G.; Bede, F.; Kollár, L. Synthesis of N-picolylcarboxamides in aminocarbonylation. *Tetrahedron* **2021**, *88*, 132128. [CrossRef]
37. Szuroczki, P.; Boros, B.; Kollár, L. Efficient synthesis of alkynyl amides via aminocarbonylation of iodoalkynes. *Tetrahedron* **2018**, *74*, 6129–6136. [CrossRef]
38. Gergely, M.; Kollár, L. Aminothiazoles and aminothiadiazoles as nucleophiles in aminocarbonylation of iodobenzene derivatives. *Tetrahedron* **2018**, *74*, 2030–2040. [CrossRef]
39. Mikle, G.; Skoda-Földes, R.; Kollár, L. Amino- and azidocarbonylation of iodoalkenes. *Tetrahedron* **2021**, *100*, 132495. [CrossRef]
40. Ismael, A.; Gevorgyan, A.; Skrydstrup, T.; Bayer, A. Renewable Solvents for Palladium-Catalyzed Carbonylation Reactions. *Org. Process. Res. Dev.* **2020**, *24*, 2665–2675. [CrossRef]
41. Fodor, D.; Kégl, T.; Tukacs, J.M.; Horváth, A.K.; Mika, L.T. Homogeneous Pd-Catalyzed Heck Coupling in γ-Valerolactone as a Green Reaction Medium: A Catalytic, Kinetic, and Computational Study. *ACS Sustain. Chem. Eng.* **2020**, *8*, 9926–9936. [CrossRef]
42. Marosvölgyi-Haskó, D.; Lengyel, B.; Tukacs, J.M.; Kollár, L.; Mika, L.T. Application of γ-Valerolactone as an Alternative Biomass-Based Medium for Aminocarbonylation Reactions. *ChemPlusChem* **2016**, *81*, 1224–1229. [CrossRef]

43. Tukacs, J.M.; Marton, B.; Albert, E.; Tóth, I.; Mika, L.T. Palladium-catalyzed aryloxy- and alkoxycarbonylation of aromatic iodides in γ-valerolactone as bio-based solvent. *J. Organomet. Chem.* **2020**, *923*, 121407. [CrossRef]
44. Ojima, I.; Commandeur, C.; Chiou, W.-H. Amidocarbonylation, Cyclohydrocarbonylation and Related Reactions. In *Comprehensive Organic Chemistry III, (COMC-III)*, 3rd ed.; Micheal, D., Mingos, P., Crabtree, R.H., Eds.; Elsevier Ltd.: Amsterdam, The Netherlands, 2007; Volume 13, pp. 511–555.
45. Tomé, V.A.; Calvete, M.J.F.; Vinagreiro, C.S.; Aroso, R.T.; Pereira, M.M. A New Tool in the Quest for Biocompatible Phthalocyanines: Palladium Catalyzed Aminocarbonylation for Amide Substituted Phthalonitriles and Illustrative Phthalocyanines Thereof. *Catalysts* **2018**, *8*, 480. [CrossRef]
46. Bousfield, T.W.; Pearce, K.P.R.; Nyamini, S.B.; Angelis-Dimakis, A.; Camp, J.E. Synthesis of amides from acid chlorides and amines in the bio-based solvent Cyrene™. *Green Chem.* **2019**, *21*, 3675–3681. [CrossRef]
47. Yan, K.; Yang, Y.; Chai, J.; Lu, Y. Catalytic reactions of gamma-valerolactone: A platform to fuels and value-added chemicals. *Appl. Catal. B Environ.* **2015**, *179*, 292–304. [CrossRef]
48. Pastore, C.; D'Ambrosio, V. Intensification of Processes for the Production of Ethyl Levulinate Using $AlCl_3·6H_2O$. *Energies* **2021**, *14*, 1273. [CrossRef]
49. Lei, P.; Mu, Y.; Wang, Y.; Wang, Y.; Ma, Z.; Feng, J.; Liu, X.; Szostak, M. Green Solvent Selection for Suzuki–Miyaura Coupling of Amides. *ACS Sustain. Chem. Eng.* **2020**, *9*, 552–559. [CrossRef]
50. Yan, K.; Jarvis, C.; Gu, J.; Yan, Y. Production and catalytic transformation of levulinic acid: A platform for speciality chemicals and fuels. *Renew. Sustain. Energy Rev.* **2015**, *51*, 986–997. [CrossRef]
51. Gao, W.; Wu, G.; Zhu, X.; Akhtar, M.A.; Lin, G.; Huang, Y.; Zhang, S.; Zhang, H. Production of methyl levulinate from cellulose over cobalt disulfide: The importance of the crystal facet (111). *Bioresour. Technol.* **2021**, *347*, 126436. [CrossRef] [PubMed]
52. Silva, J.F.L.; Mariano, A.P.; Filho, R.M. Economic potential of 2-methyltetrahydrofuran (MTHF) and ethyl levulinate (EL) produced from hemicelluloses-derived furfural. *Biomass- Bioenergy* **2018**, *119*, 492–502. [CrossRef]
53. Xu, W.; Chen, X.; Guo, H.; Li, H.; Zhang, H.; Xiong, L.; Chen, X. Conversion of levulinic acid to valuable chemicals: A review. *J. Chem. Technol. Biotechnol.* **2021**, *96*, 3009–3024. [CrossRef]
54. Sherwood, J.; Clark, J.H.; Fairlamb, I.J.S.; Slattery, J.M. Solvent effects in palladium catalysed cross-coupling reactions. *Green Chem.* **2019**, *21*, 2164–2213. [CrossRef]
55. Hao, W.; Xu, Z.; Zhou, Z.; Cai, M. Recyclable Heterogeneous Palladium-Catalyzed Cyclocarbonylation of 2-Iodoanilines with Acyl Chlorides in the Biomass-Derived Solvent 2-Methyltetrahydrofuran. *J. Org. Chem.* **2020**, *85*, 8522–8532. [CrossRef]
56. Kadam, A.; Nguyen, M.; Kopach, M.; Richardson, P.; Gallou, F.; Wan, Z.-K.; Zhang, W. Comparative performance evaluation and systematic screening of solvents in a range of Grignard reactions. *Green Chem.* **2013**, *15*, 1880–1888. [CrossRef]
57. Mondal, M.; Bora, U. Eco-friendly Suzuki–Miyaura coupling of arylboronic acids to aromatic ketones catalyzed by the oxime-palladacycle in biosolvent 2-MeTHF. *New J. Chem.* **2016**, *40*, 3119–3123. [CrossRef]
58. Santoro, S.; Marrocchi, A.; Lanari, D.; Ackermann, L.; Vaccaro, L. Towards Sustainable C−H Functionalization Reactions: The Emerging Role of Bio-Based Reaction Media. *Chem.-A Eur. J.* **2018**, *24*, 13383–13390. [CrossRef] [PubMed]
59. Přibylka, A.; Krchňák, V.; Schützenerová, E. Environmentally Friendly SPPS II: Scope of Green Fmoc Removal Protocol Using NaOH and Its Application for Synthesis of Commercial Drug Triptorelin. *J. Org. Chem.* **2020**, *85*, 8798–8811. [CrossRef] [PubMed]
60. Monticelli, S.; Castoldi, L.; Murgia, I.; Senatore, R.; Mazzeo, E.; Wackerlig, J.; Urban, E.; Langer, T.; Pace, V. Recent advancements on the use of 2-methyltetrahydrofuran in organometallic chemistry. 2016, 148, 37–48. [CrossRef]
61. Sardana, M.; Bergman, J.; Ericsson, C.; Kingston, L.P.; Schou, S.C.; Dugave, C.; Audisio, D.; Elmore, C.S. Visible-Light-Enabled Aminocarbonylation of Unactivated Alkyl Iodides with Stoichiometric Carbon Monoxide for Application on Late-Stage Carbon Isotope Labeling. *J. Org. Chem.* **2019**, *84*, 16076–16085. [CrossRef] [PubMed]
62. Barré, A.; Ţînţaş, M.-L.; Alix, F.; Gembus, V.; Papamicaël, C.; Levacher, V. Palladium-Catalyzed Carbonylation of (Hetero)Aryl, Alkenyl and Allyl Halides by Means of N-Hydroxysuccinimidyl Formate as CO Surrogate. *J. Org. Chem.* **2015**, *80*, 6537–6544. [CrossRef] [PubMed]
63. Nathel, N.F.F.; Kim, J.; Hie, L.; Jiang, X.; Garg, N.K. Nickel-Catalyzed Amination of Aryl Chlorides and Sulfamates in 2-Methyl-THF. *ACS Catal.* **2014**, *4*, 3289–3293. [CrossRef]
64. Skoda-Foldes, R. Palladium-Catalyzed Aminocarbonylation of Iodoalkenes and Iodoarenes. *Lett. Org. Chem.* **2010**, *7*, 621–633. [CrossRef]
65. Marosvölgyi-Haskó, D.; Kégl, T.; Kollár, L. Substituent effects in aminocarbonylation of para -substituted iodobenzenes. *Tetrahedron* **2016**, *72*, 7509–7516. [CrossRef]
66. Takacs, A.; Petz, A.; Jakab, B.; Kollár, L. Aminocarbonylation of 2-Iodothiophene: High-Yielding Synthesis of Thiophen-2-yl-glyoxylamides. *Lett. Org. Chem.* **2007**, *4*, 590–594. [CrossRef]
67. Takács, A.; Jakab, B.; Petz, A.; Kollár, L. Homogeneous catalytic aminocarbonylation of nitrogen-containing iodo-heteroaromatics. Synthesis of N-substituted nicotinamide related compounds. *Tetrahedron* **2007**, *63*, 10372–10378. [CrossRef]
68. Takács, A.; Marosvölgyi-Haskó, D.; Kabak-Solt, Z.; Damas, L.; Rodrigues, F.M.; Carrilho, R.M.; Pineiro, M.; Pereira, M.M.; Kollár, L. Functionalization of indole at C-5 or C-7 via palladium-catalysed double carbonylation. A facile synthesis of indole ketocarboxamides and carboxamide dimers. *Tetrahedron* **2016**, *72*, 247–256. [CrossRef]
69. Kollár, L.; Erdélyi, A.; Rasheed, H.; Takács, A. Selective Synthesis of N-Acylnortropane Derivatives in Palladium-Catalysed Aminocarbonylation. *Molecules* **2021**, *26*, 1813. [CrossRef] [PubMed]

70. Kollár, L.; Takács, A. Novel synthesis of 3-carboxamidolactam derivatives via palladium-catalysed aminocarbonylation. *Tetrahedron* **2018**, *74*, 6116–6128. [CrossRef]
71. Baburajan, P.; Elango, K.P. Co2(CO)8 as a convenient in situ CO source for the direct synthesis of benzamides from aryl halides (Br/I) via aminocarbonylation. *Tetrahedron Lett.* **2014**, *55*, 1006–1010. [CrossRef]
72. Gockel, S.N.; Hull, K.L. Chloroform as a Carbon Monoxide Precursor: *In* or *Ex Situ* Generation of CO for Pd-Catalyzed Aminocarbonylations. *Org. Lett.* **2015**, *17*, 3236–3239. [CrossRef]
73. Wang, P.; Yang, J.; Sun, K.; Neumann, H.; Beller, M. A general synthesis of aromatic amides *via* palladium-catalyzed direct aminocarbonylation of aryl chlorides. *Org. Chem. Front.* **2022**, *9*, 2491–2497. [CrossRef]

Disclaimer/Publisher's Note: The statements, opinions and data contained in all publications are solely those of the individual author(s) and contributor(s) and not of MDPI and/or the editor(s). MDPI and/or the editor(s) disclaim responsibility for any injury to people or property resulting from any ideas, methods, instructions or products referred to in the content.

Article

In Situ-Generated, Dispersed Cu Catalysts for the Catalytic Hydrogenolysis of Glycerol

Iuliana Porukova *, Vadim Samoilov, Dzhamalutdin Ramazanov, Mariia Kniazeva and Anton Maximov

A.V. Topchiev Institute of Petrochemical Synthesis, RAS, 29 Leninsky Prospect, 119991 Moscow, Russia
* Correspondence: porukova@ips.ac.ru

Abstract: The present study is dedicated to the experimental verification of a concept for the hydrogenolysis of glycerol over in situ-generated Cu dispersed particles (Cu-DP). The Cu-DP were generated by in situ reduction of a precursor salt ($Cu(OAc)_2$, $CuSO_4$, $CuCl_2$) in the presence of KOH and were active in glycerol conversion under hydrogen (T = 200–220 °C, $p(H_2)$ = 1–4 MPa), where 1,2-propylene glycol (PG) and lactic acid (LA) were detected to be the main products. The influence of the reaction conditions (temperature, hydrogen pressure, reaction time, catalyst-to-feed ratio and the KOH/Cu ratio) on the yields of the products is described. It was shown that the selectivity between the PG and LA could be tuned by changing $p(H_2)$ or by the KOH amount, i.e., higher yields of LA corresponded to lower $p(H_2)$ and higher alkalinity of the reaction media. The activity of the in situ-generated Cu-DP was found to be comparable to that of an industrial $Cu-Cr_2O_3$ catalyst. The Cu-DP catalysts were characterized by XRD, XPS, HRTEM and SEM. During the reaction, the catalyst evolved by the sintering and recrystallization of the separate Cu-DP; the crystallite sizes after 1 and 15 h reaction times amounted to 35 and 49 nm, respectively.

Keywords: dispersed copper particles; glycerol; propylene glycol; hydrogenolysis; heterogeneous catalysis

1. Introduction

The need to reduce environmental degradation and to transition to a low-carbon economy requires the search for new solutions to reduce the consumption of non-renewable carbonaceous raw materials. One such solution is the synthesis of the most important large-tonnage compounds traditionally obtained from raw mineral materials using components derived from renewable sources, i.e., so-called "petrochemical substitutes". For example, propylene glycol (1,2-propanediol) is a dihydric alcohol of great industrial importance that is currently obtained mainly from fossil-derived propylene oxide. Propylene glycol is used as a polyol in the synthesis of polyurethanes and polyesters, as a solvent and plasticizer, and as a base for technical fluids for various purposes, e.g., antifreezes, hydraulic fluids, water-based heat carriers and deicing fluids for aviation needs [1].

Due to the versatile applications of propylene glycol (PG), the development of technologies for its production from renewable raw materials is of real significance. The most promising renewable raw material for PG production is bioglycerol, the market of which experienced tenfold growth in the 2003–2012 period [2]. At present, it is known that a consortium of Oleon and BASF is producing bio-propylene glycol from bioglycerol [3]. At the same time, studies aimed at improving the technology for obtaining propylene glycol from glycerol continue in two main directions, i.e., the search for new methods of processing crude bioglycerol [4] and the development of improved process catalysts with increased activity, selectivity and stability [4–8].

Traditionally, heterogeneous catalysts based on precious metals (Ru, Rh and Pd [9–12]) and some transition metals (Co [13], Ni [14,15] and Cu [16–21]) have been used for hydrogenation and hydrogenolysis processes. The high cost of precious metals and the low selectivity of catalysts based on them for 1,2-propanediol limit their use in the production of propylene glycol, despite their high activity [22]. The low selectivity of hydrogenolysis is associated with side reactions involving the cleavage of C–C bonds in glycerol with the formation of ethylene glycol, ethanol and methanol, as is typical of both platinum group metals and cobalt and nickel.

In contrast to catalysts based on precious metals, in glycerol hydrogenolysis, copper-containing catalysts demonstrate high selectivity for the hydrogenation of C–O bonds but lower catalytic activity than precious metals. In particular, when using heterogeneous catalysts, such as Cu/Al_2O_3 [7,8] and Cu/ZnO [16,23,24], the glycerol conversion rate can be 70%, with propylene glycol selectivity of up to 90%. Other advantages of copper-based catalysts include their relative environmental friendliness and high availability. Due to the high selectivity for 1,2-PG and relative affordability of copper as the main component, copper-containing catalysts are extremely promising for industrial applications.

The main problem of copper-containing catalysts is their low activity. For example, at T = 200 °C and $p(H_2)$ = 200 psi, propylene glycol was produced on industrial catalysts 5%Pt/C and Raney copper at an equivalent weight catalyst-to-feed ratio, while the mass contents of the active components in these catalysts were 5% and 100%, respectively. Increasing the activity of copper-based catalysts may be achieved in two main ways. The first involves the promotion of copper by a second element. The most effective promotion (which is also used in the manufacture of industrial catalysts) is achieved by adding aluminum [4,18,25], chromium [5,17,26] and zinc oxides [16,18,23,24,27,28] to copper, although examples of adding Ni, Mg [21] or B [4,29] are also known. The second way to increase the activity of copper-based catalysts relies on the fact that their activity is proportional to the copper dispersion (which can be expressed in terms of the specific surface area of copper in the catalyst) [30]. In other words, catalysts with more dispersed copper particles have increased activity. The problem of using highly active heterogeneous catalysts is associated not only with their preparation but also with the configuration of the fixed-bed reactors used. The hydrogenolysis of glycerol is exothermic and, as such, the use of a fixed-bed reactor requires additional technology to control temperature gradients inside the apparatus.

An alternative approach to hydrogenolysis is dispersed-phase catalysis in a slurry reactor. The catalyst is a dispersed phase consisting of solid particles; when the average particle size approaches the sub-nanometer range in a slurry reactor, the diffusion inside the particles practically disappears and the process becomes quasi-homogeneous. To apply this approach, the following factors should be considered:

- The dispersion of catalyst particles affects the reaction rate;
- The thermal effect of the reaction complicates the use of a fixed-bed reactor;
- The catalytic dispersed phase can be formed in situ;
- The reaction medium is able to inhibit the coagulation of the particles;
- The reaction products have a boiling point lower than the raw material and can be separated from the reaction mass by distillation.

In the case of the hydrogenolysis reaction of glycerol to propylene glycol, most of these conditions are met. The polyol medium is an excellent stabilizer for dispersed metal particles and has given rise to the polyol process of obtaining dispersions of metal nanoparticles [31]. Copper is easily reduced to a metallic state from Cu^{2+} salts, and polyol itself can act as a reducing agent [32–35]; the resulting particles can have sizes in the tens of nanometers [36]. This, in turn, led us to consider the application of copper nanoparticles, the activity of which will be high in glycerol hydrogenolysis. The boiling point of propylene glycol is almost 100 °C lower than that of glycerol (190 and 290 °C, respectively), which makes it easy to separate the product by distillation. Finally, the use of a dispersed catalyst in a mixing reactor eliminates the problem of thermal gradients. Ex situ-derived copper nanoparticles were previously successfully applied as a catalyst for furfural hydrogenation [37]. For the hydrogenolysis of glycerol, Omar et al. proposed the use of Cu/ZnO nanoparticles that were also obtained by ex situ synthesis [24]. The aim of this study is to experimentally test our proposed concept for glycerol hydrogenolysis in the presence of in situ-formed, dispersed copper particles (Figure 1).

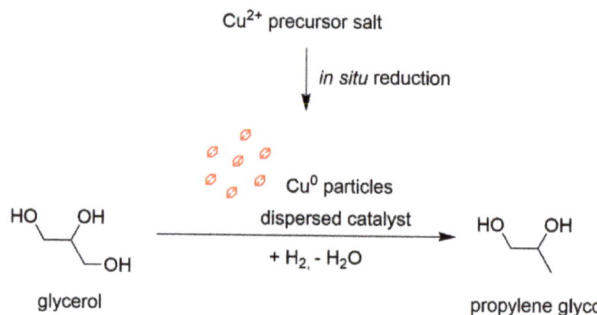

Figure 1. The concept of glycerol hydrogenolysis over in situ-generated ultrafine copper particles.

Achieving the goal requires the following:
- A description of the phenomenon of dispersed catalyst formation in situ in the reaction medium;
- Characterization of the catalytic activity of the resulting catalyst, including the dependence of product yields on the reaction conditions;
- Description of the structure and morphology of in situ-generated catalysts.

2. Results and Discussion

2.1. Phenomenon of Catalytic Activity in the Glycerol Hydrogenolysis Reaction

The initial hypothesis of our study implies that when a copper salt solution (Figure S5) is heated in glycerol, dispersed particles of metallic copper are formed, which, due to their catalytic activity, catalyze the glycerol hydrogenolysis reaction. Confirmation of this hypothesis should be based on the following observations:
- A heterogeneous catalyst must be present in the reaction mixture;
- The resulting reaction mixture must contain reaction products—in particular, the target PG;
- The conversion of glycerol and the yield of the desired products should increase with increasing temperature and reaction time;
- The conversion of glycerol and the yield of the desired products should change when the Gly/Cu ratio changes;
- Propylene glycol should be formed only in a hydrogen medium and only in the presence of a copper precursor salt.

The activity of copper-containing catalysts in glycerol hydrogenolysis is traditionally considered to be relatively low compared with that of catalysts based on platinum group metals. The results of experiments on glycerol hydrogenolysis using an industrial copper chromite catalyst (Table 1, entry 23–24) have generally confirmed this statement; at a ratio of Gly/Cu = 100 mol and T = 200 °C, the glycerol conversion for 5–10 h of reaction was 3.6–9.2%. Thus, to obtain the measured dynamics in the experiment, the catalyst must be 1–2% mol per substrate.

Table 1. Hydrogenolysis of glycerol in the presence of copper catalysts. Conditions: p(H$_2$) = 3 MPa, KOH/Cu = 5.8 mol, Gly/H$_2$O = 4.1 vol.

Entry	Precursor Salt	T, °C	Gly/Cu, mol	τ, h	X_{Gly}, %	Y_{EG}, %	Y_{PG}, %	Y_{LA}, %	Y_{GA}, %
1	-	220	-	5	2.2	0.1	0.2	2.0	-
2			50	5	15.0	0.4	7.0	7.0	0.6
3		200		10	15.2	0.5	8.9	5.0	0.8
4			100	5	7.0	0.2	4.0	2.5	0.3
5	CuSO$_4$·5H$_2$O			10	8.1	0.4	4.7	2.9	0.1
6			50	5	16.6	0.8	10.4	5.1	0.3
7		220		10	18.4	0.9	10.7	6.8	-
8			100	5	8.5	0.4	4.7	3.3	0.1
9				10	8.5	0.3	4.8	3.4	-
10			50	5	14.8	0.3	8.9	3.7	1.9
11		200		10	15.5	0.6	8.0	5.7	1.2
12			100	5	7.0	0.1	3.6	2.6	0.7
13	Cu(OAc)$_2$·H$_2$O			10	7.1	0.2	3.7	2.5	0.7
14			50	5	16.8	0.7	8.5	6.8	0.8
15		220		10	18.5	0.8	9.4	7.4	0.9
16			100	5	7.8	0.2	4.0	3.2	0.4
17				10	10.6	0.5	5.3	4.2	0.6
18		200		5	4.6	0.1	0.7	3.1	0.7
19	Nitrogen		100	5	4.4	0.1	0.7	2.8	0.8
20		220		10	4.8	0.1	0.8	3.3	0.6
21			50	5	7.4	0.1	6.0	1.3	-
22		200		10	10.6	0.1	10.1	0.4	-
23			100	5	3.6	-	3.6	-	-
24	Cu-Cr$_2$O$_3$ [1]			10	9.3	0.3	8.3	0.7	-
25			50	5	19.0	0.2	17.3	1.5	-
26		220		10	26.1	0.2	25.4	0.5	-
27			100	5	11.9	0.1	8.6	3.1	-
28				10	19.4	0.1	19.0	0.4	-
29	Cu-Cr$_2$O$_3$ + KOH	220	50	5	34.5	1.5	23.6	9.4	-
30	CuCl$_2$·2H$_2$O	200	50	5	11.0	0.3	5.8	4.3	0.6
31		220	50	5	16.2	0.5	6.7	8.3	0.7

[1] The copper chromite catalyst was used without KOH addition.

At this ratio, the precursor cannot be quickly dissolved in the initial water–glycerol mixture; therefore, to ensure homogenization, an excess of KOH was added to the solution with the precursor. In this case, the copper salts were completely dissolved, probably forming copper glycerate, as indicated by the cornflower-blue color of the solution.

When heating precursor solutions in glycerol at T = 200–220 °C and p(H$_2$) = 3 MPa, solution discoloration was observed with the simultaneous formation of a copper-colored solid phase. The analysis of reaction mixtures (Table 1) showed that these phenomena were accompanied by glycerol conversion with the formation of two main products, i.e., propylene glycol and lactic acid (or salt, depending on the pH of the reaction medium), and two by-products, i.e., ethylene glycol and glyceric acid (or salt, depending on the pH of the reaction medium). This was observed for all three tested precursors ($CuSO_4$, $Cu(OAc)_2$ and $CuCl_2$). With this in mind, the conversion of glycerol sharply decreased in the absence of a copper salt precursor, and propylene glycol was practically not formed at all. Thus, when heating glycerol with KOH at T = 220 °C (Table 1, entry 1), the glycerol conversion was 2.2%, of which, 2.0% accounted for Y_{LA} and only 0.2% for Y_{PG}. The same mixture with the addition of 2% mol (for glycerol) of copper acetate (Table 1, entry 14) was characterized by X_{Gly} = 16.8% with Y_{PG} = 8.5% and Y_{LA} = 6.8%. Thus, the formation of propylene glycol was associated with the presence of copper in the reaction mixture. It was also found that propylene glycol did not form when the reaction took place in a nitrogen atmosphere (Table 1, entry 18–20); the Y_{PG} under these conditions was only 0.7%. When the reaction was carried out under the same conditions in a hydrogen atmosphere ($p(H_2)$ = 3.0 MPa), the PG yield was 4.0% (Table 1, entry 16). These results confirm that a heterogeneous catalyst was formed from the precursor salt in the solution and that a catalytic reaction of hydrogenolysis subsequently occurred.

The conversion rate of glycerol and the yield of products varied depending on the reaction time and the catalyst-to-feed ratio. For example, in the presence of copper acetate at T = 200 and 220 °C (Table 1, entries 11 and 15), the X_{Gly} amounted to 15.5% and 18.5%, respectively. With a twofold increase in the catalyst-to-feed ratio, glycerol conversion also increased. For instance, at Gly/Cu = 100 mol and 50 mol, the X_{Gly} was 7.8% and 16.8%, respectively (Table 1, entries 16 and 14).

It is obvious that it is the alkaline nature of the reaction mixture that is responsible for the significant yield of lactic acid during the glycerol hydrogenolysis in the presence of in situ-formed copper particles. The co-catalytic performance of KOH was confirmed by a control test. The glycerol hydrogenolysis over the $Cu-Cr_2O_3$ catalyst resulted in X_{Gly} = 19.0%, Y_{PG} = 17.3% and Y_{LA} = 1.5% (Table 1, entry 25). When the fivefold molar amount of KOH was added to the $Cu-Cr_2O_3$ catalyst, the Y_{PG} was slightly increased, while the Y_{LA} rose from 1.5% to 9.4%, therefore contributing to the X_{Gly} surplus. Since the alkali acting as a co-catalyst regulates the reaction selectivity, the corresponding dependence deserves consideration.

Cu^{2+} reduction should be accompanied by partial neutralization of alkali according to the following equations:

$$Cu(OAc)_2 + H_2 = Cu^0 + 2HOAc;\ 2HOAc + 2KOH = 2KOAc + 2H_2O$$

Thus, at the ratio KOH/Cu = 2.0 mol, complete alkali neutralization should be observed during complete copper reduction. At lower KOH/Cu ratios in the initial reaction mixture, one can expect acidic medium formation; the values of KOH/Cu > 2.0 mol will correspond to the alkaline medium. As can be seen from the results obtained (Figure 2), the glycerol conversion and yields of products depended on the alkali amount in the initial reaction mixture. With a lack of alkali relative to stoichiometric amount (KOH/Cu < 2.0 mol), X_{Gly} did not exceed 3.5%, with a Y_{PG} of 0.8–1.1% and Y_{LA} of 0.6–1.1%. When the medium changed to a slightly alkaline one (KOH/Cu = 2.3 mol), the glycerol conversion increased sharply to 13.4%, and this increase was associated with an increase in the propylene glycol yield of up to 10.0%, while the lactic acid yield increased insignificantly (up to 1.7%). A further increase in the alkali amount and the medium transition to alkaline (KOH/Cu = 3.6–5.8 mol) changed the conversion selectivity at an almost constant conversion. At KOH/Cu = 3.6 mol, the glycerol conversion was 12.6%; furthermore, Y_{PG} = 6.3% and Y_{LA} = 4.7%. By-product (EG and GA) yields did not depend much on the alkali amount in the mixture. Thus, the highest selectivity of glycerol conversion in the presence of the in situ-formed copper catalyst was achieved in a slightly alkaline medium; when the dosage was increased, PG and LA alkalis were formed in comparable amounts.

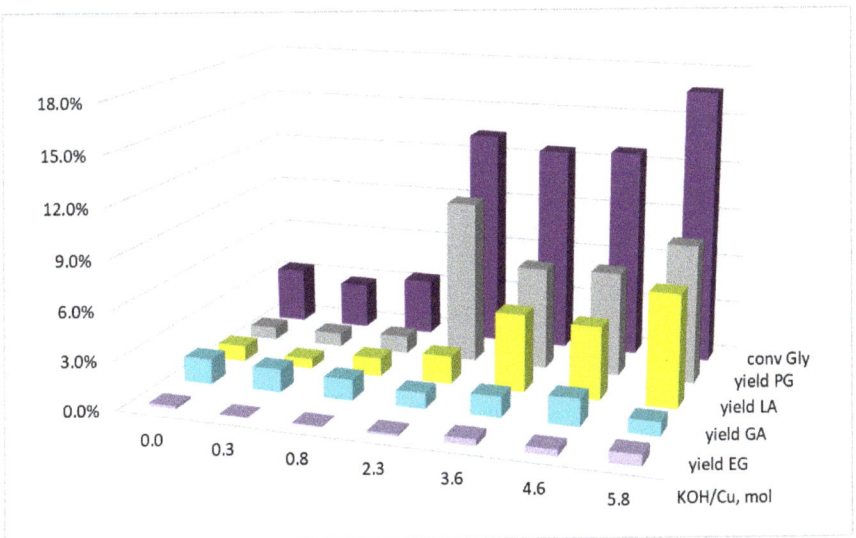

Figure 2. The influence of the KOH/Cu molar ratio on the glycerol conversion and the yields of the products. Conditions: T = 220 °C, p(H_2) = 3.0 MPa, Gly/Cu = 50 mol, τ = 5 h, precursor salt = Cu(OAc)$_2$.

Lactic acid is formed from glycerol in an alkaline medium through successive stages of dehydrogenation (catalyzed by Cu^0) and dehydration (catalyzed by OH^-). Both the rate and the equilibrium of the dehydrogenation reaction can depend on the hydrogen pressure in the system. However, the propylene glycol (hydrogenolysis product) formation rate can also depend on the hydrogen pressure. Thus, changing p(H_2) can be a way to control the reaction selectivity. The results obtained (Figure 3) indicate the validity of these assumptions: an increase in hydrogen pressure was accompanied by both an increase in glycerol conversion and an increase in selectivity between PG and LA. Thus, during glycerol hydrogenolysis in the presence of Cu(OAc)$_2$ (T = 220 °C, KOH/Cu = 5.8 mol, Gly/Cu = 50 mol, τ = 2.5 h), at p(H_2) = 1.0 MPa and 4.0 MPa, X_{Gly} amounted to 9.9% and 13.0%; Y_{PG} amounted to 3.7% and 7.2%; and Y_{LA} decreased from 4.5% to 3.7%, respectively. With that, we did not find a significant effect of hydrogen pressure on the yields of by-products of ethylene glycol and glyceric acid. Despite the lack of data on the reaction kinetics, it can be concluded from this that the glycerol hydrogenolysis reaction with the propylene glycol formation in the presence of the in situ-formed copper dispersed catalysts has a non-zero hydrogen order.

With that, both the conversion and the selectivity of the reaction changed little when the glycerol concentration in the reaction mixture was reduced by diluting it with water (Figure 4). Therefore, at Gly/H_2O = 0.4 vol, the values of $X_{Gly}/Y_{PG}/Y_{LA}$ were 13.3/4.0/7.3%, and at Gly/H_2O = 4.1 vol the values were 16.8/8.5/6.8%, respectively. Apparently, the lactic acid formation reaction has an glycerol order close to zero. Since Y_{PG} more than doubled (from 4.0 to 8.5%) with an increase in the glycerol concentration, the propylene glycol formation reaction order obviously has a non-zero value.

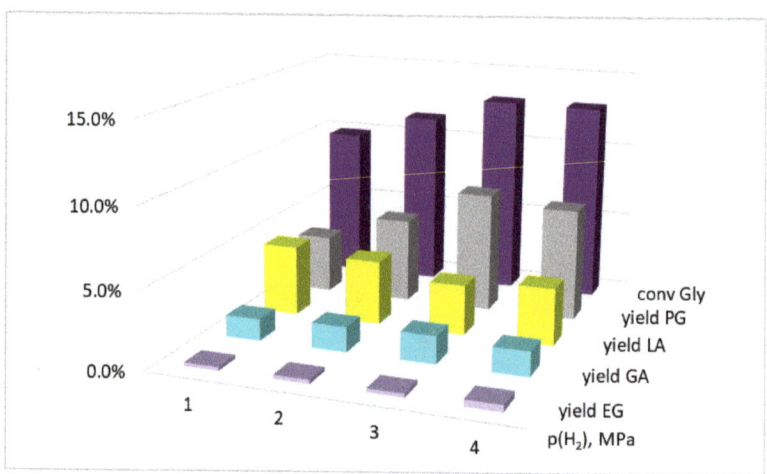

Figure 3. The influence of the hydrogen pressure on the glycerol conversion and the yields of the products. Conditions: T = 220 °C, KOH/Cu = 5.8 mol, Gly/Cu = 50 mol, Gly/H_2O = 4.1 vol, τ = 2.5 h, precursor salt = Cu(OAc)$_2$.

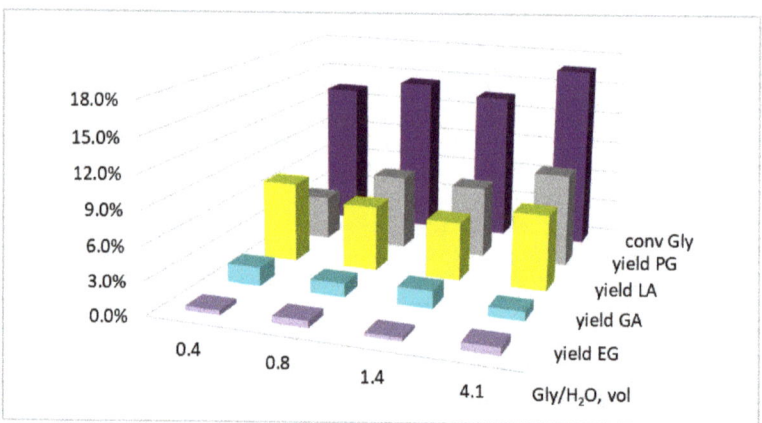

Figure 4. The influence of the Gly/H_2O volume ratio on the glycerol conversion and the yields of the products. Conditions: T = 220 °C, p(H_2) = 3.0 MPa, KOH/Cu = 5.8 mol, Gly/Cu = 50 mol, τ = 5.0 h, precursor salt = Cu(OAc)$_2$.

The dependence of glycerol conversion and product yields on the reaction time was studied for two different catalytic systems differing in their KOH/Cu molar ratio (2.3 and 5.8) (Figures 5 and 6). In both cases, the continuous growth of X_{Gly}, Y_{PG} and Y_{LA} was observed. In general, the yield of the by-product ethylene glycol increased with an increase in the PG yield, while the GA yield tended to somewhat decrease. The latter circumstance can be explained by the fact that GA formed during the copper reduction by glycerol was further converted—presumably as a result of slow hydrogenation into glycerol.

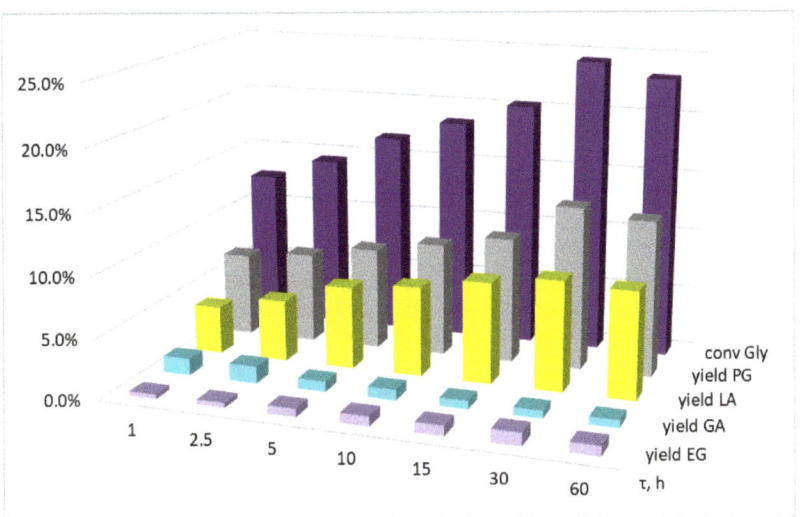

Figure 5. The influence of the reaction time on glycerol conversion and the yields of the products. Conditions: T = 220 °C, p(H$_2$) = 3.0 MPa, KOH/Cu = 5.8 mol, Gly/Cu = 50 mol, Gly/H$_2$O = 4.1 vol, precursor salt = Cu(OAc)$_2$.

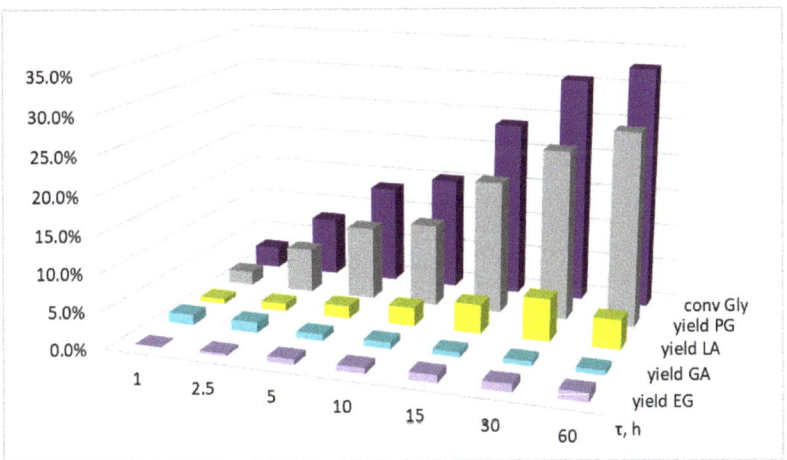

Figure 6. The influence of the reaction time on glycerol conversion and the yields of the products. Conditions: T = 220 °C, p(H$_2$) = 3.0 MPa, KOH/Cu = 2.3 mol, Gly/Cu = 50 mol, Gly/H$_2$O = 4.1 vol, precursor salt = Cu(OAc)$_2$.

With that, according to the data obtained, a sharp reaction deceleration is evident, and is more pronounced in the case of KOH/Cu = 5.8 mol (Figure 5). Thus, the change in conversion rate for the first hour of reaction was 12.6%, while for the next 29 h, the conversion rate changed by only 11.8%. In the case where KOH/Cu = 2.3 mol, the reaction deceleration was smoother: X_{Gly} at 1 and 30 h of the reaction was 2.9% and 30.6%; however, when the reaction continued for up to 60 h, the conversion increase was only 2.1% (Figure 6).

In principle, such a sharp reaction deceleration can be due to three main reasons. The first of them is the accumulation of an inhibitor compound in the reaction medium, which sharply slows down the target reaction. Theoretically, the reaction product can also act as an inhibitor, but the repeatedly described possibility of obtaining high yields of propylene glycol in the glycerol hydrogenolysis on copper catalysts indicates a low probability of such a scenario [7,25]. The second reason for reaction deceleration is to reduce the concentration of alkali by converting it to potassium lactate. When Y_{LA} reaches approximately 8%, the reaction medium becomes acidic. As is shown in Figure 2, when pH < 7, the conversion of glycerol and the yields of the main products—propylene glycol and lactic acid—are sharply reduced. The third potential cause may be the catalytic phase degradation. It is generally assumed that dispersed metal particles are well stabilized by a polyol medium; however, this does not guarantee a complete halt in agglomeration and recrystallization processes. It should be noted that the catalyst formation from the glycerol solution of the precursor salt coincides in time with the start of the catalytic reaction, and then these two processes (the catalyst evolution and the catalytic reaction) proceed in parallel. Since the main regularities of the catalytic reaction have already been described in this study, it is necessary to continue to characterize the catalyst formed in the reactor while making an attempt to describe its evolution.

2.2. In Situ-Generated Catalyst Characterization

The first questions to be answered in order to characterize the catalyst are: what is the precursor conversion degree to form a solid phase and what is the residual concentration of dissolved copper in the reaction mixture after the experiment? To answer these questions, the obtained reaction mixtures were subjected to thorough centrifugation; the copper concentration in the supernatants was determined by the XRF method (Table 2). The estimated copper concentration in the initial solution was 2.93 wt %. In the reaction mixture supernatant obtained with a reaction time of 1 h, the total residual copper content was 7×10^{-4} wt %. The reaction mixture obtained at $\tau = 5$ h was characterized by $\omega Cu = 1 \times 10^{-4}$ wt %; there was slightly more residual copper during the solid phase formation in the absence of alkali (11×10^{-4} wt %). Thus, we can speak about the complete precursor salt conversion into a heterogeneous phase under the selected reaction conditions.

Table 2. The residual copper content in the supernatant solutions measured by XRF. Conditions: T = 220 °C, p(H$_2$) = 3.0 MPa, Gly/Cu = 50 mol, Gly/H$_2$O = 4.1 vol, precursor salt = Cu(OAc)$_2$.

Reaction Conditions	ωCu^{2+} at the Beginning of the Reaction, %	ΣCu after Hydrogenolysis, 10^{-4}%
$\tau = 5$ h, Without alkali		11
$\tau = 5$ h, nKOH/nCu = 5.8	2.93	1
$\tau = 1$ h, nKOH/nCu = 5.8		7

According to the X-ray diffraction pattern obtained, the phase composition of the formed catalyst was a crystalline metallic copper Cu^0 with a typical face-centered cubic crystal lattice (Figure 7). The observed diffraction maximum at $2\theta = 43°$ corresponded to a set of Cu (1 1 1) copper planes (PDF 04-0836). Traces of phases other than metallic copper were represented by weak peaks of copper hemioxide (PDF 65-3288); the most intense peak was observed at $2\theta = 37°$, corresponding to Cu$_2$O (1 1 1), the presence of which may be due to partial oxidation of the surface of solid particles during their preparation for analysis. The phase composition was the same for samples obtained in the absence of alkali and at KOH/Cu = 5.8; no differences in the phase composition were found between samples obtained after 1, 5 and 15 h of the reaction.

In view of there being no significant differences between the XRD patterns of Cu samples, XPS patterns were obtained solely for Cu-5 and Cu-5 * with a reaction of time 5 h (Figure 8). A study of the near-surface layer of in situ-formed, dispersed catalyst using the XPS method revealed the presence of Cu^0, as evidenced by the $2p^{1/2}$–$2p^{3/2}$ spin–orbit doublet, and did not identify the CuO oxide phase. As a result of the obtained XPS spectrum deconvolution, Cu^0 peaks were identified with Cu $2p^{3/2}$ binding energies and Cu $2p^{1/2}$ equal to 932.1 eV and 951.7 eV, which is consistent with the literature data [38]. Due to the proximity of binding energies at the 2p level of Cu_2O and Cu, which leads to the superposition of peaks, it is only possible to determine the content of copper hemioxide from the Cu_2O satellite peak (E_{bin} = 945 eV). Thus, the surface layer of Cu-5 and Cu-5 * samples was represented by pure Cu^0 copper with an insignificant content of copper hemioxide in the Cu-5 sample, the formation of which could be caused by sample preparation for analysis.

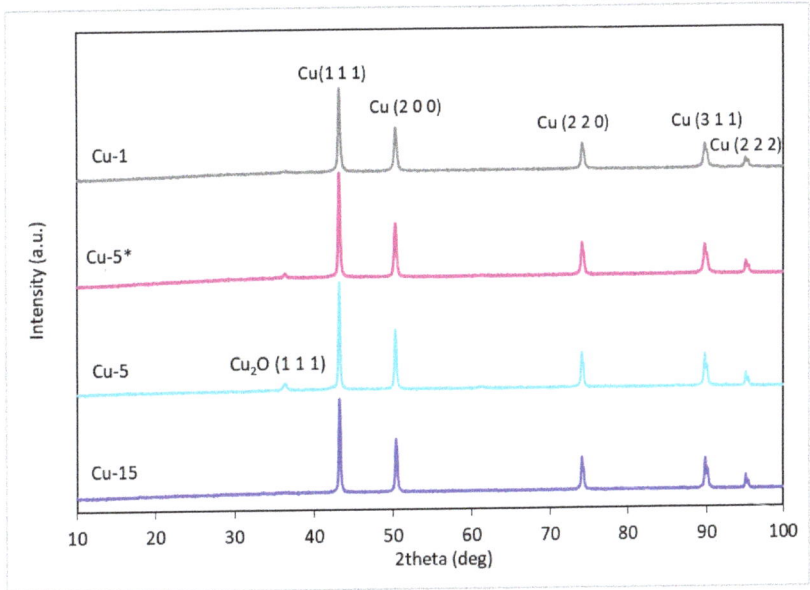

Figure 7. The XRD patterns of copper catalysts (Cu-1, Cu-5 *, Cu-5, Cu-15) generated in situ in the reaction medium during hydrogenolysis of glycerol.

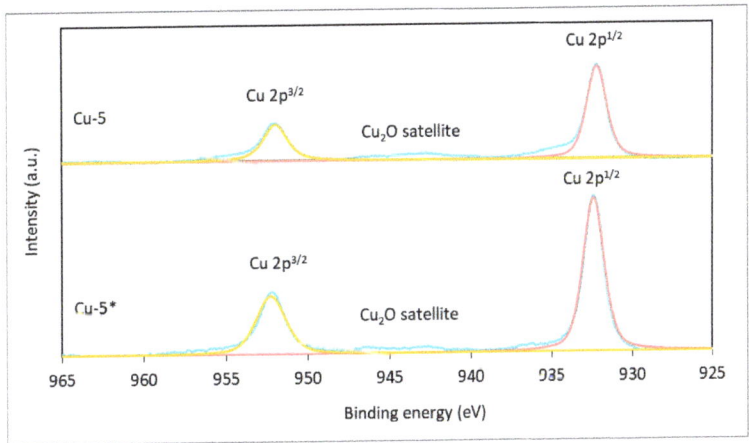

Figure 8. The XPS patterns of copper catalysts (Cu-5 *, Cu-5) generated in situ in the reaction medium during hydrogenolysis of glycerol. Blue color: the measured XPS spectrum; pink color: Cu $2p^{1/2}$ region; yellow color: Cu $2p^{3/2}$ region.

Figure 9 shows the SEM images of the catalysts Cu-1, Cu-5 and Cu-15 synthesized in situ for 1, 5 and 15 h, respectively, with KOH/Cu = 5.8 mol and Cu-5 * generated for

5 h. The morphology of all of the samples presents aggregates of various-shaped particles. According to SEM images of the catalysts Cu-1, Cu-5 and Cu-15, the particle shape changed depending on the sample formation time in situ. The average size of aggregates was 50–750 μm and the size of smaller particles was less than 5 μm (Figure S3A,B). The SEM microphotographs allowed us to make a suggestion about the evolution of the catalyst and the changes in the morphology of both the aggregates as a whole and the morphology of individual particles that made up the aggregates. For a reaction time of more than 1 h, gradual recrystallization of the copper phase resulted in the formation of agglomerates that were 50–750 μm in size, and the morphology of individual particles changed from framework- to monolith-like, probably as a result of the internal space of the framework being filled with new copper atoms (sample Cu-5, Figures 9 and S3C). Some areas of the catalyst sample Cu-15 became a monolithic surface, increasing the reaction time to 15 h (sample Cu-15, Figure 9). The fact that agglomeration of the catalysts took place during this time is also confirmed by the values of the total specific surface area obtained using the BET method. For 1 h of sample Cu-1, the value was 2.75 m^2/g; for a 15 h catalyst Cu-15, the value was more than two-times lower—0.74 m^2/g. The process of sintering continued, and after 60 h, the catalyst had the form of a monolithic sphere (Figure S4).

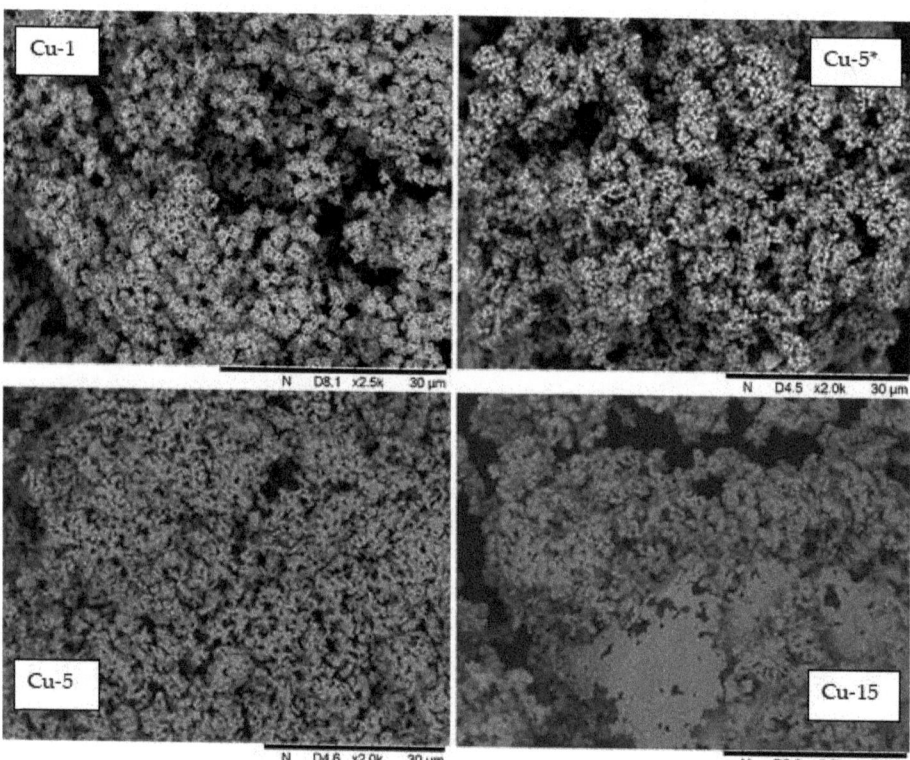

Figure 9. The SEM microphotographs of the copper catalysts (Cu-1, Cu-5 *, Cu-5, Cu-15) generated in situ in the reaction medium during hydrogenolysis of glycerol.

The absence of an alkali addition to the reaction system should be noted, and, accordingly, forming the catalyst does not lead to a significant change in morphology (sample Cu-5 *, Figure 9). Hence, sintering of the catalyst occurs during the course of the reaction time, and the active surface of the catalyst and its catalytic activity decreases, leading to a decrease in the rate of the glycerol hydrogenolysis reaction (Figures 5 and 6). Catalyst microstructure analysis using the TEM method (Figure 10) indicates the formation of cubic-shaped particle clusters. Taking into account the particle size, distribution from the HRTEM method is not evident, and one can estimate that agglomerates consist of cubic particles of approximately 50–70 nm in size.

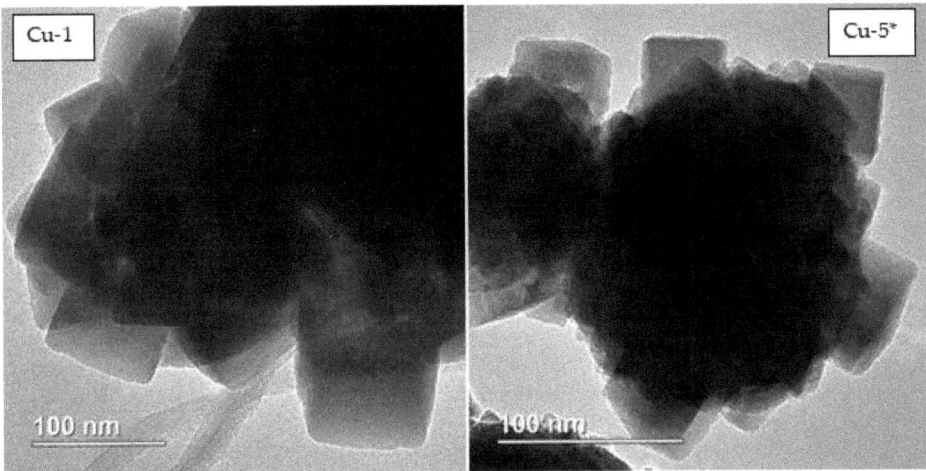

Figure 10. The HRTEM microphotographs of the copper catalysts (Cu-1, Cu-5 *) generated in situ in the reaction medium during hydrogenolysis of glycerol.

The conclusions drawn regarding the catalyst evolution from the SEM results were confirmed by the calculations of the average crystallite size from the XPA data (Table 3). The sizes of Cu crystallites were estimated by the width of the patterns, appropriate to the (111), (200), (220), (311), (222) planes in Cu according to the Hall–Williamson method. Thus, the catalyst obtained after 1, 5 and 15 h is characterized by an average crystallite size of 35 nm, 44 nm, and 49 nm, respectively.

Table 3. Phase composition and average crystallite size of in situ-generated catalysts estimated by XRD.

Catalyst	Phase Composition	Average Crystallite Size, nm
Cu-1		35
Cu-5 *	Cu	42
Cu-5		44
Cu-15		49

2.3. The Discussion Regarding the Reaction Routes and the Specific Activity

Upon glycerol hydrogenolysis of in situ-generated dispersed copper particles, four products were formed. The main reaction products were PG and LA; by-products were represented by EG and glyceric acid. The data obtained as a result of catalytic experiments in combination with published data allow us to make assumptions regarding the routes of formation of glycerol conversion products.

The glycerol conversion reaction rate increased sharply during the transition from a neutral medium to an alkaline one (Figure 2). With that, the copper catalyst activity in a weak-acid medium was several times lower compared with the specific activity of the promoted Cu/Cr_2O_3 catalyst, which confirms the previous conclusions by Nikolaev et al. [7]. This allows us to conclude that the reaction mechanism changes during the transition to an alkaline medium. It is known that in a neutral and acidic medium, the dominant conversion mechanism involves the primary glycerol dehydration with the acetol formation, which is converted to PG during hydrogenation [6,7,22,25]. During the reaction by this mechanism, intermediate acetol is always detected among the by-products. In an alkaline medium, the mechanism of glycerol conversion includes primary dehydration dehydrogenation with the formation of glyceraldehyde, followed by its dehydration into methylglyoxal (pyruvaldehyde). The latter could be either hydrogenated into PG or transformed into LA by hydration rearrangement [22,39]. Under the conditions we have chosen in an alkaline medium, the glycerol conversion proceeds according to the second mechanism through the glyceraldehyde formation; this is confirmed both by the presence of PG and LA as the main products, and by the complete acetol absence in the reaction mixtures (Figure S2A). In addition, in the presence of a Cu/Cr_2O_3 catalyst, the LA yield was insignificant in a neutral medium and sharply increased with KOH addition (Table 1, entry 29), which also indicates a change in the reaction mechanism with alkali addition.

The secondary ethylene glycol formation (Scheme 1, reaction 2) during the glyceraldehyde mechanism implementation proceeded through the retro-aldol reaction of glyceraldehyde [22,39] This mechanism is characterized by a significantly higher selectivity for EG compared with the reaction occurring through acetol formation. For example, during hydrogenolysis on a Cu/Al_2O_3 catalyst (T = 200 °C, p(H_2) = 5 MPa), the selectivity for ethylene glycol was 1.2% at X_{Gly} = 85%. Glycerol hydrogenolysis on Pt/C and Ru/C in the presence of CaO base (T = 200 °C, p(H_2) = 4 MPa) proceeded with EG selectivity from 9% to 16% [22]. During glycerol hydrogenolysis in the presence of copper DPs, ethylene glycol selectivity did not depend much on pressure and the KOH/Cu molar ratio, and in all cases, was approximately 4%, practically unchanged depending on the glycerol conversion.

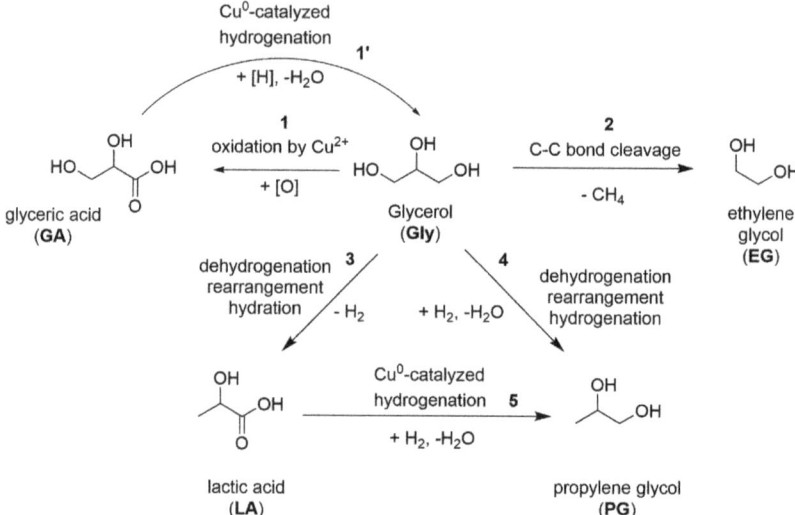

Scheme 1. The reaction routes in the catalytic hydrogenolysis of glycerol over the in situ-generated CuDPs.

Glyceric acid, detected among the reaction products in all mixtures obtained by catalysis by copper DPs, is a glycerol oxidation product (Scheme 1, reaction 1). The reaction medium was reducing; therefore, the only oxidizing agent in it could be a copper salt. Since there were no other oxidizing agents in the reaction medium, GA was formed during the copper reduction from the precursor salt at the initial stage of the reaction. As the reaction continued (Figures 4 and 5), the GA yield decreased slowly, which may indicate its hydrogenation with the glycerol formation in the presence of CuDPs (Scheme 1, reaction 1). Although glyceric acid could theoretically be generated by the Cannizzaro reaction of the intermediate glyceraldehyde, the main source of glyceric acid formation was considered the oxidation of glycerol by the Cu^{2+}.

Since glyceric acid can be hydrogenated under reaction conditions to form glycerol, it is possible to assume that the lactic acid hydrogenation with the PG formation can proceed in a similar way (Scheme 1, reaction 5). A decrease in the CuDPs activity did not allow high glycerol conversions to be achieved where, theoretically, a secondary reaction of LA hydrogenation into PG could be observed. Nevertheless, indirect confirmation of the occurrence of this reaction was provided by data on the effect of hydrogen pressure on the product yields (Figure 2); with an increase in $p(H_2)$ from 1.0 to 4.0 MPa, LA selectivity decreased from 45.8 to 28.5 mol % (T = 220 °C, Gly/Cu = 50 mol, τ = 5.0 h, precursor salt= $Cu(OAc)_2$).

The issue of comparing the activity of CuDPs and the industrial Cu/Cr_2O_3 catalyst deserves special consideration. According to Dasari et al. [17], the specific productivity for unpromoted copper heterogeneous catalysts for copper hydrogenolysis (for example, skeleton copper) is 1.5–2 times lower than for the copper chromite catalyst.

It can be assumed that the degradation rate of the catalyst, during which a decrease in dispersion and activity loss occur, is directly proportional to the concentration of copper particles in the reaction medium and temperature; as a result of which, higher relative activities of CuDPs should be observed at lower temperatures. The data obtained during the reaction at T = 200 °C show that the specific activity of the CuDPs in glycerol conversion (SP_{Gly} = 17.3–23.6 mmol g^{-1} h^{-1}) was significantly higher than the activity of the Cu/Cr_2O_3 catalyst (SP_{Gly} = 6.5 mmol g^{-1} h^{-1}) by 2.7–3.6 times (Table 4). Since 40–55% of glycerol was converted into lactic acid in the presence of the catalytic system CuDPs + KOH, the activity of dispersed particles in propylene glycol was significantly lower. For instance, for the $Cu(OAc)_2$-derived DPs, the SP_{PG} amounted to 9.4 mmol g^{-1} h^{-1}, which is still 1.8 times greater than the activity of the industrial catalyst (SP_{PG} = 5.3 mmol g^{-1} h^{-1}). With that, it is necessary to understand that in this case, KOH acts as a co-catalyst, the mass of which is ignored when calculating SP. The activity of the Cu/Cr_2O_3 catalyst was measured in a catalytic run without KOH addition.

Table 4. The SP values for the various copper catalysts in the glycerol hydrogenolysis. T = 200 °C, $p(H_2)$ = 3.0 MPa, KOH/Cu = 5.8 mol, Gly/Cu = 50 mol, τ = 5 h.

Precursor Salt	SP_{Gly}, mmol Gly $g_{catalyst}^{-1}$ h^{-1}	SP_{PG}, mmol PG $g_{catalyst}^{-1}$ h^{-1}
$CuSO_4 \cdot 5H_2O$	23.6	10.7
$Cu(OAc)_2 \cdot H_2O$	18.6	9.4
$Cu-Cr_2O_3$ [1]	6.5	5.3
$CuCl_2$	17.3	9.1

[1] The copper chromite catalyst was used without KOH addition.

The activities of CuDPs and the Cu/Cr_2O_3 catalyst were also compared in experiments without free KOH at T = 220 °C (Table 5). The activity of the copper chromite catalyst almost tripled when the temperature increased from 200 to 220 °C (SP_{Gly} = 6.5 and 16.7 mmol g^{-1} h^{-1}, respectively), while the activity of CuDPs was 26–36% higher than this value. With that, both catalysts showed similar values of propylene glycol activity SP_{PG}.

Table 5. The SP values for the CuDPs and the Cu/Cr_2O_3 catalyst in the glycerol hydrogenolysis. T = 220 °C, $p(H_2)$ = 3.0 MPa, KOH/Cu = 2.3 mol (for copper acetate) and 0 (for Cu/Cr_2O_3), Gly/Cu = 50 mol.

Precursor Salt	τ, h	SP_{Gly}, mmol Gly $g_{catalyst}^{-1}$ h^{-1}	SP_{PG}, mmol PG $g_{catalyst}^{-1}$ h^{-1}
$Cu-Cr_2O_3$	5	16.7	15.2
	1	22.8	15.0
$Cu(OAc)_2$	2.5	25.2	19.2
	5	21.1	15.7

The addition of alkali (5.8 eq) led to an increase in the activity of the Cu/Cr_2O_3 catalyst: the SP_{Gly}/SP_{PG} increased from 16.7/15.2 to 30.4/20.8 mmol $g_{catalyst}^{-1}\,h^{-1}$, respectively (Table 6). The SP values for the CuDPs derived both from $CuSO_4$ and $Cu(OAc)_2$ were somewhat lower. However, as was shown earlier (Figure 4), for the catalytic system [$Cu(OAc)_2$ + 5.8 KOH], a sharp decrease in activity was observed during the reaction, which is associated with the catalyst degradation. Obtained SP values for the reaction time of 1 and 2.5 h confirm this hypothesis; thus, according to the results after 1 h of the reaction, the SP_{Gly} and SP_{PG} for CuDPs were 3.26 and 2.61 times higher than those for the Cu/Cr_2O_3, respectively. It should be pointed out that the SP values are close for different copper salts (acetate, sulfate, chloride) under the same reaction conditions; thus, from the point of view of catalytic activity, the type of salt to be used in the hydrogenolysis of glycerol is not significant. However, it is possible that the choice of anion can have a long-term impact on reactor performance.

Table 6. The SP values for the various copper catalysts in the glycerol hydrogenolysis. T = 220 °C, p(H_2) = 3.0 MPa, KOH/Cu = 5.8, Gly/Cu = 50 mol.

Precursor Salt	τ, h	SP_{Gly}, mmol Gly $g_{catalyst}^{-1}\,h^{-1}$	SP_{PG}, mmol PG $g_{catalyst}^{-1}\,h^{-1}$
$CuSO_4 \cdot 5H_2O$	5	26.1	16.4
$Cu\text{-}Cr_2O_3$	5	30.4	20.8
$Cu(OAc)_2 \cdot H_2O$	1	99.2	54.3
	2.5	45.4	23.6
	5	26.6	13.2

It can be concluded that, for the successful application of the in situ-generated CuDPs, it is necessary to solve the problem of catalyst degradation. Obviously, the polyol medium cannot restrain processes of copper coagulation and recrystallization under the selected conditions, which are associated with both a high reaction temperature and a relatively high concentration of dispersed particles. Possible solutions may include the use of active particle dispersion stabilizers or the formation of non-monometallic particles stabilized by the second component (e.g., metal oxide).

3. Materials and Methods

3.1. Materials

Copper (II) acetate monohydrate (98%, Komponent-Reaktiv, Moscow, Russia), Copper (II) sulfate pentahydrate (98%, Ruskhim, Moscow, Russia) and Copper (II) chloride dihydrate (98%, Komponent-Reaktiv, Moscow, Russia) were used as precursors for ultrafine copper particle formation without any further purification. Potassium hydroxide (98%, Chimmed, Moscow, Russia), glycerol (≥99.3%, Komponent-Reaktiv, Moscow, Russia) and distilled water were used as reaction feeds. In addition, 1,2-Butanediol (98%, abcr, Karlsruhe, Germany) was used as the inner standard, and 1,2-propanediol (>99%, Carl Roth, Karlsruhe, Germany), ethylene glycol (≥99.5%, Komponent-Reaktiv, Moscow, Russia), DL-lactic acid in aqueous solution (80%, Komponent-Reaktiv, Moscow, Russia) and hydroxyacetol (technical, Acros organics, Austria) were used for the GC quantification method. Trimethylsilylating reagent TMS-HT (hexamethyldisilazane + trimethylchlorosilane in anhydrous pyridine; abcr, Karlsruhe, Germany) was used for derivatization. For all hydrogenation processes, hydrogen gas (grade A in accordance with GOST 3022-80, MGPZ, Moscow, Russia) was used. For comparison purposes, a commercial copper chromite catalyst, VNH-103 (Vniineftekhim-103), that contained approximately 56% copper was used. The catalyst was reduced in hydrogen flow (10 vol % H_2/Ar, 6 h at 300 °C) before use.

3.2. Catalytic Test

The liquid phase catalytic tests were performed in a stainless-steel batch reactor (50 cm^3 internal volume) equipped with a polytetrafluoroethylene (PTFE) liner, pressure gauge, a thermocouple and a magnetic stirrer. The reactor was loaded with aliquots of glycerol and water, and either a charge of copper chromite catalyst or a precursor composition consisting of a copper salt and KOH in specified quantities. After the loading of the reactor with the feed, the precursor salt/solid catalyst, the alkali and the stirrer, the reactor was purged twice with 3 MPa of hydrogen and then filled with an operating pressure of hydrogen.

The closed autoclave was placed in an electric oven and stirring was turned on; this moment was taken as the beginning of the reaction. The set temperature (200 or 220 °C) inside the reactor was reached, on average, within 30–40 min. The ratios of the glycerol volume over the water volume (0.4–4.1 vol), mass of potassium hydroxide, reaction time (1–60 h) and hydrogen pressure (1–4 MPa) were varied at the constant total volume of the reaction mixture and at the molar ratio glycerol over copper, mainly 50 or 100 mol. The complete datasets can be found in the Supplementary Materials (Tables S1–S3).

The stirrer was set to 1000 rpm (the absence of diffusion limitations at this stirring speed was proven by separate tests). At the end of the test, the reactor was taken out of the electric furnace and promptly chilled with cold water. The pressure was then gently released, and the catalyst was separated from the reaction mixture by centrifugation, washed twice with distilled water and subsequently twice with isopropyl alcohol, dried under room temperature in dry Ar flow and stored under an inert Ar atmosphere prior to the characterization analyses.

In order to check the quantity of the reaction gaseous products, the reactors were weighed twice: before the catalytic test and after the end of the test, as the reactors were cooled and the pressure was released. The carbon balance in all of the experiments was greater than $100 \pm 5\%$. The gaseous product yield was found to be negligible and therefore did was not included in further calculations.

3.3. Analysis of Products

The GC-FID analysis of the reaction products was carried out using a Krystallux 4000M chromatograph equipped with a flame ionization detector, the capillary column Optima-1 (25 m × 0.32 mm, film thickness 0.35 μm) and helium as the carrier gas. The temperature programming mode was as follows: 70 °C—holding for 1 min; from 70 to 100 °C; with a temperature rise rate of 3 deg min^{-1}; 100 °C—holding for 1 min; from 100 to 230 °C with a temperature rise rate of 30 deg min^{-1}; and 230 °C—holding for 1 min. Calibration for quantitative analysis by the internal standard method was conducted using standard samples of the compounds and 1,2-butanediol as the internal standard (Figure S1).

The feedstock and product compositions were studied using the GC–MS technique on a ThermoFocus DSQ II instrument (Figure S2) with the capillary column Varian VF-5 ms (30 m × 0.25 mm; film thickness 0.25 μm) and helium as the carrier gas. The operating mode was as follows: the injector temperature was 270 °C, the starting temperature of the chromatograph oven was 40 °C, heating was at a rate of 15 °C/min to 300 °C and isothermal holding was for 10 min. The mass spectrometer operating mode was as follows: ionization energy was 70 eV, source temperature was 230 °C and scanning was in the range of 10–800 Da at a speed of 2 scans/s, with unit resolution over the entire mass range. For the GC–MS analysis, 10 μL of the initial mixture was diluted with 1 mL of methylene chloride (analytical grade, Himmed). The GC-MS analysis did not reveal the formation of 1,3-propanediol.

Glycerol conversion (X) and product yields (Y) were calculated using the following equations.

$$X(\%) = \frac{\text{mole of consumed substrate}}{\text{initial mole of substrate}} \cdot 100\%$$

$$Y(\%) = \frac{\text{mole of formed products}}{\text{initial mole of substrate}} \cdot 100\%$$

3.4. Silylation Protocol

To obtain more volatile derivatives, the reaction mixture with 1,2-butanediol as an internal standard was derivatized with the commercially available trimethylsilylating reagent TMS-HT and analyzed by GC-FID on a non-polar column. Before GC analysis, all carboxyl-, hydroxyl groups of the obtained products during glycerol hydrogenolysis, as well as anions of lactate and glycerate, were derivatized into TMS esters. For derivatization, 5 µL of the initial mixture was mixed with 350 µL of the derivatizing agent.

3.5. Characterization of Copper Catalysts

To identify the phase composition of catalysts synthesized, X-ray diffraction (XRD) was used. The XRD spectra were obtained with a Rigaku D/Max-RC diffractometer using Cu Kα radiation. The elemental composition of the catalyst surface was studied by XPS with a PHI 5500 ESCA X-ray photoelectron spectrometer from Physical Electronics. Nonmonochromatic AlK$_a$ radiation (hn = 1486.6 eV) with a power of 300 W was used for the excitation of photoemission. Powders were pressed into an indium plate. The diameter of an analysis zone was 1.1 mm. The photoelectron peaks were calibrated based on the C^{1s} line of carbon with a binding energy of 284.9 eV. The deconvolution of spectra was performed by a nonlinear least squares method with the use of Gaussian and Lorentzian functions.

The structure and morphology of the catalyst samples were studied using high-resolution transmission electron microscopy (HRTEM) with a JEM 2100 electron microscope from JEOL at an accelerating voltage of 200 kV.

Electron microscopic images of the samples were obtained on a Hitachi TM3030 scanning electron microscope (SEM).

The surface area measurements of the samples were carried out using a Belsorp mini X device from MICROTRAC MRB. To perform the analysis, a weighed sample of ~0.2 g was placed in a preliminarily weighed glass cuvette. The stage of preliminary preparation of the samples included thermal vacuum at temperature 200 °C under pressure at 10 Pa for 8 h. The total specific surface of the samples was determined in accordance with the BET method in the range of relative pressures p/p_0 = 0.05–0.35.

To estimate the reduction completeness from the copper precursor after the reaction, the copper content in the supernatant was measured by X-ray fluorescence (XRF, Thermo ARL 4460).

In situ-generated catalysts Cu-1, Cu-5 and Cu-15 from the Cu(OAc)$_2$ precursor salt were generated under $p(H_2)$ = 3 MPa at 220 °C, KOH/Cu = 5.8 mol and Gly/H$_2$O = 4.1 vol and reaction times of 1, 5, and 15 h, respectively, and catalyst Cu-5 * was generated under the same conditions as catalyst Cu-5, except the addition of alkali. Cu-1, Cu-5, Cu-5 * and Cu-15 were characterized by XRD, HRTEM, and SEM.

In order to compare the rates of hydrogenolysis of glycerol between in situ-generated catalysts in a water–glycerol media and a commercially available Cu-Cr$_2$O$_3$ catalyst, the values of the specific productivity (SP) were calculated using the following Equation:

$$SP = \frac{n_{PG}}{m_{cat \cdot h}}, \text{ mmol g}^{-1} \text{ h}^{-1}$$

(mmol of PG formed per gram of the catalyst per hour).

4. Conclusions

In this paper, the reaction of glycerol hydrogenolysis in the presence of in situ-generated Cu dispersed particles (CuDPs) was studied. It was found that during the decomposition of precursor salts (CuSO$_4$, CuCl$_2$, Cu(OAc)$_2$) in glycerol under the conditions of a hydrogenolysis reaction (T = 200–220 °C, $p(H_2)$ = 1.0–4.0 MPa), copper dispersed particles (CuDPs) were formed and catalyzed the hydrogenolysis of glycerol to give propylene glycol and lactic acid. By-products of the conversion were ethylene glycol and glyceric acid, and the latter was formed during the copper reduction from precursor salts and was hydrogenated into glycerol as the reaction continued. The effect of reaction conditions was described (T, $p(H_2)$, KOH/Gly and Gly/H$_2$O ratios) for glycerol conversion and product yields. It was shown that an increase in the pressure and glycerol concentration increased the PG yield with practically no effect on the LA yield. The CuDPs activity at

KOH/Cu < 2 ratios was very low (Y_{PG} = 0.8–1.1%, Y_{LA} = 0.6–1.1%, T = 220 °C), increasing abruptly during the transition to a slightly alkaline medium (KOH/Cu = 2.3, Y_{PG} = 10.0%). Consequently, the lactic acid yield remained relatively low (Y_{LA} = 0.6–1.1%). A further increase in the alkali amount to KOH/Cu = 5.8 mol was accompanied by a significant increase in the selectivity for lactic acid (Y_{LA} = 6.8%) without an increase in PG yield (Y_{PG} = 8.5%). The alkalinity of the reaction media therefore allows the reaction mechanism to be tuned.

Samples of CuDPs were further characterized by XRD, XPS, SEM and HRTEM. The catalysts were found to consist of pure metallic copper with a tiny admixture of Cu_2O. According to the results of the analysis of morphology by electron microscopy, the catalyst consisted of agglomerates of an irregular shape with a mean size of 50–750 μm. Agglomerates consisted of cubic units measuring approximately 50–70 nm in size. During the reaction, the catalyst evolved, consisting of the sintering and recrystallization of copper particles, which resulted in a decrease in dispersion and activity loss. Significant activity loss had already been observed by the 5th hour of the existence of CuDPs.

Based on the results obtained, the formation methods of the main products (PG, LA) and by-products (EG, GA) were discussed. Under the selected conditions, the main reaction mechanism is presumably a glyceraldehyde-based mechanism, allowing both PG and LA to be formed. The activity of the obtained samples of CuDPs was compared with the activity of an industrial Cu/Cr_2O_3 catalyst. It was demonstrated that both in the presence of alkali and in a neutral medium, the specific activity of dispersed particles was higher or equal to the activity of an industrial catalyst. The highest activity for the [CuPDs + KOH] catalytic system was recorded at reaction time τ = 1 h (SP_{Gly} = 99.2, SP_{PG} = 54.3 mmol g^{-1} h^{-1}); under the same conditions, at τ = 5 h, the activity of the Cu/Cr_2O_3 catalyst was substantially lower (SP_{Gly} = 30.4, SP_{PG} = 20.8 mmol g^{-1} h^{-1}).

Supplementary Materials: The following supporting information can be downloaded at: https://www.mdpi.com/article/10.3390/molecules27248778/s1, Tables S1–S3: The average values of conversion of glycerol (X_{Gly}), yields of products (Y) under different reaction conditions during glycerol hydrogenolysis. Figure S1: Calibration plots of response ratio (S_{st}/S) versus mass ratio (m_{st}/m) for internal standardization by GC-FID for A: TMS-Gly derivative; B: TMS-PG derivative; C: TMS-EG derivative; D: TMS-LA derivative. Figure S2: A: The mass spectrum of a silylated liquid sample after glycerol hydrogenolysis reaction; B: The mass spectrum of TMS-Gly; C: The mass spectrum of TMS-EG; D: The mass spectrum of TMS-PG; E: The mass spectrum of TMS-LA; F: The mass spectrum of TMS-BD; G: The mass spectrum of TMS-GA. Figure S3: The SEM microphotographs of the copper catalysts generated in situ in the reaction medium during hydrogenolysis of glycerol. A, B: Cu-5; C: Cu-15. Figure S4: The appearance of the catalyst after 60 h of glycerol hydrogenolysis. Figure S5: $Cu(OAc)_2$ in a water–glycerol solution prior to adding potassium hydroxide.

Author Contributions: Conceptualization, V.S. and A.M.; methodology, I.P. and V.S.; validation, I.P., D.R. and V.S.; formal analysis, I.P.; investigation, I.P., D.R. and M.K.; resources, A.M.; data curation, I.P. and V.S.; writing—original draft preparation, I.P., V.S. and M.K. writing—review and editing, I.P. and V.S.; visualization, I.P. and V.S.; supervision, V.S. and A.M.; project administration, V.S. and A.M.; funding acquisition, V.S. All authors have read and agreed to the published version of the manuscript.

Funding: This research was funded by the Russian Science Foundation (RSF), 22-13-00252.

Institutional Review Board Statement: Not applicable.

Informed Consent Statement: Not applicable.

Data Availability Statement: Not applicable.

Conflicts of Interest: The authors declare no conflict of interest.

Sample Availability: Not applicable.

References

1. Sullivan, C.J.; Kuenz, A.; Vorlop, K.-D. Propanediols. In *Ullmann's Encyclopedia of Industrial Chemistry*; Wiley-VCH Verlag GmbH & Co., Ltd.: Weinheim, Germany, 2018; pp. 1–15.
2. Ciriminna, R.; Pina, C.D.; Rossi, M.; Pagliaro, M. Understanding the Glycerol Market. *Eur. J. Lipid Sci. Technol.* **2014**, *116*, 1432–1439. [CrossRef]
3. BASF and Oleon Celebrate Grand Opening of Propylene Glycol Production Plant. Available online: https://www.chemeurope.com/en/news/138616/basf-and-oleon-celebrate-grand-opening-of-propylene-glycol-production-plant.html (accessed on 10 November 2022).
4. Nanda, M.R.; Yuan, Z.; Shui, H.; Charles Xu, C. Selective Hydrogenolysis of Glycerol and Crude Glycerol (A by-Product Orwaste Stream from the Biodiesel Industry) to 1,2-Propanediol over B_2O_3 promoted Cu/Al_2O_3 catalysts. *Catalysts* **2017**, *7*, 196. [CrossRef]
5. Sepulveda, J.; Manuale, D.; Santiago, L.; Carrara, N.; Torres, G.; Vera, C.; Goncalves, M.; Carvalho, W.; Mandelli, D. Artigo. Selective hydrogenolysis of glycerol to propylene glycol in a continuous flow trickle bed reactor using copper chromite and Cu/Al_2O_3 catalysts. *Quim. Nova* **2018**, *41*, 926–932.
6. Vasiliadou, E.S.; Lemonidou, A.A. Kinetic Study of Liquid-Phase Glycerol Hydrogenolysis over Cu/SiO_2 Catalyst. *Chem. Eng. J.* **2013**, *231*, 103–112. [CrossRef]
7. Nikolaev, S.A.; Dmitriev, G.S.; Zanaveskin, K.L.; Egorova, T.B.; Khadzhiev, S.N. Selective Hydrogenolysis of Glycerol to 1,2-Propylene Glycol on Ultrafine Copper Particles. *Pet. Chem.* **2017**, *57*, 1074–1080. [CrossRef]
8. Dmitriev, G.S.; Khadzhiev, V.I.; Nikolaev, S.A.; Ezzhelenko, D.I.; Mel'chakov, I.S.; Zanaveskin, L.N. Copper-Containing Catalysts in the Liquid-Phase Hydrogenolysis of Glycerol. *Pet. Chem.* **2020**, *60*, 1066–1072. [CrossRef]
9. Gandarias, I.; Arias, P.L.; Requies, J.; Güemez, M.B.; Fierro, J.L.G. Hydrogenolysis of Glycerol to Propanediols over a Pt/ASA Catalyst: The Role of Acid and Metal Sites on Product Selectivity and the Reaction Mechanism. *Appl. Catal. B Environ.* **2010**, *97*, 248–256. [CrossRef]
10. Chaminand, J.; Djakovitch, L.A.; Gallezot, P.; Marion, P.; Pinel, C.; Rosier, C. Glycerol Hydrogenolysis on Heterogeneous Catalysts. *Green Chem.* **2004**, *6*, 359–361. [CrossRef]
11. Feng, J.; Fu, H.; Wang, J.; Li, R.; Chen, H.; Li, X. Hydrogenolysis of Glycerol to Glycols over Ruthenium Catalysts: Effect of Support and Catalyst Reduction Temperature. *Catal. Commun.* **2008**, *9*, 1458–1464. [CrossRef]
12. Pendem, C.; Gupta, P.; Chaudhary, N.; Singh, S.; Kumar, J.; Sasaki, T.; Datta, A.; Bal, R. Aqueous Phase Reforming of Glycerol to 1,2-Propanediol over Pt-Nanoparticles Supported on Hydrotalcite in the Absence of Hydrogen. *Green Chem.* **2012**, *14*, 3107–3113. [CrossRef]
13. Guo, X.; Li, Y.; Shi, R.; Liu, Q.; Zhan, E.; Shen, W. Co/MgO Catalysts for Hydrogenolysis of Glycerol to 1, 2-Propanediol. *Appl. Catal. A Gen.* **2009**, *371*, 108–113. [CrossRef]
14. Greish, A.A.; Finashina, E.D.; Tkachenko, O.P.; Kustov, L.M. Preparation of Propanols by Glycerol Hydrogenolysis over Bifunctional Nickel-Containing Catalysts. *Molecules* **2021**, *26*, 1565. [CrossRef] [PubMed]
15. van Ryneveld, E.; Mahomed, A.S.; van Heerden, P.S.; Green, M.J.; Friedrich, H.B. A Catalytic Route to Lower Alcohols from Glycerol Using Ni-Supported Catalysts. *Green Chem.* **2011**, *13*, 1819–1827. [CrossRef]
16. Wang, S.; Liu, H. Selective Hydrogenolysis of Glycerol to Propylene Glycol on Cu-ZnO Catalysts. *Catal. Lett.* **2007**, *117*, 62–67. [CrossRef]
17. Dasari, M.A.; Kiatsimkul, P.P.; Sutterlin, W.R.; Suppes, G.J. Low-Pressure Hydrogenolysis of Glycerol to Propylene Glycol. *Appl. Catal. A Gen.* **2005**, *281*, 225–231. [CrossRef]
18. Zhou, Z.; Li, X.; Zeng, T.; Hong, W.; Cheng, Z.; Yuan, W. Kinetics of Hydrogenolysis of Glycerol to Propylene Glycol over $Cu-ZnO-Al_2O_3$ Catalysts. *Chin. J. Chem. Eng.* **2010**, *18*, 384–390. [CrossRef]
19. Guo, L.; Zhou, J.; Mao, J.; Guo, X.; Zhang, S. Supported Cu Catalysts for the Selective Hydrogenolysis of Glycerol to Propanediols. *Appl. Catal. A Gen.* **2009**, *367*, 93–98. [CrossRef]
20. Casale, B.; Gomez, A.M. A method of hydrogenation glycerol. *Eur. Pat. Appl.* **1992**, 0523015A2.
21. Kumar, P.; Shah, A.K.; Lee, J.H.; Park, Y.H.; Štangar, U.L. Selective Hydrogenolysis of Glycerol over Bifunctional Copper-Magnesium-Supported Catalysts for Propanediol Synthesis. *Ind. Eng. Chem. Res.* **2020**, *59*, 6506–6516. [CrossRef]
22. Nakagawa, Y.; Tomishige, K. Heterogeneous Catalysis of the Glycerol Hydrogenolysis. *Catal. Sci. Technol.* **2011**, *1*, 179–190. [CrossRef]
23. Balaraju, M.; Rekha, V.; Sai Prasad, P.S.; Prasad, R.B.N.; Lingaiah, N. Selective Hydrogenolysis of Glycerol to 1, 2 Propanediol over Cu-ZnO Catalysts. *Catal. Lett.* **2008**, *126*, 119–124. [CrossRef]
24. Omar, L.; Perret, N.; Daniele, S. Self-Assembled Hybrid ZnO Nanostructures as Supports for Copper-Based Catalysts in the Hydrogenolysis of Glycerol. *Catalysts* **2021**, *11*, 516. [CrossRef]
25. Dmitriev, G.S.; Melchakov, I.S.; Samoilov, V.O.; Ramazanov, D.N.; Zanaveskin, L.N. Synthesis of 1,2-Propylene Glycol in a Continuous Down-Flow Fixed-Bed Reactor With Cu/Al_2O_3 Catalyst. *ChemistrySelect* **2022**, *7*, e202104257. [CrossRef]
26. Schmidt, S.R.; Tanielyan, S.K.; Marin, N.; Alvez, G.; Augustine, R.L. Selective Conversion of Glycerol to Propylene Glycol over Fixed Bed Raney® Cu Catalysts. In *Topics in Catalysis*; Springer: New York, NY, USA, 2010; Volume 53, pp. 1214–1216.
27. Gao, Q.; Xu, B.; Tong, Q.; Fan, Y. Selective Hydrogenolysis of Raw Glycerol to 1,2-Propanediol over Cu-ZnO Catalysts in Fixed-Bed Reactor. *Biosci. Biotechnol. Biochem.* **2015**, *80*, 215–220. [CrossRef] [PubMed]

28. Meher, L.C.; Gopinath, R.; Naik, S.N.; Dalai, A.K. Catalytic Hydrogenolysis of Glycerol to Propylene Glycol over Mixed Oxides Derived from a Hydrotalcite-Type Precursor. *Ind. Eng. Chem. Res.* **2009**, *48*, 1840–1846. [CrossRef]
29. Shesterkina, A.; Vikanova, K.; Kostyukhin, E.; Strekalova, A.; Shuvalova, E.; Kapustin, G.; Salmi, T. Microwave Synthesis of Copper Phyllosilicates as Effective Catalysts for Hydrogenation of C≡C Bonds. *Molecules* **2022**, *27*, 988. [CrossRef]
30. Bienholz, A.; Hofmann, H.; Claus, P. Selective Hydrogenolysis of Glycerol over Copper Catalysts Both in Liquid and Vapour Phase: Correlation between the Copper Surface Area and the Catalyst's Activity. *Appl. Catal. A Gen.* **2011**, *391*, 153–157. [CrossRef]
31. Favier, I.; Pla, D.; Gómez, M. Metal-Based Nanoparticles Dispersed in Glycerol: An Efficient Approach for Catalysis. *Catal. Today* **2018**, *310*, 98–106. [CrossRef]
32. Carroll, K.J.; Reveles, J.U.; Shultz, M.D.; Khanna, S.N.; Carpenter, E.E. Preparation of Elemental Cu and Ni Nanoparticles by the Polyol Method: An Experimental and Theoretical Approach. *J. Phys. Chem. C* **2011**, *115*, 2656–2664. [CrossRef]
33. Park, B.K.; Jeong, S.; Kim, D.; Moon, J.; Lim, S.; Kim, J.S. Synthesis and Size Control of Monodisperse Copper Nanoparticles by Polyol Method. *J. Colloid Interface Sci.* **2007**, *311*, 417–424. [CrossRef]
34. Sun, J.; Jing, Y.; Jia, Y.; Tillard, M.; Belin, C. Mechanism of Preparing Ultrafine Copper Powder by Polyol Process. *Mater. Lett.* **2005**, *59*, 3933–3936. [CrossRef]
35. Chokratanasombat, P.; Nisaratanaporn, E. Preparation of Ultrafine Copper Powders with Controllable Size via Polyol Process with Sodium Hydroxide Addition. *Eng. J.* **2012**, *16*, 39–46. [CrossRef]
36. Fievet, F.; Ammar-Merah, S.; Brayner, R.; Chau, F.; Giraud, M.; Mammeri, F.; Peron, J.; Piquemal, J.Y.; Sicard, L.; Viau, G. The Polyol Process: A Unique Method for Easy Access to Metal Nanoparticles with Tailored Sizes, Shapes and Compositions. *Chem. Soc. Rev.* **2018**, *47*, 5187–5233. [CrossRef] [PubMed]
37. Sun, C.; Zeng, P.; He, M.; He, X.; Xie, X. Morphological Effect of Non-Supported Copper Nanocrystals on Furfural Hydrogenation. *Catal. Commun.* **2016**, *86*, 5–8. [CrossRef]
38. Copper X-ray Photoelectron Spectra, Copper Electron Configuration, and Other Elemental Information. Available online: https://www.thermofisher.com/ru/ru/home/materials-science/learning-center/periodic-table/transition-metal/copper.html (accessed on 1 November 2022).
39. Montassier, C.; Ménézo, J.C.; Hoang, L.C.; Renaud, C.; Barbier, J. Aqueous Polyol Conversions on Ruthenium and on Sulfur-Modified Ruthenium. *J. Mol. Catal.* **1991**, *70*, 99–110. [CrossRef]

Article

TiO$_2$ Catalyzed Dihydroxyacetone (DHA) Conversion in Water: Evidence That This Model Reaction Probes Basicity in Addition to Acidity

Insaf Abdouli, Frederic Dappozze, Marion Eternot, Chantal Guillard and Nadine Essayem *

Institut de Recherches sur la Catalyse et l'Environnement de Lyon, UMR 5256, CNRS, Université Claude Bernard Lyon 1, IRCELYON, F-69626 Villeurbanne, France
* Correspondence: nadine.essayem@ircelyon.univ-lyon1.fr

Abstract: In this paper, evidence is provided that the model reaction of aqueous dihydroxyacetone (DHA) conversion is as sensitive to the TiO$_2$ catalysts' basicity as to their acidity. Two parallel pathways transformed DHA: while the pathway catalyzed by Lewis acid sites gave pyruvaldehyde (PA) and lactic acid (LA), the base-catalyzed route afforded fructose. This is demonstrated on a series of six commercial TiO$_2$ samples and further confirmed by using two reference catalysts: niobic acid (NbOH), an acid catalyst, and a hydrotalcite (MgAlO), a basic catalyst. The original acid-base properties of the six commercial TiO$_2$ with variable structure and texture were investigated first by conventional methods in gas phase (FTIR or microcalorimetry of pyridine, NH$_3$ and CO$_2$ adsorption). A linear relationship between the initial rates of DHA condensation into hexoses and the total basic sites densities is highlighted accounting for the water tolerance of the TiO$_2$ basic sites whatever their strength. Rutile TiO$_2$ samples were the most basic ones. Besides, only the strongest TiO$_2$ Lewis acid sites were shown to be water tolerant and efficient for PA and LA formation.

Keywords: titanium dioxide; dihydroxyacetone; CO$_2$ and NH$_3$ microcalorimetry; FTIR of pyridine adsorption; acidity; basicity

Citation: Abdouli, I.; Dappozze, F.; Eternot, M.; Guillard, C.; Essayem, N. TiO$_2$ Catalyzed Dihydroxyacetone (DHA) Conversion in Water: Evidence That This Model Reaction Probes Basicity in Addition to Acidity. *Molecules* **2022**, *27*, 8172. https://doi.org/10.3390/molecules27238172

Academic Editor: Lu Liu

Received: 29 October 2022
Accepted: 21 November 2022
Published: 24 November 2022

Publisher's Note: MDPI stays neutral with regard to jurisdictional claims in published maps and institutional affiliations.

Copyright: © 2022 by the authors. Licensee MDPI, Basel, Switzerland. This article is an open access article distributed under the terms and conditions of the Creative Commons Attribution (CC BY) license (https://creativecommons.org/licenses/by/4.0/).

1. Introduction

The replacement of petroleum products by biomass-derived ones has been attracting growing attention in recent years. In this regard, the valorization of biomass and its derivatives into value-added products has been largely studied by means of different processes, especially hydrothermal ones. For these processes, homogeneous and heterogeneous acid-base catalysts were used to catalyze specific reactions steps such as hydrolysis, retro-aldolization, dehydration, etc. However, in hydrothermal conditions, solid acids and bases may be inhibited by water.

Different physicochemical methods are now well established to characterize the original acid-base properties of heterogeneous catalysts. Most of them are performed in gas phase such as temperature-programmed desorption (TPD) of probe molecules [1–4]; adsorption of probe molecules monitored by microcalorimetry [1,3–5] or infrared spectroscopy (FTIR) [3,4,6], etc. Nevertheless, these techniques characterize heterogeneous catalysts in gas phase, conditions completely different from the hydrothermal medium where acid or basic sites are very often inhibited by water [7,8].

Besides, the use of model reactions is an effective method since they are conducted under conditions as similar as possible to the reactional ones. Model reactions, whose mechanisms are well known, were largely applied in the gas phase such as light alkanes or alkenes isomerization or alcohol dehydration or dehydrogenation [9–12]. Each pathway is specifically catalyzed by acid and/or basic sites. The reactions rates, the products distribution and the mechanism involved may give insight on the nature of the active sites and/or on their density or strength. In the liquid phase, the use of model reactions is less common, but it is increasing for the characterization of acid/base properties, in particular in the presence of water [13–18]. For instance, cyclohexanol dehydration was previously applied to investigate the efficiency of various solid acids [17,18], and acetone aldolization in the presence of increasing water amounts was applied to characterize the water tolerance of hydrotalcites' basic sites [8].

Recently, the production of lactic acid (LA) from dihydroxyacetone (DHA) or its derivative, pyrulvaldehyde (PA), has been extensively investigated to evaluate the Lewis acid sites water tolerance of various solid acid catalysts and was also seen to be efficient to determine their acid sites types (Lewis vs. Brønsted) [16,19–24]. It is generally accepted that the reaction proceeds in two steps. The first step, where the DHA undergoes dehydration to produce PA, is either an acid catalyzed step (Brønsted and/or Lewis) at low temperature < 100 °C or thermally induced at higher temperature [16,23]. The second step that converts PA into LA is known to be catalyzed by Lewis acid sites only (Scheme 1) [16,19,21,23].

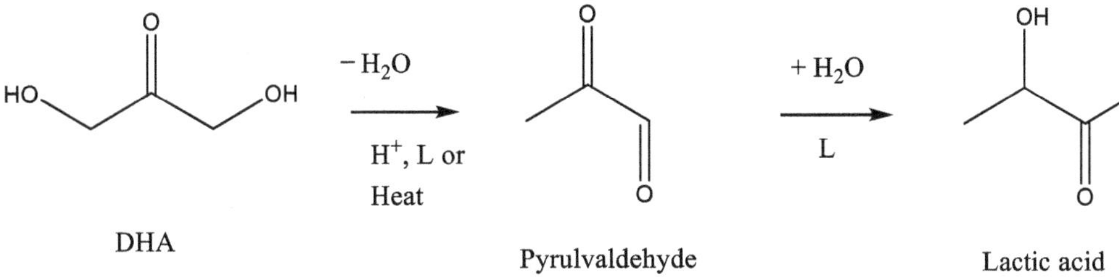

H^+: Bronsted acid sites
L: Lewis acid sites

Scheme 1. Widely accepted route for DHA conversion into lactic acid via pyruvaldehyde formation.

Many homogenous [16] or heterogeneous catalysts such as Nb_2O_5 [20,22,24], zeolites [18,19,25,26], phosphates [23], ZrO_2 [21] or doped silica [27] have been studied. Several earlier studies investigated the acid properties of TiO_2 samples using this model reaction [22,24,28] and one can note some disagreement. Anatase TiO_2, prepared by sol–gel synthesis, was shown to be a very active solid acid to produce lactic acid from pyruvaldehyde due to its high amount of Lewis acid sites, which, unlike Nb_2O_5, would not be inhibited by water adsorption [24]. Besides, Nb_2O_5 was described elsewhere to be more efficient than a commercial anatase/rutile TiO_2 [22].

It can also be noted that, despite the very broad applications of TiO_2, especially as support and in photocatalysis, there is no established agreement on the original acid base properties of a reference TiO_2 material such as TiO_2 P25 in term of nature, number of sites and strength, investigated with the usual technics in gas phase. For instance, Brønsted acid sites were not detected by FTIR of pyridine adsorption over P25 [29] or on pure anatase TiO_2 prepared by sol–gel synthesis [24,28,30,31] whereas few others mentioned the presence of Brønsted acidity over P25 or other commercial TiO_2 anatase by NH_3 adsorption monitored by FTIR [32,33]. On the other hand, the basicity of TiO_2 has been scarcely studied. One can mention the earlier ones of Watanabe et al. that revealed its bifunctionality, the presence of acid and basic sites over an anatase TiO_2, by gas phase TPD experiments [34].

More generally, there is a lack of literature on the original acid-base properties of TiO_2 as a function of its crystalline structure and/or morphology for exploring possible correlations with their (photo)catalytic activities. Therefore, probing the acid-base properties of TiO_2 samples with different crystalline structures and/or morphologies, in water, to highlight correlations with their photocatalytic behaviour, would represent a real breakthrough.

In this work, with the objective of understanding the catalytic behaviour of different TiO_2's crystalline structure (anatase/rutile) with variable specific surface areas, we studied their catalytic performances in the model reaction of DHA conversion in water. In parallel, we measured the original acidity and basicity of six selected commercial TiO_2 samples using well-established gas phase methods: namely, pyridine, NH_3 and CO_2 adsorption monitored by FTIR and microcalorimetry to get reliable data on the nature, density and strength of the original acidity and basicity of each TiO_2 samples in the absence of moisture.

In this work, we use the model reaction of DHA conversion in the aqueous phase to probe the catalyst's Bronsted/Lewis acidity, while demonstrating that this reaction can also probe the basic properties leading to DHA condensation into hexoses. This conclusion was supported by the studies of different commercial TiO_2 samples with variable acid-base properties as established by the usual gas phase methods and by the study of reference acid and base catalysts, i.e., NbOH and MgAlO

2. Results and Discussion

2.1. Catalysts Characterization

The XRD patterns of the six commercial TiO_2 catalysts are shown in Figure 1. It is confirmed that P25 and P90 are mixtures of anatase and rutile crystalline phases, UV100 and HPX-200/v2 are pure anatase and Rut160 and HPX-400C are pure rutile. From XRD measurements, the average size of TiO_2 crystallites was found to be around 8 nm for UV100 and Rut160, ~12 nm for P90 and HPX-200/v2, 16 nm for HPX-400C and around 20 nm for P25 (Table 1). The crystallite size of UV100 and P25 were in agreement with those determined by Arana et al. [32].

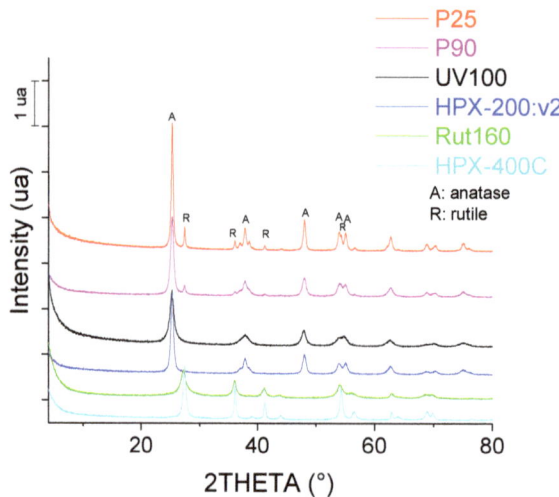

Figure 1. Diffractogram of commercial TiO_2 samples.

Table 1. Textural features and particle sizes of commercial TiO_2 samples.

Catalysts	S_{BET} [1] $(m^2 \cdot g^{-1})$	Mesopores Size [2]	S Micropore [3] $(m^2 \cdot g^{-1})$	Crystallites Size [4] (nm)
P25 (80% anatase/20% rutile)	55	>4 nm	-	18.1
P90 (92% anatase/8% rutile)	115	>5 nm	-	11.6
UV100	300	>3 nm	194	7.5
HPX-200/v2	96	>4 nm	-	13.1
Rut160	175	>4 nm	82	8.5
HPX-400C	85	>5 nm	-	16.4

[1] Pretreatment: 2 h at 300 °C under vacuum (heating rate: 5 °C.mn^{-1}), [2] BJH, [3] t-plot, [4] from XRD patterns.

The BET surface areas are roughly related to the size of the TiO_2 crystallites, which indicates that they are mainly developed by the external surfaces of the particles.

The isotherms shown in Figure 2 present the same feature, the presence of a hysteresis loop at high relative pressure, linked with the presence of large mesopores, above 3–5 nm. UV100 and Rut160 present also micropores as indicated by the significant N_2 adsorption at very low relative pressure. As shown in Table 1, the highest BET surface areas were observed for UV100 (300 $m^2.g^{-1}$) and Rut160 (175 $m^2.g^{-1}$), probably due to their smallest crystallites and the presence of micropores. The P25 showed the lowest surface area (55 $m^2.g^{-1}$), whereas the rest of the samples, i.e., P90 (mixture of anatase and rutile), HPX-200/v2 (anatase) and HPX-400C (rutile), exhibited almost similar surface areas close to 100 $m^2.g^{-1}$.

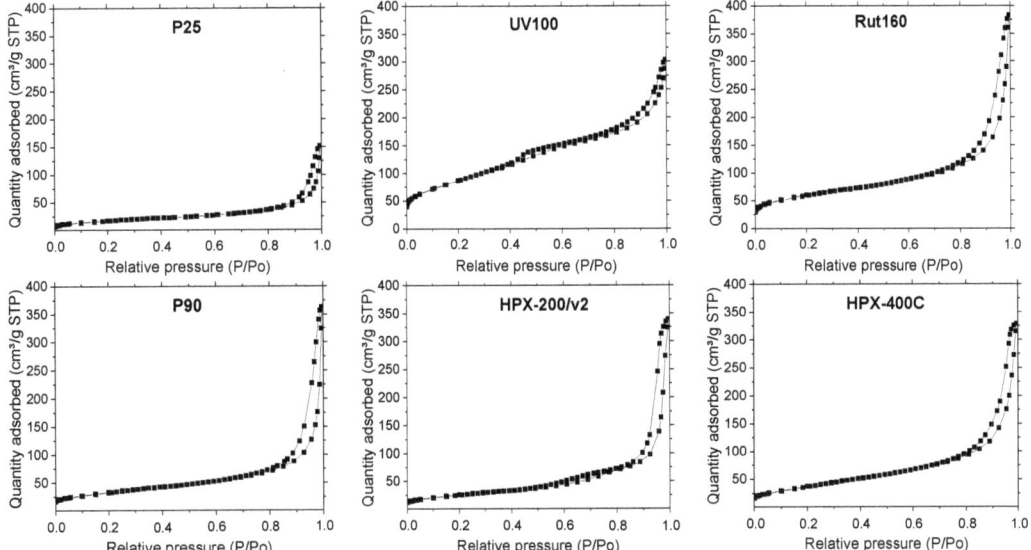

Figure 2. N_2 adsorption/desorption isotherm of the commercial TiO_2 samples. Catalyst pretreatment: 300 °C for 2 h under vacuum (heating rate: 5 °C.mn^{-1}).

2.2. Catalysts' Acidity

The catalysts' acidity was first studied by pyridine adsorption followed by FTIR to identify the nature of the acid sites, Lewis vs. Brønsted. The different FTIR spectra obtained after pyridine adsorption on TiO_2 dehydrated at 150 °C, then vacuum treated at ambient temperature and 150 °C, are shown in Figure 3. All the catalysts exhibited bands at 1445 and 1610 cm^{-1} that correspond to the vibrations characteristic of pyridine coordinated to the catalysts' Lewis acid sites. The spectra also show a band at 1490 cm^{-1} common to a vibration mode of pyridine coordinated to the catalysts' Lewis acid sites and to pyridinium ions. The band observed at 1575 cm^{-1} was previously ascribed to a vibration of pyridine linked by hydrogen bonds to the catalysts' surface [35]. However, no significant peak at 1545 cm^{-1} nor at 1640 cm^{-1}, corresponding to the vibrations characteristic of the pyridinium ions, formed in the presence of Brønsted acid sites, are clearly observed. Over Rut160 and P25, if present, this vibration is hardly detected from the baseline. Thus, from our FTIR study of pyridine adsorption on the TiO_2 samples dehydrated at 150 °C, one can conclude that all the six TiO_2 samples have only Lewis acid sites in agreement with previous investigations [24,29]. Our study does not demonstrate the presence of Brønsted acidity on these anatase and/or rutile TiO_2 samples, in contrast to others [32].

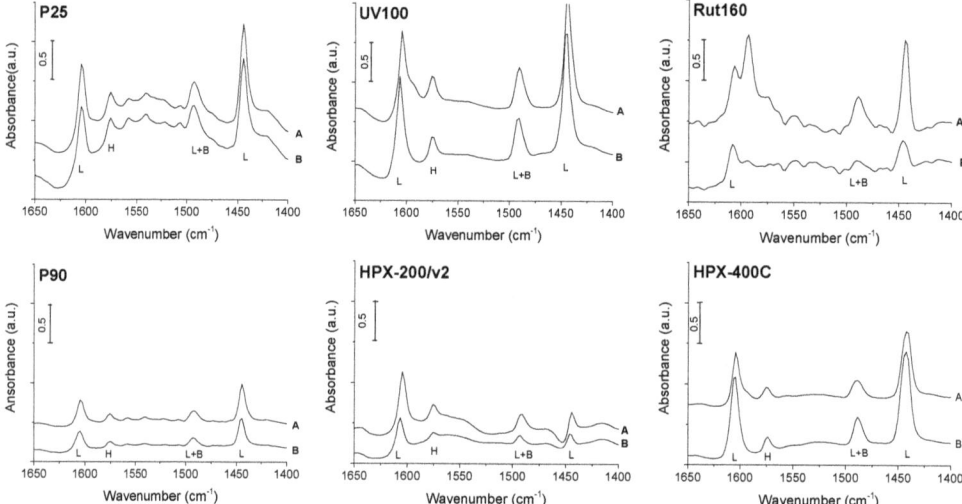

Figure 3. Infrared spectra of pyridine adsorption on P25, P90, UV100, HPX-200/v2, Rut160 and HPX-400C: (A) spectra of catalysts saturated with pyridine vapor at ambient temperature, then desorption of pyridine for 1 h at ambient temperature; (B) spectra after pyridine desorption at 150 °C for 1 h (L: characteristic vibration of Lewis acid sites, B: characteristic vibration of Brønsted acid sites, L + B: vibration common to Lewis and Brønsted acid sites, H: vibration of pyridine linked by hydrogen bonds). Conditions: Prior to pyridine adsorption, TiO_2 self-supported pellets were vacuum treated for 1 h at 150 °C (reference spectra), normalized spectra to 100 mg.

The study of the catalysts' acidity was then completed by NH_3 adsorption monitored by microcalorimetry coupled to the NH_3 isotherm measurements: a precise and reliable technique to determine the total number of acid sites and their acid strength distribution. As determined by the adsorption of pyridine followed by FTIR, all the catalysts have only Lewis acid sites; thus, the adsorption of NH_3 monitored by microcalorimetry measures the amount and the density of the Lewis acid sites.

The isotherms of NH_3 adsorption on all the samples dehydrated at 150 °C are compared in Figure 4a and the corresponding calorimetric curves are displayed in Figure 4b. The pure anatase TiO_2, UV100, which also exhibits the highest BET surface area of 300 $m^2 \cdot g^{-1}$, has more than twice the acid sites (~900 $\mu mol \cdot g^{-1}$) in comparison to other materials. On the other hand, the P25, which is made of a mixture of anatase and rutile phases with the lowest BET surface area, has the lowest amount of the acid sites, ~200 $\mu mol \cdot g^{-1}$. Total acid sites amounting to between 240 and 410 $\mu mol \cdot g^{-1}$ are measured on all other TiO_2 samples which also present intermediate BET surface areas. Since the samples have quite different specific surface areas, the calorimetric curves are compared as a function of the acid site density, expressed in μmol NH_3 per square meter (Figure 4b). The calorimetric curves of all the catalysts except Rut160 and HPX-400C show quite an equivalent pseudo plateau which indicated the presence of many acid sites of homogeneous acid strength with heat of NH_3 adsorption between 120 $kJ \cdot mol^{-1}$ and 140 $kJ \cdot mol^{-1}$ for P25, P90, HPX-200/v2 and UV100. HPX-400C presents a calorimetric curve well above the previous ones, with the presence of stronger acid sites, of homogeneous strength, characterized by a longer pseudo-plateau between 155 and 125 $kJ \cdot mol^{-1}$. In contrast, the calorimetric curve of Rut160 shows a continuous decrease of the heat of ammonia adsorption with the ammonia coverage which indicates the presence of a more heterogenous surface in terms of strength with significant lower acid strength.

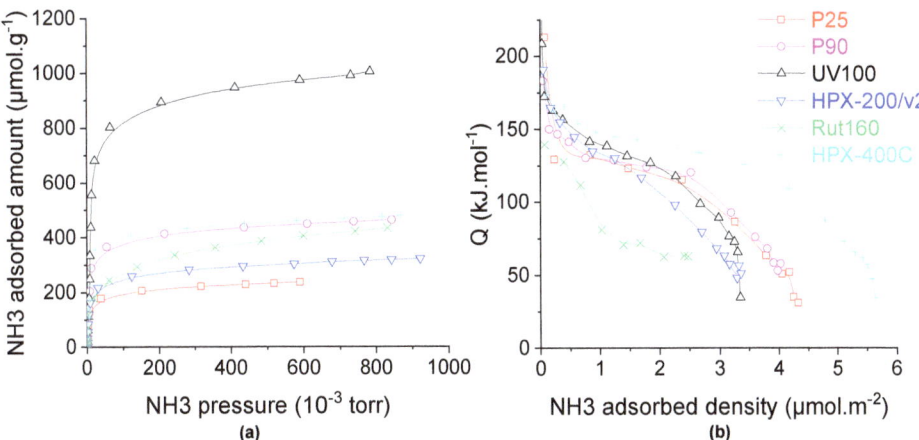

Figure 4. Ammonia adsorption on the TiO$_2$ samples: (**a**) isotherms and (**b**) calorimetric curves. Conditions: TiO$_2$ samples pre-treated at 150 °C under secondary vacuum for 5 h, NH$_3$ adsorption performed at 80 °C.

Except Rut160, all the TiO$_2$ samples present few strong acid sites with heat of ammonia adsorption higher than 150 kJ.mol^{-1}. Thus, to tentatively understand the effect of catalysts' crystalline phase and/or the crystallites' size on the TiO$_2$ acidity, the acid sites densities (acid sites amount divided by the catalyst's surface area) are compared in Table 2. As shown in Table 2, catalysts containing only anatase (UV100 and HPX-200/v2) or mainly anatase phase (80% for P25 and 92% for P90) have acid sites densities around 3 µmol.m^{-2} with close acid strength distribution. This would indicate that the Lewis acidity of anatase TiO$_2$ would not be sensitive to the structure (crystallite size). On the contrary, one can observe quite different acid site density on pure rutile TiO$_2$: around 1.4 µmol.m^{-2} and 5 µmol.m^{-2} for Rut160 and HPX-400C, respectively. Since HPX-400C with crystallites' size around 16 nm is more acidic than Rut160 with smaller particles (8.5 nm), one could suggest that the acidity of rutile TiO$_2$ would be structure sensitive and the Ti^{4+} exposed on the faces of the larger crystallites of the rutile TiO$_2$ HPX-400C would make it more acidic than the rutile TiO$_2$ Rut160.

Table 2. Catalysts' acid/basic sites amounts and densities.

Catalyst	Acid Sites Amount (µmol.g^{-1})	Acid Sites Density (µmol.m^{-2})	Basic Sites Amount (µmol.g^{-1})	Basic Sites Density (µmol.m^{-2})	Acid/Base Sites Ratio
P25	200	3.6	10	0.18	20
P90	400	3.5	15	0.13	27
UV100	900	3	120	0.4	7.5
HPX-200/v2	275	2.9	20	0.2	13.7
Rut160	240	1.4	40	0.23	6
HPX-400C	410	4.8	85	1	4.8

Conditions: TiO$_2$ samples pretreated at 150 °C under secondary vacuum for 5 h, NH$_3$ adsorption performed at 80 °C, CO$_2$ adsorption performed at 30 °C.

2.3. Catalysts Basicity

The catalysts' basicity was studied in the gas phase by carbon dioxide adsorption monitored by microcalorimetry. The CO$_2$ adsorption isotherms (Figure 5a) show that anatase UV100, which has the highest BET surface area (300 m^2.g^{-1}), has the highest amount of basic sites (120 µmol.g^{-1}), then the rutile catalysts (HPX-400C and Rut160) have an intermediate basicity (85 and 40 µmol.g^{-1}, respectively) not correlated with their surface areas (85 vs. 175 m^2.g^{-1}). Anatase HPX-200/v2 and anatase/rutile catalysts (P25 and P90) had the lowest basic sites amounts, <25 µmol.g^{-1}. Since the different TiO$_2$ samples have different surface areas, the calorimetric curves are also displayed as a function of the basic sites' densities (basic sites amount divided by the catalyst's surface area). HPX-400C (rutile) and UV100 (anatase) showed by far the highest basic sites density with a significant proportion of very weak basic sites in the case of HPX-400C, with Q$_{diffCO2}$ between 40 and 80 kJ.mol^{-1}. As shown in Table 2, the other catalysts had lower basic sites densities, <0.23 µmol.m^{-2}.

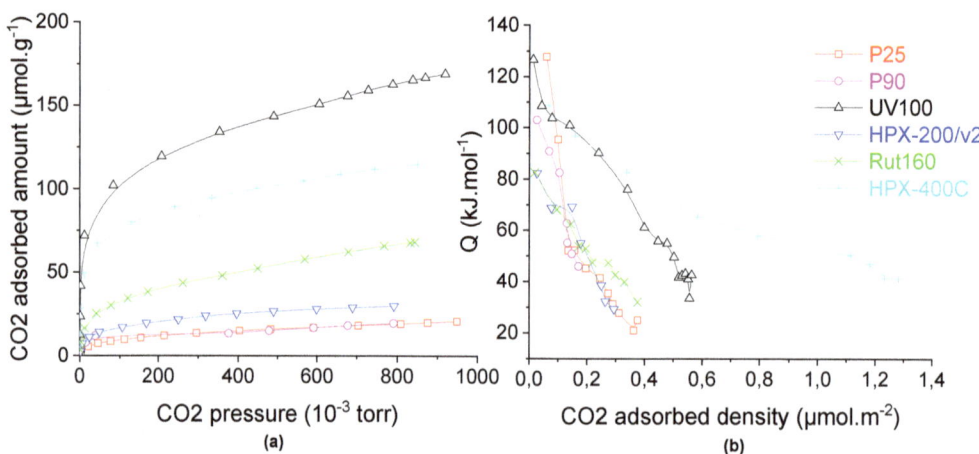

Figure 5. Carbon dioxide adsorption on TiO$_2$ (a) isotherms and (b) calorimetric curves. Conditions: TiO$_2$ samples pre-treated at 150 °C under secondary vacuum for 5 h, CO$_2$ adsorption at 30 °C.

The calorimetric curves of the six catalysts (Figure 5b) showed a progressive decrease of CO$_2$ adsorption's heat with the CO$_2$ coverage which indicates the presence of basic sites with different strength, i.e., quite heterogeneous surfaces in terms of basic strength.

From these values, one can conclude that there is no correlation between either the basic properties and the TiO$_2$ structure or the crystallites sizes and specific surface areas.

However, the superficial acid/base balance, reported in Table 2, seems to differentiate the anatase from the rutile phase: the rutile TiO$_2$ surfaces appear to exhibit the lowest acid/base balance with a ratio close to 5; while the highest acid/base ratio, between 7.5 and 27, is measured for the pure anatase or the anatase/rutile mixture. In agreement with these remarks, P25, which contains 20% of the rutile phase, presents a lower acid/base balance than P90, which contains only 8% of the rutile phase.

Note that the surface acid/base balance does not seem to depend on the morphology of the TiO$_2$ samples nor to the presence of impurities such as S or Cl traces (Table S1).

2.4. DHA Conversion in Water

As discussed above, the model reaction of DHA conversion in water was investigated in the presence of the six TiO$_2$ samples with different original acid base properties as characterized above in the gas phase. Our initial goal was to determine, a priori, their Lewis acid sites water tolerance since it is largely accepted that DHA conversion would proceed via an acid catalyzed cascade transformation producing first pyrulvaldehyde and then lactic acid in the presence of water-tolerant Lewis acid catalysts. Scheme 1 represents the largely accepted mechanism of DHA transformation.

Figure 6 shows the DHA conversion as a function of time in the presence of the six TiO$_2$ catalysts. Among the most active TiO$_2$ samples, pure rutile HPX-400C and Rut160 led to DHA conversions around 90% and 80%, respectively, after 400 min of reaction, and UV100 (pure anatase) converted also ~80% of DHA during the same reaction time. The anatase/rutile mixture, P25, appears as the less active TiO$_2$, leading to the DHA conversion of ~40% at the reaction end. The other anatase/rutile mixture (P90) is slightly more active, with ~60% DHA conversion after 400 min, similar to HPX-200/v2 (pure anatase TiO$_2$).

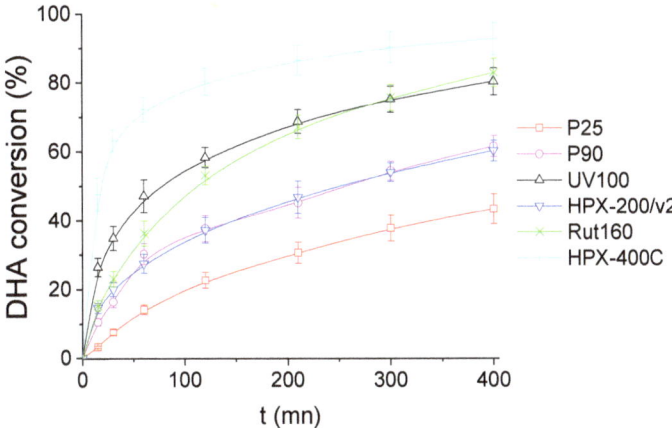

Figure 6. DHA conversion with the time course of the reaction in the presence of the six TiO$_2$. Conditions: T = 90 °C, P$_{air}$ = 1 atm, V$_{water}$ = 200 mL, (DHA) = 0.1 mol.L^{-1}, (TiO$_2$) = 10 g.L^{-1}.

From the TiO$_2$ activity ranking demonstrated in Figure 6 and the acid-base properties summarized in Table 2, the correlation with the basicity is highlighted: the TiO$_2$ samples with the lowest acid/basic sites balance, i.e., HPX-400C, Rut160 and UV100, are the most active. Moreover, one can conclude for the absence of any correlation with acidity. Indeed, we searched possible correlations between the initial rate of DHA conversion and the total amount of acid sites (µmol.g^{-1}) or the total acid sites densities (µmol.m^{-2}) in order to take into account the variability of the samples' surfaces, but no link was observed between the initial rate of DHA conversion and the original acidity of the samples determined in the gas phase (Figures S1 and S2).

We think that these results are of peculiar significance. Indeed, the absence of a correlation between the DHA conversion and the original acidity of the TiO$_2$ samples determined in the gas phase might be explained by the variability of the acid sites water tolerance as a function of the TiO$_2$ crystallinity or morphologies.

The evolutions of the products' yields with time or with the DHA conversion are presented in Figure 7 and Figure S3, respectively.

In the presence of the most active catalysts, HPX-400C, Rut160 and UV100, the formation of fructose appears as a main initial pathway in the transformation of DHA in addition to PA formation. However, whereas fructose is the main primary product in the presence of Rut160, fructose and PA are two primary products formed in equivalent amount in the presence of HPX-400C, and the formation of PA is initially slightly favoured with respect to fructose in the presence of UV100. In the presence of all the other TiO$_2$ (HPX-200/v2, P25 and P90), the initial formation of PA and its further conversion into LA is the main pathway. With the reaction progress over the first three catalysts, the fructose yield declines and we can observe a parallel increase of the glucose formation, explained by the base catalyzed fructose–glucose isomerization. The initial formation of fructose well agreed with the observed link between the TiO$_2$ samples' activity and the basicity of their surface. Indeed, fructose, which is C6 sugar, can be formed by aldolic condensation of the trioses DHA and glyceraldehyde, easily formed by DHA isomerization. Note that glyceraldehyde is detected as minor product. Both reactions, aldolic condensation and the aldo-ketose isomerization, responsible for fructose–glucose and DHA–glyceraldehyde isomerization, are base catalyzed steps [36,37]. Accordingly, a linear increase of initial rates of hexoses formation with the basic sites densities can be drawn (Figure 8). Apparently, the presence of micropores in Rut160 does not interfere with the trioses condensation reaction.

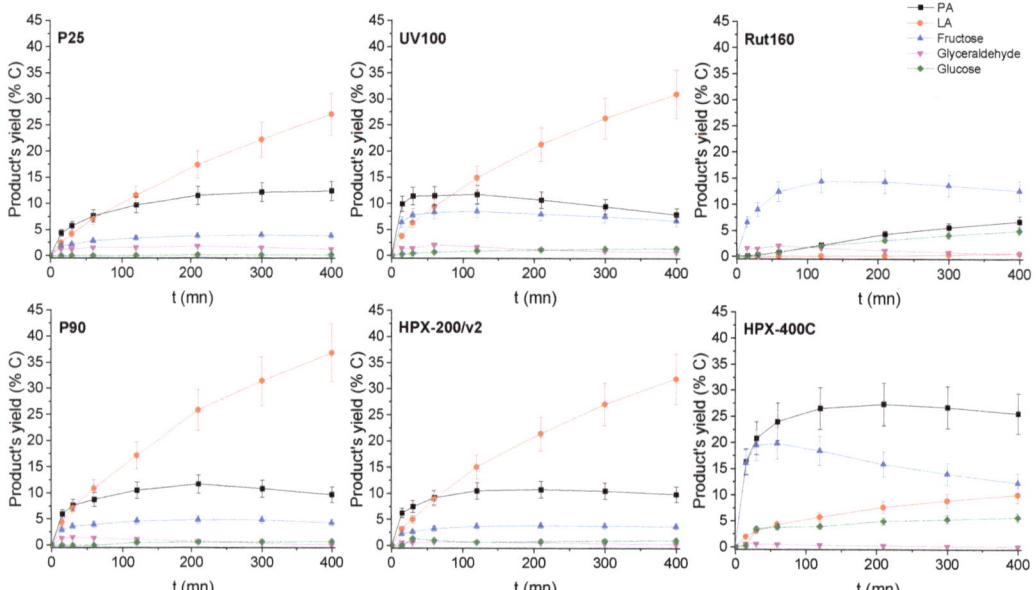

Figure 7. Products' yield time course in the presence of the six TiO$_2$s. Conditions: T = 90 °C, P$_{air}$ = 1 atm, V$_{water}$ = 200 mL, (DHA) = 0.1 mol.L^{-1}, (TiO$_2$) = 10 g.L^{-1}.

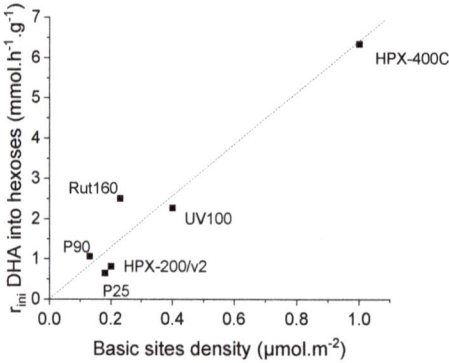

Figure 8. Increase of initial rate of hexoses formation as a function of TiO$_2$'s basic sites density.

Besides, the less active materials (P25, P90 and HPX-200/v2) presented activities consistent with the previous art: the initial formation of PA from DHA dehydration, which slows down with the reaction progress while the LA yield continuously increased (Figure 7). This fits well with the generally accepted mechanism which involves, first, an acid catalyzed (Brønsted or Lewis) DHA dehydration, and then, its rehydration into LA in the presence of water-tolerant Lewis acids sites [16,19]. However, it is interesting to note that LA is rapidly formed. At a DHA conversion lower than 10%, the amount of LA is already practically equivalent to that of PA. Such a kinetic profile was already reported over a solely water-tolerant Lewis acid catalyst such as Sc(OTf)$_3$ (OTf = triflate) [38]. Most likely, the two steps, DHA dehydration into PA and its further rehydration into LA, occur solely on the Lewis acid sites of TiO$_2$ that have kept efficiency in water. Presumably, this would suggest that the successive transformation of PA into LA would not be the rate-limiting step when only Lewis acids sites are involved. LA achieved a maximum yield of ~40% over P90 after 6 h corresponding to a selectivity of about 60% at a DHA conversion of 60% against less than 5% for pure rutile. Note that fructose, glucose, and glyceraldehyde were detected over all TiO_2 samples but in low proportions over the less active catalysts. The aldo-ketose isomerization was reported to be catalyzed as well by basic or Lewis acid sites [36,37,39].

Again, we tried to highlight an expected relationship between the initial rates of PA and LA formation and the catalyst acidity. Figure S4 evidences the absence of correlation between the initial rate of PA and LA formation with the total amount of acid sites determined in gas phase expressed per g of catalyst. The initial formation of PA and LA matches more with the total acid site density determined in gas phase and expressed in μmol per m^2: Figure 9 shows a progressive increase of the initial rates of PA and LA formation with the rise of the total acid sites densities.

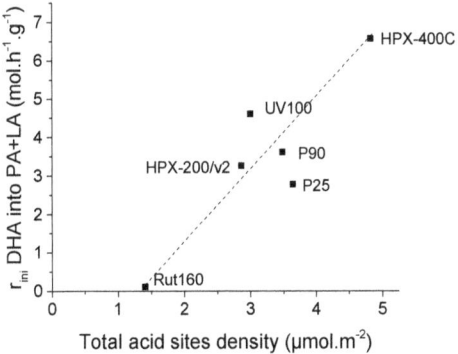

Figure 9. Increase of initial rates of pyruvaldehyde and lactic acid formation with TiO_2's total acid sites density.

This could have various explanations:

(1) A surface with enhanced proximity between the acid sites could be less inhibited by water adsorption if the DHA adsorption would require several adsorption sites. In other words, a surface with a higher density of Lewis acid sites would favor the competitive adsorption of triose sugars as regards to water, i.e., this type of surface would be more water tolerant when a triose sugar conversion is concerned. This observation is consistent with our previous observation of the improved adsorption of glucose on TiO_2 samples, which have a high density of original Lewis acid sites (determined in the gas phase) [40].

(2) Disagreeing with most of the previous literature, this could also indicate that DHA dehydration–isomerization into pyruvaldehyde and lactic acid could proceed via a bimolecular mechanism with the intermediate formation of dimeric intermediates as suggested in a recent paper [23].

It is also worth noticing that the dashed linear line, which is simply a guide for eyes to highlight the progressive increase of initial PA and LA rates formation, does not pass through the origin point but crosses the X-axis at a value ~1.5 $\mu mol.m^{-2}$. Presumably, this could be consistent with the assumption that the water tolerance of the TiO_2 surface as regards to triose sugar adsorption, and then its conversion would require a minimum density of Lewis acid sites or acid strength. In fact, when we plot the initial rate of PA and LA formation against the acid sites' density of the strongest original acid sites, characterized by a heat of ammonia adsorption higher than 130 $kJ.mol^{-1}$ as determined by microcalorimetry, the dashed line goes through the origin (Figure 10).

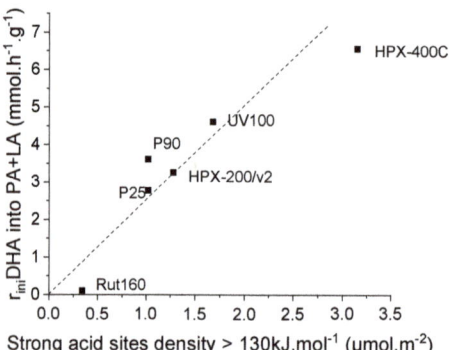

Figure 10. Linear increase of the initial rates of pyruvaldehyde and lactic acid formation with TiO$_2$'s original strong acid sites density (sites with heat of ammonia adsorption > 130 kJ.mol^{-1} as determined by calorimetry of ammonia adsorption).

This better correlation would suggest that an additional parameter would intervene for a more efficient adsorption/conversion of triose sugar in water. The higher the strength of the original Lewis acid sites, the more efficient would be the coordination of triose sugar as regards to neutral water. Below these limits, an acid sites density of ~1.5 µmol.m^{-2} and an acid strength of 130 kJ.mol^{-1}, the water/triose competitive adsorption would be favorable to water adsorption and the TiO$_2$ surface would appears as less water tolerant.

At last, LA yields as a function of DHA conversion are illustrated in Figure 11. It may be seen that, at a given DHA conversion, LA yield was the highest for P25, intermediate for the other anatase TiO$_2$ catalysts (pure or mixtures) and almost nil for rutile TiO$_2$ (Rut160 and HPX-400C).

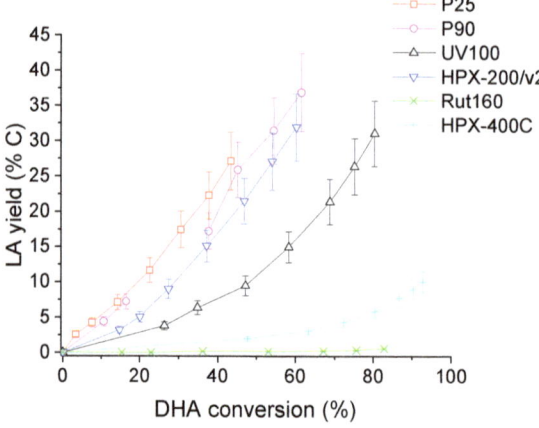

Figure 11. Evolution of lactic acid (LA) yield with DHA conversion in the presence of the six TiO$_2$. Conditions: T = 90 °C, P$_{air}$ = 1 atm, V$_{water}$ = 200 mL, (DHA) = 0.1 mol.L^{-1}, (TiO$_2$) = 10 g.L^{-1}.

To confirm the observed acidity/basicity effect on DHA transformation in the presence of the TiO$_2$ catalysts, a well-known water-tolerant solid acid (Lewis and Brønsted acid sites), NbOH [38,41] and a basic heterogeneous catalyst, MgLaO [42,43], were used to convert DHA.

NbOH (Figure 12) exhibited an activity close to the most acid TiO$_2$ sample (Figure 7) in water, where PA was initially formed from DHA dehydration and then its formation slowed down with the reaction progress while the LA yield increased continuously. NbOH produced more PA (maximum yield of 55 C%) (Figure 12) than the most acid TiO_2 catalyst (maximum yield around 15 C%) (Figure 7).

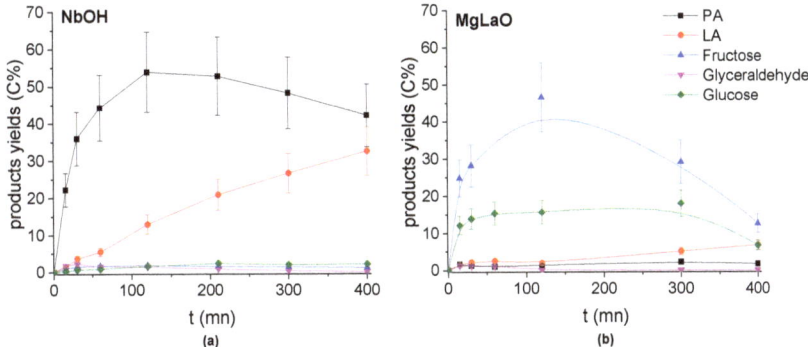

Figure 12. Evolution of the Products' yields with the reaction time in the presence of (**a**) NbOH and (**b**) MgLaO. Conditions: T = 90 °C, P_{air} = 1 atm, V_{water} = 200 mL, (DHA) = 0.1 mol.L^{-1}, (catalyst) = 10 g.L^{-1}.

On the contrary, MgLaO exhibited an activity similar to the most basic TiO_2 (Figure 12) where the main products of DHA conversion were the condensation products: fructose and glucose. PA and LA were formed with very low yields.

Finally, we plotted hexoses selectivities and the sum of PA and LA selectivities at an equivalent DHA conversion level (40%) in Figure 13. The progressive increase of hexoses selectivities against the progressive decrease of PA + LA selectivities is observed. This opposite behavior appears to be roughly linked to the original acid/base balance measured on the TiO_2 samples. The rutile TiO_2, characterized by the lowest acid/base balance (Table 2), was the most selective for hexoses formation among the TiO_2 samples with a behavior close to that of the reference basic catalyst MgLaO. On the other hand, P25 with the highest acid/base balance was the most selective for PA and LA formation close to the selectivity observed over NbOH, the reference acid catalyst.

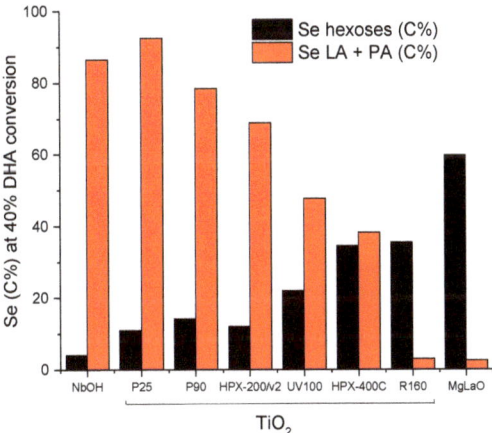

Figure 13. Evolutions of hexoses selectivity and Pyruvaldehyde/lactic acid (PA + LA) selectivities at 40% DHA conversion as a function of the TiO_2's acid/base balance. Comparison with the reference solid acid, NbOH, and the reference solid base, MgLaO.

So, these results show that DHA conversion can be converted into LA and PA by the catalysts' acid sites and/or to condensation products (fructose, glucose) by the catalysts' basic sites. This reaction appears as a suitable model reaction to characterize any catalysts' acid and basic sites in water.

To rationalize all the above reported correlations between the original acid or base properties of the TiO_2 samples and their activities/selectivities for DHA conversion in water, a mechanism is proposed (Scheme 2). This mechanism involves two parallel pathways: one catalyzed by the catalysts' basic sites leading to fructose and glucose and the other one promoted by the strongest Lewis acid sites leading to DHA isomerization into LA. The equilibrium between these two pathways depends on the superficial acid/base balance of the catalyst.

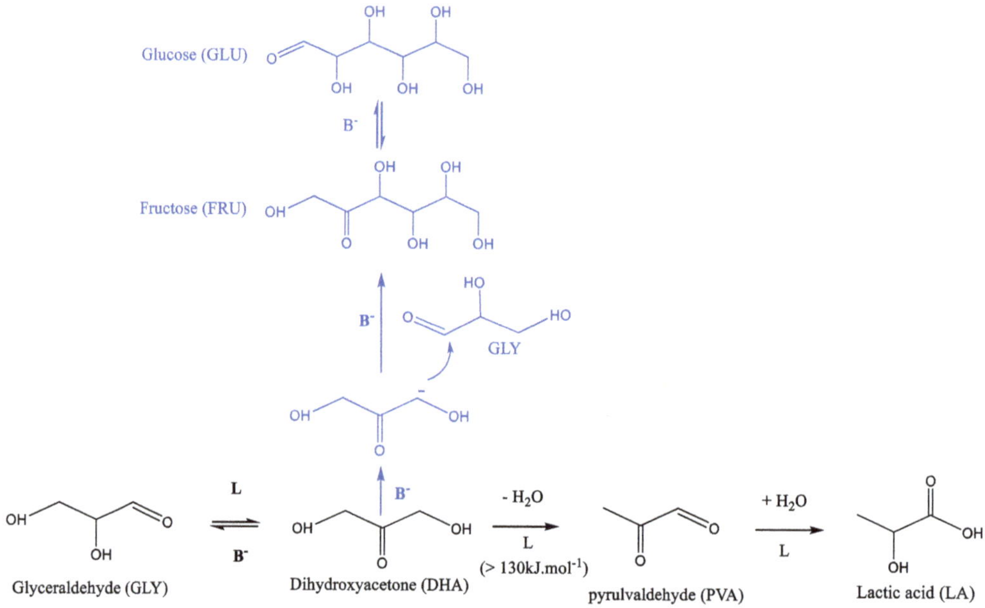

Scheme 2. Proposed mechanism which would prevail on bifunctional acid-base catalysts.

3. Materials and Methods

3.1. Commercial Titanium Dioxide (TiO_2)

Six commercial TiO_2 with different crystalline phases and different specific surface areas were tested: P25 and P90, which are mixture of anatase and rutile, provided by Evonik, two pure anatase: UV100 from Hombikat and HPX-200/v2 from Cristal and two pure rutile: Rut160 from Nanostructured and Amorphous Materials Incorporation and HPX-400C from Cristal. Rut160 was calcined in a muffle oven at 400 °C during 20 h to eliminate organic pollution from its surface. All the other catalysts were used without any treatment. The main impurities of the TiO_2 samples were summarized in the Table S1.

3.2. X-ray Diffraction (XRD)

The catalysts' crystalline structures and the sizes of their crystallites were analyzed by XRD using a Bruker D8 Advance diffractometer A25 equipped with a copper anode. Powdered samples were analyzed within a 2θ range of 4–80° at 0.02° per step and with an acquisition time of 96 s per step. The IRCELYON's XRD Service collected the diffractograms.

3.3. Brunauer–Emmett–Teller Surface Area (SBET) Analysis and Pores' Size Distribution

The catalysts' texture was characterized using nitrogen adsorption/desorption isotherm at −196 °C in a Micromeritics ASAP 2020 device. Before the analysis, the catalysts were degassed at 300 °C (heating rate 5 °C.min^{-1}) during 2 h.

3.4. Pyridine Adsorption Monitored by Fourier Transform Infrared Spectroscopy (FTIR)

The acid site types present on the catalysts (Lewis and/or Brønsted) were identified by the adsorption of pyridine monitored by FTIR. FTIR spectra were recorded with a Brucker Vector 22 spectrometer in absorption mode with a resolution of 2 cm^{-1}. Catalysts powders were pressed into self-supported pellets. The pellets were placed in an IR pyrex cell equipped with CaF_2 windows and connected to a first vacuum line allowing thermal pretreatment in the absence of pyridine pollution, then the cell was connected to a second piece of vacuum pyrex equipment to perform pyridine adsorption and desorption. All the samples were treated under flowing synthetic air (2 L.h^{-1}) for one hour at 150 °C, then under the secondary vacuum while cooling for one hour. Then, pyridine was adsorbed under saturation vapor pressure at ambient temperature. The pyridine was desorbed for one hour, firstly at ambient temperature, then at 150 °C for 1 h (to remove the physisorbed pyridine species on the catalyst). FTIR of pyridine adsorption was only used to discriminate between Lewis and Brønsted acid sites: the 19b vibration mode allows one to discriminate pyridine coordinated to Lewis acid sites (1450 cm^{-1}) from the pyridinium ion (1550 cm^{-1}). In the presence of physisorbed pyridine molecule, this vibration mode is observed at 1438 cm^{-1} [35].

3.5. NH_3 Adsorption Monitored by Microcalorimetry

The number and strength of acid sites of the catalysts were measured by ammonia adsorption at 80 °C, monitored by microcalorimetry using a Tian–Calvet calorimeter (MS80 Setaram) coupled with a volumetric equipment for isotherms measurement. This technique allows one to measure the differential heats of adsorption evolved for small and known amounts of ammonia molecules adsorbed on the catalyst surface. This energy depends on the strength of acid sites present on the catalyst surface. The total amount of acid sites is deduced from the isotherms (intersection of the tangent to the Y-axis). Thirty mg of each catalyst were placed in a glass cell and pretreated at 150 °C for 5 h under a secondary vacuum (heating rate of 2 °C.min^{-1}). The cell was then placed in the Tian–Calvet calorimeter stabilized at 80 °C for a night and coupled with a volumetric equipment. Finally, successive doses of ammonia were brought into contact with the catalyst while the heats of adsorption were recorded.

3.6. CO_2 Adsorption Monitored by Microcalorimetry

The number and strength of the basic sites of the catalysts were measured by the same calorimetric technique but using carbon dioxide as a probe, and the adsorption was done at 30 °C. The catalysts (50 mg) were treated exactly as for the acid sites quantification, following the same pretreatment procedure.

The total amount of basic sites is deduced from the isotherms (intersection of the tangent to the Y-axis).

3.7. Dihydroxyacetone (DHA) Conversion in Water

The reactions were conducted in a three-neck round bottom flask (250 mL) equipped with a water condenser, a temperature sensor and a rubber septum. At first, water and catalyst (10 g.L^{-1}) were placed in the flask (total volume of 200 mL) and heated at 90 °C in a controlled temperature oil bath and agitated by a magnetic stirrer. When the temperature stabilizes, DHA powder (1,3-dihydroxyacetone dimer, >97%, purchased from Sigma-Aldrich) was added (t = 0) and the reactions were conducted during 400 min. The DHA initial concentration was 0.1 mol.L^{-1}. Samples were taken with a 1 mL syringe through the septum and analyzed by high-performance liquid chromatography (HPLC): a Shimadzu Prominence system equipped with a refractive index detector (RID) and a COREGEL 107H column maintained at 40 °C to quantify the products and the residual substrate (DHA) (Figure S5). The mobile phase was acidified water (H_2SO_4 1.7 mM) with a flow of 0.6 mL.min^{-1}.

Two other non TiO$_2$ catalysts were tested as references for this reaction: a commercial Nobium hydroxide (NbOH) from CBMM, Brazil, and a home-made MgLaO [42].

4. Conclusions

Contrary to all expectations, the most active TiO$_2$ catalyst for DHA conversion has the most basic surface, as established by calorimetry in the gas phase. The formations of hexoses from DHA on these samples indicates the occurrence of a parallel pathway catalyzed by basic sites. In agreement with the previous literature, lactic acid formation is thought to be linked to the sample's original Lewis acidity.

This was observed on various TiO$_2$ samples with a variable superficial acid/base balance, and that was confirmed over two reference catalysts: an acid one NbOH and a basic one MgAlO.

Moreover, based on the original acid-base properties of the TiO$_2$ samples determined in the gas phase and the results obtained studying this model reaction in water, our results suggest that:

- The water tolerance of TiO$_2$ superficial acidity as regards to triose conversion would require a minimum Lewis acid sites density with enough acid strength ($Q_{diffNH3} > 130$ kJ.mol^{-1}).
- The TiO$_2$ basic sites would be water tolerant whatever their original basic strength.

Supplementary Materials: The following supporting information can be downloaded at: https://www.mdpi.com/article/10.3390/molecules27238172/s1, Table S1: TiO$_2$ samples' impurities quantified by X Fluorescence analysis (wt%); Figure S1: Initial rate of DHA conversion as a function of TiO$_2$'s acid sites amount determined in the gas phase; Figure S2: Initial rate of DHA conversion as a function of TiO$_2$'s acid sites density determined in the gas phase; Figure S3: Evolution of products' yields with DHA conversion in the presence of the six TiO$_2$s, NbOH and MgLaO; Figure S4: Initial rate of pyruvaldehyde (PA) and lactic acid (LA) formation as a function of TiO$_2$'s acid sites amount determined in the gas phase; Figure S5: HPLC Chromatograph of products after reaction for 400 min catalyzed by P25 (1 glucose, 2: fructose, 3: glyceraldehyde, 4: pyruvaldehyde, 5: lactic acid, 6: dihydroxyacetone).

Author Contributions: Investigation, I.A.; validation, I.A., C.G. and N.E.; writing—original draft preparation, I.A.; formal analysis, F.D. and M.E.; writing—review and editing, C.G. and N.E.; funding acquisition, C.G. and N.E.; supervision, C.G. and N.E. All authors have read and agreed to the published version of the manuscript.

Funding: This research was funded by the French ANR agency, grant number ANR-17-CE06-0011.

Data Availability Statement: The data presented in this study are available on request from the corresponding author.

Conflicts of Interest: The authors declare no conflict of interest.

Sample Availability: Samples are available on request from the corresponding author.

References

1. Damjanović, L.; Auroux, A. Determination of Acid/Base Properties by Temperature Programmed Desorption (TPD) and Adsorption Calorimetry. In *Zeolite Chemistry and Catalysis*; Chester, A.W., Derouane, E.G., Eds.; Springer Netherlands: Dordrecht, The Netherlands, 2009; pp. 107–167, ISBN 978-1-4020-9677-8.
2. Hunger, B.; Hoffmann, J. Temperature-Programmed Desorption (TPD) of Ammonia from H$^+$-Exchanged Zeolites with Different Structures. *J. Therm. Anal.* **1988**, *33*, 933–940. [CrossRef]
3. Che, M.; Vedrine, J.C. *Characterization of Solid Materials and Heterogeneous Catalysts: From Structure to Surface Reactivity*; John Wiley & Sons: Hoboken, NJ, USA, 2012; ISBN 978-3-527-64533-6.
4. Karge, H.G. Concepts and Analysis of Surface Acidity and Basicity. In *Handbook of Heterogeneous Catalysis*; Wiley-VCH: Weinheim, Germany, 2008; Volume 2, pp. 1096–1122, ISBN 978-3-527-61004-4.
5. Auroux, A. Acidity and Basicity: Determination by Adsorption Microcalorimetry. In *Acidity and Basicity*; Molecular Sieves; Springer Berlin Heidelberg: Berlin/Heidelberg, Germany, 2006; Volume 6, pp. 45–152, ISBN 978-3-540-73963-0.
6. Knözinger, H. Infrared Spectroscopy for the Characterization of Surface Acidity and Basicity. In *Handbook of Heterogeneous Catalysis*; American Cancer Society: Atlanta, GA, USA, 2008; pp. 1135–1163, ISBN 978-3-527-61004-4.
7. Hara, M. Heterogeneous Lewis Acid Catalysts Workable in Water. *Bull. Chem. Soc. Jpn.* **2014**, *87*, 931–941. [CrossRef]

8. Wang, Z.; Fongarland, P.; Lu, G.; Essayem, N. Reconstructed La-, Y-, Ce-Modified MgAl-Hydrotalcite as a Solid Base Catalyst for Aldol Condensation: Investigation of Water Tolerance. *J. Catal.* **2014**, *318*, 108–118. [CrossRef]
9. Guisnet, M.; Pinard, L. Characterization of Acid-Base Catalysts through Model Reactions. *Catal. Rev.* **2018**, *60*, 337–436. [CrossRef]
10. Guisnet, M.; Bichon, P.; Gnep, N.S.; Essayem, N. Transformation of Propane, n-Butane and n-Hexane over $H_3PW_{12}O_{40}$ and Cesium Salts. Comparison to Sulfated Zirconia and Mordenite Catalysts. *Top. Catal.* **2000**, *11*, 247–254. [CrossRef]
11. Essayem, N.; Ben Taârit, Y.; Feche, C.; Gayraud, P.Y.; Sapaly, G.; Naccache, C. Comparative Study of N-Pentane Isomerization over Solid Acid Catalysts, Heteropolyacid, Sulfated Zirconia, and Mordenite: Dependence on Hydrogen and Platinum Addition. *J. Catal.* **2003**, *219*, 97–106. [CrossRef]
12. Drouilly, C.; Krafft, J.-M.; Averseng, F.; Lauron-Pernot, H.; Bazer-Bachi, D.; Chizallet, C.; Lecocq, V.; Costentin, G. Role of Oxygen Vacancies in the Basicity of ZnO: From the Model Methylbutynol Conversion to the Ethanol Transformation Application. *Appl. Catal. Gen.* **2013**, *453*, 121–129. [CrossRef]
13. Frey, A.M.; Yang, J.; Feche, C.; Essayem, N.; Stellwagen, D.R.; Figueras, F.; de Jong, K.P.; Bitter, J.H. Influence of Base Strength on the Catalytic Performance of Nano-Sized Alkaline Earth Metal Oxides Supported on Carbon Nanofibers. *J. Catal.* **2013**, *305*, 1–6. [CrossRef]
14. Karaki, M.; Karout, A.; Toufaily, J.; Rataboul, F.; Essayem, N.; Lebeau, B. Synthesis and Characterization of Acidic Ordered Mesoporous Organosilica SBA-15: Application to the Hydrolysis of Cellobiose and Insight into the Stability of the Acidic Functions. *J. Catal.* **2013**, *305*, 204–216. [CrossRef]
15. Mei, D.; Lercher, J.A. Effects of Local Water Concentrations on Cyclohexanol Dehydration in H-BEA Zeolites. *J. Phys. Chem. C* **2019**, *123*, 25255–25266. [CrossRef]
16. Jolimaitre, E.; Delcroix, D.; Essayem, N.; Pinel, C.; Besson, M. Dihydroxyacetone Conversion into Lactic Acid in an Aqueous Medium in the Presence of Metal Salts: Influence of the Ionic Thermodynamic Equilibrium on the Reaction Performance. *Catal. Sci. Technol.* **2018**, *8*, 1349–1356. [CrossRef]
17. Vilcocq, L.; Koerin, R.; Cabiac, A.; Especel, C.; Lacombe, S.; Duprez, D. New Bifunctional Catalytic Systems for Sorbitol Transformation into Biofuels. *Appl. Catal. B Environ.* **2014**, *148–149*, 499–508. [CrossRef]
18. Taarning, E.; Saravanamurugan, S.; Spangsberg Holm, M.; Xiong, J.; West, R.M.; Christensen, C.H. Zeolite-Catalyzed Isomerization of Triose Sugars. *Chem. Sustain. Energy Mater.* **2009**, *2*, 625–627. [CrossRef]
19. Hossain, M.A.; Mills, K.N.; Molley, A.M.; Rahaman, M.S.; Talaphol, S.; Lalvani, S.B.; Dong, J.; Sunkara, M.K.; Sathitsuksanoh, N. Catalytic isomerization of dihydroxyacetone to lactic acid by heat treated zeolites. *Appl. Catal. A Gen.* **2021**, *611*, 117979. [CrossRef]
20. Koito, Y.; Nakajima, K.; Kitano, M.; Hara, M. Efficient Conversion of Pyruvic Aldehyde into Lactic Acid by Lewis Acid Catalyst in Water. *Chem. Lett.* **2013**, *42*, 873–875. [CrossRef]
21. Albuquerque, E.M.; Borges, L.E.P.; Fraga, M.A.; Sievers, C. Relationship between Acid-Base Properties and the Activity of ZrO_2-Based Catalysts for the Cannizzaro Reaction of Pyruvaldehyde to Lactic Acid. *ChemCatChem* **2017**, *9*, 2675–2683. [CrossRef]
22. Santos, K.M.A.; Albuquerque, E.M.; Borges, L.E.P.; Fraga, M.A. Discussing Lewis and Brønsted Acidity on Continuous Pyruvaldehyde Cannizzaro Reaction to Lactic Acid over Solid Catalysts. *Mol. Catal.* **2018**, *458*, 198–205. [CrossRef]
23. Innocenti, G.; Papadopoulos, E.; Fornasari, G.; Cavani, F.; Medford, A.J.; Sievers, C. Continuous Liquid-Phase Upgrading of Dihydroxyacetone to Lactic Acid over Metal Phosphate Catalysts. *ACS Catal.* **2020**, *10*, 11936–11950. [CrossRef]
24. Nakajima, K.; Noma, R.; Kitano, M.; Hara, M. Titania as an Early Transition Metal Oxide with a High Density of Lewis Acid Sites Workable in Water. *J. Phys. Chem. C* **2013**, *117*, 16028–16033. [CrossRef]
25. West, R.M.; Holm, M.S.; Saravanamurugan, S.; Xiong, J.; Beversdorf, Z.; Taarning, E.; Christensen, C.H. Zeolite H-USY for the Production of Lactic Acid and Methyl Lactate from C3-Sugars. *J. Catal.* **2010**, *269*, 122–130. [CrossRef]
26. Sobus, N.; Krol, M.; Piotrowski, M.; Michorczyk, B.; Czekaj, I.; Kornaus, K.; Trenczek-Zajac, A.; Komarek, S. Conversion of dihydroxyacetone to carboxylic acids on pretreated clinoptilonite modified with iron, copper, and cobalt. *Catal. Commun.* **2022**, *171*, 106509. [CrossRef]
27. Takagaki, A.; Goto, H.; Kikuchi, R.; Oyama, S.T. Silica-supported chromia-titania catalysts for selective formation of lactic acid from a triose in water. *Appl. Catal. A Gen.* **2019**, *570*, 200–208. [CrossRef]
28. Komanoya, T.; Suzuki, A.; Nakajima, K.; Kitano, M.; Kamata, K.; Hara, M. A Combined Catalyst of Pt Nanoparticles and TiO_2 with Water-Tolerant Lewis Acid Sites for One-Pot Conversion of Glycerol to Lactic Acid. *ChemCatChem* **2016**, *8*, 1094–1099. [CrossRef]
29. Kulkarni, A.P.; Muggli, D.S. The Effect of Water on the Acidity of TiO_2 and Sulfated Titania. *Appl. Catal. Gen.* **2006**, *302*, 274–282. [CrossRef]
30. Silahua-Pavón, A.A.; Espinosa-González, C.G.; Ortiz-Chi, F.; Pacheco-Sosa, J.G.; Pérez-Vidal, H.; Arévalo-Pérez, J.C.; Godavarthi, S.; Torres-Torres, J.G. Production of 5-HMF from Glucose Using TiO_2-ZrO_2 Catalysts: Effect of the Sol-Gel Synthesis Additive. *Catal. Commun.* **2019**, *129*, 105723. [CrossRef]
31. Vishwanathan, V.; Roh, H.-S.; Kim, J.-W.; Jun, K.-W. Surface Properties and Catalytic Activity of TiO_2–ZrO_2 Mixed Oxides in Dehydration of Methanol to Dimethyl Ether. *Catal. Lett.* **2004**, *96*, 23–28. [CrossRef]
32. Araña, J.; Alonso, A.P.; Rodríguez, J.M.D.; Colón, G.; Navío, J.A.; Peña, J.P. FTIR Study of Photocatalytic Degradation of 2-Propanol in Gas Phase with Different TiO_2 Catalysts. *Appl. Catal. B Environ.* **2009**, *89*, 204–213. [CrossRef]

33. Busca, G. The Surface Acidity of Solid Oxides and Its Characterization by IR Spectroscopic Methods. An Attempt at Systematization. *Phys. Chem. Chem. Phys.* **1999**, *1*, 723–736. [CrossRef]
34. Watanabe, M.; Aizawa, Y.; Iida, T.; Aida, T.M.; Levy, C.; Sue, K.; Inomata, H. Glucose Reactions with Acid and Base Catalysts in Hot Compressed Water at 473K. *Carbohydr. Res.* **2005**, *340*, 1925–1930. [CrossRef]
35. Pichat, P. *Contribution à L'étude de L'adsorption de L'ammoniac et de la Pyridine Sur des Oxydes Isolants à l'aide de la Spectrométrie Infra-Rouge*; Lyon I University: Lyon, France, 1966.
36. Souza, R.O.L.; Fabiano, D.P.; Feche, C.; Rataboul, F.; Cardoso, D.; Essayem, N. Glucose–Fructose Isomerisation Promoted by Basic Hybrid Catalysts. *Catal. Today* **2012**, *195*, 114–119. [CrossRef]
37. Moreau, C.; Lecomte, J.; Roux, A. Determination of the Basic Strength of Solid Catalysts in Water by Means of a Kinetic Tracer. *Catal. Commun.* **2006**, *7*, 941–944. [CrossRef]
38. Nakajima, K.; Hirata, J.; Kim, M.; Gupta, N.K.; Murayama, T.; Yoshida, A.; Hiyoshi, N.; Fukuoka, A.; Ueda, W. Facile Formation of Lactic Acid from a Triose Sugar in Water over Niobium Oxide with a Deformed Orthorhombic Phase. *ACS Catal.* **2018**, *8*, 283–290. [CrossRef]
39. Palai, Y.N.; Shrotri, A.; Asakawa, M.; Fukuoka, A. Silica Supported Sn Catalysts with Tetrahedral Sn Sites for Selective Isomerization of Glucose to Fructose. *Catal. Today* **2021**, *365*, 241–248. [CrossRef]
40. Abdouli, I.; Eternot, M.; Dappozze, F.; Guillard, C.; Essayem, N. Comparison of Hydrothermal and Photocatalytic Conversion of Glucose with Commercial TiO_2: Superficial Properties-Activities Relationships. *Catal. Today* **2020**, *367*, 268–277. [CrossRef]
41. Doiseau, A.C.; Rataboul, F.; Burel, L.; Essayem, N. Synergy Effect between Solid Acid Catalysts and Concentrated Carboxylic Acids Solutions for Efficient Furfural Production from Xylose. *Catal. Today* **2014**, *226*, 176–184. [CrossRef]
42. Wang, Z.; Fongarland, P.; Lu, G.; Zhan, W.; Essayem, N. Effect of Hydration on the Surface Basicity and Catalytic Activity of Mg-Rare Earth Mixed Oxides for Aldol Condensation. *J. Rare Earths* **2018**, *36*, 359–366. [CrossRef]
43. Desmartin-Chomel, A.; Hamad, B.; Palomeque, J.; Essayem, N.; Bergeret, G.; Figueras, F. Basic Properties of MgLaO Mixed Oxides as Determined by Microcalorimetry and Kinetics. *Catal. Today* **2010**, *152*, 110–114. [CrossRef]

Review

Silica-Based Supported Ionic Liquid-like Phases as Heterogeneous Catalysts

Anna Wolny and Anna Chrobok *

Department of Chemical Organic Technology and Petrochemistry, Faculty of Chemistry,
Silesian University of Technology, Krzywoustego 4, 44-100 Gliwice, Poland
* Correspondence: anna.chrobok@polsl.pl; Tel.: +48-32-237-2917

Abstract: Supported ionic liquid phases offer several advantages related with catalysis. Immobilization of ionic liquid on the solid support provides catalytic activity or efficient matrix for active phases, as enzymes or metal compounds. Ionic liquid can be physically adsorbed on the carrier (supported ionic liquid phase) or chemically grafted to the material surface (supported ionic liquid-like phase). The use of supported ionic liquid phases improves mass transport, reduces ionic amount in the process and, most importantly, enables effortless catalyst separation and recycling. Moreover, chemical modification of the surface material with ionic liquid prevents its leaching, enhancing length of catalyst life. Silica-based materials have become an effective and powerful matrix for supported ionic liquid-like phase due to its cost-efficiency, presence of hydroxyl groups on the surface enabling its functionalization, and specific material properties, such as the size and shapes of the pores. For these reasons, supported ionic liquid-like phase silica-based materials are successfully used in the organic catalysis.

Keywords: ionic liquids; acidic ionic liquids; supported ionic liquid phase; heterogeneous catalysis; silica; immobilization

1. Introduction

In recent years, responsible production and consumption has been one of the main topics of interest in both academia and industry. The chemical industry generates large amounts of hazardous waste, along with high energy consumption, use of volatile organic solvents, expensive equipment, and often harsh work conditions [1,2]. Subsequent restrictive regulations concerning health, climate, and environmental protection have forced the chemical industry to improve its existing technologies. The 2030 Agenda for Sustainable Development, adopted by all United Nations Member States in 2015, provides 17 Sustainable Development Goals. New rules for green chemistry can be a useful tool to increase the use of green technologies and achieve sustainable development in the chemical industry [3]. Green catalysis is focused on the minimization or preferably the elimination of waste, relying on the atom economy concept and the search for new effective catalysts while avoiding toxic substances. The newly developed catalysts should be characterized by high activity, selectivity, and stability under the specific process conditions [4]. Meaningful alternatives for conventional hazardous and usually expensive catalysts are enzymes and ionic liquids [5,6].

Ionic liquids (ILs), also known as low-temperature molten salts, are compounds consisting of an organic cation and an organic or inorganic anion. A major advantage of ILs is the possibility of designing their structure by selecting the proper cation and anion while projecting specific properties, meaning they have many applications in the chemical industry [7,8]. Firstly, ILs are significant alternatives for the conventional volatile organic solvents [8]. For example, in the Bayer–Villiger oxidation of ketones in the presence of ILs, lactones and esters are obtained in short reaction times (2–20 h) and in high yields (up to 95%) [9]. ILs can also stabilize enzymes in an active conformation and enhance biocatalytic processes [10]. For example, 1-butyl-3-methylimidazolium bis(trifluoromethylsulfonyl)imide ([bmim][NTf$_2$]) was used as solvent in the chemo-enzymatic oxidation of cyclobutanones and cyclohexanones to lactones with high yields (79–95%) in the presence of 30% hydrogen peroxide. In this case, the IL improved the stability of the enzyme under harsh reaction conditions [11]. The ionic nature of ILs also makes them useful as electrolytes for lithium-ion batteries and supercapacitors [12]. Furthermore, ILs are known as extractive solvents for the isolation of high-added value compounds from biomass [13], extractive solvents for analytical chemistry [14], and absorbents for gas capture, e.g., carbon dioxide [15]. ILs can also be employed as catalysts or solvent and catalyst at the same time in many reactions, e.g., Diels–Alder cycloaddition, alkylation, and acylation, as well as various types of condensations, oxidation, esterification, and transesterification reactions [16,17].

One significant group of ILs that are used as catalysts are acidic ionic liquids (AILs). AILs can be classified according to the nature of the acidic site on the Brønsted and Lewis acid types. It is possible to introduce more than one acidic function to the structure of AILs and design ILs by the combination of Brønsted and Lewis acidic types. Brønsted acidity can be introduced to ionic liquids (BAILs) as either: an acidic hydrogen in the cation (A), an anion (B) or both (C), an acidic hydrogen located in the functional group (D) or an acidic hydrogen located in the functional group and in cation/anion (E). Lewis acidic ionic liquids (LAILs) are mainly based on halometallate anions (F) and boric atom in the cation (G) (Figure 1). The formation of dual Brønsted–Lewis AILs is also presented in Figure 1 (H) [18–20].

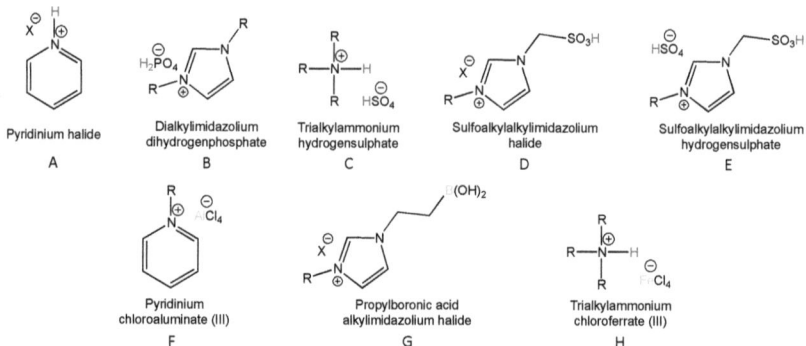

Figure 1. Examples of the structures of some acidic ionic liquids.

The most common group of BAILs are ILs with an acidic hydrogen located at the cation (A), which are also called protic acidic ionic liquids. Cations widely used for the synthesis of this type of BAIL are: 1-alkylimidazolium, 1-alkyl-2-alkylimidazolium, primary/secondary/tertiary ammonium, pyridinium, pyrolidonium, and 1,1,3,3-tetramethylguanidininium [19]. A functional group with an acidic hydrogen (e.g., -SO$_3$H, -CO$_2$H) can also be attached to the cation to obtain a BAIL [20]. An acidic site in the BAIL's anion is formed using polybasic acids such as H$_2$SO$_4$, h$_3$PO$_4$, maleic, and fumaric acids creating dialkylimidazolium, hydrogensulfate, or dihydrogenphosphate ILs [20–22]. A growing interest in green chemistry has also led to the discovery of bio-BAILs based on amino acids introduced into the structure of the cation or anion, e.g., alanine, glycine, serine, proline, and valine [23]. BAILs have been implemented in many organic reactions. For example, imidazolium-based ionic liquids functionalized with a sulfonic group were successfully employed for the hydration of alkynes under mild conditions to give ketones in high yields [24]. Some dicationic ionic liquids based on a diammonium cation and hydrogensulfate anion as environmentally benign BAILs were used for biodiesel synthesis, which was obtained with high yields and reused without significant loss of activity [25]. Imidazolium based hydrogensulfate ILs were also determined to be very efficient catalysts in the synthesis of cyclic carbonates from carbon dioxide and epoxides. high yields of cyclic carbonates (69–99%) were achieved using these ILs, which can be recycled without any loss of activity [21]. In another example for the dehydration of glycerol to acrolein using the BAIL 1-butyl-3-metylimidazolium dihydrogen phosphate, which was conducted in the liquid phase, full conversion of glycerol was achieved [22].

For the Lewis AILs, metals such as Al, Ga, Zn, Fe, In, and Sn in the form of chloride or triflate salts are used to create LAILs via complexation of the neutral IL and the metal salt in various molar ratios [18,26]. Lewis acidic cations can be formed in two ways: via a tricoordinate borenium center as a cation, or via solvation of metal cation, e.g., Li^+ as [Li(glyme)][NTf_2]/[OTf] [18]. LAILs, as well as BAILs, are readily used in organic synthesis. Water tolerant trifloaluminate ILs, synthesized from 1-alkyl-3-methylimidazolium triflates, were employed as catalysts in the cycloaddition of 2,4-dimethylphenol and isoprene to obtain a chromane. Use of the catalysts provided full conversion and high selectivity (80%) under mild reaction conditions [26]. In another example, chlorogallate(III) ILs were applied in a Bayer–Villiger oxidation of cyclic ketones to lactones. high yields (99%) in short reaction times under mild reaction conditions were also achieved [27]. Borenium LAILs used in a Diels–Alder reaction ensured good yields (90–94%) and selectivities of various dienes and dienophiles [28]. All such AILs have many applications as homogeneous catalysts [18,19], however, reducing costs and waste led to the use of heterogeneous catalysis.

Immobilization of ILs on a solid insoluble support can be performed via physical adsorption, known as supported ionic liquid phase (SILP), or via chemical bonding into the matrix, known as supported ionic liquid-like phase (SILLP) [29]. A visual representation of each can be seen in Figure 2. The IL creates a thin layer of liquid on the carrier, which decreases the amount of IL compared to the reaction in the bulk. This improves mass transfer to the catalytic centers on the fluid-fluid phase boundary and facilitates separation of the catalyst from the reaction mixture. Moreover, a heterogeneous SILP or SILLP catalyst can be successfully employed in both batch and flow processes, including fixed-bed or fluidized-bed reactors. Such applications are described later in this paper.

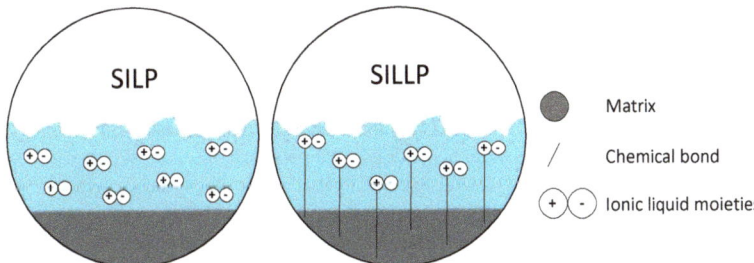

Figure 2. Supported ionic liquid phase (SILP) and supported ionic liquid-like phase (SILLP).

In this paper, achievements on the SILLP silica-based materials and their use in the organic synthesis are described. Previously, Mehnert [30] outlined the first contribution of SILPs in catalysis. Then, Sokolova et al. [31] reviewed flow processes based on catalysts immobilized on monolithic SILLPs. Next, Skoda-Földes [32] summarized the use of supported AILs in the organic synthesis, and hartmann et al. [33] characterized inorganic materials for SILLP synthesis and briefly described their input to catalysis. After that, Amarasekara [20] characterized AILs and described applications of acidic ionic liquids as SILP/SILLP, and Gruttadauria et al. described covalently-supported ionic liquid phases (SILLP) as matrices and catalysts [34], while Alinezhad et al. pointed out BAILs as SILLP in organic catalysis [35]. Then, Swadźba-Kwaśny et al. [18] briefly mentioned the applications of Lewis ILs immobilized into a solid matrix, and Leitner et al. [36] described SILP and SILLP based on nanoparticles and their applications in organic catalysis. Additionally, Vekariya [16] mentioned SILPs in the review of ILs in organic transformations. Haumann et al. [37] then presented 15 years of using SILP/SILLP catalysts in hydroformylation reactions, both in the liquid and gas phase, and Freire et al. [38] described the immobilization of ionic liquids, types of materials, and their applications. Maciejewski et al. described participation of ILs in heterogeneous catalysis, including supported IL phase catalysts (SILPC), solid catalysts with ILs (SCILL), and supported ionic liquid catalysis (SILC) techniques, as well as porous ionic liquids [39]. Moreover, Lozano et al. [40] presented applications of SILP and SILLP as supports for enzyme immobilization in organic synthesis, and Chrobok et al. [41] described SILP/SILLP biocatalysts based on nanoparticles and their applications for biocatalysis. The aim of this work is to complete the time gap and collect silica-based SILLP applications in catalysis to improve selection of the best systems for organic synthesis.

2. Immobilization of Ionic Liquids on Silica-Based Materials

The immobilization of ILs on the solid supports enables the issues related with the bulk IL systems to be overcome, such as high viscosity, mass transfer problems, IL high-cost separation, purification, regeneration, and recycling. A reduced amount of immobilized IL creates a thin layer on the matrix which, in turn, reduces costs. The possibility of creating numerous structures of ILs caused various SILPs to be designed, generating wide application potentials. Different types of materials such as silica, alumina, zeolites, polymers (e.g., polystyrene-based materials), and carbon materials (e.g., single-walled carbon nanotubes (SWCNTs), multi-walled carbon nanotubes (MWCNTs), and activated carbon) were used for such SILPs [37,41]. The most commonly used matrices are silica-based materials (e.g., silica gel, SBA-15, MCM-41 types), which are characterized by their low cost, large surface area, ordered porosity, well-defined pore geometry, and mechanical and thermal stability (except for MCM-41 type). Moreover, magnetic properties can be incorporated by coating Fe_3O_4 nanoparticles with silica, obtaining a hybrid that is even easier to separate from the reaction mixture using a magnetic field. The most important feature of silica-based materials is the presence of silanol groups (-Si-OH) on the surface, which determines the method of IL immobilization, particularly via covalent bonding (SILLP).

Physisorption is a simple method for IL immobilization that can be performed through the impregnation and adsorption from IL solution and the sol-gel procedure. The impregnation method relies on mixing the IL solution and support together before removing the solvent under vacuum conditions. The adsorption from the IL solution is accomplished by filtration, washing (to remove any excess IL), and drying under vacuum conditions. The sol-gel procedure consists of hydrolysis and polycondensation reactions of tetraethoxyorthosilicate (TEOS) in the presence of the IL, which can be described by the entrapment of the IL in the silica pores (Figure 3). The main strength of the sol-gel technique is that there is control of the molecule's growth [42]. The interactions between the IL and the silanol groups on the silica surface are based on hydrogen bonding. however, van der Waals and electrostatic interactions, as well as π-π-stacking (in the case of aromatic cation) between the IL moieties also occurs, increasing the stabilization of the SILP structure [43]. The h-bonds between -Si-OH and the IL can be confirmed via FTIR analysis, where the intensity of the characteristic peak at 952 cm^{-1} (assigned to -Si-OH) decreases if IL is present on the silica surface [44].

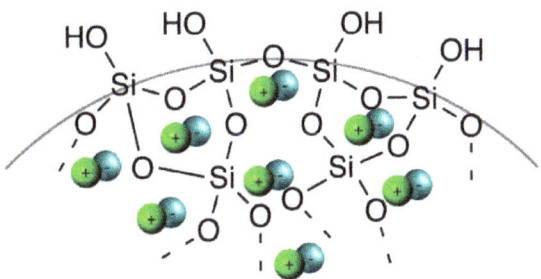

Figure 3. Ionic liquids moieties entrapped in the silica pores.

SILPs are commonly used for both chemical and biochemical processes. Entrapped triethylammonium propanesulfonate bis(trifluoromethanesulfonyl)imide [TEAPS][NTf$_2$] in the silica structure has been used for dehydration of *rac*-1-phenyl ethanol with high selectivity to styrene and recyclability (for at least 6 runs) [45]. Another BAILs, 1-methyl imidazolium hydrogen sulphate ([HMIM]HSO$_4$) and 1-methyl benzimidazolium hydrogen sulphate ([HMBIM]HSO$_4$), immobilized on silica, was applied in the isomerization of *n*-heptane and *n*-octane. The acidic SILPs showed good thermal stability high isomerization yields, were easy recyclable and environmentally friendly [46]. Then, 1-butyl-3-methylimidazolium acidic ILs with Rh-complex were immobilized on partly dehydroxylated silica surface, which created a highly active Rh/SILP catalyst dedicated for continuous hydroformylation of propene. high thermal stability, selectivity to *n*-butanal (over 95%), and TOF (turnover frequency) were observed using syn-gas and syn-gas with CO$_2$ addition [47]. For the hydrosilylation reaction, rhodium complexes immobilized in four various phosphonium based ILs anchored on silica support were applied. The amount of catalyst was reduced compared to biphasic reactions by a factor of 1000, the reaction times were shortened, and easy recycling of the Rh complexes were demonstrated [48]. The advantage of SILP catalysts in biocatalysis has also been shown. Lipase B from *Candida antarctica* (CALB) was immobilized on a SILP based on an imidazolium cation and a bis(trifluoromethanesulfonyl)imide anion used for a continuous kinetic resolution of 1-phenylethanol under supercritical CO$_2$ conditions. high enzyme activity, enantioselectivity (>99.9%), and stability (16 cycles) was achieved [49]. SILP catalysts have many advantages, such as easy and cost-efficient synthesis, where an IL multilayer on the support maintains the IL bulk properties, as well as the possibility to tailor the structure of the ILs that can be immobilized. It is worth noting that the main disadvantage is the detachment or leaching of the IL from the matrix, which is related to weak interactions between the IL and the carrier.

Covalent bonding of the IL on the surface of the support prevents its leaching and detachment. ILs immobilized as SILLPs usually create a monolayer, thus the bulk properties are lost. Methods for the preparation of SILLP silica-based materials include chemical reactions between an IL or IL precursor and hydroxyl groups present on the silica surface, or the sol-gel technique. ILs can be attached to -Si-OH group via the cation or the anion (Figure 4). Anchoring the IL into support can be obtained by direct immobilization of IL (Figure 4A) or by building the IL structure on the support (Figure 4B).

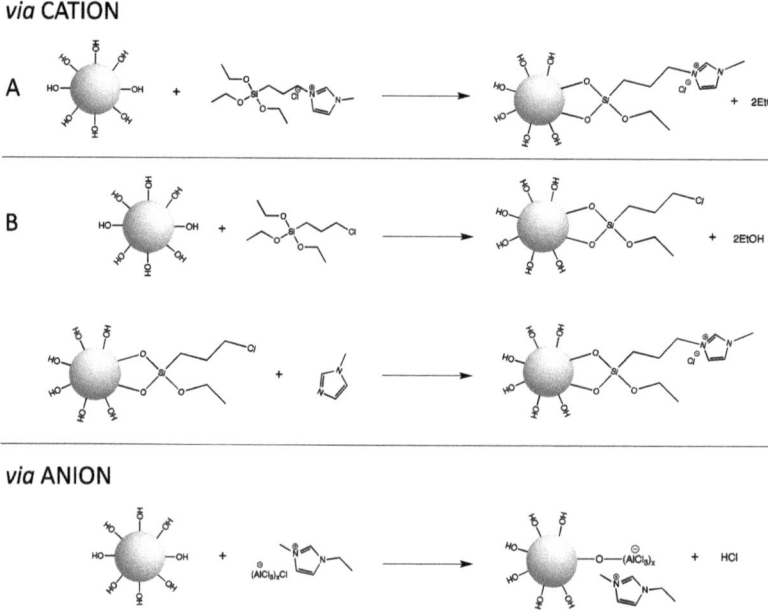

Figure 4. Covalent immobilization of ionic liquid on the silica surface via cation and via anion. Anchoring the IL into support can be obtained by direct immobilization of IL (**A**) or by building the IL structure on the support (**B**).

Typical immobilization of the IL via the cation is performed using the siliceous precursor 3-(chloropropyl)triethoxysilane, however, other precursors can also be used, e.g., 3-mercaptopropyl-trimethoxysilane [32]. As mentioned before, prepared in advance, an IL modified with ethoxysilane groups can be directly grafted to hydroxyl groups or precursors, and can be firstly anchored and be the subject of subsequent quaternization. Immobilization of IL via the cation to the siliceous surface can be confirmed by ^{29}Si MAS NMR. Peaks at -91 ppm and -101 ppm assigned to $(SiO)_2Si-(OH)_2$ and $(SiO)_3Si-OH$ groups, respectively, disappear, thus exposing the $(SiO)_4Si$ signal. Signals at -54 ppm and -66 ppm assigned to -Si-O-SiR-(OEt)$_2$ and (Si-O)$_2$-SiR-OEt, respectively, are in turn revealed [29]. If necessary, an anion exchange can be performed after the IL immobilization. Immobilization of the IL via the anion is usually observed mainly for chlorometallate ILs during the wet impregnation method where -Si-O-M bonds are obtained. For example, ^{27}Al MAS NMR spectra shows signals at 102 ppm attributed to $[Al_2Cl_7]^-$ and allows the control of the presence of $AlCl_3$ on the silica surface (1.2 ppm), which can be removed though the Soxhlet extraction [29,32]. The sol-gel method is also often used for SILLP preparation. This technique consists of polycondensation of alkoxysilane-functionalized ILs with tetrakoxysilanes, e.g., TEOS (Figure 5), and allows the control of material mesoporous character from the proper silica source/IL ratio. Besides MAS NMR spectroscopy, chemical immobilization of IL to the silica surface can be proved using FT-IR, XRD, and TEM methods.

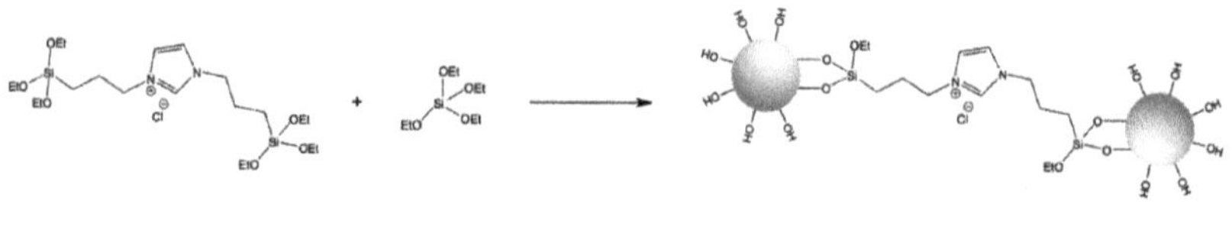

Figure 5. Covalently immobilized ionic liquid via the sol-gel method.

3. Silica-Based Supported Ionic Liquid-like Phases in Organic Catalysis
3.1. Lewis Type SILLPs

Lewis type SILLPs based on silica materials are normally synthesized via the cation method using 3-(chloropropyl)triethoxysilane, 3-(chloropropyl)trimethoxysilane, or 3-mercaptopropyltrimethoxysilane precursors, where the structures shown in Figure 6 are obtained. As can be seen in Figure 6, the material can be characterised by Lewis acidity with the Lewis center located on the alkyl chain modified with -SO_2Cl or -SO_2OH groups, or the Lewis acidity can be found in the metal halide based anion created in the complexation reaction. Table 1 presents applications of Lewis type silica based SILLP materials as catalysts in organic synthesis.

The first report on Lewis type SILLPs appeared in 2000. The presented investigations included two different immobilization methods of chloroaluminate imidazolium ILs on amorphous silica and MCM-41 supports. One of the possible SILLP synthesis routes was immobilization via the anion (Figure 4), with the second one being via the cation (**1**, Figure 6), where aluminium chloride was introduced to the IL structure in the complexation reaction. If the molar ratio of the metal halide component in the IL is more than 0.5, oligonuclear ($[Al_2Cl_7]^-$) anions are formed. The obtained heterogeneous catalysts were tested on the Friedel–Crafts alkylation, which resulted in high conversion (>90%) and selectivity (>90%) of the main product using catalyst **1** in Figure 6. In comparison, the reaction catalysed by immobilised $AlCl_3$ on the silica surface yielded only 15.7% of the main product. The better activity shown by the MCM-41 based SILLP is due to higher surface area and IL loading. Furthermore, leaching of the active phase occurred for the SILLP catalyst prepared by anion immobilization, partly due to unbonded IL moieties on the silica surface [29]. In the next report, chloroaluminate SILLP catalysts prepared via anion, cation, and sol-gel methods were used in the Friedel–Crafts alkylation of benzene with different olefins, as well as in acylation reaction. Again, leaching of the IL occurred in the SILLP prepared via anion complexation. The best activity was shown by the SILLP catalyst, where IL was grafted via the cation—almost full conversion and very high selectivity of the monoalkylated product were achieved, even at 20 °C. The lower activity of other SILLPs was most likely the result of only partly bonded acidic anions on the silica surface, which was confirmed via ^{29}Si MAS NMR analysis [50]. A tetrapropylammonium based chlorostannate (IV) IL was grafted to the silica surface via the cation (**2**, Figure 6), and used in the condensation of isobutene and formaldehyde to 3-methylbut-3-en-1-ol. Comparison of silica and MCM-41 materials resulted in better activity of the MCM-41 based SILLP catalyst in the tested reaction (α = 76%, S = 94%, Y = 71.4%, TON = 2.63·10^{-3} s^{-1}). Well-ordered and regular hexagonal pores in the MCM-41 material created micro-reactors that enhanced the SILLP catalyst activity. It is worth mentioning that the obtained heterogeneous catalyst was recyclable, and that the active phase can be used as a catalyst in the homogeneous phase as well [51]. The next report described applications of triflate Lewis type SILLP materials (**3**, Figure 6) in the synthesis of bis(indolyl)methanes [52] (Scheme 1, Figure 7), esterification of acetic or decanoic acid with various alcohols [53], nitration of aromatic compounds [53], and the addition of indole to vinyl ketones [54] (Scheme 2, Figure 7). Covalent bonding between the IL and hydroxyl groups on the silica surface was created in a radical chain transfer reaction of a 1-allylimidazolium based IL on silica gel modified with 3-mercaptopropyltrimethoxysilane. The obtained materials exhibited excellent yields, conversions, and reusability in all presented reactions. It should be pointed out that the replacement of chloroaluminate anion to triflate, and creation of Lewis centre in the cation, makes SILLP materials more resistant to water.

Table 1. Lewis type silica-based SILLP in organic catalysis.

Catalyst	Reaction Type	Reaction Conditions	Reaction Parameters	Lit.
SiO$_2$ [tespmim][Cl-AlCl$_3$] [a]	Friedel–Crafts alkylation of benzene with dodecene	6% wt. cat., benzene:dodecene (10:1; n/n), 80 °C, 1 h	A [b] > 90%, S [c] > 90%	[29]
SiO$_2$ [pmim][Cl-AlCl$_3$] [d]	Friedel–Crafts alkylation of benzene with olefins	1% wt. cat., benzene:olefin (10:1; n/n), 20 °C, 2 h	C$_6$: α = 45.3%, S = 73.8% C$_8$: α = 44.9%, S = 96.5% C$_{10}$: α = 34.1%, S = 89.6% C$_{12}$: α = 35.2%, S = 80.3% (for 6% wt. cat., 80 °C, 1 h; α = 99.4%, S = 99.7%)	[50]
SiO$_2$ [tms(p)$_4$N][Cl-SnCl$_4$] [e]	Condensation of isobutene and formaldehyde	4% mol of SnCl$_4$, isobutene:formaldehyde (1:0.1; n/n), chloroform 26 mL, 60 °C, 2 h	α = 76%, S = 94%, Y [f] = 71.4%, TON [g] = 2.63·10^{-3} s^{-1}	[51]
SiO$_2$ [p(p-SO$_2$Cl)im][OTf] [h]	Synthesis of bis(indolyl)methanes	143 mg cat., aldehyde 0.3 mmol, indole 0.5 mmol, MeCN 3 mL, rt, 1.5–9 h	Yields for: bezaldehyde 97%, p-nitrobenzaldehyde 97%, p-chlorobenzaldehyde 90%, p-acetoxybenzaldehyde 64%, p-methoxybenzaldehyde 97%, hydrocinnamaldehyde 98%	[52]
SiO$_2$ [p(p-SO$_2$Cl)im][OTf]	Esterification of acetic or decanoic acid with alcohols	Mole ratio of carboxylic acid to ionic liquid: 350, alcohol 20 mmol, carboxylic acid, 10 mmol, 100 °C, 8 h	Yields for various alcohols: (a) acetic acid: C$_8$H$_{17}$ 94.6%, C$_{10}$H$_{21}$ 95.1% (b) decanoic acid: C$_2$H$_5$ 86.3%, C$_{10}$H$_{21}$ 90.4%	[53]
SiO$_2$ [p(p-SO$_2$OH)im][OTf] [i]	Nitration of aromatic compounds	Mole ratio of aromatic compound:ionic liquid: 20, mole ratio of aromatic compound:nitric acid: 1:3, 80 °C, 4 h	Conversions for R-groups in aromatic ring: H 61.6%, Me 85.8%, Cl 10.4%, Br 22.2%	[53]
SiO$_2$ [p(p-SO$_2$Cl)im][OTf]	Addition of indole to vinyl ketones	171 mg cat., vinyl ketone 0.6 mmol, indole 0.3 mmol, Et$_2$O 0.2 mL, rt, 1.5–9 h	Yields for various ketones: 1-penten-3-one 92%, 2-cyclopentenone 88%, 3-penten-2-one 90%, benzalacetone 72%, dibenzylideneacetone 93%	[54]
SiO$_2$ [tespmim][Cl-AlCl$_3$]	Production of alkylated gasoline	0.5 g cat., iC4/C4 = 20, 80 °C, 90 min	α = 97%, S$_{C8}$ = 59.7%	[55]
SiO$_2$ [tespmim][Cl-AlCl$_3$]	Trimerization of isobutene	30% wt. cat., isobutane:isobutene molar ratio 10:1, 25 °C, 600 h^{-1}	α = 91.4%, S$_{C12}$ = 79.4%	[56]
SiO$_2$ [tespmim][Cl-FeCl$_3$] [j]	Friedel–Crafts reaction between benzene and benzyl chloride	0.05 g cat., benzene:benzyl chloride molar ratio 10:1, benzyl chloride 0.32 g, 80 °C, 45 min	α = 100%, S = 100%, 10 cycles	[57]
SiO$_2$ [(tesp)$_2$im][Cl-InCl$_3$] [k]	Friedel–Crafts reaction between benzene and benzyl chloride	0.05 g cat., benzene:benzyl chloride molar ratio 10:1, benzyl chloride 0.32 g, 80 °C, 15 min	α = 100%, S = 100%, 6 cycles	[58]
SiO$_2$ [tespmim][Cl-GaCl$_3$] [l]	Diels–Alder cycloaddition of cyclopentadiene to various dienophiles	5% mol of GaCl$_3$, cyclopentadiene:dienophile (12:8; n/n), 25 °C, 5–30 min	Methyl acrylate: α = 99%, endo:exo ratio: 95:5, 4 cycle; ethyl acrylate: α = 99%, endo:exo ratio: 93:7; diethyl maleate: α = 99%, endo:exo ratio: 93:7; methacrolein: α = 100%, endo:exo ratio: 80:20; benzoquinone: α = 83%; maleic anhydride: α = 89%	[59]
SiO$_2$ [tespmim][Cl] [m]	Cycloaddition of CO$_2$ to styrene oxide	0.5% mol cat., 0.1% mol ZnBr$_2$, styrene oxide 0.13 mol, 100 °C, P$_{CO2}$ = 1 MPa, 6 h, 700 rpm	α = 83%, Y = 72%	[60]
SiO$_2$-Zn [tespmim][Cl]	Cycloaddition of CO$_2$ to propylene oxide	S/C = 200 (PO mol per cat. mol), V$_{PO}$ = 8 mL, P$_{CO2}$ = 1.25 MPa, 100 °C, 8 h	MCM-41: α = 33%, S = 98% MSN: α = 76%, S = 97% BMMs: α = 77%, S = 98%	[61]

[a] 1-methyl-3-(triethoxysilylpropyl)imidazolium chloride—chloroaluminate (III), [b] conversion, [c] selectivity, [d] 1-propyl-3-methylimidazolium chloroaluminate (III) immobilized via anion, [e] 3-trimethoxypropyltripropylammonium chloride—chlorostannate (IV), [f] yield, [g] turnover number, [h] 1-(3-chlorosulfonylpropyl)-3-(3-trimethoxysilylmerkaptopropyl)imidazolium trifluoromethanesulfonate (triflate), [i] 1-(3-hydroxysulfonylpropyl)-3-(3-trimethoxysilylmerkaptopropyl)imidazolium triflate, [j] 1-methyl-3-(triethoxysilylpropyl)imidazolium chloride—chloroferrate (III), [k] 1-(triethoxysilylpropyl)-3-(triethoxysilylpropyl)imidazolium chloride—chloroindate (III), [l] 1-methyl-3-(triethoxysilylpropyl)imidazolium chloride—chlorogallate (III), [m] 1-methyl-3-(triethoxysilylpropyl)imidazolium chloride.

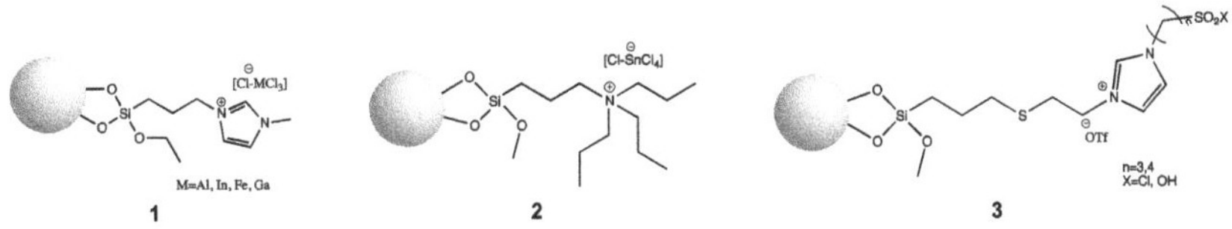

Figure 6. Structures of Lewis type silica-based SILLP materials.

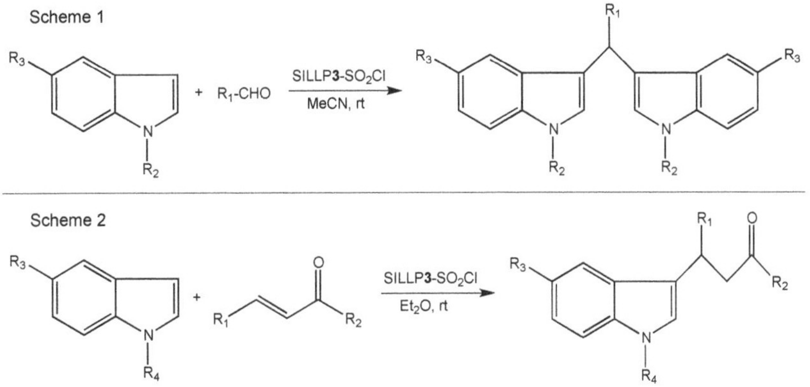

Scheme 1

Scheme 2

Figure 7. Applications of Lewis type SILLP material **3** in the synthesis of bis(indolyl)methanes (Scheme 1) and addition of indole to vinyl ketones (Scheme 2).

Further reports have presented a chloroaluminate imidazolium-based IL grafted to MCM-41 (**1**, Figure 6). High conversion (97%) and selectivity to isooctane (59.7%) in gasoline production were obtained, and the SILLP catalyst showed better activity than the IL immobilized via the anion, zeolite H-Beta, and Nafion/Silica Composite SAC 13 [55]. The same IL was immobilized on silica, MCM-41, SBA-15, active carbon, and glass materials, and the activity of the prepared SILLP was examined in the continuous gas phase trimerization of isobutene using a fixed-bed reactor under atmospheric pressure. Only silica-based SILLP catalysts enabled the trimerization reaction to occur due to synergic interactions between the IL anion and the silanol groups. In other cases, the alkylation reaction was observed. For the trimerization reaction, the MCM-41 based SILLP (α = 91.4%, S_{C12} = 79.4%) turned out to be the most active, owing to its regular hexagonal array channels that behave similar to micro-reactors, increasing the catalytic activity [56]. Apart from Al (III) and Sn (IV), other metals such as Fe, In, and Ga were used for silica-based SILLP synthesis. Chloroferrate (III) imidazolium-based IL moieties were grafted for the siliceous support of MCM-41 after complexation with $FeCl_3$ (**1**, Figure 6). The SILLP catalyst showed high efficiency and long reusability (10 cycles) in the Friedel–Crafts reaction between benzene and benzyl chloride [57]. The chloroindate (III) imidazolium-based IL was anchored to SBA-15 silica material (**1**, Figure 6), which exhibits ordered hexagonal structure, however less so than MCM-41 material. The obtained SILLP catalyst was used in the Friedel–Crafts reaction between benzene and benzyl chloride, gaining 100% conversion and 100% selectivity over 6 reaction cycles. Introducing the IL to the catalyst structure prevents $InCl_3$ from leaching [58]. A chlorogallate (III) imidazolium-based IL was covalently tethered to a multimodal silica porous silica support (**1**, Figure 6) and applied to Diels–Alder cycloaddition reactions for the synthesis of intermediates for pharmacologically active ingredients, agrochemicals, flavors, and fragrances (Scheme 1, Figure 8). The synthesized materials demonstrated a hierarchical pore structure and contained micro-, meso-, and macropores, which results in easy mass transport to and from active sites. The SILLP chlorogallate (III) catalyst showed great conversions and endo/exo selectivities in short reaction times, which is superior to other results presented in the literature. Moreover, the catalyst could be recycled five times without significant loss of activity [59].

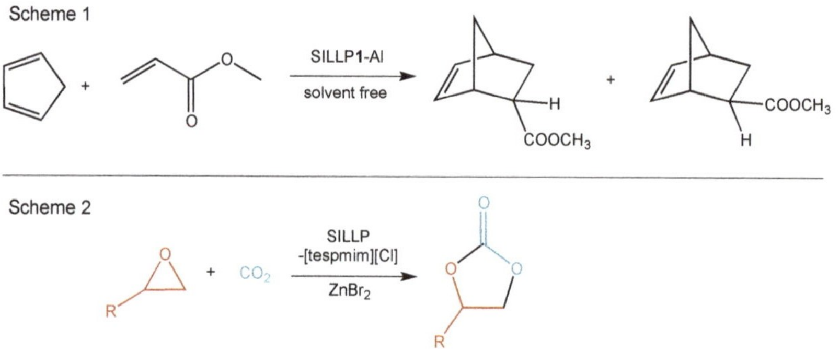

Figure 8. Diels–Alder cycloaddition (Scheme 1) and CO_2 cycloaddition for cyclic carbonates synthesis (Scheme 2) with SILP catalysts.

Other reports concern the application of ILs as co-catalysts for CO_2 cycloaddition for cyclic carbonates synthesis in the presence of a Zn Lewis centre (Scheme 2, Figure 8). It is postulated that the Zn Lewis site coordinates with the oxygen atom of epoxides, and a nucleophilic attack of the halide anion on the less sterically anion carbon atom of epoxide occurs. The first approach includes [tespmim][Cl] anchored to various silica materials such as macro/mesoporous silica, MCM-41, MSU-F (cellular foam), and MSU-H (large pore 2D hexagonal). The better results in the reaction of CO_2 with styrene oxide in the presence of $ZnBr_2$ were achieved for the silica SILLP, and the worst occurred for the MCM-41 SILLP. In this case, catalytical activity depends on pore size and not on surface area. The synthesised catalyst could be recycled four times without any loss of activity [60]. In other reports, Zn atoms and [tespmim][Cl] were grafted to the silica surface. Various materials such as MCM-41 (regular, long, hexagonal channels), MSN (nanosphere morphology, order mesopores, mainly inside pores), and BMMs (mesoporous structure, a large number accumulated inside and outside of the pores) were applied. The catalytic activity of the SILLP catalyst was examined in reaction of CO_2 with propylene oxide, where the best performance was exhibited by SILLPs with shorter and regular pore channels [61].

3.2. Brønsted-Type SILLPs

Brønsted-type SILLPS based on silica materials are synthesized via the cation method. Brønsted IL moieties are grafted to the silica surface through precursors such as 3-(chloropropyl)triethoxysilane, 3-(chloropropyl)trimethoxysilane, 3-mercaptopropyltrimethoxysilane, or (3-aminopropyl)-trimethoxysilane, creating the structures presented in Figure 9. The Brønsted acidic center located in the anion that is most often used is [HSO_4], whereas the cation Brønsted site can be found in the alkyl chain modified with an -SO_3H group through the reaction between, for example, 1,3-propanesultone and vinylimidazole. Some applications of Brønsted-type silica-based SILLP materials as catalysts in organic synthesis are presented in Table 2.

Table 2. Brønsted-type silica-based SILLP in organic catalysis.

Catalyst	Reaction Type	Reaction Conditions	Reaction Parameters	Lit.
SiO$_2$ [p(b-SO$_3$H)im][OTf] [a]	Estrification of oleic acid and methanol	0.2 mmol IL in cat., oleic acid 17.7 mmol, methanol 531.05 mmol, 100 °C, 4 h	α = 84%, 3 cycles	[62]
SiO$_2$ [p(b-SO$_3$H)im][OTf]	Transestrification of glicerol trioleate and methanol	0.2 mmol IL in cat., glicerol trioleate 17.7 mmol, methanol 531.05 mmol, 100 °C, 4 h	α = 30%, S$_{methyl\ oleate}$ = 36%	[62]
SiO$_2$ [tesp(b-SO$_3$H)im][Cl] [b]	hydrolysis of cellulose	0.02 mmol h$^+$ in cat., cellulose 0.185 mmol, 2 mL H$_2$O, 190 °C 3 h	Y$_{TRS}$ = 48.1%, Y$_{glucose}$ = 21.9%, 4 cycles	[63]
SiO$_2$ [p(p-SO$_3$H)im][OTf] [c]	Dehydration of fructose	0.175 mmol IL in cat., fructose 0.35 mmol, DMSO 2.0 g, MW: 200 W, 4 min	α = 100%, Y = 70.1% (5-hydroxymethylfurfural)	[64]
SiO$_2$ [tmsp(p-SO$_3$H)im][I] [d]	Biginelli reaction	0.8% mol cat., aldehyde 1 mmol, ethyl/methyl-acetoacetate 1 mmol, urea 1.5 mmol, 75 °C, 50–90 min	Yields for aldehydes with Ar groups: (a) ethylacetoacetate: Ph 96%, 4-OMeC$_6$H$_4$ 97%, 2-OMeC$_6$H$_4$ 86%, 4-MeC$_6$H$_4$ 96%, 4-ClC$_6$H$_4$ 90%, 3-BrC$_6$H$_4$ 96% (b) methylacetoacetate: Ph 96%, 4-OMeC$_6$H$_4$ 96%, 2-OMeC$_6$H$_4$ 89%, 4-MeC$_6$H$_4$ 95%, 4-ClC$_6$H$_4$ 92%, 3-BrC$_6$H$_4$ 95%	[65]
SiO$_2$ [tmsp(p-SO$_3$H)im][I]	Esterification of acetic acid with various alcohols	5% mol cat., alcohol 2 mmol, acetic acid 4 mmol, 60–70 °C, 12–24 h	Yields for alcohols: PhCH$_2$OH 95%, PhCH(OH)CH$_3$ 86%, PhCH(OH)CH$_2$CH$_3$ 85%, PhCH$_2$CH$_2$OH 88%, CH$_3$CH$_2$OH 93%, C$_8$H$_{17}$OH 93%, C$_9$H$_{19}$OH 92%, C$_{10}$H$_{21}$OH 92%	[66]
SiO$_2$ [tesp(p-SO$_3$H)im][OTf] [e]	Self-condensation of pentanal	10% wt. cat, 120 °C, 6 h	α = 77.4%, Y = 69.4%, S = 89.6%, TON = 230.5	[67]
SiO$_2$ [tesp(p-SO$_3$H)bim][Cl] [f]	Synthesis of fatty acid methyl esters	3% wt. cat., castor/jatropha/neem oil:methanol 1:12 molar ratio, 70 °C, 6–7 h	Castor oil Y = 94.9% Jatropha oil Y = 95.7% Neem oil Y = 94.4%	[68]
SiO$_2$ [tesp(p-SO$_3$H)bim][Cl]	Synthesis of 1-amidoalkyl naphthols	80 mg cat., aldehyde 20 mmol, 2-naphthol 20 mmol, acetamide 24 mmol, 100 °C, 7–10 h	Yields for benzaldehydes with R group: H 90%, 3-NO$_2$ 95%, 4-OH 87%, 4-OMe 89%, 4-Cl 92%, 4-Cl 93%, 4-NO$_2$ 89%	[69]
SiO$_2$ [tespmim][HSO$_4$] [g]	Bayer–Villiger oxidation of cyclic ketones	0.4 g cat., ketone 1 mmol, 68% H$_2$O$_{2(aq.)}$ 3 mmol, dichloromethane 4 mL, 50 °C, 5–15 h	cyclobutanone: α = 100%, Y = 96%; cyclopentanone: α = 98%, Y = 75%; cyclohexanone: α = 86%, Y = 64%; 2-adamantanone: α = 95%, Y = 89%; 1-indanone: α = 81%, Y = 78%; 1-tetralone: α = 78%, Y = 77%	[70]
SiO$_2$ [tespmim][HSO$_4$]	Esterification of acetic acid and butanol	4% wt. cat., 96°C, butanol 0.12 mol, acetic acid 0.10 mol, cyclohexane 6 mL, 3 h	α = 99.4%, 6 cycles	[71]
SiO$_2$ [tespmim][HSO$_4$]	Synthesis of 1-(benzothiazolylamino) phenylmethyl-2-naphthols	150 mg cat., aldehyde 1 mmol, 2-aminobenzothiazole 1 mmol, 2-naphthol 1 mmol, 110 °C, 3–5 h	Yields for various aryl aldehydes with R-groups: H 93%, 2-Cl 89%, 4-Cl 92%, 3-Br 93%, 4-Br 91%, 3-Me 93%, 2-OMe 90%, 3-OMe 92%, 4-OMe 93%, 2-NO$_2$ 90%, 3-NO$_2$ 92%, 4-NO$_2$ 91%	[72]
SiO$_2$ [tespmim][HSO$_4$]	Formylation of amines	0.8% mol cat., amine 1 mmol, formic acid 2 mmol, 60 °C, 1–15 h	Yields and TOF for amines: aniline 94%, 1428 h^{-1}; 4-methoxy aniline 97%, 7275 h^{-1}; benzyl amine 93%, 465 h^{-1}	[73]
SiO$_2$ [tespmim][HSO$_4$]	Knoevenagel–Michael cyclization for polyhydroquinolines synthesis	2% mol cat., aldehyde 1 mmol, dimedone 1 mmol, enaminone 1.2 mmol, NH$_4$OAc 1.5 mmol, H$_2$O 3 mL, 45 °C, 2–3 h	Yields for enaminone-COOMe with aldehydes with Ar-group: C$_6$H$_5$ 90%, 4-ClC$_6$H$_5$ 92%, 2-C$_6$H$_5$ 92%, 4-OMeC$_6$H$_5$ 88%, 2-MeC$_6$H$_5$ 90%	[74]
SiO$_2$ [tespmim][HSO$_4$]	Synthesis of 3,4-dihydropyrano[c] chromenes	0.15 g cat., 4-hydroxycoumarin 1 mmol, malononitrile 1 mmol, Ar-aldehyde 1 mmol, 100 °C, 30 min	Yields for aldehydes with Ar-groups: C$_6$H$_5$ 94%, 4-ClC$_6$H$_4$ 95%, 3-ClC$_6$H$_4$ 93%, 4-BrC$_6$H$_4$ 94%, 2,4-(Cl)$_2$C$_6$H$_3$ 90%, 3-O$_2$NC$_6$H$_4$ 93%, 4-O$_2$NC$_6$H$_4$ 90%, 2-O$_2$NC$_6$H$_4$ 89%, 4-MeC$_6$H$_4$ 94%, 3,4,5-(CH$_3$O)$_3$C$_6$H$_2$ 89%, 4-HO-C$_6$H$_4$ 93%	[75]

Table 2. Cont.

Catalyst	Reaction Type	Reaction Conditions	Reaction Parameters	Lit.
SiO_2 [tespmim][HSO_4]	Synthesis of pyrano[3,2-b]indole derivatives	10% mol cat., 3-hydroxypyrrole 1 mol, benzaldehyde, 1 mol, malononitrile 1 mol, acetonitrille 8 mL, 80 °C, 6–8 h	Yields for aldehydes with Ar groups: 4-$CH_3OC_6H_4$ 84%, C_6H_5 90%, 4-$CH_3C_6H_4$ 85%, 4-BrC_6H_4 90%, 2-BrC_6H_4 86%, 4-ClC_6H_4 90%, 2-ClC_6H_4 86%, 4-CNC_6H_4 90%, 4-$NO_2C_6H_4$ 85%, 2-$NO_2C_6H_4$ 88%	[76]
SiO_2 [tespmim][HSO_4]	Synthesis of pyrano[2,3-b]pyrrole derivatives	10% mol cat., 2-hydroxypyrrole 1 mol, benzaldehyde 1 mol, malonoitrile 1 mol, acetonitrille 4 mL, 60 °C, 2–8 h	Yields for aldehydes with Ar groups: 4-$CH_3OC_6H_4$ 76%, C_6H_5 90%, 4-$CH_3C_6H_4$ 82%, 4-BrC_6H_4 90%, 2-BrC_6H_4 88%, 4-ClC_6H_4 90%, 2-ClC_6H_4 86%, 4-CNC_6H_4 73%, 2-CNC_6H_4 70%, 4-$NO_2C_6H_4$ 64%, 2-$NO_2C_6H_4$ 62%	[77]
SiO_2 [tespmim][HSO_4]	Synthesis of benzo[f]chromene compounds	15% mol cat., 2-naphthol 1 mol, benzaldehyde 1 mol, triethyl orthobenzoate 1 mol, acetonitrille 4 mL, 65 °C, 4–8 h	Yields for benzaldehydes with 4-group: H 85%, Br 85%, Cl 88%, NO_2 80%, Me 88%, OMe 90%, OH 84%	[78]
SiO_2 [tespmim][HSO_4]	Synthesis of 2,9-dihydro-9-methyl-2-oxo-4-aryl-1H-pyrido[2,3-b]indole-3-carbonitrile compounds	15% mol cat., 1-methyl-1H-indol-2-ol 1 mol, (triethoxymethyl)arene 1 mol, cyanoacetamide 1 mol, DMF 6 mL, 100 °C, 2–7 h	Yields for (triethoxymethyl)arene with groups: 4-OMe 73%, h 65%, 4-Me 65%, 4-Br 61%, 2-Br 56%, 4-Cl 61%, 2-Cl 55%, 4-F 53%	[79]
SiO_2 [tespmim][HSO_4]	Synthesis of acenaphtho[1,2b]pyrroles.	10% mol cat., silyl enol of acenaphthylen-1(2H)-one 1 mol, 2,4-dimethoxybenzaldehyde 1 mol, isocyanocyclohexane 1 mol, DMF 50 mL, reflux, 10 h	Y = 97%	[80]
SiO_2 [tespmim][HSO_4]	Synthesis of 5-Amino-7-aryl-6-cyano-4H-pyrano[3,2-b]pyrroles	10% mol cat., 3-hydroxypyrrole 1 mol, aldehyde 1 mol, malononitrile 1 mol, acetonitrille 4 mL, 50 °C, 1–8 h	Yields for aldehydes with Ar groups: 4-$CH_3OC_6H_4$ 62%, C_6H_5 89%, 4-$CH_3C_6H_4$ 80%, 4-BrC_6H_4 91%, 2-BrC_6H_4 89%, 4-ClC_6H_4 88%, 2-ClC_6H_4 88%, 4-CNC_6H_4 70%, 2-CNC_6H_4 67%, 4-$NO_2C_6H_4$ 61%, 2-$NO_2C_6H_4$ 69%	[81]
SiO_2 [tespmim][$H_2PW_{12}O_{40}$] [h]	Oxidation of dibenzothiophene	0.01 g cat., O/S molar ratio: 3:1 (H_2O_2 0.8 mmol), 60 °C, 40 min	α = 100%, 4 cycles	[82]
SiO_2 [p(p-SO_3H)im][HSO_4] [i]	Esterification of acetate acid and n-butanol	6% wt. cat., n-butanol:acetic acid (2:1, n/n), 94 °C, 3 h	Y = 99.5%	[83]
SiO_2 [tesp(b-SO_3H)im][HSO_4] [j]	Synthesis of amidoalkyl naphtols	80 mg cat., aldehyde:2-naphtol:acetamide (2:2:2.4; n/n/n), 85 °C, 5–15 min	Yields and TOF for different aldehydes with R-groups: Ph 90%, 6.43 min^{-1}; 4-Cl-C_6H_4 89%, 3.18 min^{-1}; 2,4-Cl_2-C_6H_4 86%, 3.84 min^{-1}; 4-BrC_6H_4 88%, 3.15 min^{-1}; 3-NO_2-C_6H_4 92%, 6.59 min^{-1}; 4-$NO_2C_6H_4$ 93%, 6.65 min^{-1}; 3-MeO-C_6H_4 86%, 3.07 min^{-1}; 4-MeO-C_6H_4 80%, 1.91 min^{-1}; 4-Me-C_6H_4 87%, 3.11 min^{-1}	[84]
SiO_2 [tesp(b-SO_3H)im][HSO_4]	Thioacetalization of carbonyl compounds	5% mol cat., 4-methoxybenzaldehyde with thiophenol, rt, 5 h	Y = 96%, 6 cycles	[85]
SiO_2 [tesp(b-SO_3H)im][HSO_4]	Acetalization of benzaldehyde or furfural with diols	4% wt. cat., benzaldehyde 70 mmol, ethanediol 126 mmol, cyclohexane 8 mL, reflux, 1.5–3 h	Yields: (a) benzaldehyde: ethanediol 95.2%, 1,2-propanediol 93%, 1,4-Butanediol 87.1% (b) furfural: ethanediol 85%, 1,2-propanediol 95.9%	[86]
SiO_2 [tesp(p-SO_3H)im][HSO_4]	Synthesis of 2H-indazolo[1,2-b]phthalazine-triones	30 mg cat., benzaldehyde 1 mmol, dimedone 1 mmol, phthalhydrazide 1 mmol, 80 °C, 10 min	Y = 94%, 8 cycles	[87]
SiO_2 [tesp(p-SO_3H)im][HSO_4]	Synthesis of polyoxymethylene dimethyl ethers	4% wt. cat., molar ratio of methylal to trioxane 3, 105 °C, 1 h	α = 92%, S = 52%, 6 cycles	[88]
SiO_2 [tesp(p-SO_3H)im][HSO_4] [k]	Lignin depolymerization	0.5 g cat., dealkaline lignin 2% wt., 30 mL H_2O:C_2H_5OH (1:5, v/v), 200 °C, 1 h	Yields for THF soluble products 90%,	[89]
SiO_2 [p(p-SO_3H)im][HSO_4]	Esterification of acetic acid and n-butanol	8% wt. cat., acetic acid 4.8 g, n-butanol 7.12 g, cyclohexane 8 mL, 89 °C, 3 h	Y = 99.2%, S = 100%, 7 cycles; yields for other alcohols: C_6H_{13} 99.4%, C_2H_5 84.1%, $C_6H_5CH_2$ 98.5%	[90]

Table 2. Cont.

Catalyst	Reaction Type	Reaction Conditions	Reaction Parameters	Lit.
SiO$_2$ [tesp(p-SO$_3$H)a][HSO$_4$] [l]	Biodiesel synthesis	0.05 g cat., rapeseed oil 5 g, methanol 2.33 g, 70 °C, 9 h	Y = 99%, 6 cycles	[91]
SiO$_2$ [tesp(p-SO$_3$H)a][HSO$_4$]	Acetalization of benzaldehyde and 1,2-ethanediol	0.05 g cat., benzaldehyde 0.1 mol, 1,2-ethanediol 0.15 mol, 25 °C, 12 h	Y = 98%	[91]
SiO$_2$ [tespim][H$_2$PW$_{12}$O$_{40}$] [m]	Oxidations of alkenes	0.05 g cat., alkene 5 mmol, hydrogen peroxide (30%) 5 mmol, acetonitrile 4.5 mL, 60 °C, 4 h	Conversion, selectivity and TOF for alkenes: cyclooctene 90%, 99%, 162 h^{-1}; 1-octene 34%, 99%, 61 h^{-1}; norbornene 85%, 99%, 153 h^{-1}; limonene 76%, 29%, 137 h^{-1}	[92]
SiO$_2$ [p(p-SO$_3$H)im] [H$_2$PW$_{12}$O$_{40}$] [n]	Esterification of palmitic acid	15% wt. cat., methanol:palmitic acid molar ratio 9, 65 °C, 8 h	Y = 88.1%, 5 cycles	[93]

[a] 3-(4-sulfobutyl)-1-(3-trimethoxysilylmerkaptopropyl)imidazolium triflate, [b] 3-(4-sulfobutyl)-1-(3-propyltriethoxysilane)imidazolium chloride, [c] 3-(3-sulfopropyl)-1-(3-propyltriethoxysilane)imidazolium triflate, [d] 3-(3-sulfopropyl)-1-(3-propyltrimethoxysilane)imidazolium iodide, [e] 3-(3-sulfopropyl)-1-(3-propyltriethoxysilane)benzimidazolium chloride, [g] 1-methyl-3-(3-propyltriethoxysilane)imidazolium hydrogensulfate, [h] 1-methyl-3-(3-propyltriethoxysilane)imidazolium dihydrogenphosphotungstate, [i] 3-(3-sulfopropyl)-1-(3-trimethoxysilylmerkaptopropyl)imidazolium hydrogensulfate, [j] 3-(4-sulfobutyl)-1-(3-propyltriethoxysilane)imidazolium hydrogensulfate, [k] 3-(3-sulfobutyl)-1-(3-propyltriethoxysilane)imidazolium hydrogensulfate, [l] N-(3-sulfopropyl)-N-(3-propyltriethoxysilane)ammonium hydrogensulfate, [m] 1-(3-propyltriethoxysilane)imidazolium dihydrogenphosphotungstate, [n] 3-(3-sulfopropyl)-1-(3-trimethoxysilylmerkaptopropyl)imidazolium dihydrogenphosphotungstate.

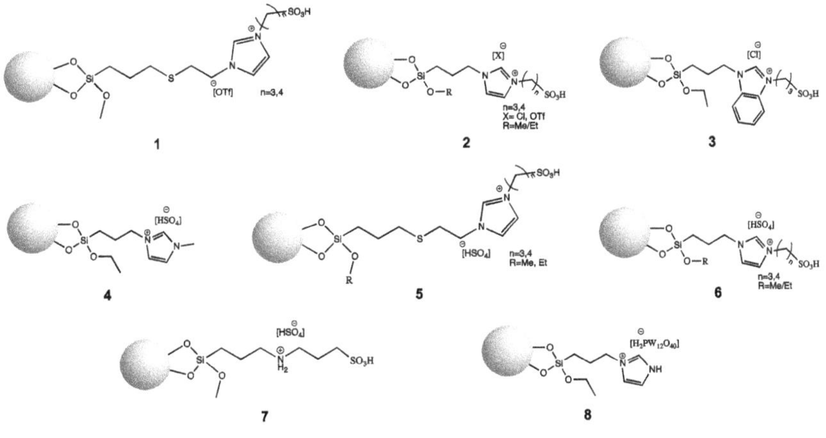

Figure 9. Structures of Brønsted-type silica-based SILLP materials.

Brønsted acidic vinylimidazolium-based IL moieties modified with -SO$_3$H groups were grafted to a sulfhydryl group-modified silica surface through free radical addition obtaining SILLP catalyst 1 (Figure 9). The prepared material was used in the esterification of oleic acid with methanol and the transesterification of glycerol trioleate with methanol. It was reported that the loading density of the IL influenced both reactions. Increasing the loading of the IL on the support induced the conversion of oleic acid. However, for the conversion of glycerol trioleate, the opposite effect is observed. This is due to the size of glycerol trioleate molecules and the decreasing pore size of the SILLP. The SILLP catalyst could be used in the esterification for three cycles, after which the catalytic activity dropped and the hydrolysis or alcoholysis of the -Si-O-Si- bonds occurred [62]. In the following report, the structure of a Brønsted imidazolium IL modified with -SO$_3$H groups was produced in three stages: first, 3-(chloropropyl)trimethoxysilane was anchored to the silica surface. Next, imidazole moieties were introduced to the structure before 1,4-

butanesultone was used to modify the imidazole ring with a -C$_4$H$_9$SO$_3$H group (**2**, Figure 9). The catalytic activity of the obtained SILLP material was tested in a cellulose hydrolysis and yielded a 48.1% reduction of sugar and 21.9% of glucose. In this case, the catalyst maintained activity for three cycles. Due to the use of the SILLP catalyst, the total yields in the reduction of sugar and glucose were higher than using SO$_3$HC$_3$H$_7$-SiO$_2$ or SO$_3$H-SiO$_2$, keeping the same -SO$_3$H group loading. This effect results from the interaction between the imidazolium IL and the hydroxyl groups in cellulose [63]. The SILLP catalyst **1** (Figure 9) was also used in the dehydration of fructose to 5-hydroxymethylfurfural (HMF) with 100% conversion and 70.1% yield of the main product (Scheme 1, Figure 10). In this reaction, the catalyst was reused without significant loss of activity for 7 cycles. Simultaneously, the same SILLP material with a Lewis -SO$_2$Cl center was tested for this reaction. However, the catalyst exhibited inferior efficiency compared with the Brønsted-type SILLP [64]. The next report described the functionalization of bifunctional periodic mesoporous organosilica with IL and -SO$_3$H groups, as well as its application in a Biginelli condensation reaction for the synthesis of pharmacological and biological activities compounds. The novel material assured high yields of various products (Table 2) and could be recycled over 10 times without any decrease in efficiency [65]. The same catalyst was also used in the esterification of acetic acid with various alcohols and, again, high yields of the main products were reached and the SILLP material could be reused several times [66]. Another report mentioned a triflate imidazolium-based IL with a -SO$_3$H group anchored to the silica, MCM-41, and SBA-15 materials (**2**, Figure 9). Activity tests were performed for the self-condensation of pentanal to 2-propyl-2-heptenal, where the best results were achieved for the silica based SILLP (69.4% yield, 89.6% selectivity), which was due to the highest IL loading on the surface [67]. A benzimidazolium IL with a -SO$_3$H group in alkyl chain was grafted to silica surface (**3**, Figure 9) in stages (which were described above). The SILLP catalyst was employed in the transesterification of non-edible oils with high free fatty acids, as well as for the synthesis of 1-amidoalkyl naphthols from 2-naphthol, amides, and aldehydes (Scheme 2, Figure 10). This eco-friendly and efficient catalyst for transesterification provided 95% yield of fatty acid methyl esters and catalytic stability over 5 runs [68]. Moreover, the SILLP material in the synthesis of 1-amidoalkyl naphthols exhibited high yield of the obtained products (Table 2), high product quality, short reaction times, and reusability for five reaction cycles, which makes the catalyst very useful for industrial practices [69].

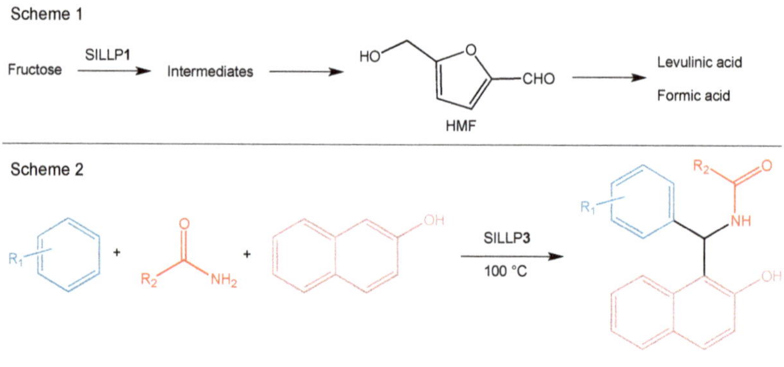

Figure 10. Brønsted-type SILLP based on silica material in the dehydration of fructose (Scheme 1) and synthesis of 1-amidoalkyl naphthols (Scheme 2).

Further reports concern the application of covalently immobilized imidazolium-based ILs with a hydrogensulfate anion on silica materials (**4**, Figure 9). The Bayer–Villiger oxidation of cyclic ketones to lactones (Figure 11) is one example of numerous reactions catalyzed by SILLP **4** (Figure 9). For that purpose, a silica material with the extensive system of meso- and macropores was used. Here, the catalyst showed great activity, which resulted in high conversions of ketones and yields of lactones (60–91%), short reaction times, and good reusability (three cycles) [70]. The same SILLP catalyst was used in the esterification of acetic acid with butanol with a 99.4% conversion, and a reusability of six catalytic cycles with a slight decrease of conversion were observed [71]. The synthesis of 1-(benzothiazolylamino)phenylmethyl-2-naphthols catalyzed by SILLP **4** (Figure 9) was also reported. In this case, IL was anchored to rice husk ash, which is a natural source of amorphous silica. High yields for various aldehydes (90–93%) and high TOF (92 h^{-1}) were achieved (Table 2) in very short reaction times. Furthermore, the catalyst could be reused six times without activity loss [72]. The same catalytic system was examined for the formylation of amines. Again, the catalyst proved to be simple, stable, and efficient, since high yields (93–97%), TOF (465–7275 h^{-1}) and reusability over 10 cycles were reached [73]. In another report, an IL containing a hydrogensulfate anion was immobilized on nanoporous silica SBA-15 and used in the synthesis of hexahydroquinolines via the Knoevenagel–Michael cyclization as an alternative to conventional catalysts. Excellent yields (90–93%), short reaction times, aqueous conditions, and reusability (seven runs) made the process more environmentally friendly [74]. The synthesis of 3,4-dihydropyrano[c]chromenes (Scheme 1, Figure 12) and pyrano[2,3-c]pyrazoles were also proceeded in the presence of SILLP **4** (Figure 9). Various ILs with anions, such as $[HSO_4]^-$, $[H_2PO_4]^-$, $[Br]^-$, and $[OTf]^-$ were tested, with the best results gained for the hydrogensulfate anion. The catalyst exhibited very good yields (89–95%) and reusability (five cycles) [75]. The use of SILLP **4** (Figure 9) was also successful for the synthesis of pyrano[3,2-b]indoles (Scheme 2, Figure 12) [76], pyrano[2,3-b]pyrroles (Scheme 3, Figure 12) [77], benzo[f]chromenes [78], 2,9-dihydro-9-methyl-2-oxo-4-aryl-1H-pyrido[2, 3-b]indole-3-carbonitriles [79], acenaphtho[1,2b]pyrroles [80], and 5-amino-7-aryl-6-cyano-4H-pyrano[3,2-b]pyrroles [81]. As shown in Table 2, satisfying yields for different aldehydes, arenes, and components were achieved, which indicates the versatility of SILLP **4** (Figure 9) catalyst, as well as the developed methods. Moreover, the catalyst could be reused several times [76–81]. The dihydrogenphosphotungstate anion ($[H_2PW_{12}O_{40}]^-$) was reported in an SBA-15 based SILLP, exhibiting well-ordered, mesoporous specific high surface area. This novel catalyst found application in the oxidation of dibenzothiophene, 4,6-dimethylbenzothiophene, and benzothiophene for fuel desulfurization. This SILLP showed excellent efficiency, with 100% conversion of dibenzothiophene and 4,6-dimethylbenzothiophene, which means the total ability of removal of toxic compounds from the fuel. Furthermore, the catalyst could be successfully reused four times [82].

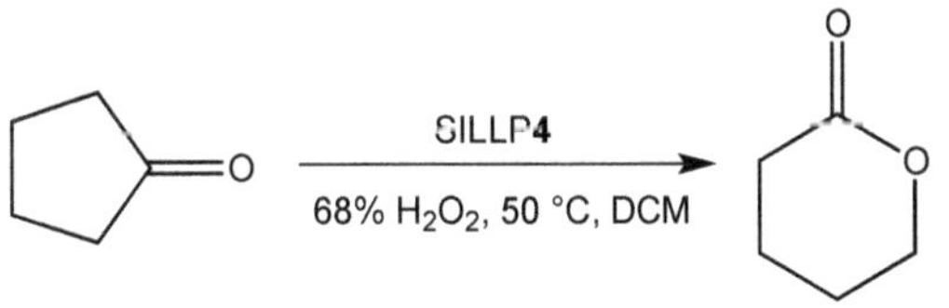

Figure 11. Brønsted-type SILLP based on silica material in Bayer–Villiger oxidation.

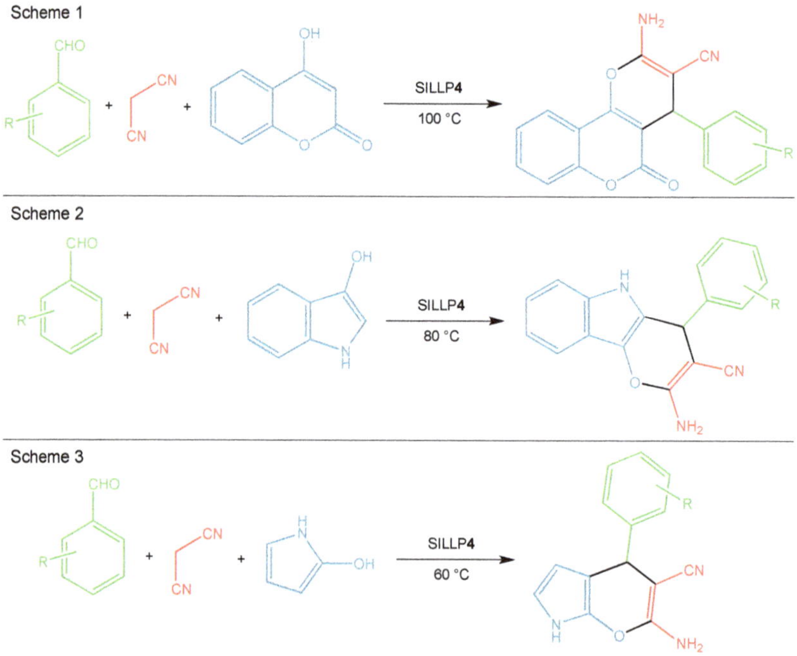

Figure 12. Brønsted-type SILLP based on silica material in the synthesis of 3,4-dihydropyrano[c]chromenes (Scheme 1), pyrano[3,2-b]indoles (Scheme 2), and pyrano[2,3-b]pyrroles (Scheme 3).

Dual Brønsted acidic ILs immobilized on silica materials are another group of Brønsted-type SILLPs. In this case, the Brønsted centers are located both in the cation and anion, like the -SO$_3$H groups grafted to the alkyl chain in the cation and like the [HSO$_4$]$^-$ in the anion (**5, 6**, Figure 9). This kind of catalyst is quite often used according to the literature. One use is the esterification of acetic acid and n-butanol. SILLP **5** (Figure 9) material caused 99.5% yield of n-butyl acetate, where it could be recycled eight times with only a slight decrease in conversion to 90.1% [83]. The next report presents 3-sulfopropyl-1-(3-propyltrimethoxysilane)imidazolium hydrogensulfate IL anchored to silica gel forming SILLP **6** (Figure 9) in the synthesis of amidoalkyl naphthols by the multicomponent condensation. high yields and TOFs (Table 2) were obtained, and the catalyst kept activity for seven cycles without significant loss [84]. SILLP **6** (Figure 9) was also used as a catalyst in the thioacetalization of carbonyl compounds, providing high yields (85–96%). The reaction between 4-methoxybenzaldehyde with thiophenol (Scheme 1, Figure 13) was characterized by 96% yield, mild reaction conditions, and short reaction times, with the catalyst efficiently being recycled six times [85]. Furthermore, the same catalyst was employed in acetalization of benzaldehyde or furfural with various diols. High catalytic activity (yields 85–96%) for 10 reaction runs was reached for the synthesis of benzaldehyde ethanediol acetal [86]. Again, SILLP **6** (Figure 9) was used as a catalyst in the synthesis of 2H-indazolo[1,2-b]phthalazine-triones (Scheme 2, Figure 13) [87] and polyoxymethylene dimethyl ethers [88]. In the first case, nano-silica formed a matrix for IL immobilization. The synthesized material showed high catalytic activity, gaining 81–96% yield of indazolophthalazine-triones and bisindazolophthalazine-triones, while maintaining activity over seven reaction cycles [87]. Various types of such silica gels used in SILLP synthesis are described widely throughout the literature. In order to reduce the ratio of the catalyst in the reactants, the SILLP with the highest surface area and IL loading was selected as the catalyst. This resulted in a 52% trioxane conversion and 92% polyoxymethylene dimethyl ethers selectivity [88]. Next,

SILLP **6** (Figure 9) was also employed in a lignin depolymerization. This highly thermally stable catalyst allowed a 90% yield of tetrahydrofuran soluble products to be obtained in 1 h at 200 °C [89]. It was also found that SILLP **5** (Figure 9) catalyzed the esterification of acetic acid and n-butanol. The catalyst provided a 99.2% yield and 100% selectivity and was active for seven cycles. Moreover, high yields for reactions with various alcohols were achieved (Table 2) [90]. The novel silica-based SILLP **7** (Figure 9) material was also synthesized from the (3-aminopropyl)-trimethoxysilane precursor. The prepared catalyst was used in the acetalization of benzaldehyde with 1,2-ethanedioland and biodiesel synthesis. In both cases, the SILLP exhibited high catalytic activity, assuring 99% yields of the main product and reusability over six cycles. In comparison with conventional biodiesel synthesis, SILLP catalysts are a very promising alternative [91]. SILLP **8** (Figure 9) with Brønsted acidic sites introduced with an imidazolium cation and dihydrogenphosphotungstate anion was tested in the oxidation of alkenes. The catalyst proved to be efficient in this reaction, providing high selectivities, conversions, and TOF (Table 2) [92]. The dihydrogenphosphotungstate anion was involved in the synthesis of SILLP **5** (Figure 9) instead of the hydrogensulfate anion. The SBA-15 SILLP catalyst was applied in the biodiesel synthesis from palmitic acid. In comparison to other anions such as hydrogensulfate and triflate, the dihydrogenphosphotungstate anion performed the highest IL loading and catalytic activity, giving an 88.1% yield and reusability of over five times [93].

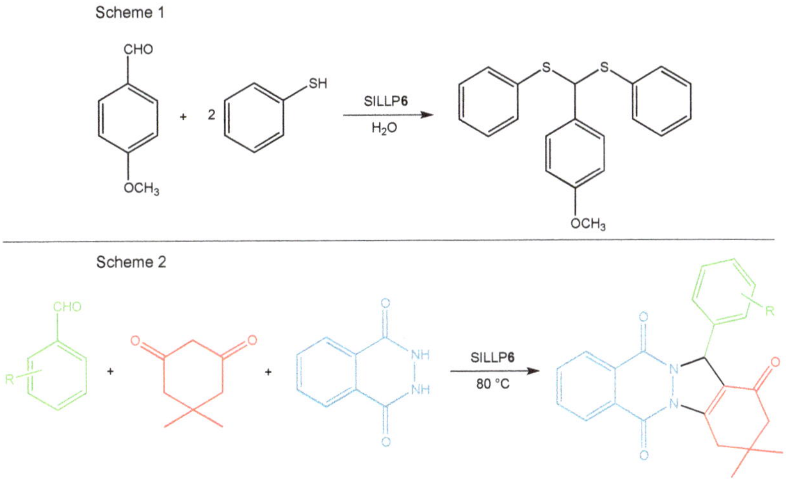

Figure 13. Brønsted-type SILLP based on silica material in the synthesis of thioacetalization of (Scheme 1)), 2H-indazolo[1,2-b]phthalazine-triones (Scheme 2).

3.3. Fe_3O_4-Silica hybrid Based SILLPs

Immobilization of an ionic liquid on a solid matrix provides easy catalyst separation from the reaction mixture, as well as its recycling. Doping silica materials with Fe_3O_4 offers new features, such as magnetic properties, for example. A silica-Fe_3O_4 hybrid could be even faster and more easily separated from the reaction mixture using an external magnetic field, making it an attractive support. Table 3 shows applications of silica-Fe_3O_4-based SILLP in organic catalysis, and Figure 14 presents chosen structures of silica-Fe_3O_4-based SILLPs.

Table 3. Silica-Fe$_3$O$_4$-based SILLPs in organic catalysis.

Catalyst	Reaction Type	Reaction Conditions	Technological Parameters	Lit.
SiO$_2$·Fe$_3$O$_4$ [tmspmim][Cl-AlCl$_3$] [a]	Synthesis of β-keto enol ethers	0.27 g cat., 5,5-dimethylcyclohexane-1,3-dione 1 mmol, alcohol 3 mL, rt, 50–95 min	Yields for alcohols: methanol 94%, ethanol 93%, n-butanol 89%, n-pentanol 87%, 2-propanol 88%, cyclohexanol 86%	[94]
SiO$_2$·Fe$_3$O$_4$ [tmspmim][Cl-ZnCl$_2$] [b]	Synthesis of benzoxanthenes	15 mg cat., benzaldehyde 1 mmol, 2-naphthol 1 mmol, dimedone 1 mmol, sonication, 80 °C, 30 min	Yields for benzaldehydes with R-groups: H 96%, 4-Me 84%, 2-OH 81%, 4-F 81%, 4-Cl 72%, 4-Br 76%, 2-F 70%, 2-Cl 75%, 2-Br 79%, 2-NO$_2$ 90%	[95]
SiO$_2$·Fe$_3$O$_4$ [tmspmim][Cl-ZnCl$_2$]	Synthesis of pyrroles	15 mg cat., aniline 1 mmol, acetonylacetone 1.2 mmol, sonication, 30–90 min	Yields for anilines with R-groups: H 91%, 4-I 98%, 4-OH 95%, 2-OH, 5-Me 78%, 3,5-Cl 77%	[95]
SiO$_2$·Fe$_3$O$_4$ [tmspmim][HSO$_4$] [c]	Synthesis of 1,8-dioxodecahydro-acridines	cyclic diketones: amines:aldehydes: catalyst (2:1:1:0.01), 80 °C, 10–30 min	Yields 87–97%	[96]
SiO$_2$·Fe$_3$O$_4$ [tmspim][HSO$_4$] [d]	Synthesis of 3-thiocyanato-1H-indole	5 mg cat., indole:H$_2$O$_2$:KSCN (1:3:3; n/n), water:ethanol (1:4; v/v), rt	Y = 95%; for various substrates 88–98%	[97]
SiO$_2$·Fe$_3$O$_4$ [tmsptetrazole-SO$_3$H][Cl] [e]	Synthesis of arylcyanamide	0.2 g cat., arylcyanamide 1 mmol, NaOCN 1 mmol, H$_2$O 10 mL, reflux	Yields for arylcyanamide with R-groups: 3-Br 90%, 4-Cl 89%, 4-Me 92%, 4-OMe 93%	[98]
SiO$_2$·Fe$_3$O$_4$ [tmspim][HSO$_4$]	Synthesis of 1-carbamoyl-1-phenylureas	50 mg cat., benzaldehyde 1 mmol, acetic anhydride 5 mmol, rt, 10–120 min	Yields for benzaldehydes with R-groups: H 91%, 4-Cl 95%, 4-Me 93%, 4-OH 91%, 2-OH 97%, 4-MeO 90%, 2-MeO 87%, 4-COOH 90%, 4-CN 88%, 4-NO$_2$ 98%	[99]
SiO$_2$·Fe$_3$O$_4$ [tmspdabco(SO$_3$H)][OTf]$_2$ [f]	Acetylation of aldehydes with acetic anhydride	50 mg cat., isatin 0.5 mmol, indole 1 mmol, H$_2$O 2 mL, 90 °C, 2 h	Y = 85–96%, 8 cycles	[100]
SiO$_2$·Fe$_3$O$_4$ [tespmim][H$_2$PW$_{12}$O$_{40}$] [g]	Synthesis of 3,3-di(indolyl)indolin-2-ones	0.1 mg cat., hydrazine hydrate 2 mmol, ethyl acetoacetate 2 mmol, aryl aldehydes 1 mmol, ammonium acetate 3 mmol, water 15 mL, rt, 30 min.	Yields for different Ar-aldehydes: H 96%, Cl 95%, F 97%, NO$_2$ 98%, OMe 92%, Me 93%, OH 90%, CN 95%	[101]
SiO$_2$·Fe$_3$O$_4$ [tesp(b-SO$_3$H)im][HSO$_4$] [h]	Synthesis of tetrahydrodipyrazolopyridines	55 mg cat., aldehyde 2 mmol, 2-naphthol 2 mmol, dimedone 2.4 mmol, 90 °C, 35–65 min	Yields for aldehydes with Ar groups: C$_5$H$_6$ 89%, 4-MeC$_6$H$_4$ 86%, 4-OMeC$_6$H$_4$ 84%, 4-ClC$_6$H$_4$ 91%, 3-ClC$_6$H$_4$ 84%, 4-BrC$_6$H$_4$ 90%, 3-BrC$_6$H$_4$ 88%, 4-NO$_2$C$_6$H$_4$ 93%, 3-NO$_2$C$_6$H$_4$ 90%, 2-NO$_2$C$_6$H$_4$ 85%	[102]
SiO$_2$·Fe$_3$O$_4$ [tesp(b-SO$_3$H)im][HSO$_4$]	Synthesis of benzoxanthenes	50 mg cat., isatin 1 mmol, 1,3-dimethyl-2-amino uracil 1 mmol, barbituric acid 1 mmol, H$_2$O, 1 mL, rt, 4–8 h	Y = 81–90%, 5 cycles	[103]
SiO$_2$·Fe$_3$O$_4$ [tmsp(p-SO$_3$H)im][HSO$_4$] [i]	Synthesis of spirooxindoles	0.2 g cat., oleic acid 10 mmol, alcohol 60 mmol, 373K, 4 h	Methanol: Y = 89.6% Ethanol: Y = 93.5% n-propanol: Y = 92% n-butanol: Y = 91.5%	[104]
SiO$_2$·Fe$_3$O$_4$ [tesp(p-SO$_3$H)im][HSO$_4$]	Biodiesel production from oleic acid	10.8% wt. cat., methanol:oleic acid molar ratio 6, 110 °C, 4 h	α = 92.9%, 8 cycles	[105]
SiO$_2$·Fe$_3$O$_4$ [tesp(Ph-SO$_3$H)$_3$P][Cl] [j]	Biodiesel production from oleic acid	0.06 g cat., benzaldehyde 30 mmol, ethylene glycol 90 mmol, cyclohexane 185 mmol, reflux, 2 h	Yields for: benzaldehyde 97% (5 cycles), propionaldehyde 96%, butanone 95%, cyclohexanone 94%	[106]

Table 3. Cont.

Catalyst	Reaction Type	Reaction Conditions	Technological Parameters	Lit.
$SiO_2 \cdot Fe_3O_4$ [Cl][diammonium] $[HSO_4]$ [k]	Acetalization of aldehyde or ketone with ethylene glycol	0.048 cat., dimedone 1 mmol, benzaldehyde 1 mmol, 6-amino-1,3-dimethyluracil 1 mmol, 120 °C, 15–30 min	Yields for various benzaldehydes with R-group: H 94%, 3-Br 92%, 4-Br 90%, 2-Cl 88%, 4-Cl 96%, 4-Me 93%, 4-OMe 94%, 4-OH 81%	[107]
$SiO_2 \cdot Fe_3O_4$ [tesp2pyr][HSO_4] [l]	Synthesis of pyrimido[4,5-b]quinolines.	200 mg cat., aromatic amine 1 mmol, $NaNO_2$ 2.5 mmol, NaI 2.5 mmol, rt, 12–15 min	Yields for aromatic amines: $C_6H_5NH_2$ 73%, 4-$H_2NC_6H_4$COOH 95%, 4-$NO_2C_6H_4NH_2$ 83%, 4-Br$C_6H_4NH_2$ 78%, 4-Cl$C_6H_4NH_2$ 82%, 4-Me$C_6H_4NH_2$ 62%	[108]
$SiO_2 \cdot CoFe_2O_4$ [p(b-SO_3H)im] [OTf] [m]	Diazotization–iodination of the aromatic amines	1:30 equimolar amount of oleic acid and the catalyst, alcohol 17.02 g, 100 °C, 4 h	CH_3: α = 75%, C_4H_9: α = 40%, C_6H_{13}: α = 20%, C_8H_{17}: α = 16%	[109]
$SiO_2 \cdot Fe_3O_4$ [tmsptetrazole-SO_3H][HSO_4] [n]	Esterification of oleic acid with straight-chain alcohols	20 mg cat., benzaldehyde 1 mmol, 2-thiobarbituric acid 2 mmol, acetate ammonium 1 mmol, H_2O 5 mL, rt, 35–60 min	Yields for benzaldehydes with R-groups: H 89%, 4-Cl 91%, 4-NO_2 95%, 4-Me 87%, 4-OMe 84%, 2-NO_2 93%, 2-OH 82%, 2-OMe 85%, 2–80%, 3-OMe 90%	[110]
$SiO_2 \cdot Fe_3O_4$ [OH-etNH_3][b-SO_3] [o]	Synthesis of pyrimidine derivatives	Aldehyde:malononitrile: thiophenol:catalyst (1/2/1/0.012; n/n/n/n), 50 °C, 5–20 min	Y = 81–91%; 5 cycles (benzaldehyde, malononitrile and thiophenol)	[111]
$SiO_2 \cdot Fe_3O_4$ [tmspdabco][Cl] [p]	Synthesis of 2-amino-3,5-dicarbonitrile-6-thio-pyridines	Aldehyde, ethyl cyanoacetate, H_2O-polyethylene glycol	8 cycles, high yields	[112]
$SiO_2 \cdot Fe_3O_4$ [tespmim][Cl] [r]	Knoevenagel condensation	0.0007 g cat., aromatic aldehyde 1 mmol, anilines 1 mmol, thioglycolic acid 1 mmol, 70 °C, 55–70 min	(a) aniline + aromatic aldehydes Yields for R-groups in aldehydes: H 94% (10 cycles), 4-Me 88%, 4-Cl 95%, 4-NO_2 92%, 3-NO_2 89% (b) p-methylaniline + aromatic aldehydes Yields for R-groups in aldehydes: H 90%, Me 93%, 90%	[113]
$SiO_2 \cdot Fe_3O_4$ [tespmim][Cl]	Synthesis of 1,3-thiazolidin-4-ones	20% mol cat., 6-amino-N,N-dimethyuracil 1 mmol, 3-(2-methyl-1H-indol-3-yl)-3-oxopropanenitrile 1 mmol, arylaldehydes 1 mmol, DMF 10 mL, 120 °C, 55–120 min	Yields for aldehydes with Ar-groups: 4-FC_6H_4 90% (3 cycles), 4-ClC_6H_4 90%, 4-BrC_6H_4 85%, 4-CNC_6H_4 90%, 4-$CF_3C_6H_4$ 90%, C_6H_5 80%, 3-ClC_6H_4 90%, 3-OMeC_6H_4 75%	[114]
$SiO_2 \cdot Fe_3O_4$ [tespmim][Cl]	Synthesis of indole-substituted pyrido[2,3-d]pyrimidines	1% mol cat., epoxide 10 mmol, P_{CO2} = 1 Mpa, 140 °C, 4–12 h	Styrene oxide Y = 93% (11 cycles), propylene oxide Y = 99%, epichlorohydrin Y= 99%	[115]
$SiO_2 \cdot Fe_3O_4$ [tespmim][Cl]	Cycloaddition of CO_2 to epoxides	0.05 g cat., aromatic aldehyde 2 mmol, ethyl acetoacetate 2 mmol, urea/thiourea 3 mmol, 100 °C, 25–40min	Yields for aldehydes: (a) urea: Ph 95%, 3-ClC_6H_4 97%, 3-$NO_2C_6H_4$ 97%, 2-tiophen 98%, 3-FC_6H_4 92% (b) thiourea: Ph 96%, 4-OMeC_6H_4 90%, 2-tiophen 95%	[116]

Table 3. Cont.

Catalyst	Reaction Type	Reaction Conditions	Technological Parameters	Lit.
$SiO_2 \cdot Fe_3O_4$ [tespmim][Cl]	Synthesis of 3,4-dihydropyrimidin-2(1H)-ones/thiones	7 mg cat., aniline 1 mmol), formic acid 3 mmol, rt, 5–10 min	Yields for anilines with R-groups: H 99% (5 cycles), 4-Me 98%, 4-OMe 98%, 4-Cl-90%, 4-NO_2 98%	[117]
$SiO_2 \cdot Fe_3O_4$ [tesptriazinium][Cl] [s]	N-formylation	0.02 g cat., aromatic aldehyde 1 mmol, malononitrile 1 mmol, 5-hydroxy-2-hydroxymethyl-4H-pyran-4-one (kojic acid) 1 mmol H_2O 5 mL, reflux, 30–45 min	Yields for benzaldehydes with R-groups: H 94%, 2,3-Cl_2 94%, 2,6-Cl_2 97%, 4-NO_2 98%, 3-NO_2 97%, 4-OH 85%	[118]
$SiO_2 \cdot Fe_3O_4$ [tesampmim][Cl] [t]	of amines	10 mg cat., benzylalcohol 1 mmol, anhydride 2 mmol, rt, 20–60 min	Yields for various benzylalcoholes with R-groups: 4-Br 96% (9 cycles), 4-OMe 94%, 4-F 94%, i-C_3H_7 93%	[119]
$SiO_2 \cdot Fe_3O_4$ [tmsp(alanine)im][Cl] [u]	Synthesis of 4H-dihydropyrano	0.001 g cat., arylaldehyde 2.5 mol, arylamine 2.5 mol cyclohexanon 3 mol, EtOH 20 mL, sonication (70 W)	Yields and selectivity (anti:syn) for aniline+ benzaldehydes with R-groups: H 92%, 99:1; 2-Cl 91%, 97:3; 4-Me 88%, 99:1; 4-Cl 92%, 99:1; 4-Br 92%, 99:1; 4-OMe 89%, 99:1; 2-OMe 86%, 99:1	[120]
$SiO_2 \cdot Fe_3O_4$ [tespdeaim][PF_6] [w]	[3,2-b]pyran-3-carbonitrile	25 mg cat., aldehyde or ketone 2 mmol, malononitrile 2 mmol, water 10 mL, 30 °C, 1 h	α for aldehydes/ketones: cyclohexanone >99%, furfural >99%, benzaldehyde >99%, 4-nitrobenzaldehyde 91.6%, 4-hydroxybenzaldehyde 89.4%, 2-hydroxybenzaldehyde 80.3%, 2-methylpropanal 92%	[121]
$Fe_{3-x}Ti_xO_4-SiO_2$ [TrpEt$_3$][I] [x]	Derivatives	0.12 g cat., anilines 1 mmol, dialkyl acetylenedicarboxylates 1 mmol, terminal alkynes or acetophenones 1.2 mmol, 100 °C, 15–18 h	Methyl 4-propylquinoline2-carboxylate: Y = 75% ethyl 6-hydroxy4-propylquinoline-2-carboxylate: Y = 92%	[122]
$Fe_{3-x}Ti_xO_4-SiO_2$ [TrpEt$_3$][I]	Acetylation of alcohols	0.12 g cat., anilines 1 mmol, dialkyl acetylenedicarboxylates 2.2 mmol, 100 °C, 10–22 h	Ethyl 4-(4-bromophenyl)benzo quinoline-2-carboxylate: Y = 77% dimethyl 8-nitroquinoline2,4-dicarboxylate: Y = 82%	[122]

[a] 1-methyl-3-(trimethoxysilylpropyl)imidazolium chloride—chloroaluminate (III), [b] N-(trimethoxysilylpropyl)imidazolium chloride—chlorozincate (II), [c] 1-methyl-3-(trimethoxysilylpropyl)imidazolium hydrogensulfate, [d] 1-methyl-3-(trimethoxysilylpropyl)imidazolium hydrogensulfate, [5] N-(trimethoxysilylpropyl)-5-phenyl-1H-tetrazolium-SO_3H chloride, [e] N-(3-sulfopropyl)-N-(3-propyltrimethoxysilane)triethylenediammonium ditriflate, [f] 1-methyl-3-(trimethoxysilylpropyl)imidazolium dihydrogenphosphotungstate, [g] 3-(4-sulfobutyl)-1-(3-propyltriethoxysilane)imidazolium hydrogensulfate, [h] 3-(3-sulfopropyl)-1-(3-propyltriethoxysilane)imidazolium hydrogensulfate, [i] P-(trimethoxysilylpropyl)-P,P,P-tri(4-sulfophenyl)phosphonium chloride, [j] N-(trimethoxysilylpropyl)-N,N-dimethyl-N-(dimethylammonium)ammonium chloride hydrosulfate, [k] N-(propyl-triethoxysilane)-2-pyrrolidinium hydrogensulfate, [l] 3-(4-sulfobutyl)-1-(3-trimethoxysilylmerkaptopropyl)imidazolium triflate, [m] N-(trimethoxysilylpropyl)-5-phenyl-1H-tetrazolium-sulfobutyl hydrogensulfate, [n] 2-hydroxyethylammonium butylsulphonate, [o] N-(3-propyltrimethoxysilane)triethylenediammonium chloride, [p] 1-methyl-3-(triethoxysilylpropyl)imidazolium chloride, [r] N-(triethoxysilylpropyl)triazinium hydrosulfate, [s] 3-((3-(trisilyloxy)propyl)propionamide)-1-methylimidazolium chloride, [t] 3-(trimethoxysilylpropyl)-1-(2-aminopropanoate)imidazolium trimethylethanolammonium chloride, [u] imidazolium alanine based IL, [w] 3-(trimethoxysilylpropyl)-1-(triethylamine)imidazolium hexafluorophosphate, [x] triethyltryptophanium iodide.

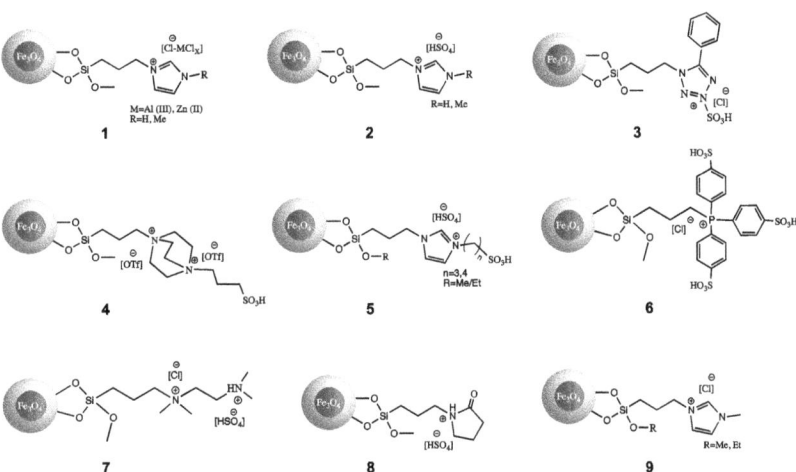

Figure 14. Structures of silica-Fe$_3$O$_4$-based SILLPs.

A Lewis chloroaluminate IL was grafted to a SiO$_2$·Fe$_3$O$_4$ nanomaterial, and the catalytic activity of obtained SILLP **1** (Figure 14) was tested in the synthesis of β-keto enol ethers. The magnetic catalyst showed proper efficiency and provided high yields under mild reaction conditions. Moreover, the SILLP maintained activity for six reaction cycles, and its recovery through external magnetic field was very effective [94]. The next report examined Lewis magnetic SILLP **1** (Figure 14) based on the chlorozincate (II) anion in the synthesis of benzoxanthenes (Scheme 1, Figure 15) and pyrroles (Scheme 2, Figure 15). In both reactions, the SILLP presented excellent activity, reusability for 5 runs, and achieved 76–96% yields of benzoxanthenes and pyrroles. In comparison with the described catalysts, the magnetic SILLP is a promising alternative due to its versatility [95]. Brønsted hydrogensulfate IL was anchored to magnetic silica-based material, where SILLP **2** (Figure 14) exhibited excellent activity and achieved 87–97% yields in the condensation reaction of cyclic diketones with aromatic aldehydes and ammonium acetate or primary amines. The catalyst could be reused nine times, which additionally proves the wide applicability of this nanomaterial [96]. In another report, SILLP **2** (Figure 14) found application as a catalyst for the thiocyanation of aromatic and heteroaromatic compounds. High yields of 88–98%, regioselectivity, short reaction times, and reusability (seven runs) were achieved [97]. The next report described the magnetic SILLP **3** (Figure 14) based on the phenyltetrazole cation. The SILLP nanocatalyst examined its efficiency in the synthesis of antibacterially active 1-carbamoyl-1-phenylureas in water. The magnetic nanomaterial gave 89–93% yields of the main products and kept good catalytic activity during five reaction cycles [98]. Hydrogensulfate poly(ionic liquid) was grafted to silica magnetic nanoparticles via the polymerization of vinylimidazolium moieties. The catalytic activity of the prepared heterogeneous catalyst was checked in the acetylation of aldehydes with acetic anhydride, which resulted in 90–98% yields and 10 reaction cycles without activity loss. Moreover, the SILLP also showed good efficiency in the deprotection reaction of acyl [99]. The Brønsted triethylenediammonium ditriflate based magnetic SILLP **4** (Figure 14) was found to be a great catalyst in the synthesis of 3,3-di(indolyl)indolin-2-ones. A yield of 85–96% of various indolines compounds with medical properties and eight efficient catalytic cycles were achieved with SILLP **4** [100]. In other work, Fe$_3$O$_4$ nanoparticles coated with silica SILLP **2** (Figure 14) based on the dihydrogenphosphotungstate anion catalyzed the synthesis of tetrahydrodipyrazolo-pyridines. This catalytic system could be reused several times using magnetic external forces and high loadings of the IL, providing excellent yields (90–98%) under mild conditions [101]. Further reports present Dual Brønsted acidic ILs immobilized on silica coated magnetic nanoparticles. SILLP **5** (Figure 14) found applications as a catalyst in the synthesis of benzoxanthenes [102], spirooxindoles [103], and biodiesel production from oleic acid [104,105]. This novel catalyst demonstrated great versatility and activity in all mentioned processes, achieving high yields for benzoxanthenes (84–91%), spirooxindoles (81–90%), biodiesel (90–94%) synthesis, as well as short reaction times, high products quality, easy catalyst recovery via magnetic field, and great reusability, which makes SILLP **5** very attractive for industrial use [102–105]. In other work, a phosphonium-SO$_3$H based IL was anchored to the magnetic silica nanomaterial, creating the Brønsted-type SILLP **6** (Figure 14) catalyst. Its activity was tested in the acetalization of aldehyde or ketone with ethylene glycol, which resulted in high product yields of 94–97% with various substrates, and the possibility of SILLP catalyst recycling five times without significant loss of activity [106]. Interestingly, the dicationic IL grafted to magnetic nanoparticles (**7**, Figure 14) found application in the synthesis of pyrimido[4,5-b]quinolines (Scheme 3, Figure 15). The novel SILLP **7** hydrogensulfate anion provides one acidic hydrogen and one weakly basic (negative oxygen) site, and was successfully used in the synthesis that requires an acidic and a basic catalyst. This magnetic catalytic system performed well, with yields of 81–96%, short reaction times, and recovery for four reaction cycles, with only a slight decrease in activity [107].

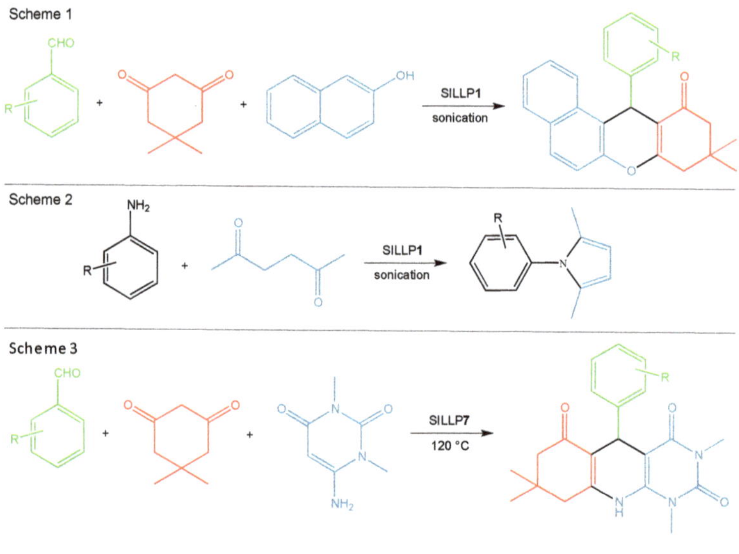

Figure 15. Silica-ferrite hybrid-based SILLP as the catalyst in synthesis of benzoxanthenes (Scheme 1), pyrroles (Scheme 2), pyrimido[4,5-b]quinolines (Scheme 3).

A follow-up report described *N*-(propyl-triethoxysilane)-2-pyrrolidinium hydrogensulfate immobilized on Fe$_3$O$_4$ silica nanoparticles (**8**, Figure 14) as an efficient catalyst for the one-pot diazotization–halogenation of the aromatic amines. Utilization of SILLP **8** as a green catalyst turned out to provide satisfying yields and short reaction times [108]. Silica coated cobalt ferrite nanoparticles were modified with a 3-(4-sulfobutyl)-1-(3-trimethoxysilylmerkaptopropyl)imidazolium triflate IL, and were used in the esterification of oleic acid with straight-chain alcohols. higher SH-group loading on the silica surface resulted in a decreasing pore diameter and surface area. On the other hand, however, less IL moieties could be immobilized on the surface of the nanomaterial. Increasing the alkyl chain in the alcohol caused mass transfer resistance, which resulted in a decreased conversion. This kind of SILLP could find application in shape-selective catalysis [109]. Other work described a sulfo-tetrazolium hydrogensulfate based IL anchored to magnetic nanoparticles. Its activity was tested in the one-pot synthesis of pyrimidine derivatives under mild conditions. The catalyst provided 80–95% yields, an easy separation method using magnetic forces, and could be recycled for six reaction cycles without any activity loss. Moreover, in comparison with another catalyst described in literature, this SILLP is an outstanding green alternative [110]. 2-hydroxyethylammonium sulphonate IL was immobilized via the anion on a magnetic silica-based material. The catalyst possesses basic sites such as hydroxyl groups and acidic sites such as ammonium moieties, and was therefore successfully used in the one-pot three-component synthesis of 2-amino-3,5-dicarbonitrile-6-thio-pyridines. Satisfying yields of 81–91% of various pyridines were achieved, as well as a reusability of five reaction cycles in the reaction between benzaldehyde, malononitrile, and thiophenol characterized this catalyst as very efficient [111]. More recent work was also carried out on a 1,4-diazabicyclo[2.2.2]octane-based basic IL immobilized on silica coated ferrite nanomolecules for a Knoevenagel condensation. The SILLP showed excellent catalytic performance, high yields, short reaction times, and could be reused for eight times. Specific activity could be explained with synergistic action of the tertiary amine, IL, and nanoparticles [112]. Further reports concern the applications of Fe_3O_4-silica nanoparticles modified with 1-methyl-3-(triethoxysilylpropyl)imidazolium chloride. SILLP 9 (Figure 14) catalytic activity was investigated in the synthesis of 1,3-thiazolidin-4-ones [113], indole-substituted pyrido[2,3-d]pyrimidines [114], 3,4-dihydropyrimidin-2(1H)-ones/thiones [116], cycloaddition of CO_2 to epoxides [115], and N-formylation of amines [117]. As shown in Table 3, high reaction yields, easy catalyst recovery, and the possibility of recycling make SILLP 9 not only versatile, but also very efficient. The same catalyst type, but with a triazinium cation, was tested for the synthesis of 4H-dihydropyrano[3,2-b]pyran-3-carbonitrile derivatives. High yields of 85–98% were achieved for various benzaldehydes (Table 3), and newly synthesized compounds indicate potential antioxidant and antifungal properties. Furthermore, this catalyst could be reused four times without any loss of activity [118]. Next, a 3-((3-(trisilyloxy)propyl)propionamide)-1-methylimidazolium chloride IL anchored to silica magnetic nanoparticles was used in the acetylation of alcohols with acetic anhydride under mild conditions. Good yields (93–96%), simple separation by magnetic decantation, and reusability for nine cycles without activity loss were reported for this SILLP [119]. The SILLP magnetic nanoparticles formed from imidazolium-aniline based IL were applied in the Mannich reaction between arylaldehydes, anilines, and cyclohexanone under ultrasound irradiation. The catalyst provided high yields of the main product, high diastereoselectivity (anti:syn), short reaction times, and could be easy reused six times without activity loss, which makes it competitive to previous achievements in this field [120]. In another report, a basic 1-triethylamineimidazolium based IL immobilized on silica coated magnetic nanoparticles was tested for Knoevenagel condensation between various aldehydes and malononitrile. As shown in Table 3, high yields and five reaction cycles with this magnetic SILLP were achieved. In comparison, the IL was immobilized on polystyrene-divinylbenzene resin, but the magnetic silica-based SILLP showed better activity than the polymeric one, presumably due to a more basic character of the silica-ferrite matrix [121]. Studies on immobilized triethyltryptophanium iodide IL on titanomagnetite silica matrix as the catalyst in the synthesis of 6-substituted quinolinedialkyl-2,4-dicarboxylates showed that the library of compounds achieved good yields, the possibility of convenient catalyst recovery, and reusability for three reaction runs were reached in the presence of the SILLP [122].

3.4. SILLP as Matrix for Metals, Organocatalysts, and Enzymes

The catalytic features of the developed SILLPs applications as a matrix or co-catalyst are known, and there are many reports of the use of an SILLP as a matrix/co-catalyst for metal particles, organocatalysts, or enzymes. In Table 4, only examples of silica-based SILLP applications as a matrix are shown due to existing accurate reviews on this topic [34,39–41,123].

Table 4. Examples of silica-based SILLP as a matrix or co-catalyst in organic catalysis.

Catalyst	Reaction Type	Reaction Conditions	Reaction Parameters	Lit.
SiO_2/Rh [tespbim][BF_4] [a]/(tppti) [b]	Hydroformylation of 1-hexene	CO/H_2 (1:1; v/v), Rh/P (1:10, n/n), 100 °C, 5 h	α = 33%, S = 2.4 (n/i-heptanal ratio), TOF = 65 min^{-1}	[124–126]
SiO_2/Ni [tesp(p-SO_3H)im][OTf] [c]	Hydrogenation of n-valeraldehyde	4.5 g cat., n-valeraldehyde 30 mL, P_{H2} = 3 MPa, 200 °C 8 h	α = 100%, S = 98.6%	[127]
SiO_2/PbS [tespmim][Cl] [d]	Dehydrogenation of formic acid	0.0007 g cat., HCOOH/HCOONa 9.00 mmol, 8:1; n/n, H_2O 2.5 mL, 40 °C, 750 rpm	Y = 97% (formic acid decomposition), S_{H2} = 78%, TOF = 604 h^{-1}	[128]
SiO_2/Pd [bvim][Br] [e]	Suzuki coupling	1% mol. cat. phenylboronic acid:aryl halide (1.1:1; n/n), H_2O/EtOH (1.2 mL; 1:1; v/v), K_2CO_3 (0.6 mol), 50 °C, 19 h	Yields for aryl bromides with R-groups: 4-CHO 81%, 4-OMe 89%, 3-OMe 85%, 4-NO_2 80%, 2-CHO 95%, 4-$COCH_3$ 88%, 3-$COCH_3$ 70%, 4-COOH 88%, 2-CH_3 86%, 2-CN 88%, 3,5-(CF_3)$_2$ 89%, H 78%, 1-naphthyl 85%)	[129]
SiO_2/POSS [f]/Pd [tesppim][Cl] [g]/[tespmim][Cl]	Suzuki coupling	0.07% mol. cat. phenylboronic acid:aryl halide (1.1:1; n/n), H_2O/EtOH (1.2 mL; 1:1; v/v), K_2CO_3 (0.6 mol), 50 °C, 19 h	Yields and TOF for aryl bromides with R-groups: 4-CHO 99%, 1429 h^{-1}; 4-OMe 95%, 1327 h^{-1}; 3-OMe 75%, 1071 h^{-1}; 4-NO_2 99%, 1429 h^{-1}; 4-$COCH_3$ 99%, 1429 h^{-1}; 3-$COCH_3$ 99%, 1429 h^{-1}; 3-CH_3 99%, 1414 h^{-1}; 4-CH_3 93%, 1329 h^{-1}; 4-CN 99%, 1429 h^{-1}	[130]
SiO_2/POSS/Pd [tesppim][Cl]/[tespmim][Cl]	Heck reaction	0.07% mol. cat. aryl halide, 0.5 mmol, methyl acrylate 0.75 mmol, triethylamine 1 mmol, DMF 1 mL, 120 °C, 3 h	Yields and TOF for aryl iodides with R-groups: H >99%, 476 h^{-1}; 4-CH_3 >99%, 476 h^{-1}; 4-$COCH_3$ 99%, 471 h^{-1}; 4-OCH_3 99%, 471 h^{-1}; 3-OCH_3 99%, 471 h^{-1}; 4-NO_2 >99%, 476 h^{-1}; 2-C_4H_3S 91%, 433 h^{-1}; 4-CHO >99%, 286 h^{-1}	[130]
SiO_2/Pd [bvim][Br]	Suzuki coupling	0.1% mol. cat. phenylboronic acid:aryl halide (45.2:40; n/n mmol), 0.33 M EtOH (121.2 mL), K_2CO_3 (48 mmol), 50 °C, 1.5 $mLmin^{-1}$, 36 h	Yields for different aryl bromides (H 96%, CH_3 96%, CHO 98%) TON = 3800	[131]
SiO_2/Proline [bvim][NTf_2] [h]	Asymmetric aldol reaction	5% mol cat., aldehyde 1 mmol, cyclohexanone 5 mmol, 1.2 mmol H_2O, rt, 2.5 h	Yields and enantiomeric excess (ee) for aldehydes: 4-NO_2Ph Y = 99%, ee = 98%; 4-ClPh Y = 92%, ee = 99%; 4-BrPh Y = 95%, ee = 97%, 4-CNPh Y = 99%, ee = 92%	[132]
SiO_2/CALB [i] [tespmim][BF_4] [j]	Diacylglycerol production	5% wt. cat., corn oil 4.4 g, glycerol 0.23 g, tert-pentanol 17 mL, 50 °C, 12 h	α = 70.94%, 5 cycles	[133]
SiO_2/PPL [k] [tmspmim][BF_4] [l]	Triacetin hydrolysis	6.83 g of glyceryl triacetate, pH = 7, 45 °C, 10 min	5 cycles	[134]
$SiO_2 \cdot Fe_3O_4$/CRL [m] [tespmim][Cl]	Production of trans-free plastic fats	Palm stearin or liquid rice bran oil, 45 °C, 48 h	4 cycles	[135]

[a] 1-butyl-3-(triethoxysilylpropyl)imidazolium tetrafluoroborate, [b] tri(m-sulfonyl)triphenyl phosphine tris(1-butyl-3-methyl-imidazolium) salt as a ligand, [c] 3-(3-sulfopropyl)-1-(3-propyltriethoxysilane)imidazolium triflate, [d] 1-methyl-3-(triethoxysilylpropyl)imidazolium chloride, [e] 1,4-bis(3-vinylimidazolium-1-yl) bromide, [f] polyhedral oligomeric silsesquioxanes, [g] 1propyl-3-(triethoxysilylpropyl)imidazolium chloride, [h] 1,4-bis(3-vinylimidazolium-1-yl) bis(trifluoromethane)sulfonimide, [i] *Candida antarctica* lipase B, [j] 1-methyl-3-(triethoxysilylpropyl)imidazolium tetrafluoroborate, [k] *Porcine pancreas* lipase, [l] 1-methyl-3-(trimethoxysilylpropyl)imidazolium tetrafluoroborate, [m] *Candida rugosa* lipase.

The hydroformylation reaction is one of the firsts reports on the application of silica based-SILLP as a matrix for metal-based catalysts. Rh particles were introduced to the SILLP with a ligand (to prevent any leaching of Rh) and used for n,i-heptanal production with TOF = 65 min^{-1}, which, compared to the typical biphasic IL approach (TOF = 23 min^{-1}), was a major accomplishment. The catalyst owes its higher activity to a higher concentration of Rh particles on the surface, as well as a larger surface area. Further studies on this topic included physical adsorption of ILs on solid supports (SILP) as Rh particles matrix and continuous-flow processes, which is more accurately described by haumann in the review [37,124–126]. Next, research shows a novel bifunctional Ni-IL/SiO$_2$ in 2-propylheptanol synthesis through a one-pot, self-condensation and hydrogenation from n-valeraldehyde. Ils possess a Brønsted -SO$_3$H group and act like both a matrix and a co-catalyst. With the Ni-SILLP catalytic system, 100% conversion, 75.4% selectivity of the main product, and 98.6% production of 2-propylheptanol and pentanol were achieved [127]. Other reports present SILLP as effective matrices for PbS nanoparticles. A high surface area and IL presence enabled high loading of PdS molecules without aggregation. Additionally, synergistic effects between metal-based particles and Ils provided great catalytic activity in the dehydrogenation of formic acid, with 100% degradation of the acid, 78% selectivity to hydrogen, and TOF = 604 h^{-1} [128]. Further studies present immobilization and stabilization of Pd particles via SBA-15-based SILLP. New versatile and efficient catalytic systems were tested in the Suzuki coupling and heck reactions. As shown in Table 4, the library of compounds was synthesized with high yields, and a catalyst could be recycled several times. Applications of SBA-15 with hexagonal pores as a matrix, which behaved as nanoreactors, assured excellent catalytic activity [129–131]. Moreover, transition from a batch to a continuous process provided conversion of 27 g of substrate to the main product using only 42 mg of the Pd-SILLP catalyst as well as reducing waste, which significantly reduced the E-factor [131]. The next report shows a silica-based SILLP as a carrier for the cis-ion-tagged proline. Proline moieties dissolved in covalently immobilized multilayered IL film performed with excellent activity in an asymmetric aldol reaction. The catalyst provided high yields and enantioselectivity of the main products (Table 4) and could also be recycled up to 15 times [132]. SILLPs can also be used for enzyme immobilization. Moreover, many reports confirmed an IL stabilizing effect on three-dimensional structures of enzymes, increasing protein activity. For example, the catalytic activity of lipase from *Candida antarctica* (CALB) adsorbed on an imidazolium silica-based SILLP was examined in corn oil glycerolysis to diacylglycerol production. The presence of the IL resulted in increasing the catalytic activity from 1855 to 5044 U/g and selectivity from 3.72 to 11.99 (ratio of diacylglycerols/monoacylglycerols). Additionally, the biocatalyst could be recycled for five reaction cycles, and even retained its activity at 50 °C [133]. Another lipase from *Porcine pancreas* immobilized on the same SILLP matrix was used in triacetin hydrolysis. Immobilized enzyme exhibited extremely high thermal stability, where even at 65 °C activity loss did not occur [134]. Lipase from *Candida rugosa* (CRL) was adsorbed on magnetic silica nanoparticles and used in the production of trans-free plastic fats. CRL-SILLP bionanomaterial catalyzed interesterifications of solid palm stearin and liquid rice bran oil for product possesses desirable physicochemical properties. In this case, convenient separation of the biocatalyst enabled its recycling up to four times [135].

4. Conclusions

In summary, achievements in the use of silica-based supported ionic liquid-like phases in heterogeneous organic catalysis were presented. Many Lewis and Brønsted acidic ionic liquids were found to be extremely active as heterogeneous catalysts. For the synthesis of Lewis type silica-based SILLP, chloroaluminate (III), chlorogallate (III), chloroferrate (III), chloroindate (III), chlorostannate (II), chlorozincate (II) anions or hydroxysulfonyl/chlorosulfonyl groups in the cation alkyl chain were used. In case of the forming of Brønsted-type silica-based SILLP, hydrogensulfate, dihydrogenphosphate, dihydrogenphosphotungstate anions, and/or sulfoalkyl group on the cation were found.

Replacement of the halogen anions should be further investigated to prevent hydrolysis and the formation of hazardous acids as hCl. This would result in the reduction of costs, toxic waste, specific equipment, and apparatus corrosion. The anion of the ionic liquid has a crucial influence on the SILLP properties—the synthesized catalyst could be more or less acidic depending on the specific requirements, therefore application of SILLP as a catalyst is very convenient. For the SILLP synthesis, various cations such as imidazolium, alkylammonium, phosphonium, pyrrolidinium, tetrazolium, diammonium, triazinium, and tryptophanium were used, though their selection depended mainly on the substrate or nature of the reaction. It was mainly acidic ILs that were anchored to the silica surface that exhibited great catalytic activity and reusability, and from this the heterogeneous catalyst recovery was very easy. By cross-referencing the presence of the homogeneous and heterogeneous catalysis in the ionic liquids, it can be concluded that IL immobilization increases its catalytic activity due to enhanced mass transfer and availability of active sites. The selection of the silica material also brings many options in terms of size, shape, and density of pores, and hydroxyl groups on the surface. Silica materials such as SBA-15 and MCM-41, with their well-ordered, regular, and hexagonal array of pores, form microreactors that enhance the process efficiency. Obviously, the most important feature is the simplicity of the chemical modification of the surface via trimethoxysilyl/triethoxysilyl groups present as IL precursors, e.g., (3-(chloropropyl)triethoxysilane, 3-(chloropropyl)trimethoxysilane, 3-mercaptopropyltrimethoxysilane, or (3-aminopropyl)-trimethoxysilane). Moreover, the silica can be doped with ferrate nanoparticles, giving the surface magnetic properties. These magnetic-silica nano-catalysts can be removed and recycled by applying external magnetic forces, which is a very convenient approach. Additionally, the magnetic separation also increases the product purity and quality. It should be noted, however, that few examples of the continuous processes with silica-based SILLP have been developed. Flow catalysis offers many advantages compared to batch processes, for example: waste reduction, optimization of pure product synthesis and isolation, reduction of the amount of solvent required, and optimization of the catalyst recovery and recycling. Continuous catalysis simply means an efficient process, as well as green and environmentally friendly production, which is very attractive to the chemical industry. Silica-based SILLPs are versatile, stable catalysts, easy to synthesize, and reusable, with big potential for continuous-flow processes. SILLPs are also potential candidates for the development of sustainable and green chemical processes.

Author Contributions: Conceptualization, A.C. and A.W.; literature survey A.W.; writing—original draft preparation, A.W.; writing—review and editing, A.C. and A.W.; visualization, A.W.; supervision, A.C. All authors have read and agreed to the published version of the manuscript.

Funding: This work was financed by the National Science Centre, Poland (grant no. UMO-2020/39/B/ST8/00693) and Silesian University of Technology (Poland), grant No. 04/050/BKM22/0151.

Institutional Review Board Statement: Not applicable.

Informed Consent Statement: Not applicable.

Data Availability Statement: Data sharing is not applicable for this article.

Conflicts of Interest: The authors declare no conflict of interest. The funders had no role in the design of the study; in the writing of the manuscript, or in the decision to publish.

References

1. Thore, S.; Tarverdyan, R. *Measuring Sustainable Development Goals Performance*; Elsevier: Amsterdam, The Netherlands, 2021.
2. Scheldon, R.A. The E factor 25 years on: The rise of green chemistry and sustainability. *Green Chem.* **2017**, *19*, 18–43. [CrossRef]
3. Ratti, R. Industrial applications of green chemistry: Status, Challenges and Prospects. *SN Appl. Sci.* **2020**, *2*, 263. [CrossRef]
4. Sheldon, R.A.; Arends, I.; Hanefeld, U. *Green Chemistry and Catalysis*; Wiley-VCH Verlag GmbH: Berlin, Germany, 2020.
5. Scheldon, R.A.; Woodley, J.M. Role of biocatalysis in sustainable chemistry. *Chem. Rev.* **2018**, *118*, 801–838. [CrossRef]
6. De los Ríos, A.P.; Irabien, A.; Hollmann, F.; Fernández, F.J.H. Ionic Liquids: Green Solvents for Chemical Processing. *J. Chem.* **2013**, *2013*, 402172. [CrossRef]

7. Greer, A.J.; Jacquemin, J.; Hardacre, C. Industrial applications of ionic liquids. *Molecules* **2020**, *25*, 5207. [CrossRef]
8. Padvi, S.A.; Dalal, D.S. Task-Specific Ionic Liquids as a Green Catalysts and Solvents for Organic Synthesis. *Curr. Green Chem.* **2020**, *7*, 105–119. [CrossRef]
9. Chrobok, A. The Baeyer–Villiger oxidation of ketones with Oxone® in the presence of ionic liquids as solvents. *Tetrahedron* **2010**, *66*, 6212–6216. [CrossRef]
10. Domínguez de María, P. *Ionic Liquids in Biotransformations and Organocatalysis: Solvents and Beyond*; John Wiley & Sons, Inc.: Hoboken, NJ, USA, 2012; pp. 1–435.
11. Drożdż, A.; Erfurt, K.; Bielas, R.; Chrobok, A. Chemo-enzymatic Baeyer–Villiger oxidation in the presence of Candida antarctica lipase B and ionic liquids. *New J. Chem.* **2015**, *39*, 1315–1321. [CrossRef]
12. Karuppasamy, K.; Theerthagiri, J.; Vikraman, D.; Yim, C.-J.; Hussain, S.; Sharma, R.; Kim, H.S. Ionic Liquid-Based Electrolytes for Energy Storage Devices: A Brief Review on Their Limits and Applications. *Polymers* **2020**, *12*, 918. [CrossRef]
13. Passos, H.; Freire, M.G.; Coutinho, J.A.P. Ionic liquid solutions as extractive solvents for value-added compounds from biomass. *Green Chem.* **2014**, *16*, 4786–4815. [CrossRef]
14. Bajkacz, S.; Rusin, K.; Wolny, A.; Adamek, J.; Erfurt, K.; Chrobok, A. Highly efficient extraction procedures based on natural deep eutectic solvents or ionic liquids for determination of 20-Hydroxyecdysone in Spinach. *Molecules* **2020**, *25*, 4736. [CrossRef] [PubMed]
15. Shukla, S.K.; Khokarale, S.G.; Bui, T.Q.; Mikkola, J.-P.T. Ionic Liquids: Potential Materials for Carbon Dioxide Capture and Utilization. *Front. Mater.* **2019**, *6*, 42. [CrossRef]
16. Vekariya, R.L. A review of ionic liquids: Applications towards catalytic organic transformations. *J. Mol. Liq.* **2017**, *227*, 44–60. [CrossRef]
17. Welton, T. Ionic liquids in catalysis. *Coord. Chem. Rev.* **2004**, *248*, 2459–2477. [CrossRef]
18. Brown, L.C.; Hogg, J.M.; Swadźba-Kwaśny, M. Lewis Acidic Ionic Liquids. *Top Curr. Chem.* **2017**, *375*, 78. [CrossRef]
19. Greaves, T.L.; Drummond, C.J. Protic Ionic Liquids: Properties and Applications. *Chem. Rev.* **2008**, *108*, 206–237. [CrossRef] [PubMed]
20. Amarasekara, A.S. Acidic Ionic Liquids. *Chem. Rev.* **2016**, *116*, 6133–6183. [CrossRef]
21. Jasiak, K.; Siewniak, A.; Kopczyńska, K.; Chrobok, A.; Baj, S. Hydrogensulphate ionic liquids as an efficient catalyst for the synthesis of cyclic carbonates from carbon dioxide and epoxides. *J. Chem. Technol. Biotechnol.* **2016**, *91*, 2827–2833. [CrossRef]
22. Shen, L.; Yin, H.; Wang, A.; Lu, X.; Zhang, C.; Chen, F.; Chen, H. Liquid phase catalytic dehydration of glycerol to acrolein over Brønsted acidic ionic liquid catalysts. *J. Ind. Eng. Chem.* **2014**, *20*, 759–766. [CrossRef]
23. Zhang, L.; He, L.; Hong, C.-B.; Qin, S.; Tao, G.-H. Brønsted acidity of bio-protic ionic liquids: The acidic scale of [AA]X amino acid ionic liquids. *Green Chem.* **2015**, *17*, 5154–5163. [CrossRef]
24. Kore, R.; Kumar, T.J.D.; Srivastava, R. Hydration of alkynes using Brönsted acidic ionic liquids in the absence of Nobel metal catalyst/H$_2$SO$_4$. *J. Mol. Cat. A Chem.* **2012**, *360*, 61–70. [CrossRef]
25. Fang, D.; Yang, J.; Jiao, C. Dicationic Ionic Liquids as Environmentally Benign Catalysts for Biodiesel Synthesis. *ACS Catal.* **2010**, *1*, 42–47. [CrossRef]
26. Latos, P.; Culkin, A.; Barteczko, N.; Boncel, S.; Jurczyk, S.; Brown, L.C.; Nockemann, P.; Chrobok, A.; Swadźba-Kwaśny, M. Water-Tolerant Trifloaluminate Ionic Liquids: New and Unique Lewis Acidic Catalysts for the Synthesis of Chromane. *Front. Chem.* **2018**, *6*, 535. [CrossRef]
27. Markiton, M.; Chrobok, A.; Matuszek, K.; Seddon, K.R.; Swadźba-Kwaśny, M. Exceptional activity of gallium(iii) chloride and chlorogallate(iii) ionic liquids for Baeyer–Villiger oxidation. *RSC Adv.* **2016**, *6*, 30460–30467. [CrossRef]
28. Matuszek, K.; Coffie, S.; Chrobok, A.; Swadźba-Kwaśny, M. Borenium ionic liquids as catalysts for Diels–Alder reaction: Tuneable Lewis superacids for catalytic applications. *Catal. Sci. Technol.* **2017**, *7*, 1045–1049. [CrossRef]
29. Valkenberg, M.H.; deCastro, C.; Hölderich, W.F. Immobilisation of chloroaluminate ionic liquids on silica materials. *Top. Catal.* **2000**, *14*, 139–144. [CrossRef]
30. Mehnert, C.P. Supported Ionic Liquid Catalysis. *Eur. J. Chem.* **2005**, *11*, 50–56. [CrossRef]
31. Burguete, M.I.; García-Verdugo, E.; Karbass, N.; Luis, S.V.; Sans, V.; Sokolova, M. Development of efficient processes under flow conditions based on catalysts immobilized onto monolithic supported ionic liquid-like phases. *Pure Appl. Chem.* **2009**, *81*, 1991–2000. [CrossRef]
32. Skoda-Földes, R. The Use of Supported Acidic Ionic Liquids in Organic Synthesis. *Molecules* **2014**, *19*, 8840–8884. [CrossRef]
33. Schwieger, W.; Selvam, T.; Klumpp, M.; Hartmann, M. Porous Inorganic Materials as Potential Supports for Ionic Liquids. In *Supported Ionic Liquids: Fundamental and Applications*, 1st ed.; Fehrmann, R., Riisager, A., Haumann, M., Eds.; Wiley-VCH Verlag GmbH: Berlin, Germany, 2014; pp. 37–74.
34. Giacalone, F.; Gruttadauria, M. Covalently Supported Ionic Liquid Phases: An Advanced Class of Recyclable Catalytic Systems. *ChemCatChem* **2016**, *8*, 664–684. [CrossRef]
35. Vafaeezadeh, M.; Alinezhad, H. Brønsted acidic ionic liquids: Green catalysts for essential organic reactions. *J. Mol. Liq.* **2016**, *218*, 95–105. [CrossRef]
36. Migowski, P.; Luska, K.L.; Leitner, W. Nanoparticles on Supported Ionic Liquid Phases - Opportunities for Application in Catalysis. Nanocatalysis in Ionic Liquids. In *Nanocatalysis in Ionic Liquids*, 1st ed.; Martin, H., Prechtl, G., Eds.; Wiley-VCH Verlag GmbH: Berlin, Germany, 2017; pp. 249–273.

37. Marinkovic, J.M.; Riisager, A.; Franke, R.; Wasserscheid, P.; Haumann, M. Fifteen Years of Supported Ionic Liquid Phase-Catalyzed Hydroformylation: Material and Process Developments. *Ind. Eng. Chem. Res.* **2019**, *58*, 2409–2420. [CrossRef]
38. Pedro, A.Q.; Coutinho, J.A.P.; Freire, M.G. Immobilization of Ionic Liquids, Types of Materials, and Applications. In *Encyclopedia of Ionic Liquids*; Zhang, S., Ed.; Springer Nature: Singapore, 2019; pp. 1–12.
39. Bartlewicz, O.; Dąbek, I.; Szymańska, A.; Maciejewski, H. Heterogeneous Catalysis with the Participation of Ionic Liquids. *Catalysts* **2020**, *10*, 1227. [CrossRef]
40. Garcia-Verdugo, E.; Lozano, P.; Luis, S.V. Biocatalytic Processes Based on Supported Ionic Liquids. In *Supported Ionic Liquids: Fundamental and Applications*, 1st ed.; Fehrmann, R., Riisager, A., Haumann, M., Eds.; Wiley-VCH Verlag GmbH: Berlin, Germany, 2014; pp. 351–368.
41. Wolny, A.; Chrobok, A. Ionic Liquids for Development of Heterogeneous Catalysts Based on Nanomaterials for Biocatalysis. *Nanomaterials* **2021**, *11*, 2030. [CrossRef] [PubMed]
42. Stöber, W.; Fink, A.; Bohn, E. Controlled growth of monodisperse silica spheres in the micron size range. *J. Colloid Interface Sci.* **1968**, *26*, 62–69. [CrossRef]
43. Donato, K.Z.; Matějka, L.; Mauler, R.S.; Donato, R.K. Recent Applications of Ionic Liquids in the Sol-Gel Process for Polymer–Silica Nanocomposites with Ionic Interfaces. *Colloids Interfaces* **2017**, *1*, 5. [CrossRef]
44. Du, A.; Wang, Z.; Shang, Y.; Sun, X. Interactions Between an Ionic Liquid and Silica, Silica and Silica, and Rubber and Silica and Their Effects on the Properties of Styrene-Butadiene Rubber Composites. *J. Macromol. Sci. Phys.* **2019**, *58*, 99–112. [CrossRef]
45. Wang, Y.-M.; Ulrich, V.; Donnelly, G.F.; Lorenzini, F.; Marr, A.C.; Marr, P.C. A Recyclable Acidic Ionic Liquid Gel Catalyst for Dehydration: Comparison with an Analogous SILP Catalyst. *ACS Sustain. Chem. Eng.* **2015**, *3*, 792–796. [CrossRef]
46. Dhar, A.; Siva Kumar, N.; Khimani, M.; Al-Fatesh, A.S.; Ibrahim, A.A.; Fakeeha, A.H.; Patel, H.; Vekariya, R.L. Silica-immobilized ionic liquid Brønsted acids as Highly effective Heterogeneous catalysts for the isomerization of n-heptane and n-octane. *RSC Adv* **2020**, *10*, 15282. [CrossRef]
47. Riisager, A.; Fehrmann, R.; Haumann, M.; Wasserscheid, P. Supported Ionic Liquid Phase (SILP) Catalysis: An Innovative Concept for Homogeneous Catalysis in Continuous Fixed-Bed Reactors. *Eur. J. Inorg. Chem.* **2006**, *4*, 695–706. [CrossRef]
48. Kukawka, R.; Pawlowska-Zygarowicz, A.; Dzialkowska, J.; Pietrowski, M.; Maciejewski, H.; Bica, K.; Smiglak, M. A highly effective supported ionic liquid phase (SILP) catalysts—Characterization and application to the hydrosilylation reaction. *ACS Sustain. Chem. Eng.* **2019**, *7*, 4699–4706. [CrossRef]
49. Lozano, P.; Diego, T.; de Carrié, D.; Vaultier, M.; Iborra, J.L. Continuous green biocatalytic processes using ionic liquids and supercritical carbon dioxide. *Chem. Commun.* **2002**, *7*, 692–693. [CrossRef] [PubMed]
50. Valkenberg, M.H.; deCastro, C.; Hölderich, W.F. Immobilisation of ionic liquids on solid supports. *Green Chem.* **2001**, *4*, 88–93. [CrossRef]
51. Jyothi, T.M.; Kaliya, M.L.; Herskowitz, M.; Landau, M.V. A comparative study of an MCM-41 anchored quaternary ammonium chloride/SnCl4 catalyst and its silica gel analogue. *Chem. Commun.* **2001**, *11*, 992–993. [CrossRef]
52. Hagiwara, H.; Sekifuji, M.; Hoshi, T.; Qiao, K.; Yokoyama, C. Synthesis of Bis(indolyl)methanes Catalyzed by Acidic Ionic Liquid Immobilized on Silica (ILIS). *Synlett* **2007**, *8*, 1320–1322. [CrossRef]
53. Qiao, K.; Hagiwara, H.; Yokoyama, C. Acidic ionic liquid modified silica gel as novel solid catalysts for esterification and nitration reactions. *J. Mol. Catal. A Chem.* **2006**, *246*, 65–69. [CrossRef]
54. Hagiwara, H.; Sekifuji, M.; Hoshi, T.; Suzuki, T.; Quanxi, B.; Qiao, K.; Yokoyama, C. Sustainable Conjugate Addition of Indoles Catalyzed by Acidic Ionic Liquid Immobilized on Silica. *Synlett* **2008**, *4*, 608–610. [CrossRef]
55. Kumar, P.; Vermeiren, W.; Dath, J.-P.; Hoelderich, W.F. Production of alkylated gasoline using ionic liquids and immobilized ionic liquids. *Appl. Catal. A Gen.* **2006**, *304*, 131–141. [CrossRef]
56. Liu, S.; Shang, J.; Zhang, S.; Yang, B.; Deng, Y. Highly Efficient Trimerization of Isobutene Over Silica Supported Chloroaluminate Ionic Liquid Using C4 Feed. *Catal. Today* **2013**, *200*, 41–48. [CrossRef]
57. Wang, G.; Yu, N.; Peng, L.; Tan, R.; Zhao, H.; Yin, D.; Yin, D. Immobilized Chloroferrate Ionic Liquid: An Efficient and Reusable Catalyst for Synthesis of Diphenylmethane and its Derivatives. *Catal. Lett.* **2008**, *123*, 252–258. [CrossRef]
58. Zhao, H.; Yu, N.; Wang, J.; Zhuang, D.; Ding, Y.; Tan, R.; Yin, D. Preparation and catalytic activity of periodic mesoporous organosilica incorporating Lewis acidic chloroindate(III) ionic liquid moieties. *Microporous Mesoporous Mater.* **2009**, *122*, 240–246. [CrossRef]
59. Matuszek, K.; Chrobok, A.; Latos, P.; Markiton, M.; Szymańska, K.; Jarzębski, A.; Swadźba-Kwaśny, M. Silica-supported chlorometallate(iii) ionic liquids as recyclable catalysts for Diels–Alder reaction under solventless conditions. *Catal. Sci. Technol.* **2016**, *6*, 8129–8137. [CrossRef]
60. Siewniak, A.; Forajter, A.; Szymańska, K. Mesoporous Silica-Supported Ionic Liquids as Catalysts for Styrene Carbonate Synthesis from CO_2. *Catalysts* **2020**, *10*, 1363. [CrossRef]
61. Yao, J.; Sheng, M.; Bai, S.; Su, H.; Shang, H.; Deng, H.; Sun, J. Ionic Liquids Grafted Mesoporous Silica for Chemical Fixation of CO_2 to Cyclic Carbonate: Morphology Effect. *Catal. Lett.* **2021**, *152*, 781–790. [CrossRef]
62. Zhen, B.; Jiao, Q.; Wu, Q.; Li, H. Catalytic performance of acidic ionic liquid-functionalized silica in biodiesel production. *J. Energy Chem.* **2014**, *23*, 97–104. [CrossRef]
63. Wiredu, B.; Amarasekara, A.S. Synthesis of a silica-immobilized Brönsted acidic ionic liquid catalyst and Hydrolysis of cellulose in water under mild conditions. *Catal. Commun.* **2014**, *48*, 41–44. [CrossRef]

64. Bao, Q.; Qiao, K.; Tomida, D.; Yokoyama, C. Preparation of 5-hydroymethylfurfural by dehydration of fructose in the presence of acidic ionic liquid. *Catal. Commun.* **2008**, *9*, 1383–1388. [CrossRef]
65. Elhamifar, D.; Nasr-Esfahani, M.; Karimi, B.; Moshkelgosha, R.; Shábani, A. Ionic Liquid and Sulfonic Acid Based Bifunctional Periodic Mesoporous Organosilica (BPMO-IL-SO$_3$H) as a Highly Efficient and Reusable Nanocatalyst for the Biginelli Reaction. *ChemCatChem* **2014**, *6*, 2593–2599. [CrossRef]
66. Elhamifar, D.; Karimi, B.; Moradi, A.; Rastegar, J. Synthesis of Sulfonic Acid Containing Ionic-Liquid-Based Periodic Mesoporous Organosilica and Study of Its Catalytic Performance in the Esterification of Carboxylic Acids. *ChemPlusChem* **2014**, *79*, 1147–1152. [CrossRef]
67. Wang, W.; Wang, D.; Yang, Q.; An, H.; Zhao, X.; Wang, Y. Silica-immobilized acid ionic liquid: An efficient catalyst for pentanal self-condensation. *J. Chem. Technol. Biotechnol.* **2020**, *95*, 2964–2972. [CrossRef]
68. Kotadia, D.A.; Soni, S.S. Sulfonic acid functionalized solid acid: An alternative eco-friendly approach for transesterification of non-edible oils with high free fatty acids. *Monatsh. Chem.* **2013**, *144*, 1735–1741. [CrossRef]
69. Kotadia, D.A.; Soni, S.S. Silica gel supported –SO$_3$H functionalised benzimidazolium based ionic liquid as a mild and effective catalyst for rapid synthesis of 1-amidoalkyl naphthols. *J. Mol. Catal. A Chem.* **2012**, *353–354*, 44–49. [CrossRef]
70. Chrobok, A.; Baj, S.; Pudło, W.; Jarzębski, A. Supported hydrogensulfate ionic liquid catalysis in Baeyer–Villiger reaction. *Appl. Catal. A Gen.* **2009**, *366*, 22–28. [CrossRef]
71. Zhang, J.; Wan, H.; Guan, G. Preparation and Catalytic Performance of Silica Gel Immobilized Acidic Ionic Liquid Catalyst. *Reaction. Eng. Technol.* **2008**, *24*, 503–508.
72. Seddighi, M.; Shirini, F.; Mamaghani, M. Brönsted acidic ionic liquid supported on rice husk ash (RHA-[pmim]HSO$_4$): A highly efficient and reusable catalyst for the synthesis of 1-(benzothiazolylamino)phenylmethyl-2-naphthols. *Comptes Rendus Chim.* **2015**, *18*, 573–580. [CrossRef]
73. Shirini, F.; Seddighi, M.; Mamaghani, M. Brönsted acidic ionic liquid supported on rice husk ash (RHA–[pmim]HSO$_4$): A highly efficient and reusable catalyst for the formylation of amines and alcohols. *RSC Adv.* **2014**, *4*, 50631–50638. [CrossRef]
74. Rostamnia, S.; Hassankhani, A.; Hossieni, H.G.; Gholipour, B.; Xin, H. Brønsted acidic Hydrogensulfate ionic liquid immobilized SBA-15: [MPIm][HSO$_4$]@SBA-15 as an environmentally friendly, metal- and halogen-free recyclable catalyst for Knoevenagel–Michael-cyclization processes. *J. Mol. Catal. A Chem.* **2014**, *395*, 463–469. [CrossRef]
75. Niknam, K.; Piran, A. Silica-Grafted Ionic Liquids as Recyclable Catalysts for the Synthesis of 3,4-Dihydropyrano[c]chromenes and Pyra-no [2,3-c]pyrazoles. *Green Sustain. Chem.* **2013**, *3*, 31420. [CrossRef]
76. Damavandi, S. Immobilized Ionic Liquid-Catalyzed Synthesis of Pyrano[3,2-b]indole Derivatives. *E-J. Chem.* **2012**, *9*, 1490–1493. [CrossRef]
77. Damavandi, S.; Sandaroos, R. Novel Synthetic Route to Pyrano[2,3-b]pyrrole Derivatives. *Syn. React. Inorg. Metal Org. Nano Metal Chem.* **2012**, *42*, 621–627. [CrossRef]
78. Eshghi, H.; Zohuri, G.H.; Sandaroos, R.; Damavandi, S. Synthesis of novel benzo[f]chromene compounds catalyzed by ionic liquid. *Heterocycl. Commun.* **2012**, *18*, 67–70. [CrossRef]
79. Damavandi, S.; Sandaroos, R. Novel Multicomponent Synthesis of 2,9-Dihydro-9-methyl-2-oxo-4-aryl-1H-pyrido [2, 3-b] indole-3-carbonitrile Compounds. *J. Chem. Sci.* **2013**, *125*, 95–100. [CrossRef]
80. Goldani, M.T.; Sandaroos, R.; Damavandi, S. Efficient Polymeric Catalyst for One-pot Synthesis of Acenaphtho [1, 2-b] Pyrroles. *Res. Chem. Intermed.* **2014**, *40*, 139–147. [CrossRef]
81. Sandaroos, R.; Damavandi, S.; Salimi, M. Facile one-pot synthesis of 5-amino-7-aryl-6-cyano-4H-pyrano[3,2-b]pyrroles using supported hydrogen sulfate ionic liquid. *Monatsh. Chem.* **2012**, *143*, 1655–1661. [CrossRef]
82. Xiong, J.; Zhu, W.; Ding, W.; Yang, L.; Chao, Y.; Li, H. Phosphotungstic Acid Immobilized on Ionic Liquid-Modified SBA-15: Efficient hydrophobic heterogeneous Catalyst for Oxidative Desulfurization in Fuel. *Ind. Eng. Chem. Res.* **2014**, *53*, 19895–19904. [CrossRef]
83. Wan, H.; Zhang, J.; Guan, G. Preparation of Supported Acidic Ionic Liquid by Covalent Bond Grafting and its Catalysis in Synthesis of n-Butyl Acetate. *Shiyou Huagong/Petrochem. Technol.* **2009**, *38*, 134–138.
84. Zhang, Q.; Luo, J.; Wei, Y. A silica gel supported dual acidic ionic liquid: An efficient and recyclable Heterogeneous catalyst for the one-pot synthesis of amidoalkyl naphthols. *Green Chem.* **2010**, *12*, 2246–2254. [CrossRef]
85. Vafaeezadeh, M.; Dizicheh, Z.B.; Hashemi, M.M. Mesoporous silica-functionalized dual Brønsted acidic ionic liquid as an efficient catalyst for thioacetalization of carbonyl compounds in water. *Catal. Commun.* **2013**, *41*, 96–100. [CrossRef]
86. Miao, J.; Wan, H.; Shao, Y.; Guan, G.; Xu, B. Acetalization of carbonyl compounds catalyzed by acidic ionic liquid immobilized on silica gel. *J. Mol. Catal. A Chem.* **2011**, *348*, 77–82. [CrossRef]
87. Safaei, S.; Mohammadpoor-Baltork, I.; Khosropour, A.R.; Moghadam, M.; Tangestaninejad, S.; Mirkhani, V. Nano-silica supported acidic ionic liquid as an efficient catalyst for the multi-component synthesis of indazolophthalazine-triones and bis-indazolophthalazine-triones. *Catal. Sci. Technol.* **2013**, *3*, 2717. [CrossRef]
88. Wu, Y.; Li, Z.; Xia, C. Silica-Gel-Supported Dual Acidic Ionic Liquids as Efficient Catalysts for the Synthesis of Polyoxymethylene Dimethyl Ethers. *Ind. Eng. Chem. Res.* **2016**, *55*, 1859–1865. [CrossRef]
89. Singh, S.K.; Dhepe, P.L. Novel Synthesis of Immobilized Brønsted-Acidic Ionic Liquid: Application in Lignin Depolymerization. *ChemistrySelect* **2018**, *3*, 5461–5470. [CrossRef]

90. Miao, J.; Wan, H.; Guan, G. Synthesis of immobilized Brønsted acidic ionic liquid on silica gel as heterogeneous catalyst for esterification. *Catal. Commun.* **2011**, *12*, 353–356. [CrossRef]
91. Ma, W.; Wang, W.; Liang, Z.; Hu, S.; Shen, R.; Wu, C. Synthesis of novel acidic ionic liquid immobilized on silica. *Kinet. Catal.* **2014**, *55*, 665–670. [CrossRef]
92. Sofia, L.T.A.; Krishnan, A.; Sankar, M.; Kala Raj, N.K.; Manikandan, P.; Rajamohanan, P.R.; Ajithkumar, T.G. Immobilization of Phosphotungstic Acid (PTA) on Imidazole Functionalized Silica: Evidence for the Nature of PTA Binding by Solid State NMR and Reaction Studies. *J. Phys. Chem. C* **2009**, *113*, 21114–21122. [CrossRef]
93. Wang, Y.; Zhao, D.; Wang, L.; Wang, X.; Li, L.; Xing, Z.; Ding, H. Immobilized phosphotungstic acid based ionic liquid: Application for heterogeneous esterification of palmitic acid. *Fuel* **2018**, *216*, 364–370. [CrossRef]
94. Li, P.-H.; Li, B.-L.; Hu, H.-C.; Zhao, X.-N.; Zhang, Z.-H. Ionic liquid supported on magnetic nanoparticles as highly efficient and recyclable catalyst for the synthesis of β-keto enol ethers. *Catal. Commun.* **2014**, *46*, 118–122. [CrossRef]
95. Nguyen, H.T.; Thi Le, N.-P.; Nguyen Chau, D.-K.; Tran, P.H. New nano-Fe_3O_4-supported Lewis acidic ionic liquid as a highly effective and recyclable catalyst for the preparation of benzoxanthenes and pyrroles under solvent-free sonication. *RSC Adv.* **2018**, *8*, 35681–35688. [CrossRef]
96. Alinezhad, H.; Tajbakhsh, M.; Ghobadi, N. Ionic liquid immobilized on Fe_3O_4 nanoparticles: A magnetically recyclable heterogeneous catalyst for one-pot three-component synthesis of 1,8-dioxodecahydroacridines. *Res. Chem. Intermed.* **2015**, *41*, 9979–9992. [CrossRef]
97. Nezhad, E.R.; Karimian, S.; Sajjadifar, S. Imidazole functionalized magnetic Fe_3O_4 nanoparticles a highly efficient and reusable Brønsted acid catalyst for the regioselective thiocyanation of aromatic and heteroaromatic compounds at room temperature in water:ethanol. *J. Sci.* **2015**, *26*, 233–240.
98. Nasrollahzadeh, M.; Issaabadi, Z.; Sajadi, S.M. Fe_3O_4@SiO_2 nanoparticle supported ionic liquid for green synthesis of antibacterially active 1-carbamoyl-1-phenylureas in water. *RSC Adv.* **2018**, *8*, 27631–27644. [CrossRef] [PubMed]
99. Pourjavadi, A.; Hosseini, S.H.; Doulabi, M.; Fakoorpoor, S.M.; Seidi, F. Multi-Layer Functionalized Poly(Ionic Liquid) Coated Magnetic Nanoparticles: Highly Recoverable and Magnetically Separable Brønsted Acid Catalyst. *ACS Catal.* **2012**, *2*, 1259–1266. [CrossRef]
100. Gupta, R.; Yadav, M.; Gaur, R.; Arora, G.; Rana, P.; Yadav, P.; Sharma, R.K. Silica-Coated Magnetic-Nanoparticle-Supported DABCO-Derived Acidic Ionic Liquid for the Efficient Synthesis of Bioactive 3,3-Di(indolyl)indolin-2-ones. *ACS Omega* **2019**, *4*, 21529–21539. [CrossRef] [PubMed]
101. Sadeghzadeh, S.M. A heteropolyacid-based ionic liquid immobilized onto magnetic fibrous nano-silica as robust and recyclable heterogeneous catalysts for the synthesis of tetrahydrodipyrazolopyridines in water. *RSC Adv.* **2016**, *6*, 75973–75980. [CrossRef]
102. Zhang, Q.; Su, H.; Luo, J.; Wei, Y. A Magnetic Nanoparticle Supported Dual Acidic Ionic Liquid: A "Quasi-Homogeneous" Catalyst for the One-pot Synthesis of Benzoxanthenes. *Green Chem.* **2012**, *14*, 201–208. [CrossRef]
103. Khalafi-Nezhad, A.; Mohammadi, S. Magnetic, Acidic, Ionic Liquid-catalyzed One-pot Synthesis of Spirooxindoles. *ACS Comb. Sci.* **2013**, *15*, 512–518. [CrossRef] [PubMed]
104. Wan, H.; Wu, Z.; Chen, W.; Guan, G.; Cai, Y.; Chen, C.; Liu, X. Heterogenization of ionic liquid based on mesoporous material as magnetically recyclable catalyst for biodiesel production. *J. Mol. Catal. A Chem.* **2015**, *398*, 127–132. [CrossRef]
105. Wu, Z.; Li, Z.; Wu, G.; Wang, L.; Lu, S.; Wang, L.; Guan, G. Brønsted Acidic Ionic Liquid Modified Magnetic Nanoparticle: An Efficient and Green Catalyst for Biodiesel Production. *Ind. Eng. Chem. Res.* **2014**, *53*, 3040–3046. [CrossRef]
106. Wang, P.; Kong, A.; Wang, W.; Zhu, H.; Shan, Y. Facile Preparation of Ionic liquid Functionalized Magnetic Nano-solid Acid Catalysts for Acetalization Reaction. *Catal. Lett.* **2010**, *135*, 159–164. [CrossRef]
107. Zare, A.; Barzegar, M. Dicationic ionic liquid grafted with silica-coated nano-Fe_3O_4 as a novel and efficient catalyst for the preparation of uracil-containing Heterocycles. *Res. Chem. Intermed.* **2020**, *46*, 3727–3740. [CrossRef]
108. Isaad, J. Acidic ionic liquid supported on silica-coated magnetite nanoparticles as a green catalyst for one-pot diazotization–halogenation of the aromatic amines. *RSC Adv.* **2014**, *4*, 49333–49341. [CrossRef]
109. Zhen, B.; Jiao, Q.; Zhang, Y.; Wu, Q.; Li, H. Acidic ionic liquid immobilized on magnetic mesoporous silica: Preparation and catalytic performance in esterification. *Appl. Catal. A Gen.* **2012**, *445–446*, 239–245. [CrossRef]
110. Naeimi, H.; Nejadshafiee, V.; Islami, M.R. Iron (III)-doped, ionic liquid matrix-immobilized, mesoporous silica nanoparticles: Application as recyclable catalyst for synthesis of pyrimidines in water. *Microporous Mesoporous Mater.* **2016**, *227*, 23–30. [CrossRef]
111. Sobhani, S.; Honarmand, M. Ionic liquid immobilized on γ-Fe_2O_3 nanoparticles: A new magnetically recyclable heterogeneous catalyst for one-pot three-component synthesis of 2-amino-3,5-dicarbonitrile-6-thio-pyridines. *Appl. Catal. A Gen.* **2013**, *467*, 456–462. [CrossRef]
112. Jia, X.; Zhang, X.; Wang, Z.; Zhao, S. Tertiary amine ionic liquid incorporated Fe_3O_4 nanoparticles as a versatile catalyst for the Knoevenagel reaction. *Synth. Commun.* **2022**, *52*, 774–786. [CrossRef]
113. Azgomi, N.; Mokhtary, M. Nano-Fe_3O_4@SiO_2 supported ionic liquid as an efficient catalyst for the synthesis of 1,3-thiazolidin-4-ones under solvent-free conditions. *J. Mol. Catal. A Chem.* **2015**, *398*, 58–64. [CrossRef]
114. Mamaghani, M.; Sheykhan, M.; Sadeghpour, M.; Tavakoli, F. An expeditious one-pot synthesis of novel bioactive indole-substituted pyrido[2,3-d]pyrimidines using Fe_3O_4@SiO_2-supported ionic liquid nanocatalyst. *Monatsh. Fur Chem.* **2018**, *149*, 1437–1446. [CrossRef]

115. Zheng, X.; Luo, S.; Zhang, L.; Cheng, J.-P. Magnetic nanoparticle supported ionic liquid catalysts for CO_2 cycloaddition reactions. *Green Chem.* **2009**, *11*, 455. [CrossRef]
116. Safari, J.; Zarnegar, Z. Brønsted Acidic Ionic Liquid based Magnetic Nanoparticles: A New Promoter for the Biginelli Synthesis of 3, 4-Dihydropyrimidin-2 (1 h)-ones/thiones. *New J. Chem.* **2014**, *38*, 358–365. [CrossRef]
117. Garkoti, C.; Shabir, J.; Mozumdar, S. An imidazolium based ionic liquid supported on Fe_3O_4@SiO_2 nanoparticles as an efficient heterogeneous catalyst for N-formylation of amines. *New J. Chem.* **2017**, *41*, 9291–9298. [CrossRef]
118. Azarifar, D.; Ebrahimiasl, H.; Karamian, R.; Ahmadi-Khoei, M. s-Triazinium-based ionic liquid immobilized on silica-coated Fe_3O_4 magnetic nanoparticles: An efficient and magnetically separable heterogeneous catalyst for synthesis of 2-amino-4,8-dihydropyrano[3,2-b]pyran-3-carbonitrile derivatives for antioxidant and antifungal evaluation studies. *J. Iran. Chem. Soc.* **2018**, *16*, 341–354.
119. Ghorbani-Choghamarani, A.; Norouzi, M. Synthesis and characterization of ionic liquid immobilized on magnetic nanoparticles: A recyclable heterogeneous organocatalyst for the acetylation of alcohols. *J. Magn. Magn. Mater.* **2016**, *401*, 832–840. [CrossRef]
120. Safaei Ghomi, J.; Zahedi, S. Novel ionic liquid supported on Fe_3O_4 nanoparticles and its application as a catalyst in Mannich reaction under ultrasonic irradiation. *Sonochemistry* **2017**, *34*, 916–923. [CrossRef] [PubMed]
121. Zhang, Y.; Xia, C. Magnetic hydroxyapatite-encapsulated γ-Fe_2O_3 nanoparticles functionalized with basic ionic liquids for aqueous Knoevenagel condensation. *Appl. Catal. A Gen.* **2009**, *366*, 141–147. [CrossRef]
122. Nikoofar, K.; molaei Yielzoleh, F. Cascade embedding triethyltryptophanium iodide ionic liquid ($TrpEt_3^+I^-$) on silicated titano-magnetite core ($Fe_{3-x}Ti_xO_4$-SiO_2@$TrpEt_3^+I^-$): A novel nano organic–inorganic hybrid to prepare a library of 4-substituted quinoline-2-carboxylates and 4,6-disubstituted quinoline-2-carboxylates. *J. Chin. Chem. Soc.* **2021**, *68*, 1549–1562.
123. Gu, Y.; Li, G. Ionic Liquids-Based Catalysis with Solids: State of the Art. *Adv. Synth. Catal.* **2009**, *351*, 817–847. [CrossRef]
124. Mehnert, C.P.; Cook, R.A.; Dispenziere, N.C.; Afeworki, M. Supported Ionic Liquid Catalysis—A New Concept for homogeneous hydroformylation Catalysis. *J. Am. Chem. Soc.* **2002**, *124*, 12932–12933. [CrossRef]
125. Riisager, A. Continuous fixed-bed gas-phase hydroformylation using supported ionic liquid-phase (SILP) Rh catalysts. *J. Catal.* **2003**, *219*, 452–455. [CrossRef]
126. Riisager, A.; Eriksen, K.M.; Wasserscheid, P. Propene and 1-Octene hydroformylation with Silica-Supported, Ionic Liquid-Phase (SILP) Rh-Phosphine Catalysts in Continuous Fixed-Bed Mode. *Catal. Lett.* **2003**, *90*, 149–153. [CrossRef]
127. An, H.; Wang, D.; Miao, S.; Yang, Q.; Zhao, X.; Wang, Y. Preparation of Ni-IL/SiO_2 and its catalytic performance for one-pot sequential synthesis of 2-propylheptanol from n-valeraldehyde. *RSC Adv.* **2020**, *10*, 28100–28105. [CrossRef]
128. Sadeghzadeh, S.M. PbS based ionic liquid immobilized onto fibrous nano-silica as robust and recyclable heterogeneous catalysts for the hydrogen production by dehydrogenation of formic acid. *Microporous Mesoporous Mater.* **2016**, *234*, 310–316. [CrossRef]
129. Gruttadauria, M.; Liotta, L.F.; Salvo, A.M.P.; Giacalone, F.; La Parola, V.; Aprile, C.; Noto, R. Multi-Layered, Covalently Supported Ionic Liquid Phase (mlc-SILP) as highly Cross-Linked Support for Recyclable Palladium Catalysts for the Suzuki Reaction in Aqueous Medium. *Adv. Synth. Catal.* **2011**, *353*, 2119–2130. [CrossRef]
130. Calabrese, C.; Campisciano, V.; Siragusa, F.; Liotta, L.; Aprile, C.; Gruttadauria, M.; Giacalone, F. SBA-15/POSS-Imidazolium hybrids as Catalytic Nanoreactor: The role of the support in the stabilization of Palladium species for C−C Cross Coupling Reactions. *Adv. Synth. Catal.* **2019**, *361*, 3758–3767. [CrossRef]
131. Pavia, C.; Ballerini, E.; Bivona, L.A.; Giacalone, F.; Aprile, C.; Vaccaro, L.; Gruttadauria, M. Palladium Supported on Cross-Linked Imidazolium Network on Silica as highly Sustainable Catalysts for the Suzuki Reaction under Flow Conditions. *Adv. Synth. Catal.* **2013**, *355*, 2007–2018. [CrossRef]
132. Montroni, E.; Lombardo, M.; Quintavalla, A.; Trombini, C.; Gruttadauria, M.; Giacalone, F. A Liquid-Liquid Biphasic homogeneous Organocatalytic Aldol Protocol Based on the Use of a Silica Gel Bound Multilayered Ionic Liquid Phase. *ChemCatChem* **2012**, *4*, 1000–1006. [CrossRef]
133. Zhong, N.; Li, Y.; Cai, C.; Gao, Y.; Liu, N.; Liu, G.; Tan, W.; Zeng, Y. Enhancing the catalytic performance of Candida antarctica lipase B by immobilization onto the ionic liquids modified SBA-15. *Eur. J. Lipid. Sci. Tech.* **2018**, *120*, 1700357. [CrossRef]
134. Zou, B.; Chu, Y.; Xia, J.; Chen, X.; Huo, S. Immobilization of lipase by ionic liquid-modified mesoporous SiO_2 adsorption and calcium alginate-embedding method. *Appl. Biochem. Biotechnol.* **2018**, *185*, 606–618. [CrossRef]
135. Xie, W.; Zang, X. Lipase immobilized on ionic liquid-functionalized magnetic silica composites as a magnetic biocatalyst for production of trans-free plastic fats. *Food Chem.* **2018**, *257*, 15–22. [CrossRef]

MDPI
St. Alban-Anlage 66
4052 Basel
Switzerland
www.mdpi.com

Molecules Editorial Office
E-mail: molecules@mdpi.com
www.mdpi.com/journal/molecules

Disclaimer/Publisher's Note: The statements, opinions and data contained in all publications are solely those of the individual author(s) and contributor(s) and not of MDPI and/or the editor(s). MDPI and/or the editor(s) disclaim responsibility for any injury to people or property resulting from any ideas, methods, instructions or products referred to in the content.

www.ingramcontent.com/pod-product-compliance
Lightning Source LLC
LaVergne TN
LVHW070442100526
838202LV00014B/1648